Springer Series in
MATERIALS SCIENCE 91

Springer Series in
MATERIALS SCIENCE

Editors: R. Hull   R. M. Osgood, Jr.   J. Parisi   H. Warlimont

The Springer Series in Materials Science covers the complete spectrum of materials physics, including fundamental principles, physical properties, materials theory and design. Recognizing the increasing importance of materials science in future device technologies, the book titles in this series reflect the state-of-the-art in understanding and controlling the structure and properties of all important classes of materials.

98 **Physics of Negative Refraction and Negative Index Materials**
Optical and Electronic Aspects and Diversified Approaches
Editors: C.M. Krowne and Y. Zhang

99 **Self-Organized Morphology in Nanostructured Materials**
Editors: K. Al-Shamery and J. Parisi

100 **Self Healing Materials**
An Alternative Approach to 20 Centuries of Materials Science
Editor: S. van der Zwaag

101 **New Organic Nanostructures for Next Generation Devices**
Editors: K. Al-Shamery, H.-G. Rubahn, and H. Sitter

102 **Photonic Crystal Fibers**
Properties and Applications
By F. Poli, A. Cucinotta, and S. Selleri

103 **Polarons in Advanced Materials**
Editor: A.S. Alexandrov

104 **Transparent Conductive Zinc Oxide**
Basics and Applications in Thin Film Solar Cells
Editors: K. Ellmer, A. Klein, and B. Rech

105 **Dilute III-V Nitride Semiconductors and Material Systems**
Physics and Technology
Editor: A. Erol

106 **Into The Nano Era**
Moore's Law Beyond Planar Silicon CMOS
Editor: H.R. Huff

107 **Organic Semiconductors in Sensor Applications**
Editors: D.A. Bernards, R.M. Ownes, and G.G. Malliaras

108 **Evolution of Thin-Film Morphology**
Modeling and Simulations
By M. Pelliccione and T.-M. Lu

109 **Reactive Sputter Deposition**
Editors: D. Depla and S. Mahieu

110 **The Physics of Organic Superconductors and Conductors**
Editor: A. Lebed

111 **Molecular Catalysts for Energy Conversion**
Editors: T. Okada and M. Kaneko

112 **Atomistic and Continuum Modeling of Nanocrystalline Materials**
Deformation Mechanisms and Scale Transition
By M. Cherkaoui and L. Capolungo

113 **Crystallography and the World of Symmetry**
By S.K. Chatterjee

114 **Piezoelectricity**
Evolution and Future of a Technology
Editors: W. Heywang, K. Lubitz, and W. Wersing

115 **Defects, Photorefraction and Ferroelectric Switching in Lithium Niobate**
By T. Volk and M. Wöhlecke

116 **Einstein Relation in Compound Semiconductors and Their Nanostructures**
By K.P. Ghatak, S. Bhattacharya, and D. De

117 **From Bulk to Nano**
The Many Sides of Magnetism
By C.G. Stefanita

118 **Fundamental and Technological Aspects of Extended Defects in Germanium**
By C. Claeys and E. Simoen

Volumes 50–97 are listed at the end of the book.

Pietro Ferraro
Simonetta Grilli
Paolo De Natale
Editors

Ferroelectric Crystals for Photonic Applications

Including Nanoscale Fabrication
and Characterization Techniques

With 262 Figures

Dr. Pietro Ferraro
Dr. Simonetta Grilli
Dr. Paolo De Natale
CNR, Istituto Nazionale di Ottica Applicata
Via Campi Flegrei 34, 80078 Pozzuali, Italy
E-mail: pietro.ferraro@inoa.it, simonetta.grilli@inoa.it, paolo.denatale@inoa.it

Series Editors:

Professor Robert Hull
University of Virginia
Dept. of Materials Science and Engineering
Thornton Hall
Charlottesville, VA 22903-2442, USA

Professor R. M. Osgood, Jr.
Microelectronics Science Laboratory
Department of Electrical Engineering
Columbia University
Seeley W. Mudd Building
New York, NY 10027, USA

Professor Jürgen Parisi
Universität Oldenburg, Fachbereich Physik
Abt. Energie- und Halbleiterforschung
Carl-von-Ossietzky-Strasse 9–11
26129 Oldenburg, Germany

Professor Hans Warlimont
Institut für Festkörper-
und Werkstoffforschung,
Helmholtzstrasse 20
01069 Dresden, Germany

Springer Series in Materials Science   ISSN 0933-033X

ISBN 978-3-540-77963-6   e-ISBN 978-3-540-70548-2

Library of Congress Control Number: 2008930219

© Springer-Verlag Berlin Heidelberg 2009

This work is subject to copyright. All rights are reserved, whether the whole or part of the material is concerned, specifically the rights of translation, reprinting, reuse of illustrations, recitation, broadcasting, reproduction on microfilm or in any other way, and storage in data banks. Duplication of this publication or parts thereof is permitted only under the provisions of the German Copyright Law of September 9, 1965, in its current version, and permission for use must always be obtained from Springer-Verlag. Violations are liable to prosecution under the German Copyright Law.

The use of general descriptive names, registered names, trademarks, etc. in this publication does not imply, even in the absence of a specific statement, that such names are exempt from the relevant protective laws and regulations and therefore free for general use.

Typesetting: Data prepared by VTEX using a Springer T$_{\!E}$X macro package
Cover concept: eStudio Calamar Steinen
Cover production: WMX Design GmbH, Heidelberg

SPIN: 11935704   57/3180/Vtex
Printed on acid-free paper

9 8 7 6 5 4 3 2 1

springer.com

Preface

The idea to write a new book in the field of ferroelectric crystals arose from some considerations reported in the following. In the last 5 years, several groups all around the world in the field of engineering and characterization of ferroelectric crystals have published more than 300 papers. The motivation for such an intense research activity is referable to the fact that the ferroelectric crystals are a key element for the most attractive and useful photonic and optoelectronic devices. In fact, during the 60ies, the scientists realized that the ferroelectric crystals could have been efficiently used to generate new, unavailable frequencies, taking advantage of the freshly proposed birefringent phase-matching method. The synchronized rush for the development of novel coherent sources and for the discovery of the best-suited nonlinear crystals for mixing and generation had started. Consequently, the range of applications of ferroelectric crystals has enormously widened in the last years, especially based on the use of periodically poled structures (i.e., PPLN, PPLT, PPKTP, or PPKTA) to quasi-phase-match optical interactions. A new generation of sources is finding increasing applications in various fields, including high sensitivity trace gas monitoring and any kind of advanced spectroscopic set-ups, thus replacing "old style" gas lasers like Argon-ion or dye lasers. New possibilities are also being explored to engineer ferroelectric crystals with two- or three-dimensional geometries. Results from this field will allow developing photonic devices combining photonic band-gap properties and nonlinear conversion processes, i.e., nonlinear photonic crystals.

Moreover new micro-devices have been developed and built, based on domain-engineering processes, for telecom or sensing applications (filters, ring resonators, whispering gallery mode based sensing devices, etc.). Other interesting and emerging topics are growing rapidly, and one of the most promising is related to the possibility of fabricating structures on micrometre and/or nanometre scale in ferroelectric crystals in order to realize photonic band-gap devices. Several papers appeared on this subject presenting different fabrication approaches (i.e., e-beam processing, poling by atomic force microscope scanning and interference lithography, and subsequent electric poling). Many new configurations for photonic devices would become possible taking advantage of the photonic band-gap related physics with two- and three-dimensional geometries. The appeal of ferroelectric crystals for similar applications also arises from the additional properties they exhibit, e.g., electro-optic, piezoelectric, pyroelectric effects, with respect to other materials, thus making possible unique performances.

Simultaneously several papers appeared during the last years in the most authoritative journals, where new characterization methods have been developed and proposed to investigate the basic properties of ferroelectric crystals (optical microscopy, interferometric, scanning probe microscopy, X-ray diffraction, etc.). Specific methods and procedures have been invented to investigate the structures during the engineering process as well as after the fabrication and while operating into the photonic and optoelectronic devices.

The aim of the present volume is definitely to give an up-to-date source of information in this scientific and technological field of increasing interest and not covered yet by other books. The book gathers the latest achievements in the field of ferroelectric domain engineering and characterization at micron and nano scale dimensions and periods. The results obtained in the last years by the main scientific groups all over the world recognized as the most experts in this area are presented in this book, thus providing, we hope, a valid and precious overview on the last developments and moreover on the future innovative applications of those engineered materials in the field of photonics, for scientists working in this area. The text is aimed at researchers and PhD students who wish to be introduced rapidly in the last achievements in the field of material processing and photonic applications of ferroelectric materials.

The book is organized in 15 chapters grouped into three parts: *Fabrication*; *Characterization*; *Applications*. The first part focuses on the development of advanced methods for micron- and nano-scale engineering of ferroelectric crystals, while the second one deals with the most widely used techniques for the characterization of material and engineering related properties of the crystals. The last part provides an overview of the most important current and future applications of the new ferroelectric structure devices in the field of photonics.

Pozzuoli, August 2008

Pietro Ferraro
Simonetta Grilli
Paolo De Natale

Contents

Preface v

Part I Fabrication

1 Micro-Structuring and Ferroelectric Domain Engineering of Single Crystal Lithium Niobate
S. Mailis, C.L. Sones, R.W. Eason 3
   1.1 Introduction 3
   1.2 Other Methods 4
   1.3 Differential Chemical Etching 5
      1.3.1 z-Faces 5
      1.3.2 y-Faces 9
      1.3.3 Microstructures 11
   1.4 Summary and Future Work 17
   References 18

2 Fabrication and Characterization of Self-Assembled Ferroelectric Linear and Nonlinear Photonic Crystals: GaN and LiNbO$_3$
L.-H. Peng, H.-M. Wu, A.H. Kung, C.-M. Lai 21
   2.1 Introduction 21
   2.2 Micro-Domain Engineering with Conventional Poling Electrode Design 23
      2.2.1 Internal Field Effect in the Poling of Congruent-Grown LiNbO$_3$ or LiTaO$_3$ 23
      2.2.2 Origin of the Fringe Field 24
      2.2.3 Poling Issues with Doped or Stoichiometric LiNbO$_3$ and LiTaO$_3$ 26
   2.3 From Micron to Submicron Domain Engineering with Improved Electrode Design 27
      2.3.1 Charged Potential Barrier Method 27
      2.3.2 Stack of High-k Dielectric Poling Electrode Method 36

2.3.3 Submicron Domain Engineering with Self-Assembly Type
　　　　　of Poling Electrodes 40
　　　2.3.4 Submicron Domain Engineering in Ferroelectric
　　　　　Semiconductors 44
　2.4 Conclusion 47
　　　References 48

3　Sub-Micron Structuring of LiNbO$_3$ Crystals with Multi-Period and Complex Geometries
S. Grilli, P. Ferraro 53
　3.1 Introduction 53
　3.2 Overview of the Etching Techniques Applied to Lithium Niobate 53
　3.3 Electric Field Poling and Overpoling 59
　3.4 Holographic Lithography 61
　3.5 Periodic Sub-Micron Structuring 63
　　　3.5.1 Overpoling Applied to One-Dimensional Michelson Resist
　　　　　Gratings 63
　　　3.5.2 Overpoling Applied to Two-Dimensional Michelson Resist
　　　　　Gratings 65
　　　3.5.3 Overpoling Applied to Two-Beams Resist Gratings
　　　　　at Sub-Micron Scale 66
　　　3.5.4 Complex Surface Structures by Moiré HL 67
　3.6 Double-Face Sub-Micron Surface Structures 72
　3.7 Possible Applications for Novel Photonic Crystal Devices 73
　　　References 76

4　Nonlinear Optical Waveguides in Stoichiometric Lithium Tantalate
M. Marangoni, R. Ramponi 79
　4.1 Material Properties 81
　　　4.1.1 Physical Properties 81
　　　4.1.2 Optical Properties 83
　4.2 Waveguide Fabrication through Reverse-Proton-Exchange 85
　　　4.2.1 Fabrication and Characterization Procedures 87
　　　4.2.2 Modelling 88
　4.3 Second-Harmonic Generation in RPE-PPSLT Waveguides 93
　　　4.3.1 Highly Confining Waveguides 93
　　　4.3.2 Weakly-Confining Waveguides 95
　　　References 97

5　3-D Integrated Optical Microcircuits in Lithium Niobate Written by Spatial Solitons
E. Fazio, M. Chauvet, V.I. Vlad, A. Petris, F. Pettazzi, V. Coda, M. Alonzo 101
　5.1 Review of Waveguide Fabrication Techniques 101
　5.2 Theory of Photorefractive–Photovoltaic Spatial Solitons
　　　in Biased LiNbO$_3$ 102

  5.2.1 Photorefractive Model 102
  5.2.2 Time Dependent Electric Field Distribution 103
  5.2.3 PR Space Charge Field 104
  5.2.4 Soliton Solutions 105
  5.2.5 Dark Solitons 106
  5.2.6 Bright Solitons 110
 5.3 Photorefractive Bright Soliton Observation....................... 112
 5.4 Waveguiding in Soliton Channels/Strips 115
  5.4.1 Experimental Observation 115
  5.4.2 Fixing Soliton Waveguides and Circuits in Lithium Niobate
     Crystals 116
  5.4.3 Waveguide Characteristics............................... 117
 5.5 Optical Microcircuits with Soliton Waveguides 117
  5.5.1 Passive 117
 5.6 Optical Microcircuits with Solitons Waveguides 119
  5.6.1 Passive 119
  5.6.2 Active.. 122
 5.7 Three-Dimensional Optical Micro-Circuits with SWGs 129
 References.. 132

Part II Characterization

6 Light Aided Domain Patterning and Rare Earth Emission Based Imaging of Ferroelectric Domains
V. Dierolf, C. Sandmann 137
 6.1 Introduction and Background 137
  6.1.1 Overview 137
  6.1.2 Rare Earth Ions in $LiNbO_3$ 137
  6.1.3 Combined Excitation Emission Spectroscopy 139
  6.1.4 Confocal Microscopy and Spectroscopy 141
 6.2 Application of RE Spectroscopy to the Imaging of Integrated
   Optical Devices in Lithium Niobate............................ 143
  6.2.1 Rare Earth Ions as Probes 144
  6.2.2 Imaging of Waveguides 144
  6.2.3 Imaging of Ferroelectric Domains and Domain Wall Regions .. 147
  6.2.4 Imaging of Periodically Poled Waveguide Structures.......... 151
 6.3 Light Induced Domain Inversion 153
  6.3.1 Methods... 153
  6.3.2 Build-Up of Charge under Focussed Laser Irradiation 154
  6.3.3 Influence of Light on Domain Inversion and Growth 155
  6.3.4 Direct Writing of Domain Patterns 158
 6.4 Summary and Conclusions 162
 References.. 163

7 Visual and Quantitative Characterization of Ferroelectric Crystals and Related Domain Engineering Processes by Interferometric Techniques

P. Ferraro, S. Grilli, M. Paturzo, S. Nicola 165
    7.1 Introduction 165
    7.2 Measuring the Refractive Indices and Thickness of Lithium Niobate Wafers 166
    7.3 Visualization and *In-Situ* Monitoring of Domains Formation 171
        7.3.1 Digital Holography and Experimental Configuration for *In-Situ* Investigation of Poling 174
        7.3.2 Investigation of the Electro-Optic Effect and Internal Fields 192
        7.3.3 Evaluation of Optical Birefringence at Ferroelectric Domain Wall in $LiNbO_3$ 201
    References 204

8 New Insights into Ferroelectric Domain Imaging with Piezoresponse Force Microscopy

T. Jungk, A. Hoffmann, E. Soergel 209
    8.1 Introduction 209
        8.1.1 Ferroelectrics 209
        8.1.2 Lithium Niobate ($LiNbO_3$) 210
    8.2 Principles of Scanning Force Microscopy (SFM) 211
        8.2.1 Tip-Cantilever-Surface Interactions 211
        8.2.2 Cantilever Movements 212
        8.2.3 Cross-Talk 213
        8.2.4 Calibration 213
    8.3 Principles of Piezoresponse Force Microscopy (PFM) 214
        8.3.1 PFM Setup & Standard Settings 214
        8.3.2 System-Inherent Background in PFM Measurements 216
        8.3.3 Vectorial Description 216
    8.4 Consequences of the System-Inherent Background 218
        8.4.1 Background-Induced Misinterpretations 218
        8.4.2 Background-Free PFM Imaging 220
    8.5 Quantitative Piezoresponse Force Microscopy 221
        8.5.1 Amplitude of the PFM Signal 222
        8.5.2 Domain Wall Width 222
    8.6 Ferroelectric Domain Imaging by Lateral Force Microscopy 223
        8.6.1 Origin of the Lateral Signal 224
        8.6.2 Application to PPLN 224
    8.7 Conclusions 226
    References 226

9 Structural Characterization of Periodically Poled Lithium Niobate Crystals by High Resolution X-Ray Diffraction

M. Bazzan, N. Argiolas, C. Sada, P. Mazzoldi 229
- 9.1 Introduction... 229
- 9.2 The Principle of the XRD Technique 232
  - 9.2.1 The Theory of High Resolution X-Ray Diffraction 233
  - 9.2.2 The HRXRD Applied to PPLN Crystals 239
- 9.3 Experimental Set-Up for Structural Characterization by HRXRD 242
- 9.4 Applications 246
  - 9.4.1 Investigation of Sub-Micrometric PPLN Crystals............. 246
  - 9.4.2 Investigation of Micrometric PPLN Crystals with Bent Domain Walls 249
- 9.5 Conclusions... 253
- References.. 254

Part III Applications

10 Nonlinear Interactions in Periodic and Quasi-Periodic Nonlinear Photonic Crystals

A. Arie, A. Bahabad, N. Habshoosh 259
- 10.1 Introduction.. 259
- 10.2 Wave Equations in NLPC 261
- 10.3 Analysis of a Periodic Nonlinear Photonic Crystal................ 263
  - 10.3.1 The Real Lattice... 263
  - 10.3.2 The Reciprocal Lattice 265
  - 10.3.3 Conversion Efficiency for Specific Types of 2D Periodic Structures 266
- 10.4 Analysis of a Quasi-Periodic Nonlinear Photonic Crystal 272
  - 10.4.1 Statement of the Problem 272
  - 10.4.2 Solution by Quasiperiodic Lattices 273
  - 10.4.3 Establishing an Orthogonality Condition 274
  - 10.4.4 Tiling the Quasi-Periodic Lattice by the Dual Grid Construction 275
  - 10.4.5 The Fourier Transform of the Quasi-Periodic Lattice 276
  - 10.4.6 From Lattice to a Nonlinear Photonic Crystal 277
  - 10.4.7 A One-Dimensional Example – The Three Wave Doubler ... 278
- 10.5 Discussion and Summary 282
- References.. 283

11 Domain-Engineered Ferroelectric Crystals for Nonlinear and Quantum Optics

M. Bellini, P. Cancio, G. Gagliardi, G. Giusfredi, P. Maddaloni,
D. Mazzotti, P. Natale 285
- 11.1 Introduction.. 285

	11.1.1 Classification of Nonlinear Processes	285
	11.1.2 Phase Matching	286
11.2	Nonlinear Optics for Spectroscopic Applications	287
	11.2.1 Coherent Sources for mid-IR Spectroscopy and Metrology	287
	11.2.2 OFCS Extension to the mid-IR	288
	11.2.3 Future Perspectives	295
11.3	Structured Nonlinear Crystals for Quantum Optics	296
	11.3.1 Quantum Light Sources	297
	11.3.2 Single-Photon Detectors	301
	References	303

12 Photonic and Phononic Band Gap Properties of Lithium Niobate

M.P. Bernal, M. Roussey, F. Baida, S. Benchabane, A. Khelif, V. Laude 307

- 12.1 Introduction ... 307
- 12.2 Photonic Crystals ... 309
  - 12.2.1 Band Structure Theory and Slow Light ... 309
  - 12.2.2 Fabrication and Examples ... 313
  - 12.2.3 Experimental Procedure ... 314
  - 12.2.4 Measurement of a PBG in a LN Photonic Crystal ... 318
  - 12.2.5 LN PtC Waveguides: Transmission and SNOM Characterization ... 319
  - 12.2.6 A LN PtC Intensity Modulator ... 321
- 12.3 Phononic crystals ... 323
  - 12.3.1 Theory ... 323
  - 12.3.2 Fabrication and Examples ... 326
- 12.4 Conclusion ... 332
- References ... 334

13 Lithium Niobate Whispering Gallery Resonators: Applications and Fundamental Studies

L. Maleki, A.B. Matsko 337

- 13.1 Introduction ... 337
- 13.2 Modulators ... 338
  - 13.2.1 Principle of Operation ... 340
  - 13.2.2 Performance ... 340
- 13.3 Tunable Filters ... 341
  - 13.3.1 First-Order Filter ... 341
  - 13.3.2 Third-Order Filter ... 342
  - 13.3.3 Fifth-Order Filter ... 344
  - 13.3.4 Insertion Loss ... 345
- 13.4 WGRs Made of Periodically Poled Lithium Niobate ... 346
  - 13.4.1 Optical Frequency Doubling ... 347
  - 13.4.2 Calligraphic Poling ... 350
  - 13.4.3 Reconfigurable Filters ... 351
- 13.5 Photorefractive Damage ... 352

          13.5.1 Congruent LiNbO$_3$ 354
          13.5.2 Magnesium Doped Congruent LiNbO$_3$ 356
          13.5.3 Crossings and Anticrossings of the Modes 358
          13.5.4 Holographic Engineering of the WGM Spectra 359
     13.6 Infrared Transparency and Photorefractivity of Lithium Niobate
          Crystals: Theory 359
          13.6.1 Rate Equations 362
          13.6.2 Solution of the Rate Equations 365
          13.6.3 Absorption of the Light and Initial Concentration
                 of the Filled Traps 368
     Appendix A: Basic Properties of WGMs 371
     Appendix B: Lithium Niobate Impurities: A Short Review
     of Existing Results 372
          B.1 Small Polarons 373
          B.2 Bipolarons 374
          B.3 Iron 375
     Appendix C: Photorefractivity in Red: A Short Review
     of the Existing Results 376
          C.1 Light Induced Change of Refractive Index 376
          C.2 Light Induced Change of Absorption 377
     Appendix D: Numerical Values of the Basic Rates
     Characterizing the Impurities 377
     References 380

**14 Applications of Domain Engineering in Ferroelectrics
   for Photonic Applications**
D.A. Scrymgeour 385
     14.1 Introduction 385
     14.2 Ferroelectrics and Domain Engineering 385
     14.3 Applications of Domain Engineered Structures 387
          14.3.1 Frequency Conversion 387
          14.3.2 Electro-Optic Devices 390
     14.4 Challenges of Domain Engineered Ferroelectric Devices 396
     14.5 Conclusions 397
     References 398

**15 Electro-Optics Effect in Periodically Domain-Inverted Ferroelectrics
   Crystals: Principles and Applications**
J. Shi, X. Chen 401
     15.1 Introduction 401
     15.2 Basic principle 402
          15.2.1 Electro-Optic Effect in Crystals 402
          15.2.2 Electro-Optical Effect for Crystals of 3m Symmetry Group .. 404
          15.2.3 Electro-Optical Effect in Periodically Domain-Inverted
                 Crystals with 3m Symmetry 407

15.3 Applications 413
    15.3.1 Devices Based on Bragg Diffraction Grating Structure 413
    15.3.2 Devices Based on Solc-Layered Structure 416
    15.3.3 Other Application Devices 418
    References.. 419

List of Contributors

M. Alonzo
Dipartimento di Energetica
Universitá La Sapienza Rome
Roma, Italy

N. Argiolas
Physics Department
University of Padova
Via Marzolo 8
35131 Padova, Italy

A. Arie
Dept. of Physical Electronics
School of Electrical Engineering
Tel-Aviv University
Tel-Aviv 69978, Israel
ady@eng.tau.ac.il

A. Bahabad
Dept. of Physical Electronics
School of Electrical Engineering
Tel-Aviv University
Tel-Aviv 69978, Israel

Fadi Baida
Département d'Optique P.-M. Duffieux
Institut FEMTO-ST
CNRS UMR 6174
Université de Franche-Comté
Besancon, France

M. Bazzan
Physics Department
University of Padova
Via Marzolo 8
35131 Padova, Italy

Marco Bellini
Istituto Nazionale di Ottica
Applicata (CNR-INOA)
Largo Fermi 6
50125 Firenze, Italy;

European Laboratory for Nonlinear
Spectroscopy (LENS)
Via Carrara 1
50019 Sesto Fiorentino, FI, Italy

Maria Pilar Bernal
Département d'Optique P.-M. Duffieux
Institut FEMTO-ST
CNRS UMR 6174
Université de Franche-Comté
Besancon, France
maria-pilar.bernal@femto-st.fr

Sarah Benchabane
Département de Physique
et Métrologie des Oscillateurs
Institut FEMTO-ST
CNRS UMR 6174
Université de Franche-Comté
Besancon, France

Pablo Cancio
Istituto Nazionale di Ottica
Applicata (CNR-INOA)
Largo Fermi 6
50125 Firenze, Italy;

European Laboratory for Nonlinear
Spectroscopy (LENS)
Via Carrara 1
50019 Sesto Fiorentino, FI, Italy

M. Chauvet
FEMTO-ST Institute
Université de Franche-Comté Besançon
France

Xianfeng Chen
Institute of Optics and Photonics
Department of Physics
The State Key Laboratory
on Fiber-Optic
Local Area Network and Advanced
Optical Communication Systems
Shanghai Jiaotong University
800 Dongchuan Road
Shanghai 200240, China
xfchen@sjtu.edu.cn

V. Coda
FEMTO-ST Institute
Université de Franche-Comté Besançon
France

Paolo De Natale
Istituto Nazionale di Ottica
Applicata (CNR-INOA)
Largo Fermi 6
50125 Firenze, Italy
paolo.denatale@inoa.it;

European Laboratory for Nonlinear
Spectroscopy (LENS)
Via Carrara 1
50019 Sesto Fiorentino, FI, Italy

Sergio De Nicola
Istituto di Cibernetica
"E. Cainiello" del CNR

Viale Campi Flegrei 34
80078 Pozzuoli, NA, Italy

Volkmar Dierolf
Physics Department
Lehigh University
16 Memorial Drive East
Bethlehem, PA 18015, USA
vod2@lehigh.edu

Robert W. Eason
Optoelectronics Research Centre
University of Southampton, UK

E. Fazio
Dipartimento di Energetica
Universitá La Sapienza
Roma, Italy

Pietro Ferraro
Istituto Nazionale di Ottica Applicata
del CNR (CNR-INOA)
Pozzuoli, Napoli, Italy

Gianluca Gagliardi
Istituto Nazionale di Ottica
Applicata (CNR-INOA)
Largo Fermi 6
50125 Firenze, Italy

Giovanni Giusfredi
Istituto Nazionale di Ottica
Applicata (CNR-INOA)
Largo Fermi 6
50125 Firenze, Italy;

European Laboratory for Nonlinear
Spectroscopy (LENS)
Via Carrara 1
50019 Sesto Fiorentino, FI, Italy

Simonetta Grilli
Istituto Nazionale di Ottica Applicata
del CNR, CNR-INOA
Pozzuoli, Napoli, Italy

List of Contributors xvii

N. Habshoosh
Dept. of Physical Electronics
School of Electrical Engineering
Tel-Aviv University
Tel-Aviv 69978, Israel

Akos Hoffmann
Institute of Physics
University of Bonn
Wegelerstr. 8
53115 Bonn, Germany

Tobias Jungk
Institute of Physics
University of Bonn
Wegelerstr. 8
53115 Bonn, Germany

Abdelkrim Khelif
Département de Physique
et Métrologie des Oscillateurs
Institut FEMTO-ST
CNRS UMR 6174
Université de Franche-Comté
Besanton, France

A.H. Kung
Institute of Atomic and Molecular
Sciences
Academia Sinica
Taipei, Taiwan
People's Republic of China

Chih-Ming Lai
Department of Electronic Engineering
Ming Chuan University
Taoyuan, Taiwan
People's Republic of China

Vincent Laude
Département de Physique
et Métrologie des Oscillateurs
Institut FEMTO-ST
CNRS UMR 6174
Université de Franche-Comté
Besanton, France
vincent.laude@femto-st.fr

Pasquale Maddaloni
Istituto Nazionale di Ottica
Applicata (CNR-INOA)
Largo Fermi 6
50125 Firenze, Italy

Sakellaris Mailis
Optoelectronics Research Centre
University of Southampton, UK

Lute Maleki
OEwaves, Inc.
1010 East Union St Pasadena
CA 91106 USA;

Jet Propulsion Laboratory
California Institute of Technology
4800 Oak Grove Drive
Pasadena, CL 91109-8099, USA
Lute.Maleki@jpl.nasa.gov

Marco Marangoni
Dipartimento di Fisica Politecnico
di Milano and INFM-CNR
Piazza Leonardo da Vinci 32
20133 Milan, Italy
marco.marangoni@polimi.it

Andrey B. Matsko
Jet Propulsion Laboratory
California Institute of Technology
4800 Oak Grove Drive
Pasadena, CA 91109-8099, USA
Andrey.Matsko@jpl.nasa.gov

P. Mazzoldi
Physics Department
University of Padova
Via Marzolo 8
35131 Padova, Italy

Davide Mazzotti
Istituto Nazionale di Ottica
Applicata (CNR-INOA)
Largo Fermi 6
50125 Firenze, Italy;

European Laboratory for Nonlinear
Spectroscopy (LENS)
Via Carrara 1
50019 Sesto Fiorentino, FI, Italy

Melania Paturzo
Istituto Nazionale di Ottica
Applicata del CNR, CNR-INOA
Puzzuoli, Napoli, Italy

Lung-Han Peng
Institute of Electro-Optical Engineering
National Taiwan University
Taipei, Taiwan
People's Republic of China

A. Petris
Institute of Atomic Physics
NILPRP
Bucharest, Romania

F. Pettazzi
Dipartimento di Energetica
Universitá La Sapienza Rome
Rome, Italy;
FEMTO-ST Institute
Université de Franche-Comté Besançon
France

Roberta Ramponi
Dipartimento di Fisica
Politecnico di Milano and Istituto
di Fotonica e Nanotecnologie del CNR
Piazza Leonardo da Vinci 32
20133 Milan, Italy
roberta.ramponi@fisi.polimi.it

C. Sada
Physics Department
University of Padova
Via Marzolo 8
35131 Padova, Italy
cinzia.sada@unipd.it

Christian Sandmann
Physics Department
Lehigh University
16 Memorial Drive East
Bethlehem, PA 18015, USA;
BD Technologies
21 Davis Drive
Research Triangle Park
NC 27709, USA
Christian_Sandmann@bd.com

David A. Scrymgeour
Sandia National Laboratories
P.O. Box 5800, Albuquerque
NM 87185-1415, USA
dscrymg@sandia.gov

Jianhong Shi
Institute of Optics and Photonics
Department of Physics
The State Key Laboratory
on Fiber-Optic
Local Area Network and Advanced
Optical Communication Systems
Shanghai Jiaotong University
800 Dongchuan Road
Shanghai 200240, China
purewater@sjtu.edu.cn

Elisabeth Soergel
Institute of Physics
University of Bonn
Wegelerstr. 8
53115 Bonn, Germany
soergel@uni-bonn.de

Collin L. Sones
Optoelectronics Research Centre
University of Southampton, UK

V.I. Vlad
Institute of Atomic Physics
NILPRP
Bucharest, Romania

Han-Ming Wu
Institute of Electro-Optical Engineering
National Taiwan University
Taipei, Taiwan
People's Republic of China

Part I

Fabrication

1 Micro-Structuring and Ferroelectric Domain Engineering of Single Crystal Lithium Niobate

S. Mailis, C.L. Sones, and R.W. Eason

1.1 Introduction

The ability to microstructure specific materials is always associated with the ability to selectively remove material over small scale-lengths. Localized etching whether it is chemical or physical, wet or dry, parallel or sequential is central to every modern microstructuring method. For example a beam of accelerated ions is scanned on the surface of interest removing material along its trajectory. Alternatively the surface is prepared/treated in a manner that changes its "quality" locally making it more susceptible or more resistive to a particular etching agent. The whole surface is subsequently exposed to the etching agent which can be a uniform accelerated ion beam, a laser beam or an acid. The etching agent preferentially attacks the pretreated (or the untreated) portion of the surface removing material.

In the case of silicon crystals there are anisotropic chemical etching agents which attack preferentially specific crystallographic planes [1]. Etching using these agents produces 3D structures while selective oxidization followed by HF etching is used for undercutting and slicing. It is fair to say that micro-structuring in the case of silicon crystals has managed to revolutionize silicon based devices and to produce a novel class of technology i.e. Micro-Electro-Mechanical-Systems (MEMS) [2–4].

Lithium niobate is a well established material which is widely used in photonics and, similar to silicon, excellent quality material is available at low cost. As in the case of silicon, microstructuring of lithium niobate is also expected to add functionality and value to the material by both improving the performance of existing devices and also by creating new ones.

In the rest of the chapter we will demonstrate how ferroelectric domain engineering, achieved mainly by the application of an external electric field at room temperature [5, 6], can be combined with subsequent chemical etching to provide a powerful method for the microstructuring of lithium niobate.

However, before embarking into the description of differential chemical etching of domain-engineered structures we present other methods that have been used for the microstructuring of lithium niobate. The methods listed here have been developed for the microelectronics industry hence they are considered standard and are very well characterized.

1.2 Other Methods

A high-precision method for micro-etching is *electron beam milling*. In this method a beam of accelerated electrons is scanned across the sample to remove material from the surface. The resolution of the features which can be fabricated with this method can be very high and is comparable to the effective wavelength of the accelerated electrons however, although very precise, this method is slow when etching of large surfaces is required. Electron beam milling has been used not only for the fabrication of microstructures in lithium niobate [7] but also for ferroelectric domain inversion [8, 9].

A method similar to the electron beam milling is *ion beam milling* where accelerated ions are used instead of electrons to selectively remove material. Focussed ion beam milling has been used successfully for the fabrication of ultra fine photonic structures in lithium niobate [10, 11]. However, this method is again limited due to its sequential nature and can only be sensibly applied for small areas (≤ 1 mm). In another arrangement a broad beam of accelerated argon ions is used to bombard large sample areas that carry photolithographic patterning. Surface relief patterns are produced as the argon ions etch preferentially the parts of the material that are not covered by the photoresist. This arrangement is more suitable for the fabrication of extended structures across full wafers but its resolution is limited to the resolution of the photolithography.

Reactive ion etching is used in conjunction with photolithography, as in the case of ion beam milling to produce surface relief structures. In this method etching is aided by reaction of the ion species with the material to be etched. This method has been used for the fabrication of ridge waveguides and other photonic structures in lithium niobate [12–14].

Laser ablation or laser sputtering is a straightforward method whereby a highly absorbed (usually UV) laser beam is used to induce explosive evaporation of material from the surface. Part of the absorbed energy becomes kinetic energy of the part of the removed material which is ejected form the illuminated area. Etching of small features (suitable for photonic applications) can be achieved either by simply focusing of the laser beam, by using an absorption mask either in contact or projection mode and finally by interference [15–17]. Laser ablation is a "dirty" method because usually there is a large quantity of debris caused by re-deposition of the ejected material around the etched feature. However, subsequent cleaning or even brief wet etching can significantly improve the quality of the features.

Finally *wet etching* is another microstructuring method (one that is discussed at length here) whereby use of chemical reagents which interact with the surface results in material removal at the interface between the surface and the reagent. Spatially selective etching is achieved by covering the surface with an inert film that inhibits the contact of the acid with that particular part of the surface or, as in the case of lithium niobate, by modifying the "type" of the interacting surface by inversion of the ferroelectric domain orientation.

1.3 Differential Chemical Etching

1.3.1 z-Faces

Chemical etching is often used for cleaning, polishing and purification of optical surfaces. For this purpose a number of potential etch/polish chemical reagents were examined in the comprehensive work by Nassau et al. [18] in a variety of experimental conditions to assess their ability to polish/etch lithium niobate single crystal surfaces. This study showed that certain chemical reagents were able to induce an optical contrast between opposite ferroelectric domains thus making them visible.

More specifically it was observed that optical contrast appeared between opposite ferroelectric domains on the z faces of the crystal after 15 sec to 60 sec immersion in molten KOH or 15 minutes in a hot (50°C) mixture of H_2O_2 and NaOH. However, the highest contrast between opposite ferroelectric domains was observed for the $HF:2HNO_3$ mixture ratio [18]. This particular mixture of acids attacks preferentially the $-z$ face (defined as the z face that becomes negative upon cooling of the crystal [19]) removing material from the surface while the opposite $+z$ face remains unaffected. As a result of this effect any ferroelectric domain pattern will be transformed into a surface relief pattern after etching.

Brief (few minutes) etching in a mixture of HF and HNO_3 acid mixtures is today common practice for the quality assessment of nonlinear optical periodically poled lithium niobate (PPLN) devices. The relief pattern which is the result of differential etching corresponds precisely to the inverted ferroelectric domain shape and increases dramatically the optical contrast of the domain walls. Although destructive, it is still a very popular method for visualization of inverted domains in lithium niobate because it can show structural details of the domain boundary down to the nm scale.

Figure 1.1 shows two scanning electron microscopy (SEM) images corresponding to the opposite faces of a briefly (10 min) etched, hexagonally poled lithium

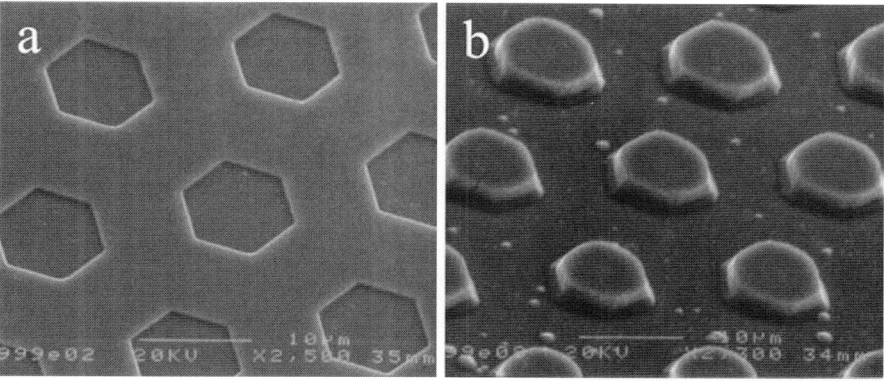

Fig. 1.1. SEM of briefly (10 min) etched hexagonally poled lithium niobate crystal: (**a**) $-z$ face terminating hexagonal domains, (**b**) $+z$ face terminating hexagonal domains

niobate crystal (HexLN) which is a z-cut crystal that carries hexagonally shaped inverted ferroelectric domains arranged in a 2D hexagonal lattice [20]. The recessed hexagons shown in Fig. 1.1(a) correspond to the $-z$ face of the inverted ferroelectric domains that have been attacked preferentially by the acid. The surrounding area corresponds to a $+z$ face and hence stays unaffected, resulting in the observed difference in height. As the polarity of the domain terminating surfaces is inverted on the opposite face of the sample the surface topography after etching is complementary to that of Fig. 1.1(a), and appears as an inverted pattern as shown in Fig. 1.1(b).

Etch-Rates

In the context of fabricating physical (high aspect ratio) structures it is necessary to etch the crystal for longer time (few hours to a few days depending on etching temperature and target etch depth) and to be able to predict the depth of the etching accurately. Hence, for any practical application it is useful to know the etching characteristics (etch-rate and quality) prolonged etching performed under different conditions. In a simple experiment, domain-engineered samples are etched for a fixed duration (15 hrs) at a temperature of 60°C in HF:xHNO$_3$ ratios where $x = 0$, 1, 2, 4. The initial concentrations of the electronic grade HF and HNO$_3$ as supplied were 48% and 78% respectively. Upon etching a difference in height is produced between anti-parallel domain termination areas on opposite z faces of the crystal because the $-z$ termination area is preferentially attacked by the acids. A schematic of the etch behaviour of the opposite domain surfaces is shown in Fig. 1.2 where arrows are used to indicate the direction of the permanent dipole. The sign of the charge that appears on the polar surfaces upon cooling of the crystal is also indicated in the schematic.

The difference in step height "d" along a single interface between opposite domains corresponds to the etch depth of the $-z$ face. The topography of the interface between opposite domains etched with mixtures of different HF and HNO$_3$ acid ratios at 60°C for 15 hrs is shown in the SEM images of Fig. 1.3. The step which is formed at the interface between opposite domains was measured using an α-step

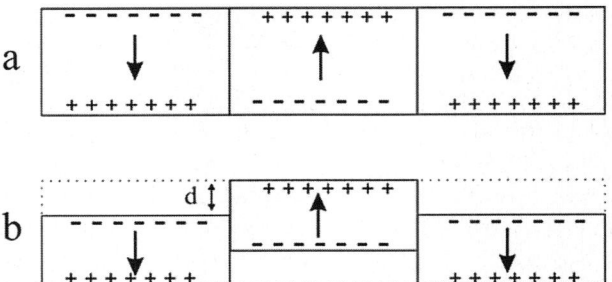

Fig. 1.2. Schematic showing the dependence of the preferential acid attack on the polarity of the electric dipole terminating surface, as indicated by the *arrows*

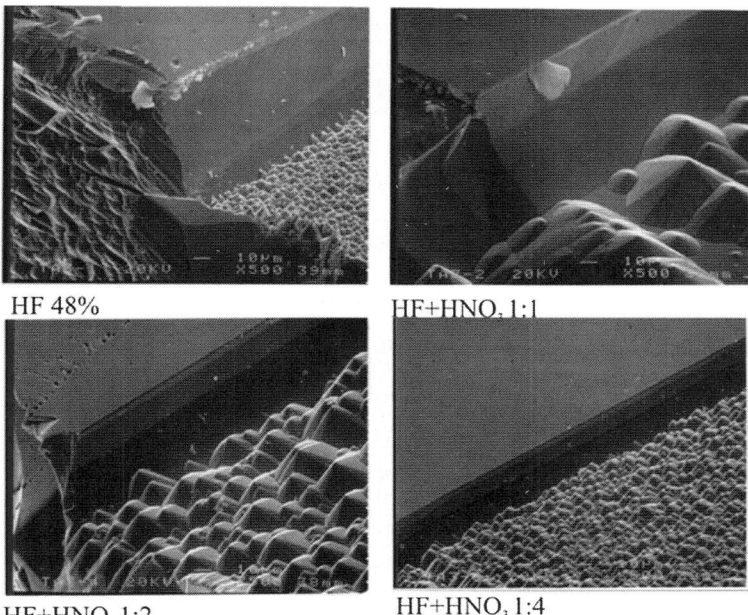

Fig. 1.3. SEM scans of the boundary between two inverted domains after etching for 15 hours in different HF:HNO$_3$ acid mixture ratios. The magnification is the same in all the scans

Table 1.1. Etch-rate as a function of HF:HNO$_3$ ratio at 60°C

HF:HNO$_3$	% HF in mixture	d (μm)	Etch-rate (μm/hr)
0:1	0	0	0
1:4	9.6	26.2 ± 0.7	1.7
1:2	16	48.3 ± 2.4	3.2
1:1	24	70.9 ± 1.8	4.7
1:0	48	81.9 ± 20.0	5.5

surface profiler, and the deduced etch-rates values (in μm/hr) as a function of the HF:HNO$_3$ mixture ratio are presented in Table 1.1.

The plot of the etch-rate as a function of the HF concentration in HF:HNO$_3$ mixture is shown in Fig. 1.4 where the ratio of acids is indicated in brackets next to the data points. The etch-rate increases with increasing temperature of the acid following an Arrhenius law [21].

This experiment concludes that pure HF provides the fastest etching-rate for the $-z$ face of lithium niobate and also that pure HNO$_3$ acid is not capable of etching the crystal, at least by any measurable amount, for the etching conditions employed in the experiment. In another experiment the HNO$_3$ acid in the mixture is substituted with water and again a domain engineered sample was etched for the same duration and temperature with different concentrations of HF. The etch-rates of the

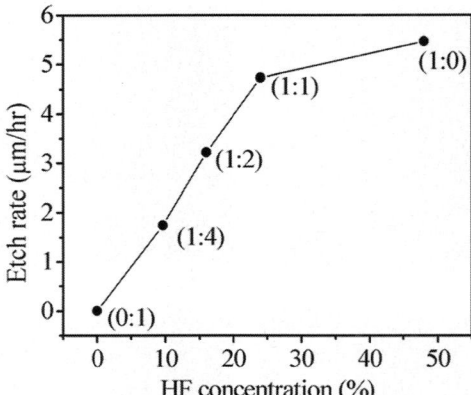

Fig. 1.4. Etch-rate at 60°C for various ratios of HF and HNO_3 acids. The *horizontal axis* shows the concentration of HF in the mixture while the *brackets* next to the experimental points indicate the acid ratio ($HF:HNO_3$)

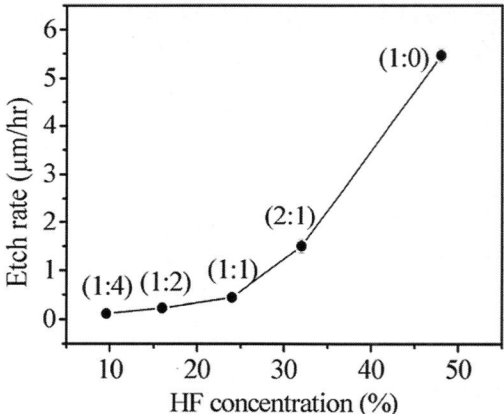

Fig. 1.5. Etch-rate of the $-z$ face at 60°C as a function of the concentration of HF acid. The *horizontal axis* shows the concentration of HF in the mixture while the *brackets* next to the experimental points indicate the acid ratio ($HF:HNO_3$)

$-z$ face were deduced and are shown in the plot of Fig. 1.5 as a function of the HF concentration in the etching solution.

It can be observed in this plot that the etch-rates were slower for water diluted HF which indicates that HNO_3 does play an auxiliary role in the etch process. More details about the chemistry of differential etching of lithium niobate in HF can be found in [22, 23].

Prolonged etching of the $-z$ face of the crystal affects the quality of the etched surface usually resulting in significant surface roughness, which is not desirable of course in any micro-fabrication processes. However, the quality of the etched sur-

1 Micro-Structuring and Ferroelectric Domain 9

Fig. 1.6. SEM scan of the $-z$ face after 15 hours of etching at 60°C with different HF:HNO$_3$ ratios. The magnification is the same in all four scans

face after prolonged etching can be controlled to some extent by the ratio of HF and HNO$_3$ mixture. Figure 1.6 shows a sequence of SEM images corresponding to the crystal surface after etching for the same duration and temperature but with different acid mixture ratios. In order to exclude the possibility of material quality variation, the samples used in this experiment came from the same wafer. It is obvious that the HF:HNO$_3$ mixture ratio can dramatically change the surface topography and furthermore, it seems that pure HF (48%) etching not only is faster but gives the best etched surface quality.

There are however several indications that the chemical etching induced surface roughness can be further controlled e.g. by changing the temperature of the process or by other pre-treatment of the surface such as UV illumination.

1.3.2 y-Faces

Careful observation of etched single inverted ferroelectric domains shows that the original shape of the inverted domain is not preserved during the etch process. This is indicative of simultaneous etching of more than one crystal planes/surfaces. To demonstrate this more clearly we have etched a sample similar to the one shown in Fig. 1.1 for 24 hrs. The initial inverted domain distribution consists of hexagonally shaped domains arranged in a hexagonal close packed 2D lattice. The result of the prolonged etching is shown in Fig. 1.7. These SEM images correspond to the two

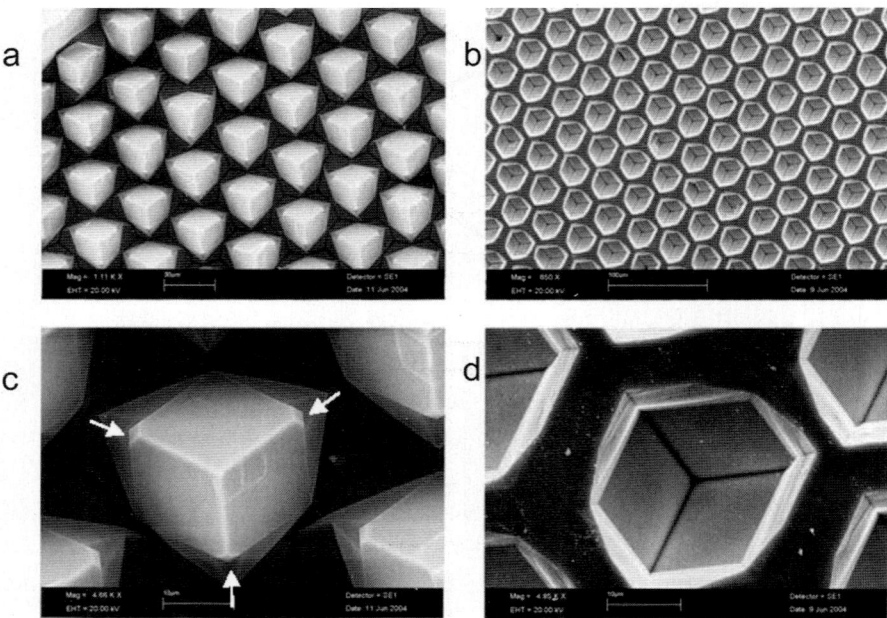

Fig. 1.7. SEM scans of opposite faces of a prolonged HF etched HexLN crystal. (**a**) Pyramids corresponding to $+z$ terminating domains, (**b**) cavities corresponding to $-z$ face terminating domains, (**c**), (**d**) higher magnification scans of (**a**) and (**b**) respectively

opposite faces of the sample show upstanding features (Fig. 1.7(a)) and cavities (Fig. 1.7(b)) arranged in the original 2D hexagonal lattice.

However, the cross section of these features transforms from hexagonal at the base to triangular at the top and finally degenerates to a point at the very top as the faces of the triangle merge. This is more obvious in the higher magnification images shown in Fig. 1.7(c), (d).

This transformation of the shape of the etched feature can be attributed to a lateral etching taking place simultaneously with the vertical etching process (along the z direction). Since the lateral etching is not pronounced after brief etching we can conclude that it is much slower than the primary etching along the z direction. Lateral etching is also not uniform because if it was, the cross section of the etched structures would tend to become circular rather than triangular. From the symmetry of the resulting pyramidal structures we can deduce that the lateral etching occurs between opposite y-face pairs and the etch-rate depends on the "polarity" of the y-face.

As shown in Fig. 1.8, the three indistinguishable y-directions in lithium niobate single crystals, contained within the three mirror planes, and normal to the z axis, clearly explains the transformation from hexagonal to a triangular shape. The schematic illustrates the shape transformation of a hexagonal domain as a re-

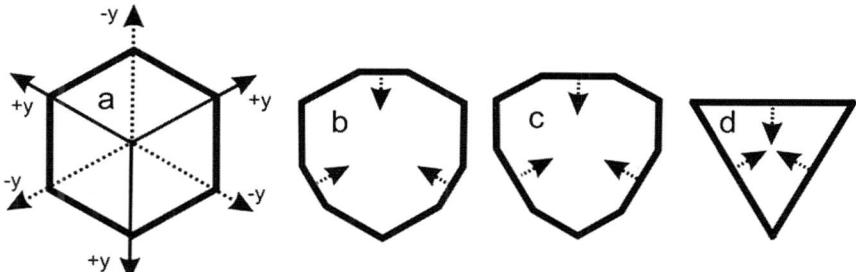

Fig. 1.8. Schematic illustrating the shape transformation of the top surface of an isolated hexagonal domain due to differential y-face etching. From left to right: (**a**) initial shape of the etched domain, the *axes* indicate the three y symmetry directions and their polarity, (**b**) and (**c**) intermediate stages where the *arrows* show the etch directions, (**d**) the final triangular shape

sult of lateral y-face etching. One can notice that the apices of the initial hexagonal domain that correspond to $+y$ faces stay pinned in the schematic. Further etching reduces the size of the hexagon hence the apices start to move until the triangular shape degenerates to a single point (not shown in the schematic).

This shape transformation is common for both faces of the domain and only the etch direction is different. The etching process of opposite y-faces unlike the differential etching between opposite z-faces, affects both faces. This can be observed near the base of the triangular pyramid in the areas which are indicated by the arrows in Fig. 1.7(c). In these areas the tips of the triangular cross section (which correspond to the $+y$-face that etches slower) are flattened as a result of etching.

The direction of the y-axes is inverted during ferroelectric domain inversion [24] hence, the depth profile of inverted ferroelectric domains can be visualized by etching of the y-face [25]. Finally, there is no observation in the prolonged etching of inverted domains structures to suggest that the x-faces of the crystal etch at all in HF and HNO_3 mixtures and if there is any x-face etching at all the rate should be remarkably slow. At this point it is worth mentioning that the pyramids which are produced by etching of individual hexagonal domains like the ones shown in Fig. 1.7 terminate in ultra-sharp tips which could be used in scanning probe microscopy. A side view of a pyramid array and a high magnification of the ultra-sharp tip are shown in the SEM images of Fig. 1.9.

1.3.3 Microstructures

The next step now will be to show how this very important feature i.e. differential etching between opposite domains can be used to fabricate 2D and 3D microstructures in lithium niobate single crystals.

Fig. 1.9. SEM images of: (*left*) side view of an array of pyramids and (*right*) high magnification side-view of the terminating tip

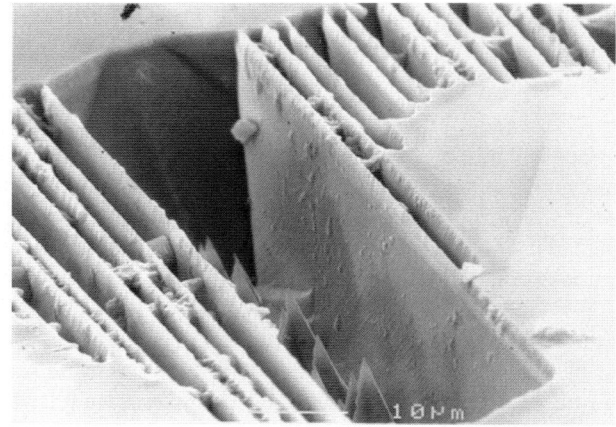

Fig. 1.10. SEM scan of a high aspect ratio structure fabricated by long etching of a periodically poled lithium niobate crystal

Direct Etching of Domain-Inverted Crystals

From the previous discussion it is clear that direct chemical etching of a ferroelectric domain distribution results in the generation of a surface relief pattern. Prolonged etching can indeed result in very high aspect ratio structures an example of which is illustrated in Fig. 1.10 where an SEM image of a periodically poled crystal which has been etched for a few hours is shown. The inverted domain slabs in this structure have a width of approximately 2 μm and a depth of ~20 μm. The quality and depth of the vertical wall which corresponds to the inverted domain sheet can be clearly observed through the gap left by missing PPLN. There are numerous applications that can benefit from structures like these.

Figure 1.11 shows SEM images of a set of waveguide ridges fabricated by the chemical HF etching of a set of stripe inverted domains. The vertical confinement of light is achieved by subsequent titanium indiffusion [26]. Waveguiding takes place in the top ~2–3 μm of the ridge which is away from the rough surface of the etched

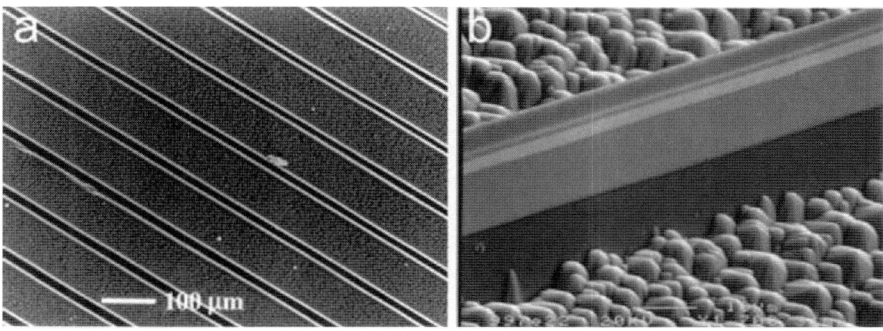

Fig. 1.11. (a) set of ridge waveguides; (b) high magnification side view of a single ridge waveguide

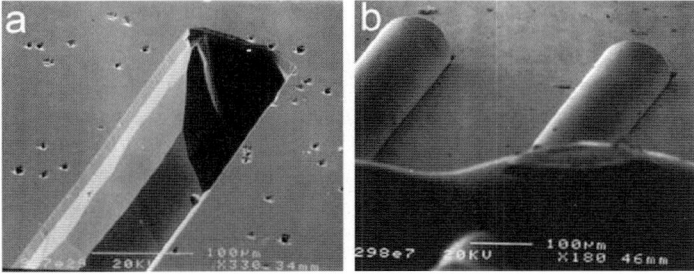

Fig. 1.12. (a) SEM image of a fiber alignment groove; (b) standard telecom fibers mounted into a pair of alignment grooves

$-z$ face as shown in Fig. 1.11(b). The side wall of the ridge structure is very smooth, free from any roughness that could induce optical loss.

Waveguides like the ones shown in Fig. 1.11 have been subsequently periodically poled and used to demonstrate nonlinear second harmonic generation [29].

Etched cavities can also be used for the precise alignment of optical components on a photonic chip. Shown in the SEM image of Fig. 1.12(a) is a set of etched cavities which are used for the mounting of two optical fibres onto a common substrate. This arrangement could be used for pigtailing of a photonic device pre- or post-fabricated on the substrate.

Other applications could include surface structuring for impedance matching in fast optical switches and acousto-optic waveguides. Additionally if the ferroelectric domain distributions become comparable to the optical wavelength of a guided wave in the crystal then etched periodic poled structures can be used as Bragg gratings [27] or photonic band-gaps for 2D structures [28].

Complex 3D Structures

Fabrication of 3D structures, especially free standing structures, involves a substantial complication as it is necessary to undercut (etch below) an existing structure, a task which cannot be achieved merely by domain engineering and etching. One way to fabricate such structures is by employing the method of contact bonding [29–32]. This method relies on the Van Der Waals forces between two ultra-clean flat surfaces to maintain contact between them when brought to close proximity and is used routinely in the microelectronics industry.

This approach for fabrication of complex structures will be illustrated by following the steps for the fabrication of a set of free standing structure cantilevers which is a typical MEMS structure. For this purpose we will use an already processed substrate which carries all the undercut sections. This substrate is subsequently bonded to a second, structure-carrying, substrate forming a composite which can be further processed (polished, etched) as a whole.

The fabrication steps of the two separate substrates and the composite are outlined in the schematic which is shown in Fig. 1.13. The substrate shown on the left in the schematic is the one carrying the cantilever structures while the substrate on the right will form the basis of the structure.

In steps 1–3, as shown in the schematic, the two substrates are domain engineered in order to encode, in domain inverted patterns, the desired physical structures to be revealed after etching. The substrate on the left hand side of the schematic carries the cantilever structures while the substrate on the right hand side carries a large domain inverted square that provides the undercut section that allows the cantilevers to be free standing. The areas appearing grey in the schematic correspond to $-z$ face therefore material can be removed from this part of the surface by chemical HF etching.

In step 5 the composite is polished in order to reduce the thickness of the cantilever structure-carrying side. In this step the thickness of the cantilevers is defined and finally in the final step 6 the composite is chemically etched in HF to reveal the free standing cantilevers.

An example of the finished structure is shown in the SEM images of Fig. 1.14. The SEM image on the left shows the tip of the cantilever while the image on the right shows the point of support. The structure shown in Fig. 1.14 is $\sim$1 mm long, and has a top surface width of $\sim$50 μm and a height of $\sim$100 μm.

The cross-section of the cantilever however is not square because it is subject to lateral etching. In the image shown in Fig. 1.15 focussed gallium ions were used to cut off a part of the tip of the cantilever beam revealing its cross section which shows significant back etching as this particular sample has been over-etched.

Longer etching leads to more severe deformation of the original structure where in extreme cases the single cantilever can split into two beams with a common support point forming a structure that looks very much like a tuning fork.

The image shown in Fig. 1.16, taken using an optical microscope, shows that it is possible to focus through the free-standing beam. This indicates that the surfaces of the cantilever maintain their good optical quality during the fabrication process which is very important for any optical waveguide arrangement superimposed onto the physical structure.

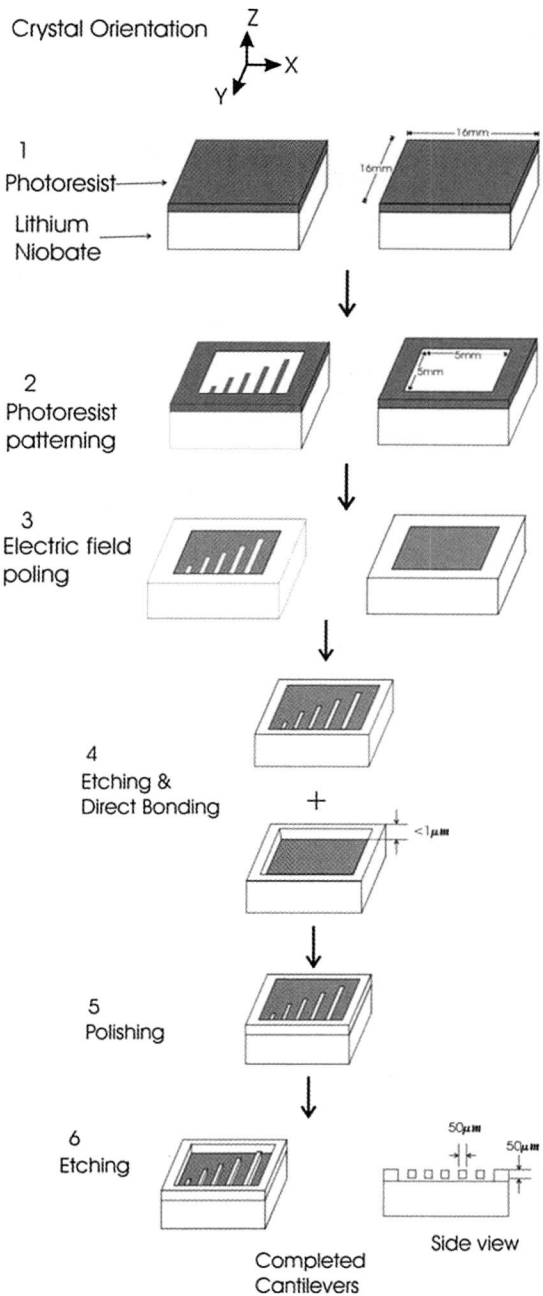

Fig. 1.13. Schematic illustrating the steps for the fabrication of free standing single crystal cantilevers

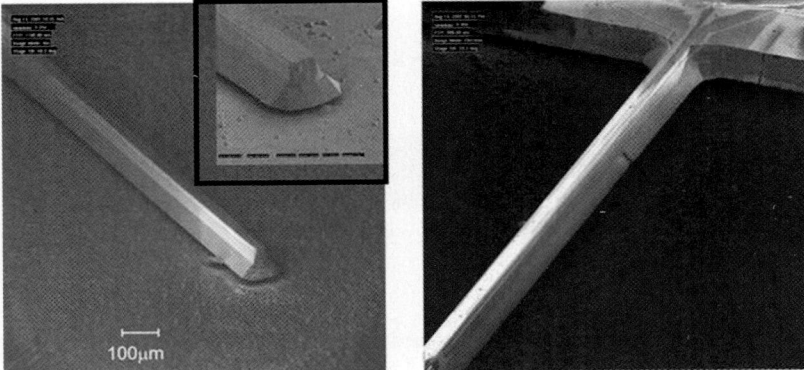

Fig. 1.14. SEM scans of a single crystal lithium niobate micro cantilever. *Left*: the tip of the structure, *right*: the support point, *inset*: high magnification image of the cantilever tip

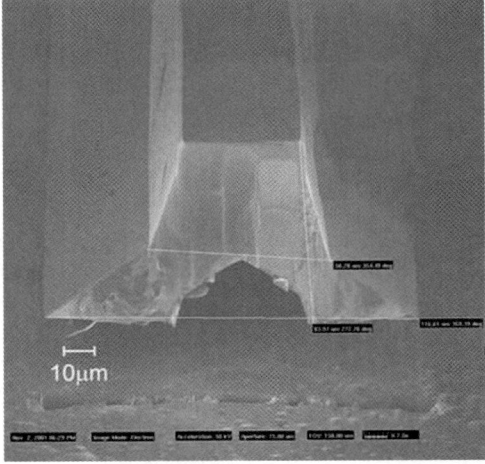

Fig. 1.15. Cross-section of a lithium niobate micro-cantilever

Tuneability is one unique feature of lithium niobate crystals that makes this material particularly attractive for use in optical devices. The optical and mechanical properties of the crystal can be modified by a small amount by (among others) the application of an electric field. Hence, any device will need to be equipped with electrodes that deliver the appropriate voltages locally. One way to deposit metals with precision is by using a focused ion beam to break down metalorganic precursors in the vicinity of the surface.

Although time consuming (depending on the size of the electrode) this is a very precise and straightforward method for fabricating electrical contacts. Deposition of tungsten (W) electrodes along the sides of a single crystal pyramid and a cantilever beam is shown in the SEM images of Fig. 1.17.

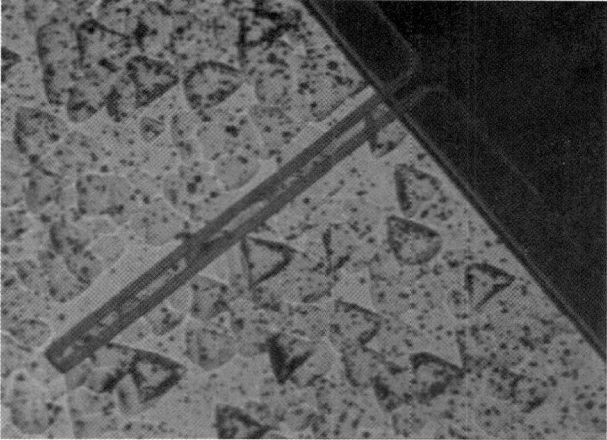

Fig. 1.16. Optical microscope image of a free-standing cantilever. The rougher undercut substrate is visible through the bulk of the cantilever

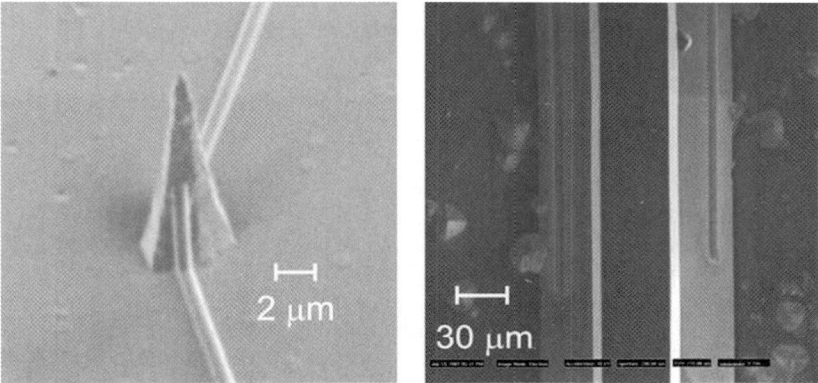

Fig. 1.17. Tungsten electrodes deposited by FIB induced CVD along the side surface of a single crystal tip (*left*) and a cantilever beam (*right*)

1.4 Summary and Future Work

Differential chemical etching between opposite z faces of lithium niobate crystals has been known to researchers since the discovery and initial experimental growth of crystals. While this effect has been extensively used for the assessment of the crystal quality in this chapter we have shown how differential etching between opposite z faces can be used for microstructuring of domain-engineered crystals. We have shown studies of the etch-rates under different conditions and discussed the simultaneous lateral differential etching between opposite y-faces and its impact on the topography of the resulting structures.

The better understanding of the differential etching process has helped to optimize it and furthermore to employ it in the fabrication of a variety of structures ranging from simple 2D surface structures to high aspect ratio 3D microstructures and finally, incorporation of the standard fabrication technique of contact bonding has enabled the realization of free standing structures such as a single crystal microcantilever. Structures which are fabricated using differential etching of ferroelectric domains benefit from the high optical quality of the etched facets which enables potential integration of optical structures onto the physical structures.

Finally, we have identified the method of chemical vapour deposition induced by FIB for the application of metal electrodes onto delicate micro-structures.

The micro-fabrication method that we have presented here however does not represent a panacea. There are still a number of processing issues that have to be addressed, especially if there is a scope for industrial scale fabrication. In reality it is necessary to consider all microfabrication methods and apply the suitable one for different intended structures. The message that we would like to emphasize with this chapter however is that there is indeed an extensive toolbox available for the microfabrication of lithium niobate and that it is feasible to fabricate complex structures thus expanding the application range, increasing the degree of integration and adding value to this very important nonlinear optical ferroelectric material.

References

1. K.E. Petersen, Proc. IEEE **70**, 420–456 (1982)
2. W.P. Eaton, J.H. Smith, Smart Mater. Struct. **6**, 530–539 (1997)
3. T. Frank, J. Micromech. Microeng. **8**, 114–118 (1998)
4. M. Koch, D. Chatelain, A.G.R. Evans, A. Brunnschweiler, J. Micromech. Microeng. **8**, 123–126 (1998)
5. M. Yamada, N. Nada, M. Saitoh, K. Watanabe, Appl. Phys. Lett. **62**, 435 (1993)
6. J. Webjorn, V. Pruneri, St.P.J. Russell, J.R.M. Barr, D.C. Hanna, Electron. Lett. **30**, 894 (1994)
7. C. Restoin, S. Massy, C. Darraud-Taupiac, A. Barthelemy, Opt. Mater. **22**, 193 (2003)
8. C. Restoin, C. Darraud-Taupiac, J.L. Decossas, J.C. Vareille, J. Hauden, Mater. Sci. Semicon. Proc. **3**, 405–407 (2000)
9. A.C.G. Nutt, V. Gopalan, M.C. Gupta, Appl. Phys. Lett. **60**, 2828 (1992)
10. S.Z. Yin, Microwave Opt. Technol. Lett. **22**, 396 (1999)
11. F. Lacour, N. Courjal, M.P. Bernal, A. Sabac, C. Bainier, M. Spajer, Opt. Mater. **27**, 1421 (2005)
12. H. Hu, A.P. Milenin, R.B. Wehrspohn, H. Hermann, W. Sohler, J. Vac. Sci. Technol. A **24**, 1012–1015 (2006)
13. V. Foglietti, E. Cianci, D. Pezzeta, C. Sibilia, M. Marangoni, R. Osellame, R. Ramponi, Microelectron. Eng. **67–68**, 742 (2003)
14. W.S. Yang, H.Y. Lee, W.K. Kim, D.H. Yoon, Opt. Mater. **27**, 1642–1646 (2005)
15. H.W. Chong, A. Mitchell, J.P. Hayes, M.W. Austin, Appl. Surf. Sci. **201**(1–4), 196–203 (2002)
16. L. Gui, B.X. Xu, T.C. Chong, IEEE Photonics Technol. Lett. **16**, 1337–1339 (2004)

17. S. Mailis, G.W. Ross, L. Reekie, J.A. Abernethy, R.W. Eason, Electron. Lett. **36**, 1801–1803 (2000)
18. K. Nassau, H.J. Levinstein, G.M. Loiacono, J. Phys. Chem. Solids **27**, 983 (1966)
19. R.S. Weis, T.K. Gaylord, Appl. Phys. A **37**, 191–203 (1985)
20. N.G.R. Broderick, G.W. Ross, H.L. Offerhaus, D.J. Richardson, D.C. Hanna, Phys. Rev. Lett. **84**, 4345 (2000)
21. I.E. Barry, G.W. Ross, P.G.R. Smith, R.W. Eason, G. Cook, Mater. Lett. **37**, 246 (1998)
22. C.L. Sones, S. Mailis, W.S. Brocklesby, R.W. Eason, J.R. Owen, J. Mater. Chem. **12**, 295 (2002)
23. D. Xue, K. Kitamura, Ferroelectrics **29**, 89–93 (2002) (Letters section)
24. T.J. Sono, J.G. Scott, C.L. Sones, C.E. Valdivia, S. Mailis, R.W. Eason, J.G. Frey, L. Danos, Phys. Rev. B **74**, 205424 (2006)
25. N. Niizeki, T. Yamada, H. Toyoda, Jpn. J. Appl. Phys. **6**(3), 318–326 (1967)
26. I.E. Barry, G.W. Ross, P.G.R. Smith, R.W. Eason, Appl. Phys. Lett. **74**, 1487 (1999)
27. A.C. Busacca, C.L. Sones, V. Apostolopoulos, R.W. Eason, S. Mailis, Appl. Phys. Lett. **81**, 4946 (2002)
28. S. Grilli, P. Ferraro, L. Sansone, M. Paturzo, S. De Nicola, G. Plerattini, P. De Natale, IEEE Photonics Technol. Lett. **18**, 541–543 (2006)
29. K. Mitane, U. Goselr, J. Electron. Mater. **21**, 669 (1992)
30. F. Gray, K. Hermansson, Appl. Phys. Lett. **71**, 3400 (1997)
31. K. Ljungberg, A. Soderbarg, Y. Backlund, Appl. Phys. Lett. **78**, 1035 (1993)
32. H. Himi, H. Matsui, S. Fujino, T. Hattori, Jpn. J. Appl. Phys. Part I **33**, 6 (1994)

2 Fabrication and Characterization of Self-Assembled Ferroelectric Linear and Nonlinear Photonic Crystals: GaN and LiNbO$_3$

L.-H. Peng, H.-M. Wu, A.H. Kung, and C.-M. Lai

2.1 Introduction

Ferroelectrics distinguish themselves from the commonly-known dielectrics or semiconductors by having a homogeneous spontaneous polarization ($\mathbf{P}_s$) due to the crystal's symmetry. The direction of $\mathbf{P}_s$ of a ferroelectric crystal can be changed at a temperature higher than the material's Curie point T_c or by applying an external field above a threshold value related to the crystal's coercive field $\mathbf{E}_c$ [1]. Single domain bulk ferroelectric crystals can be conveniently produced by using the technique of external electric field poling during the crystal growth [2]. Periodically inverted ferroelectric domains can be obtained by applying a series of short electrical pulses with a suitable waveform across a pair of electrodes on a proper ferroelectric substrate [3]. For the latter approach, caution should be exercised to avoid the back-switching of $\mathbf{P}_s$ due to an internal field that arises from the crystal's non-stoichiometric defects. The capability for a ferroelectric crystal/thin film to locally switch its polarization direction and to retain a structure with an alternating sign of polarization states can substantially modify the material's tensor properties and result in novel device applications.

According to tensoranalysis, all the odd higher-rank tensors such as the nonlinear susceptibility $\chi^{(2)}$, the piezoelectric modulus $\mathbf{d}_{ijk}$, and the electro-optic coefficient $\mathbf{r}_{ijk}$, which are of the third rank, will have their sign changed with the flipping of $\mathbf{P}_s$, whereas the sign of the even-ranked tensors such as the linear susceptibility $\chi^{(1)}$ will remain unchanged. This unique property makes ferroelectrics a promising material for applications in the field of quantum electronics and optoelectronics [4]. For example, high-density data storage can utilize the reversibility of $\mathbf{P}_s$ by fast switching the ferroelectric domain between opposite polarization states [5]. Applications related to energy storage and charge amplification can benefit from ferroelectric devices characterized with spatially graded polarizations [6].

When a periodic sign reversal of $\mathbf{P}_s$ appears in a "*head-to-head*" configuration, the discontinuity of $\mathbf{P}_s$ can generate a periodic distribution of localized charge density whose sign also will change periodically at the domain boundaries. The

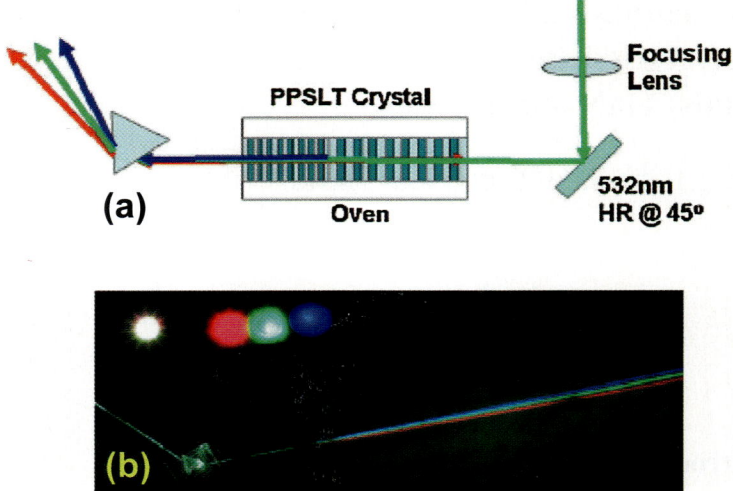

Fig. 2.1. (a) Schematic draw, and (b) generation of three basic colors from a green 532 nm pump periodically poled stoichiometric LiTaO$_3$ which invokes a cascaded parametric oscillation and sum frequency mixing

coupling of lattice vibrations with the electromagnetic waves can render various long-wavelength optical properties such as microwave resonance, dielectric abnormality, and polariton excitation [7]. Moreover, when the nonlinear susceptibility tensor $\chi^{(2)}$ changes sign with a spatial periodicity equal to the coherent length $l_c = \pi/\Delta k$ with a "*top and down*" configuration of $\mathbf{P}_s$, it imposes on the ferroelectric a structure-related lattice vector $\mathbf{G} = 2\pi/\Lambda$, where Λ is the structure period [8]. This vector provides a mechanism to satisfy the momentum conservation condition, $\mathbf{k}_1 + \mathbf{k}_2 + \mathbf{G} = \mathbf{k}_3$, in a three-wave nonlinear optical process to allow an efficient energy exchange. This effect therefore can compensate destructive interference in the nonlinear interacting waves caused by the material's dispersion as the waves propagate along the crystal. Such a ferroelectric domain engineering process is known as quasi-phase-matching (QPM) and is now widely used to achieve efficient wavelength conversion in harmonic [9] or parametric [10] generation. Compared to conventional angle- or temperature-tuned birefringent phase matching, QPM offers a higher conversion efficiency due to access to the largest nonlinear susceptibility tensor $\chi^{(2)}_{33}$ and absence of beam walk-off [11].

Figure 2.1 is a monolithic white light laser that highlights a promising application of such QPM-devices in display technology [12]. Here the white light laser is composed of coherent red, green, and blue beams generated by cascaded parametric oscillation and sum frequency mixing.

When the inverted domains are made to periodically reside in two-dimensional (2D) lattice sites, they form a new type of $\chi^{(2)}$ nonlinear photonic crystal ($\chi^{(2)}$-NPC) [13]. A wealth of nonlinear optical phenomena such as cascaded harmonic generation [14, 15] and enhanced phonon-polariton Raman scattering [16] can be observed due to the availability of plural lattice vectors involved in the QPM processes.

When the ferroelectric domain is made to have a periodic structure at a pitch around λ/n, it forms another new type of ferroelectric photonic crystal that can be used to disperse and control light-matter interaction [17]. It also enables novel optical processes of SHG and OPO in which the pump beam and the generated waves propagate in opposite directions [18, 19]. In these applications, the domain structure has to be engineered into the submicron regime. There are many other interesting applications such as high-density ferroelectric memory and electronic devices that also benefit from ferroelectric domain engineering into the nanometer regime. References [20] and [21] provide a good review concerning these latter applications. Owing to these accounts, it is of much scientific interest to establish versatile polarization switching techniques that are effective in manipulating the domain reversal to a desired dimensionality and configuration.

The organization of this chapter is as follows. In Sect. 2.2 we address the challenges commonly encountered in the fabrication of periodically poled ferroelectric domain structures whose pitch size is in the range of a few micrometers. In particular, we focus on the issue of domain control in the material systems of undoped and doped congruent-grown and stoichiometric lithium niobate ($LiNbO_3$) and lithium tantalate ($LiTaO_3$) that are widely used in academic research and industrial applications. In Sect. 2.3 we discuss the fabrication methods developed in our labs that allow the realization of ferroelectric domain engineering in the micron and submicron regime. We address the issues in domain engineering by manipulating the electrostatic field by the methods of (a) charged potential barrier, (b) high-k, and (c) self-assembled electrode. Examples of efficient nonlinear harmonic generation and the characterization of temperature and wavelength tuning bandwidth will be given for the 2D $\chi^{(2)}$-NPC. Single mode stimulated emission from submicron cavities of the ferroelectric semiconductor (gallium nitride, GaN) will be described. The compatibility of the ferroelectric nano-domain engineering techniques with the existing processing facility will be discussed.

2.2 Micro-Domain Engineering with Conventional Poling Electrode Design

2.2.1 Internal Field Effect in the Poling of Congruent-Grown $LiNbO_3$ or $LiTaO_3$

Based upon early studies of domain reversal in $BaTiO_3$, it is suggested that the surface nucleation rate, sidewise expansion, and coalescence of domain constitute three major steps in the kinetics of polarization switching [22, 23]. For domain nucleation to take place under a *uniform* field, one applies an external voltage to overcome the activation barrier—a factor determined from the minimization of the ferroelectric free energy [24]. For realizing a periodically poled ferroelectric structure, one takes advantage of carrier injection from an external circuit to compensate the depolarization field underneath the corrugated metallic electrodes and to initiate the polarization switching process. In the latter, domain reversal takes place underneath the edge of the metal electrode, where the distorted field has a peak value that increases the nucleation rate of the inverted domain according to $\sim\exp(-E_a/E_z)$, where E_a is the phenomenological activation energy [25].

For poling ferroelectrics such as congruent-grown LiNbO$_3$ or LiTaO$_3$ that have a high coercive field ($E_c \sim 21$ kV/mm), a typical procedure is to work with a "liquid electrode" by immersing the high voltage leads in a salt solution such as lithium chloride (LiCl) or in oil. The remaining tasks in the development of high-quality periodically poled ferroelectric structures are to control the material's internal field E_{int} and a structure-related fringe field $E_{x,y}$ in the tangential plane of the ferroelectric surface. The former arises from the material's stoichiometry, whereas the latter is due to field distortion caused by dielectric discontinuity at the ferroelectric corrugated surface and the poling electrode [26].

For the widely-used ferroelectric crystals of congruent-grown LiNbO$_3$ and LiTaO$_3$, there exists a large internal field $E_{int} \sim (2-4\,\text{kV/mm})$ due to non-stoichiometric defects [27, 28]. The direction of E_{int} points to the crystal's $+C$ axis and can cause instability in the inverted domain such that the latter has a tendency to flip the corresponding P_s back to its original polarization status upon termination of the poling field. Improved waveform designs in the poling field such as to utilize a low field to stabilize the inverted domains or to apply a reverse field to compensate the domain wall movement have been demonstrated [29]. An alternative solution is to put a fast turn-on rectifying diode in series with the poling apparatus such that the relaxation of the inverted domains driven by E_{int} is inhibited when terminating the poling field [30].

2.2.2 Origin of the Fringe Field

Irregularity [31] and broadening [32] of inverted domains, however, are troublesome issues commonly observed in the conventional poling procedure that can result in a great loss in the nonlinear optical conversion efficiency. This is related to a fundamental problem due to the unwanted lateral domain motion caused by the fringe field ($E_{x,y}$) in the tangential plane of the ferroelectric. The origin of the fringe field can be ascribed to the bending of equal potential line across an abrupt dielectric boundary in the corrugated ferroelectric surface which typically is coated with an insulating layer of photo-resist, SiO$_2$ [33] or spin-on-glass [34]. The fringe field causes current to spread into the unpatterned region and results in an unwanted domain wall motion into this area. To quantitatively evaluate such a fringe field effect, we develop a coding algorithm based upon a finite difference method [35] to solve the Poisson equation that considers a spatial modulation in the dielectric function and a distribution of fixed charge density as well.

Let us assume the following structure parameters for evaluating the fringe field: an applied voltage $V/d = E_c$, a QPM periodicity Λ of 20 μm, and an insulating layer of photo-resist with 1 μm thickness and a dielectric constant $\varepsilon_r = 3$ over

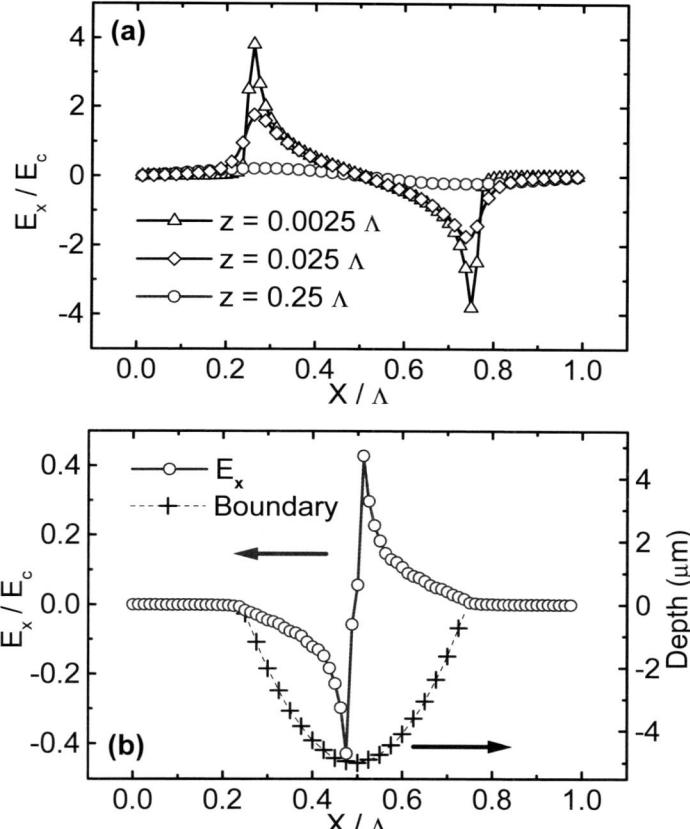

Fig. 2.2. Calculated field line distribution for the fringe field E_x in LiNbO$_3$ with (a) corrugated 1D electrode design covered with photo-resist, (b) flat 1D electrode design embedded with positive charge density $\rho = 10^{-3}$ **P**$_s$ underneath the surface

a congruent-grown LiNbO$_3$ of thickness $d = 0.5$ mm and a $\varepsilon_r = 30$. With such a conventional poling electrode design, our calculation results in Fig. 2.2(a) indicate that a relatively large $\mathbf{E}_x \sim 4E_c$ can exist at $z = 0.0025\Lambda$ underneath the insulating layer, manifesting itself to a dramatic distortion of the local switching field ($\mathbf{E}_z$ not shown). Even though $\mathbf{E}_x$ only affects a thin surface region up to $z \sim 0.25\Lambda$ and fades away at a larger distance, what is important is the direction of its field line pointing *inward* from both sides underneath the insulating layer. This creates field screening and field compensation in the unpatterned regime and provides a current flow path responsible for domain wall movement into the unpatterned regime. The fringe field becomes one of the most troublesome issues related to domain broadening and has to be solved to retain a high-fidelity in periodically poling of ferroelectric structures.

2.2.3 Poling Issues with Doped or Stoichiometric LiNbO$_3$ and LiTaO$_3$

From the material's point of view, selective doping of congruent-grown LiNbO$_3$ (CLN) with zinc oxide (ZnO) [30, 36] or magnesium oxide (MgO) [37] can dramatically decrease the crystal's coercive and internal field due to the suppression of the non-stoichiometric defects. The underlying mechanism can be understood to result from the replacement of the crystal's point defects such as Li vacancy (V_{Li}) and Nb antisite (Nb$_{\text{Li}}$) by dopants (such as Zn or Mg) with a suitable atomic radius that matches well to the virgin crystal [38, 39]. Alternatively, stoichiometric ferroelectric crystals of lithium niobate (SLN) or lithium tantalate (SLT), grown by an improved double-crucible Czochralski method, are also shown to have similar effects in reducing the coercive and internal field due to the improvement of the material's composition [40, 41]. For the doped congruent-grown or stoichiometric-grown LN/LT, the coercive field can be reduced by one-order of magnitude with respect to the value in the undoped congruent-grown counterparts. The doped LN/LT crystals are also found to have a higher resistance to the photo-refractive effect [42], i.e., higher optical damage resistance, thus making them a suitable material for high-power laser applications [43]. However, uniformity issues and crystal quality of doped congruent-grown and stoichiometric-grown LN/LT become a major concern from the user's (poling) point of view.

On the other hand, for the more mature MgO:CLN material, it was recently noted that the surface resistance of a ferroelectric domain changed by a factor $\sim 10^6$ upon the flipping of $\mathbf{P}_s$, whereas no resistance change was observed in the case of polarization switching of undoped CLN [44]. Such a wide range of resistance change causes a serious problem, which is, it forms a path of current leakage upon performing ferroelectric poling on the doped CLN. A random distribution of such nucleation sites of inverted domain at the initial stage of polarization switching can evolve into a formation of leakage paths that can quickly drain the external bias current to produce an area A of inverted domains according to $i_{\text{pol}} = 2A\,d\mathbf{P}_s/dt$. Since the majority of the poling current would flow into these "leaky" paths which have a much lower resistance, the electrostatic interaction between the field and the spontaneous polarization, i.e., $-\int (2\mathbf{P}_s \cdot \mathbf{E}_z)\,dr^3$, would speed up the expansion of inverted domains from these randomly distributed sites and the coalescence of nearby neighboring inverted domains to minimize the total energy. As a result, it becomes difficult to fabricate periodically poled QPM-structures with a high-fidelity with doped CLN of a standard 0.5 mm-thick substrate when the domain periodicity is less than 10 μm.

Another report identifies that thermal history prior to poling (such as cooling in the postbake stage) can significantly affect the poling quality of the MgO:CLN crystal due to charge induced pyroelectric effect on the surface [45]. Since MgO:CLN has an $\mathbf{E}_c$ that is only $\sim 1/6 \sim 1/8$ of the undoped case (21 kV/mm), the pyroelectric field can produce randomly distributed nucleation sites in the form of shallow inverted domains that essentially play the role of "leakage path" in the poling procedure.

Special techniques, including the use of multi-pulse poling on relatively thick (2–3 mm) substrates at elevated temperature, have been developed to improve the poling quality of MgO:CLN [46, 47]. The idea is to prevent the formation of current leakage paths by using short pulses to prevent the penetration of the inverted domains from reaching the opposite side of the thick substrate, thus to mitigate the issues of domain broadening and irregularity [48]. However, low yield and a high amount of wasted unused wafer material of the expensive doped substrate constitute a major concern in such an approach.

2.3 From Micron to Submicron Domain Engineering with Improved Electrode Design

In this section, we provide three proof-of-concept mechanisms, viz. the use of charged potential barriers and the use of self-assembled dielectric spheres in conjunction with high-k dielectric layer to provide spatial modulation of the poling field and to mitigate the fringe field effect. A major advantage of these methods is their compatibility with conventional semiconductor processing procedures such as photolithography and thin film coating, thus leading to affordability of submicron ferroelectric domain engineering in a cost-effective manner. In addition, our approaches provide a versatile means of fabricating practical device-size (1 cm^2) ferroelectric optoelectronic and electronic devices with a domain size varying from a few microns to submicron feature sizes. Examples will be illustrated on 1D/2D periodically poled LiNbO$_3$, LiTaO$_3$, and the ferroelectric semiconductor gallium nitride (GaN).

2.3.1 Charged Potential Barrier Method

Example of Forming 1D Periodically Poled LiNbO$_3$

Let us assume there is a distribution of positively charged potential parabola periodically underneath a ferroelectric surface which has a pitch corresponding to a QPM periodicity. Such a potential can arise from a distribution of localized ions due to an implantation process or from the fixed depolarization charges due to a spatial divergence of $\mathbf{P}_s$ ($-\nabla \cdot \mathbf{P}_s = \rho$). As an example, we illustrate in Fig. 2.2(b) a positively charged parabola representing the case of polarization-induced depolarization charge due to a distribution of $\mathbf{P}_s$ in a *head-to-head* configuration along the crystal's $+C$ axis [49]. We now evaluate the horizontal tangential field $\mathbf{E}_x$ at a depth $z = 0.0025\Lambda$ ($\Lambda = 20\,\mu\text{m}$) along the parabolic domain boundary according to the finite-difference algorithm outlined in Sect. 2.2.2. Compared with the case shown in Fig. 2.2(a) of a conventional poling electrode over the ferroelectric surface, we note the replacement of the *corrugated* insulating pattern by a positively charged parabola of $\rho = 10^{-3}\mathbf{P}_s$ underneath a *flat* surface not only can significantly reduce the value of $\mathbf{E}_x$ by about *one order of magnitude* to $0.4E_c$ in the peak value but also can *reverse* the direction of the $\mathbf{E}_x$ field line to point *outward* underneath the intentionally unpoled region.

The existence of such a positively charged parabola underneath the ferroelectric surface essentially forms a potential barrier (by exerting repelling field line) to inhibit the migration of inverted domain wall to reach the charged boundary. This signifies a promising mechanism to oppose the current spreading effect discussed in Fig. 2.2(a) and offers a solution to mitigate the issues of domain broadening. To provide a concrete evidence of a charged barrier effect in micron-domain engineering, we tested this technique on a 1D QPM structure with a periodicity of 20 μm on a Z-cut, 0.5 mm thick, congruent-grown undoped LiNbO$_3$ substrate. To introduce a positively charged parabola on the subsurface of LiNbO$_3$, we perturbed the subsurface domain structure in a *head to head* configuration along the $+C$ axis. To avoid generating new crystal phases on the polar surface, we adopted a high temperature induced out-diffusion process [50] instead of the commonly used ion-exchange method [51]. We placed the LiNbO$_3$ substrate patterned with a thin (~50 nm) aluminum (Al) electrode inside a quartz tube furnace and allowed it to undergo a heat treatment at 1050°C for 5 hours in an air ambience. After the heat treatment, the Al-electrode transformed to micro-pore like aluminum oxide (Al$_2$O$_3$) electrode. The sample was then transferred to a poling apparatus consisting of a high voltage amplifier (Trek 20/20 A for 20 kV and 20 mA output) in series with a fast turn-on diode to execute pulsed-field poling using liquid electrodes.

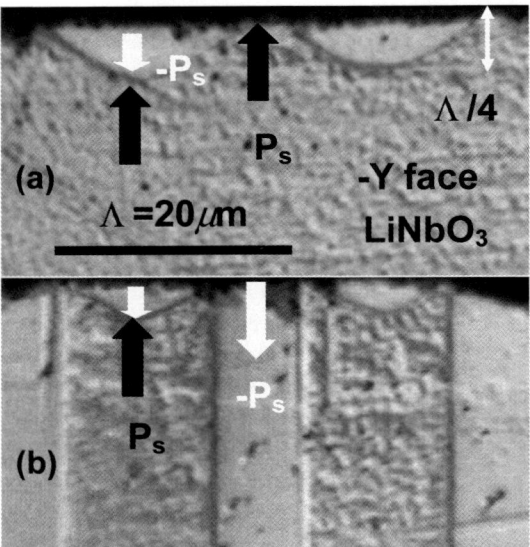

Fig. 2.3. $-Y$ face micrographs of LiNbO$_3$ illustrating the effects of (**a**) forming head to head domain configuration after the first high-temperature treatment at 1050°C, and (**b**) confined lateral domain to the charged boundary. The domain period was 20 μm and poled at 21 kV/mm across a 0.5 mm-thick substrate

Our poling results are displayed in Fig. 2.3, where (a) is the $-y$ face micrograph of an etched 20 μm-period QPM structure after the high-temperature treatment at 1050°C. We denote a formation of shallow surface inverted domain of a rounded triangular shape at a depth of 0.25Λ (5 μm) below the uncovered LiNbO$_3$ surface. This phenomenon differs from a previous study on SiO$_2$ coated LiNbO$_3$[52] upon which the stress induced piezoelectric effect is responsible for causing domain inversion in region covered with SiO$_2$. Our data suggest that the origin of surface domain inversion is attributable to the space charge effect that arises from out-diffusion or the defect gradient associated with the high-temperature process [53–55].

The *head-to-head* distribution of $\mathbf{P}_s$ in Fig. 2.3(a) indicates the existence of fixed depolarization charge according to the basics of electrostatics by $-\nabla \cdot \mathbf{P}_s = \rho$. It points out a possibility of using this potential "barrier" to manipulate the nucleation of inverted domains and the movement of the charged domain walls in the kinetic response of ferroelectric domain engineering. As a pedagogical illustration of this principle, we show in Fig. 2.3(b) the etched $-y$ face micrograph of a periodically poled LiNbO$_3$ structure that has undergone both the heat treatment and pulsed field poling at 21 kV/mm. We note that the inverted domain shown in Fig. 2.3(a) did not resume a forward growth along the existing 180° triangular domain boundary. Indeed the volume of such shallow surface inverted domain formed in the first heat treatment remains unchanged after the pulsed field poling.

Instead, we observe from Fig. 2.3(b) that the nucleation of inverted domain and its forward growth takes place within the region where the original Al electrode turns into micro-pore like electric contacts spaced by the self-assembled Al$_2$O$_3$ micro-electrode. The lateral motion of such domain walls is found to be *confined* to the edge of the charged triangular domain boundary.

It is to be noted that our observation differs from recent studies of using the proton-exchange process to destroy the ferroelectric phase so as to perform periodic domain reversal in LiTaO$_3$ [56] and LiNbO$_3$ [57]. In the latter, by periodically inserting a non-ferroelectric shallow layer into the ferroelectric host crystal, one can suppress the sidewise motion of the inverted domains.

Here our work suggests the following interesting aspects:

(a) Positive depolarization charge, arisen from $\rho = -\nabla \cdot \mathbf{P}_s$, constitutes a potential barrier to reduce the displacement (switching) current in the sidewise and the vertical direction.
(b) From (a) such a charged boundary can stop the forward growth of inverted domain along the existing 180° domain boundary, where the head to head distribution of $\mathbf{P}_s$ leads to a much larger formation energy of inverted domain.
(c) From (a) such a charged boundary can introduce a repelling field line to inhibit the lateral inverted domain motion.
(d) The electric contact to the ferroelectric surface, hence the nucleation site of the inverted domain, is established via the micro-pore like fissure in the self-assembled Al$_2$O$_3$ electrode (high-k dielectric insulator).

Example of Forming 2D Periodically Poled LiTaO$_3$

To further apply the aforementioned technique for making 2D periodical poled ferroelectrics, we simulate in Fig. 2.4 the $\mathbf{E}_y$ field line distribution in the tangential plane of LiTaO$_3$ for the case of (a) a corrugated surface with photo-resist patterning and (b) a planar surface but periodically embedded with a 2D distributed positively charged parabola [58]. The charged parabola is again assumed to have a density of $10^{-3}\mathbf{P}_s$ placed at a maximum depth of 0.25Λ below the surface and to have a periodicity equal to that of a given QPM structure. We evaluate the $\mathbf{E}_y$ field distribution at a depth of $z = 0.05\Lambda$ below the $+Z$ face according to the algorithm of the finite-difference method discussed in Sect. 2.2.2. The relative strength of $\mathbf{E}_y$ has been normalized with respect to the coercive field $\mathbf{E}_c = V/d$, where V is the applied voltage across a planar LiTaO$_3$ substrate of thickness d. We first note from Fig. 2.4(a) a peak value of $\mathbf{E}_y \sim 4E_c$ with the filed line pointing *inward* underneath the photo-resist coverage region. A *saddle*-like field distribution of $\mathbf{E}_x$, similar to $\mathbf{E}_y$ but having an inverse symmetry with respect to the mid point, can be obtained by making a 90° coordinate transformation of letting $x \leftrightarrow y$. These effects make $\mathbf{E}_{x,y}$ act as a driving force, causing current to spread away from the four corners of the poling electrodes. As a result, fast carrier redistribution underneath the photoresist region accelerates the sidewise motion of the inverted domains. This becomes the source of domain irregularity and broadening in the 2D periodically poled device.

In comparison, for the case of flat ferroelectric surface but embedded a 2D positively charged parabola as shown in Fig. 2.4(b), we note that a compensation of the fringe field $\mathbf{E}_y$ by the positive charge with respect to the external applied field can significantly reduce its peak value by one order of magnitude down to $0.1\mathbf{E}_c$. In addition, the field line of $\mathbf{E}_{x,y}$ in Fig. 2.4(b) characteristically points in an *outward* direction with respect to the charged parabola, resembling its 1D counterpart shown in Fig. 2.2. Such a group of distributed potentials indeed form confinement barriers that block lateral current flow and offer a mechanism to inhibit lateral motion of the inverted domain.

To provide a more concise demonstration of such a self-assembled 2D poling process, we utilize a nickel (Ni) in-diffusion process [59] to realize a positively charged potential barrier in the Z-cut congruent-grown LiTaO$_3$ substrate of 0.5 mm thickness. At an annealing temperature around 580°C which is below the Curie temperature ($\sim$600°C) of LiTaO$_3$, the Ni diffusion process results in a shallow surface domain reversal, thus providing a 2D charged potential parabola according to $-\nabla \cdot \mathbf{P}_s = \rho$ underneath the Z-face of LiTaO$_3$. Here we denote that nucleation of the inverted domain in annealed LiTaO$_3$ begins in regions not covered by the Ni pad via contact with the liquid electrode, and the lateral motion of the invert domains is prohibited by the 2D positively charged potential barrier.

Data shown in Fig. 2.5 are the (a) $-Z$ face optical micrograph of 2D periodically poled LiTaO$_3$, which has a 2D QPM periodicity of $7.8 \times 13.6\,\mu\text{m}^2$, and (b) the 2D SHG pattern generated by a 10 ns pulsed 1064 nm Nd:YAG pump laser at a temperature of 30, 40, 45, 50, and 55°C respectively. We first note there are two kinds of inverted domains in Fig. 2.5(a), viz., the larger one ($\sim$6 μm in size) has a *triangu-*

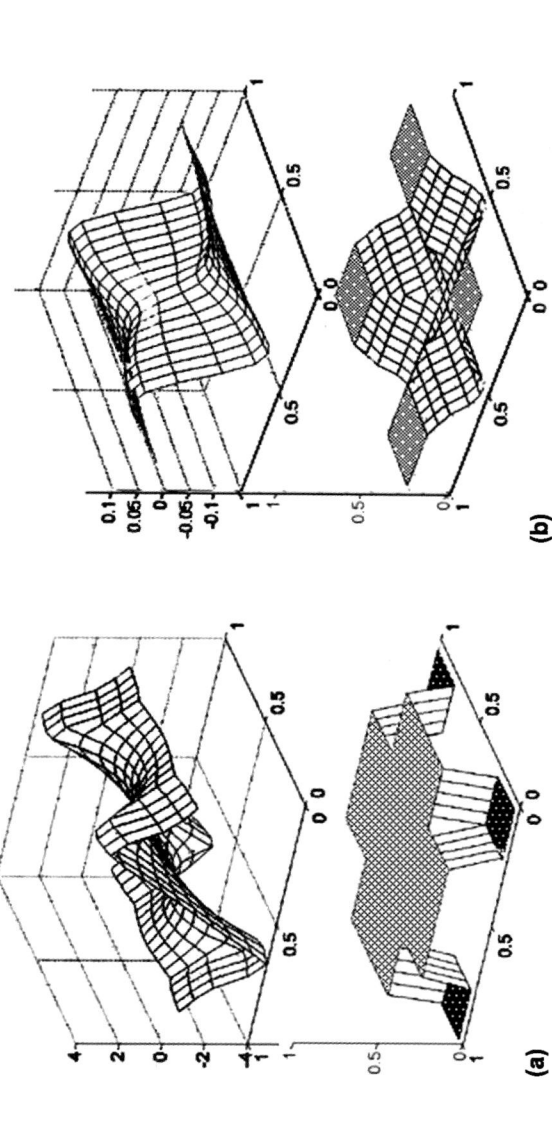

Fig. 2.4. Calculated Field line distribution in the x–y plane for the fringe field $\mathbf{E}_y$ in LiTaO$_3$ with (**a**) corrugated 2D electrode design covered with photo-resist, (**b**) flat 2D electrode design embedded with positive charge density $\rho = 10^{-3}\,\mathbf{P}_s$ underneath the surface

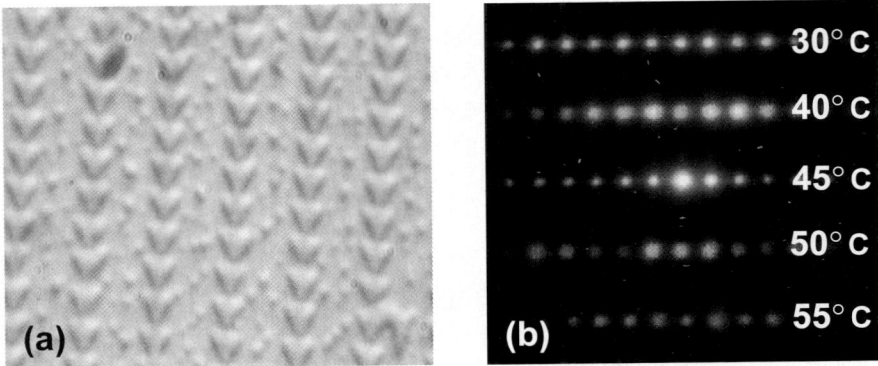

Fig. 2.5. (a) $-Z$ face micrographs of 2D poled LiTaO$_3$ with a periodicity of $7.8 \times 13.6\,\mu m^2$, (b) far field pattern of 2D SHG signals generated by normal incident 10 ns Nd:YAG laser at temperatures of 30, 40, 45, 50, 55°C

lar shape and obeys the crystal symmetry of LiTaO$_3$, and the smaller one (∼1 μm in size) of circularly poled crystallites. The triangular shape of inverted LiTaO$_3$ domain results from fast expansion and movement of domain wall along the 3-fold degenerate axes once the forward domain growth reaches the opposite side of the substrate, reminiscent of a case for incomplete bulk screening [60]. On the other hand, the nucleation of the scattered smaller (∼1 μm) rounded crystallites is due to the localized high-field action through a micro-pore contact at the NiO$_x$/LiTaO$_3$ interface. The charged potentials surrounding this nucleation site further constrain the lateral expansion of the inverted domain and lead to the appearance of rounded shape. A more careful examination of the green emission pattern in Fig. 2.5(b) indicates that at $T = 45°C$, QPM-SHG takes place due to phase matching with the reciprocal lattice vector $\mathbf{G}_{10}$. This begins the appearance of an array of high-brightness SHG spots in the far field of the 2D NPC.

Optical Characterization of 2D Periodically Poled LiNbO$_3$

(a) Temperature/Azimuth Angle Dependent $\mathbf{G}_{mn}(T, \phi)$ of the QPM-SHG Process

The plurality of reciprocal lattice vectors in a 2D periodically poled QPM ferroelectric nonlinear photonic crystal (NPC) offers a wide tuning range in the phase matching temperature and rotation angle for second harmonic and cascaded harmonics generation. To illustrate one such capability, we investigated temperature tuning of a QPM-SHG process in a 2D poled LiNbO$_3$ with a periodicity of $6.6 \times 13.6\,\mu m^2$. This device can be pumped and phase matched either by a 1064 nm Nd:YAG or a 1.3 μm-band pump laser from one of the x/y axis of the 2D-NPC. With such a structure modification, one would expect to have the generation and propagation of the nonlinear interacting waves different from those in the case of bulk birefringent matching. We schematically show these optical processes in Fig. 2.6(a), with the

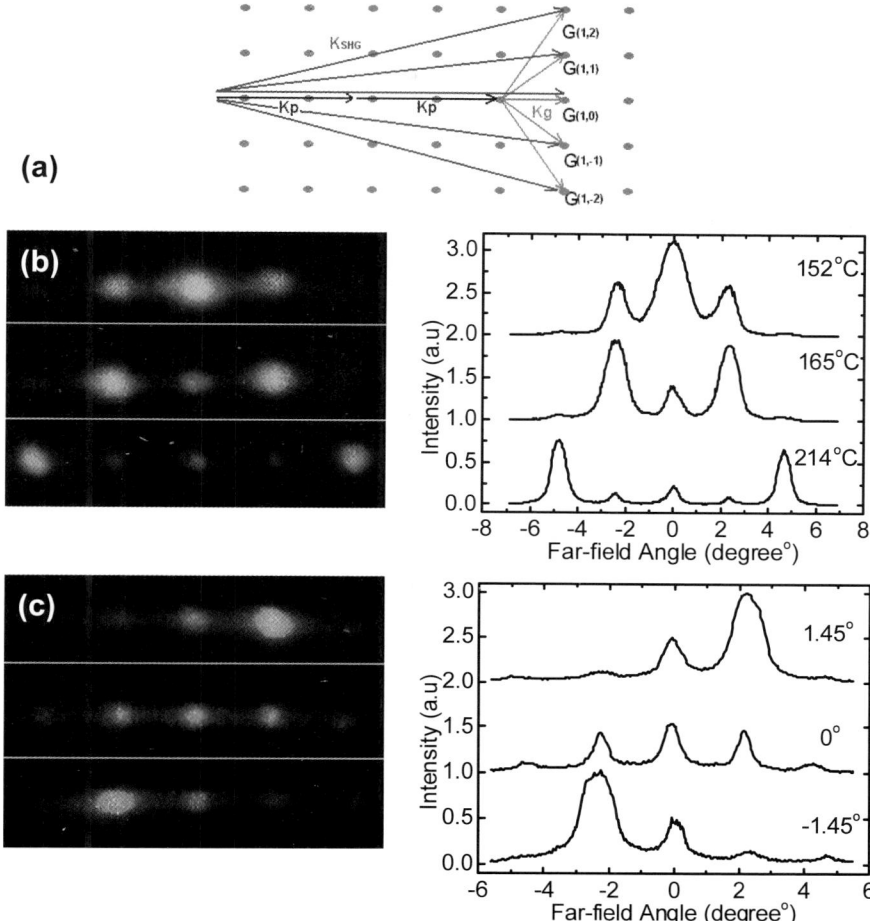

Fig. 2.6. (a) Schematic show of 2D QPM-SHG processes involved various $\mathbf{G}_{mn}$ reciprocal lattice vectors, (b) far-field intensity and image of the $\mathbf{G}_{mn}$-assisted SHG process (at normal incidence) measured at various temperatures for 2D poled LiNbO$_3$ with a periodicity of 6.6 × 13.6 μm^2, (c) far-field intensity and image of the $\mathbf{G}_{mn}$-assisted SHG process (at $T = 10°C$) at various incident angles for 2D poled LiNbO$_3$ with a periodicity of 6.9 × 13.6 μm^2

image/intensity distribution of the far-field SHG pattern presented as a function of the crystal rotation angle and the crystal temperature in Figs. 2.6(b) and (c), respectively [61].

Data in Fig. 2.6(b) were taken at a crystal temperature of 152, 165, and 214°C, respectively. They were measured for a *normal* incident configuration with a 10 ns pulsed Nd:YAG laser at a peak intensity of ∼2 MW/cm^2 and propagating along the short period (i.e., 6.6 μm) side of the sample. The appearance of a series of

SHG signals with *unequal* intensity clarifies the participation of *transverse* reciprocal lattice vectors that are absent in the 1D QPM-SHG case. These characteristic distributions of the far-field angle dependent SHG spectra reveal a QPM condition $\mathbf{k}_{2\omega} - 2\mathbf{k}_\omega - \mathbf{G}_{mn}(n, T) = 0$ that can be sequentially fulfilled by the participation of the reciprocal lattice vectors $\mathbf{G}_{1,0}$, $\mathbf{G}_{1,\pm 1}$, and $\mathbf{G}_{1,\pm 2}$ at the respective temperatures.

The reciprocal lattice vector $\mathbf{G}_{1,\pm m}$-related QPM conditions are further explored in Fig. 2.6(c), where the SHG signals are taken from another 2D poled LiNbO$_3$ with a periodicity of $6.9 \times 13.6\,\mu\text{m}^2$ at $T = 10°C$. The data were measured with an azimuth rotation angle of $\phi = 0°$ and $\pm 1.45°$, respectively. Note the far-field patterns for the SHG signals taken at $\phi = \pm 1.45°$ have a mirror image with respect to each other and have a peak SHG signal appear at a far-field angle of $\pm 2.3°$, but overlaid with discrete green spots that are weak in intensity and not phase-matched. In the *normal* incident condition ($\phi = 0°$), only weak SHG signals are observed, indicating that there is no phase-matching in this configuration.

These temperature/angle dependent SHG signals observed in Figs. 2.6(b) and (c) can be understood by considering the $\mathbf{G}_{mn}(T, \phi)$-related QPM process, where the reciprocal lattice vector $\mathbf{G}_{mn}$ for phase-matching is a function of the temperature T and the azimuth rotation angle ϕ in the x–y plane. Taking the material dispersion in the refractive index $n(\omega, T)$ of LiNbO$_3$, [62] and the QPM condition $\mathbf{k}_{2\omega} - 2\mathbf{k}_\omega - \mathbf{G}_{mn} = 0$, one can understand the effect of 2D $\chi^{(2)}$ structure on the dispersive generation of 2D SHG green lasers. Figure 2.7 shows the calculated 2D dispersion curves for those two periodically poled LiNbO$_3$ whose SHG data were presented in Fig. 2.6. Here we detect a general trend of a monotonic increase in the phase-matching temperature with the rotation angle. This finding agrees with most of the experimental observations. The exception being that the $\mathbf{G}_{10}(T, \phi)$ assisted QPM-SHG process is less sensitive to the variation of the incident angle. We also observe an increase of the phase-matching temperature with the magnitude of $\mathbf{G}_{mn}$ involved in the nonlinear wave interaction. This effect is also correlated to the increase of the far-field emission angle in the $\mathbf{G}_{mn}$-related QPM-SHG signal. The

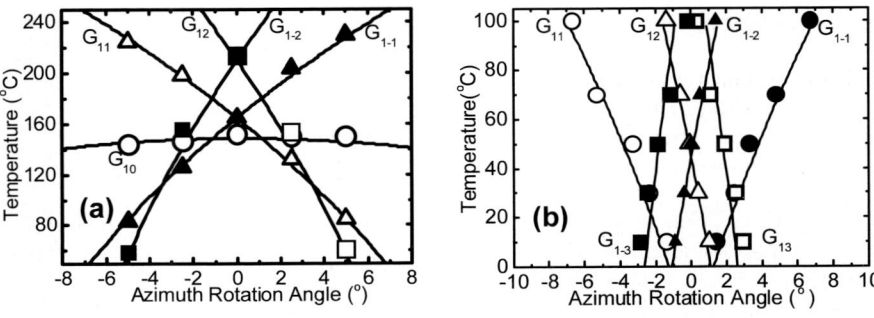

Fig. 2.7. Calculated temperature dispersion curves overlaid with the experimental data to show the acceptance bandwidth for the $\mathbf{G}_{mn}$-assisted SHG processes for 2D poled LiNbO$_3$ with a periodicity of (a) $6.6 \times 13.6\,\mu\text{m}^2$, and (b) $6.9 \times 13.6\,\mu\text{m}^2$

$G_{m\pm n}(T,\phi)$ dispersion curves show an axial symmetry as one varies the rotation angle in the x–y plane. This can be related to the fact that the $\chi^{(2)}$ nonlinearity in the QPM-NPC is characterized by an orthorhombic lattice structure that exhibits a C_{2v} symmetry in our calculation.

(b) Wavelength/Azimuth Angle Dependent $G_{mn}(\lambda,\phi)$ of the QPM-SHG Process

The wavelength tunability of a QPM-SHG process was investigated by using another two sets of 2D poled LiNbO$_3$ with a periodicity of 20 × 20 and 29.5 × 29.5 μm^2 [63]. These two samples were designed to phase-match to the 1.5 μm- and 1.9 μm-wavelength bands for the second harmonic and cascaded harmonic generation processes. Here a home-built PPLN-based narrow-band ns-OPO was used as the fundamental pump source in a grazing incidence configuration with a grating cavity [64]. The peak power at 1.5 μm was 2 kW, while that at 1.9 μm was 1.2 kW. The effective length of the 20 × 20 and 29.5 × 29.5 μm^2 sample was 6 mm and 18 mm, respectively.

Figures 2.8(a) and (b) show the images and the far-field intensity distribution taken at wavelengths that match the QPM-SHG process with the $G_{1,0}$, $G_{1,\pm1}$, $G_{1,\pm2}$ reciprocal lattice vectors for the two 2D periodically LiNbO$_3$ crystals. The SHG data were measured with the fundamental beam propagating along the crystal's x-axis (i.e., 0° incident angle). The images were taken by placing an IR sensitive florescence card at 10 cm away from the sample. Data of Fig. 2.8 exhibit a symmetrical distribution of the SHG patterns in a plane *transverse* to the propagation direction of the fundamental beam. By rotating the crystal 90° to let the pump beam propagate along the crystal's y-axis, similar transverse distributions of the SHG signal can also be observed. Such a four-fold rotational symmetry indeed reflects the tetragonal structural effect of the 2D NPC on the SHG process.

By rotating the 2D poled LiNbO$_3$ samples to vary the effective internal incident angle of the pump beam relative to the x (or y) axis of the 2D NPC, one equivalently changes the period of the QPM structure and hence can use it to examine the wavelength dependence of the QPM-SHG processes. This procedure allows us to analyze the lattice spacing effect on the wavelength span of QPM-SHG in the 2D NPC. Data in Fig. 2.8(c) represent the measured and calculated phase-matching wavelength as a function of the azimuth rotation angle and the reciprocal lattice vector G_{mn}. We note an increase in the spacing between the QPM wavelengths with the azimuth rotation angle ϕ. The wavelength span of the QPM-SHG process is shown to increase with the lattice spacing of the 2D NPC, changing from 150 to 200 nm as the 2D lattice spacing increases from 20 to 29.5 μm.

We further illustrate in Figs. 2.9(a) and (b) the conversion efficiency of the $G_{1,0}$-assisted QPM-SHG process in the 20 × 20 and 29.5 × 29.5 μm^2 2D poled LiNbO$_3$. The data were taken at a QPM pump beam of 1589 and 1998 nm wavelength, respectively [65]. Here we note a saturation at an internal SHG efficiency at ~50%, occurring at a pump intensity ~8 MW/cm^2. We also observed cascaded generation of the third- and fourth-harmonics in Fig. 2.9(c) with the fundamental pump laser that accompanied the SHG saturation. The latter reflects an interesting

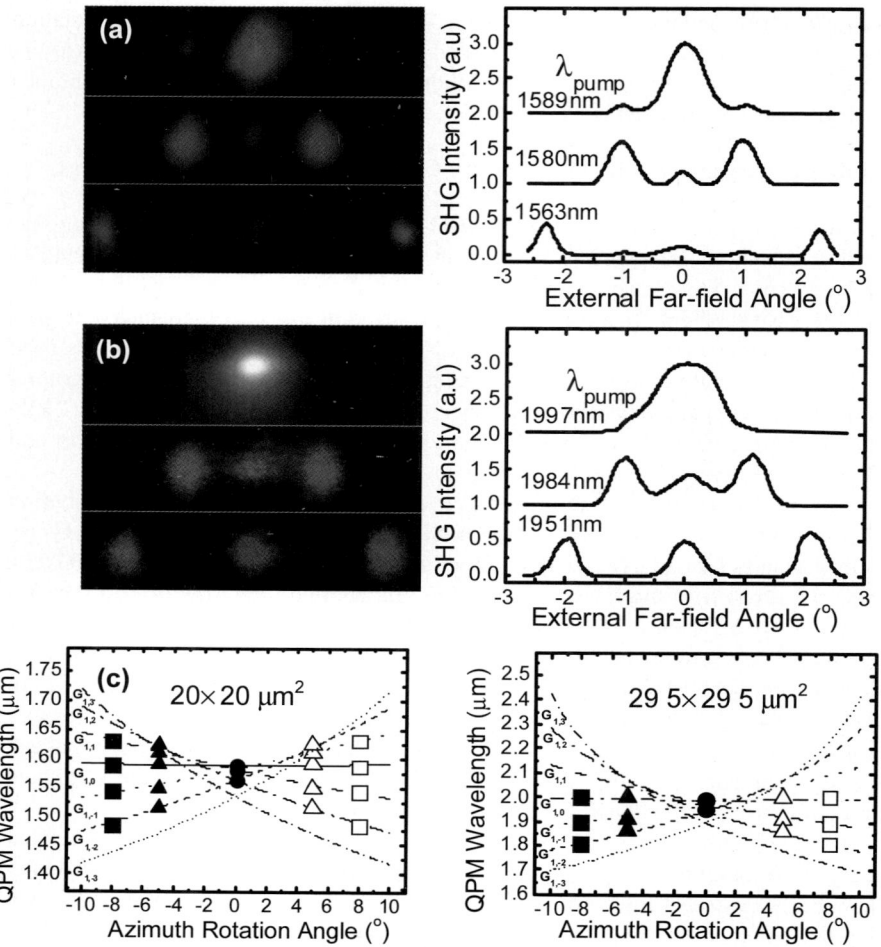

Fig. 2.8. (a) Images, (b) far-field intensity, and (c) calculated dispersion curves showing the wavelength dependent G_{mn}-assisted SHG processes for 2D poled LiNbO$_3$ with a periodicity of 20×20 and $29.5 \times 29.5\,\mu m^2$, respectively

process of energy distribution into cascaded high-order harmonic processes at high incident intensity condition due to the availability of plural reciprocal lattice vectors.

2.3.2 Stack of High-*k* Dielectric Poling Electrode Method

In this section we explore the possibilities of using stacks of high-*k* dielectric as the poling electrodes in order to reach the goals of (i) *increasing* the spatial homogeneity of the poling field $\mathbf{E}_z$ and (ii) *reducing* the fringe field $\mathbf{E}_{x,y}$. When modification in the composite/geometry of the corrugated current blocking layers matches these

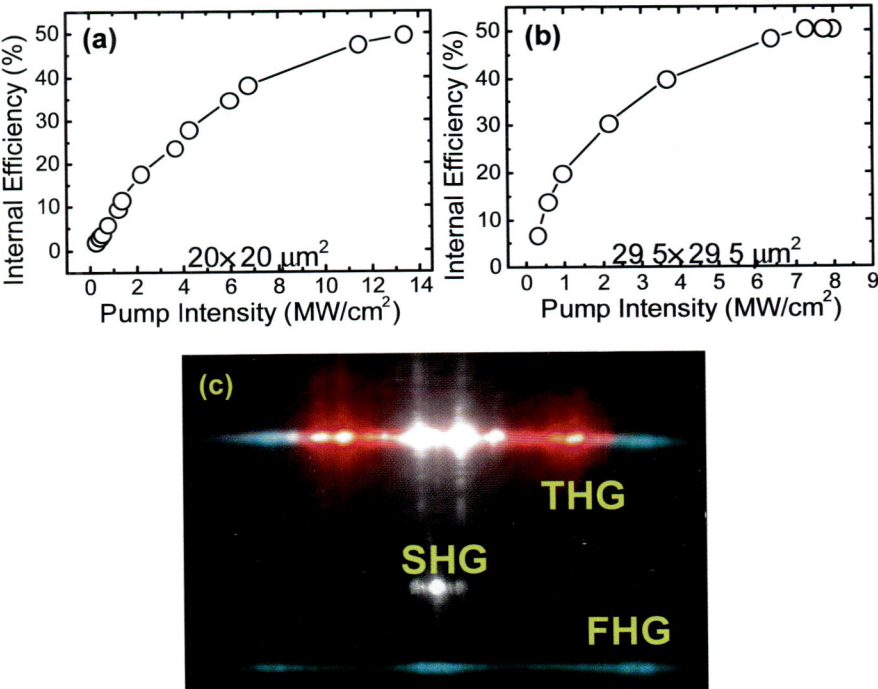

Fig. 2.9. SHG conversion efficiency for 2D poled LiNbO$_3$ with a periodicity of (**a**) 20 × 20, (**b**) 29.5 × 29.5 μm^2, and (**c**) cascaded harmonic generations of the fourth and fifth harmonics for sample in (**b**) measured at a normal incidence configuration

purposes, it increases the uniformity of the domain nucleation sites and favorably constrains their lateral motion. To give a pedagogical illustration of this principle, we schematically draw in Fig. 2.10(a) two sets of poling electrodes on LiNbO$_3$ ($\varepsilon_r = 30$). Case A is a conventional design with a corrugated layer of photo-resist ($\varepsilon_r = 2.5$), whereas Case B differs from A by depositing an additional layer of high-k ($\varepsilon_r = 40$) dielectric on top of the LiNbO$_3$ substrate. The corresponding poling electric field $\mathbf{E}_z$ and the accompanying fringe field $\mathbf{E}_x$, normalized to the material's coercive field $\mathbf{E}_c$, are displayed side by side in Fig. 2.10(b) for comparison. Here we emphasize again that at a shallow depth (0.1 μm or less) below the ferroelectric surface, the local distribution of the poling field $\mathbf{E}_z$ determines the inverted domain nucleation rate and follows an exponential dependence $\sim\exp(-\mathbf{E}_a/\mathbf{E}_z)$. On the other hand, the extent of lateral domain motion would be largely determined by the accompanying fringe field $\mathbf{E}_x$. For Case A with a conventional design of the poling electrode, we note that the poling field $\mathbf{E}_z$ has a peak value localized near the edges ($\pm x_0$) of the corrugated photo-resist pattern and follows an asymptotic form of $\sim\exp[-(x \pm x_0)/\Lambda]$ at a depth of 0.01Λ (and less) below the surface. This would result in a preferential nucleation of inverted domains along the edges of the poling

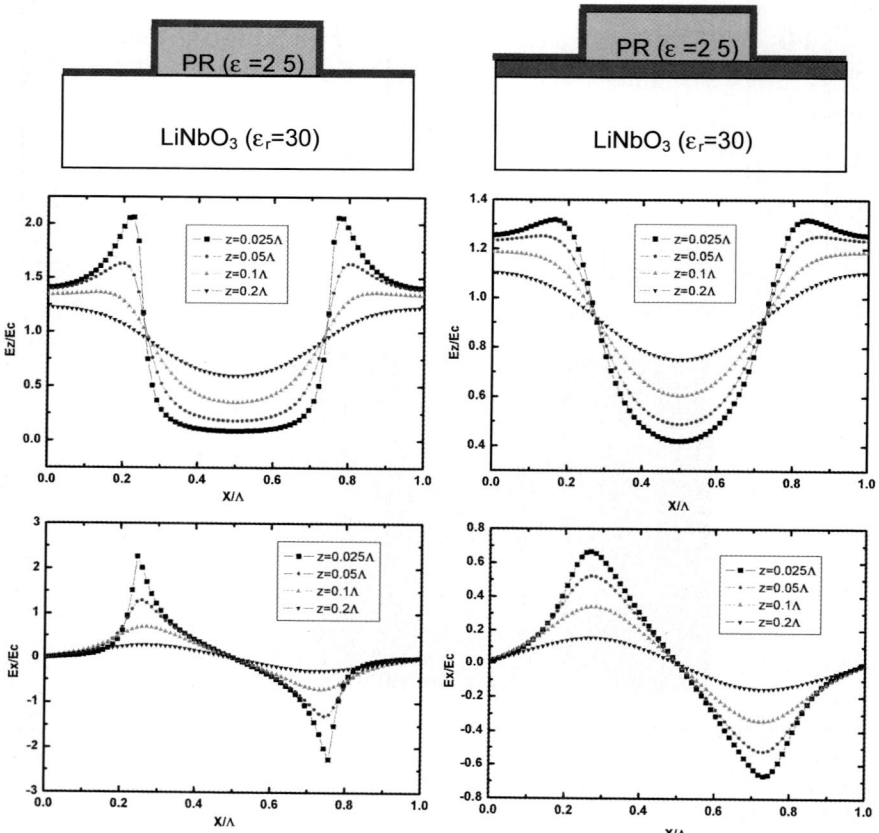

Fig. 2.10. (a) Schematic draw, and calculated field line distribution for (b) E_z and (c) E_x in two kinds of poling electrode design, i.e., conventional one with the ferroelectric surface covered with a corrugated electrode and the new one by inserting an additional high-k ($\varepsilon_r = 40$) dielectric into the ferroelectric surface

electrode. However, a large companying fringe field $\mathbf{E}_x$, which in this case has a magnitude $>2E_c$ and is comparable to its vertical poling component $\mathbf{E}_z$, can result in a fast redistribution of the injected carriers to spread to the regions covered by the photo-resist and thus lead to the broadening of the inverted domains. In comparison, for Case B where a high k-dielectric is inserted into the ferroelectric surface, one can detect a transformation of the poling field $\mathbf{E}_z$ into a much more uniform and smooth function, whereas the magnitude of the accompanying fringe field $\mathbf{E}_x$ is reduced to $\sim 1/4$ of its counterpart in Case A. The combination of these observations suggests a solution for better control in domain nucleation density and reduction of the fringe field effect. It can lead to the fabrication of ferroelectric domain structures with a higher fidelity than the original electrode design.

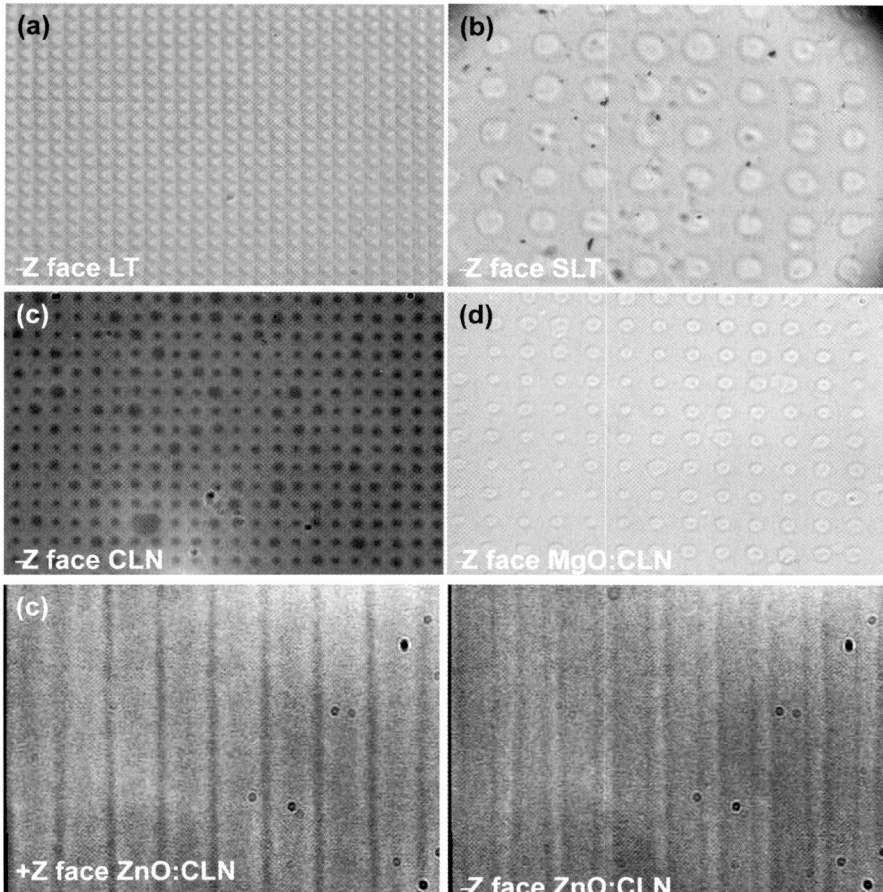

Fig. 2.11. 1D/2D samples of 0.5 mm-thickness poled by the high-k electrode method. (**a**) $-Z$ face micrograph of 2D poled CLT with a periodicity of $5.4 \times 5.4\,\mu m^2$, (**b**) $-Z$ face micrograph of 2D poled SLT with a periodicity of $13.6 \times 13.6\,\mu m^2$, (**c**) $-Z$ face micrograph of 2D poled CLN with a periodicity of $11 \times 11\,\mu m^2$, (**d**) $-Z$ face micrograph of 2D poled MgO:CLN with a periodicity of $11 \times 11\,\mu m^2$, and (**e**) $+Z$ and $-Z$ face micrographs of 1D poled ZnO:CLN with a periodicity of $11\,\mu m$

Example of Forming 1D/2D Periodically Poled LiNbO$_3$ and LiTaO$_3$

Using the proposed method discussed in Sect. 2.3.2, we fabricated five sets of 1D/2D periodically poled structures on 0.5 mm-thick LiNbO$_3$ and LiTaO$_3$ substrates under the conditions of different doping and composition. Figures 2.11(a) and (b) illustrate the $-Z$ micrographs of 2D poled Samples A and B made on the undoped congruent-grown and stoichiometric LiTaO$_3$ substrates with a periodicity of 5.4×5.4 and $13.6 \times 13.6\,\mu m^2$, respectively, Figs. 2.11(c) and (d) show the $-Z$ face micrographs of 2D poled Samples C and D made on the congruent-grown undoped and MgO:LiNbO$_3$ substrates at a periodicity of $11 \times 11\,\mu m^2$, and Sample E for a 1D poled ZnO:CLN structure with a periodicity of $11\,\mu m$. Here we seek the high-k dielectric from a combination of gallium oxide Ga$_2$O$_3$ ($\varepsilon_r \sim 15$) or silicon dioxide SiO$_2$ ($\varepsilon_r \sim 4$) that can be conveniently deposited by the RF-sputtering technique. We note that these devices have at a domain periodicity ranging from 5.4 to $11\,\mu m$ and a working area of $6 \times 6\,mm^2$, which are suitable for making themselves as the QPM-SHG red, green, and blue lasers. Data for the red QPM-SHG, including the far-field emission pattern, wavelength span, and the efficiency curve, are shown in Fig. 2.12 for the undoped 2D poled LiNbO$_3$ Sample C of a periodicity $11 \times 11\,\mu m^2$. Here contributions to the QPM-SHG processes from reciprocal lattice vector up to $G_{1,\pm 5}$ are observed. A saturated external SHG conversion efficiency of 35% and a wavelength bandwidth of $\sim 10\,nm$ are detected.

2.3.3 Submicron Domain Engineering with Self-Assembly Type of Poling Electrodes

Example of Domain Poling Along the Perimeter of a Ring and the Corners of a Rectangle

The manipulation of the electrostatic poling field E_z, as discussed in Sect. 2.3.2, with the insertion of a high-k dielectric on top of the ferroelectric surface can be further improved by periodically modulating the dielectric constant in the region defined by the photo-resist opening [66]. This field modulation technique allows us to fabricate submicron size ferroelectric domains over thick ($\sim 0.5\,mm$) substrates. To illustrate this principle, we add 2D arrays of self-assembled dielectric spheres to the high-k dielectric region in Fig. 2.13(a) and reevaluate the poling field E_z. Here the parameters used in the calculation are: high-k ($\varepsilon_r = 40$) dielectric with a $0.36\,\mu m$ thickness, photo-resist opening ($\varepsilon_r = 2.5$) with a $2.5\,\mu m$ thickness, and self-assembled dielectric spheres of diameter $2.5\,\mu m$ and $\varepsilon_r = 2.5$, and the QPM periodicity is $5.4\,\mu m$. These are the adjustable parameters that can be used to retain an optimum condition for a QPM structure with a known periodicity.

Our calculation reveals a dramatic change of E_z from an even and smooth function of Case B in Fig. 2.10 to a spatially modulated function in Figs. 2.13(b) and (c) whose periodicity is determined by the spacing of the self-assembled spheres. In particular, the modulation depth of E_z critically depends on the thickness of the high-k dielectric when the composition/thickness of the self-assembled spheres is fixed. For a special case shown in Fig. 2.13(b), we find that field modulation by the self-assembled dielectric spheres can result in a 20% enhancement of E_z near the "corrugated" photo-resist edge compared with those lying in the inter-sphere "flat" region. This suggests a preferential nucleation of the inverted domains along the photo-resist edge. Simultaneously, underneath the region of the spheres one also finds a screening effect such that the E_z field value would drop below E_c. The probability for domain to nucleate in this region becomes negligibly small. We further note from Fig. 2.13(c) that the fringe field in the "flat" inter-sphere regime is essentially zero, thus it has no effect on the lateral domain motion once lost the nu-

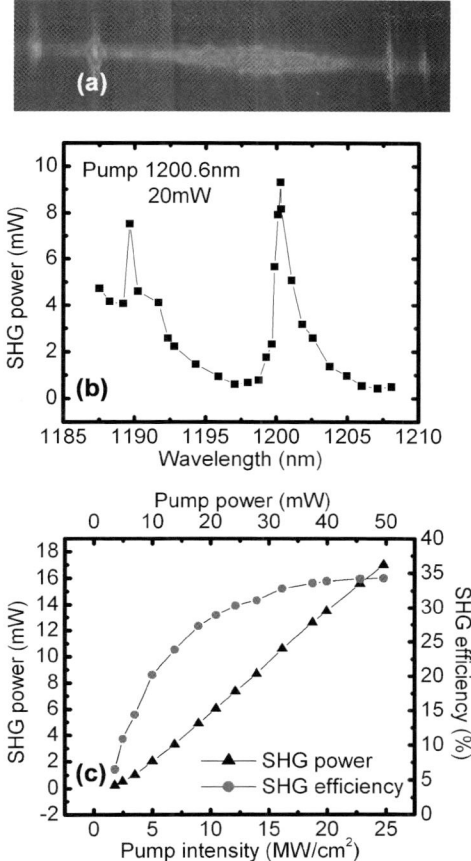

Fig. 2.12. (a) Far-field red SHG images, (b) wavelength acceptance width, and (c) external SHG conversion efficiency for Sample C of Fig. 2.11 using a 1200 nm ns pump laser

cleation preference to its edge component as discussed in Fig. 2.13(b). Although a small fringe field ~0.25E_c can be found underneath the region of the self-assembled sphere, there the corresponding E_z value is well below E_c and thus cannot initiate the domain reversal process. These localized field distributions in E_z and $E_{x,y}$ with a sufficient modulation depth allow us to purse a preferential nucleation of inverted domains along the boundary of the photo-resist opening with a periodicity (spacing) determined by the size of the self-assembled dielectric spheres.

The $+Z$ face micrographs in Fig. 2.14 demonstrate the versatility of submicron size structures that can be periodically poled by the proposed field modulation technique on 0.5 mm-thick congruent-grown undoped LiNbO$_3$ substrates. We show in Fig. 2.14(a) a periodic distribution of inverted domains consisting of pyramidal shaped nucleation sites along the perimeter of a 10 μm-dia LiNbO$_3$ micro-

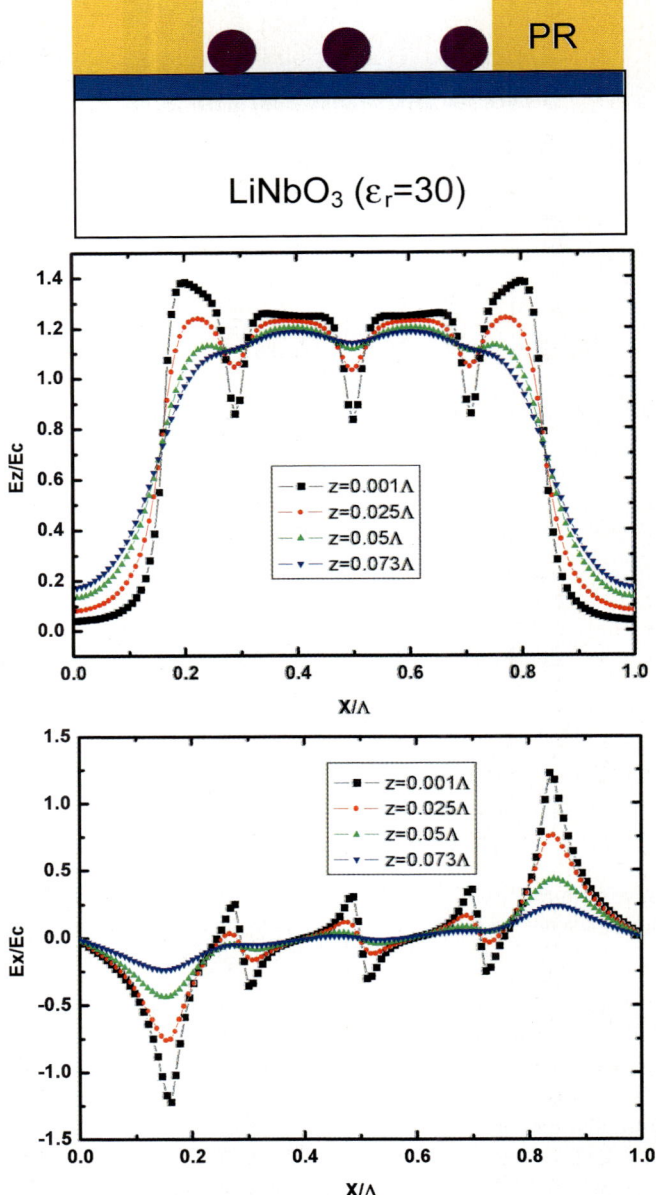

Fig. 2.13. (a) Schematic draw, and calculated field line distribution for (b) $\mathbf{E}_z$ and (c) $\mathbf{E}_x$ in a self-assembly type of poling electrode design, i.e., use of dielectric spheres over the high-k dielectric to provide modulation in the poling $\mathbf{E}_z$ and the fringe $\mathbf{E}_x$ field

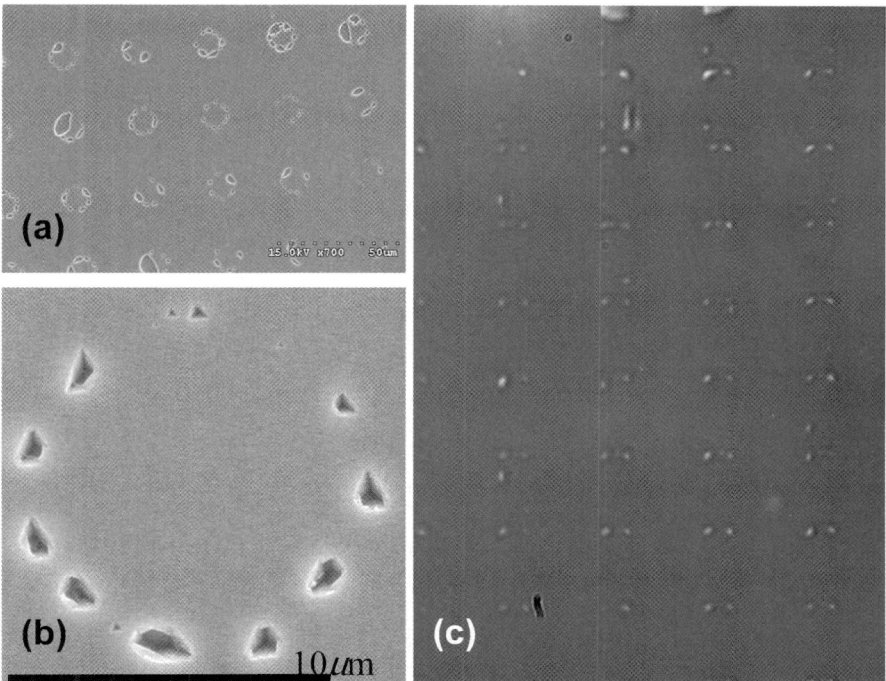

Fig. 2.14. $+Z$ face micrographs showing (**a**) periodically poled submicron domains along the perimeter of a LiNbO$_3$ ring of 10 μm diameter, (**b**) magnified picture of (**a**) showing the inverted pyramidal shape of nucleation sites, and (**c**) dot-like inverted domain on the four corners of a LiNbO$_3$ rectangle of 14×5 μm^2 in size

ring. The magnified picture in Fig. 2.14(b) indicates that a periodic spacing between these inverted domain sites marks a spatial modulation of the poling field by the self-assembled spheres with an average diameter ∼2 μm. The unpoled region corresponds to areas covered by the self-assembled dielectric spheres, where the underlying $\mathbf{E}_z$ field was screened to below the value of $\mathbf{E}_c$ and unable to activate the domain reversal process. Similarly in Fig. 2.14(c) we observe a 2D distribution of dot-like inverted domains residing on the four corners of a LiNbO$_3$ rectangle of 14×5 μm^2 in size. We further note that the smallest domain appearing on the $+Z$ face of LiNbO$_3$ corresponds to a nucleation site with an inverted pyramid shape of size ∼0.2 μm. It expands to a hexagon (not shown) with a side length ∼0.5 μm on the $-Z$ face of LiNbO$_3$ due to a finite lateral domain motion.

We should comment on this field modulation technique with a recent report that employs thickness modulation of a 2D photo-resist to fabricate submicron reversed domains in LiNbO$_3$ [67]. In the latter case, the thickness modulation in the photo-resist was realized by moiré interference lithography to have the resist pattern size and thickness to follow the intensity modulation. By applying a technique of electric field overpoling, submicron *surface* domain structures can be realized. However, in our approach, the equivalent modulation in the resistance thickness consists of adding self-assembled dielectric spheres over a high-k dielectric inserted into a conventional photoresist opening. Our approach not only is compatible with the standard material processing technology, it also offers the advantages of creating versatile geometric structures such as periodically poled rings or rectangles with submicron feature size of inverted domains over a much *thicker* substrate. This latter possibility becomes advantageous when high power laser applications are of interest.

2.3.4 Submicron Domain Engineering in Ferroelectric Semiconductors

Example of Forming Prism-Like GaN Submicron Cavities

Gallium nitride (GaN) is an emerging ferroelectric semiconductor that has received wide applications ranging from light emitters to high-power electronics, and more recently to nonlinear optics [68]. GaN-based materials, like its fellow ferroelectrics LiNbO$_3$ or LiTaO$_3$, have a spontaneous polarization $\mathbf{P}_s$ and are subject to an (internal) crystal field due to the crystal's wurtzite structure symmetry and a discontinuity of $\mathbf{P}_s$ at the hetero-structure interface. The fabrication of GaN-based photonic crystals typically involves the use of e-beam lithography followed by dry-etching or by direct ion-beam writing/etching to create air holes. By introducing a periodic discontinuity/modulation in the dielectric function, one can open the photonic bandgap to manipulate light-matter interaction [69]. However, recent investigations of a thin GaN membrane [70] or a bulk GaN [71] equipped with photonic crystal structures only reveals moderate ($\sim$300) to low ($\sim$185) value of a quality (Q)-factor in the light emission properties. The deterioration of the Q factor in these GaN photonic cavities can be ascribed to material damage and imperfect structure encountered in the device processing [72]. For example, dry etch of GaN can result in a rough circumference [73] or preferential loss of the nitrogen element [74]. Theoretical analysis also showed that geometric imperfections generated in the fabrication process can significantly reduce the cavity Q-factors [75].

It is generally known that, for an unintentionally n^- doped GaN layer grown on a c-plane sapphire substrate, there occurs a surface band bending effect which would dominate the electro-optic response [76]. This observation allows us to take advantage of the effect of photo-excitation with a self-assembled mask to modulate the internal field associated with the photo-enhanced chemical (PEC) wet etching of GaN. This approach enables us to realize prism-like GaN submicron cavities with a low scattering loss and leads to the observation of single-mode stimulated emission from GaN [77].

The principle of PEC wet oxidation and etching of GaN can be understood from an oxidative dissolution of GaN with the photo-generated carriers when immersed in an appropriate electrolyte such as potassium hydroxide (KOH) or phosphoric acid (H$_3$PO$_4$) [78, 79]. In our approach, modulation of the surface field-activated PEC process was realized by shining a UV light source (a 254 nm Hg lamp) over a 2D

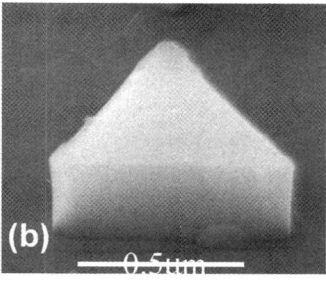

Fig. 2.15. $+Z$ face micrographs showing (**a**) prism-like GaN submicron cavity (side length $\sim$0.7 µm) with $\{11\bar{2}0\}$ facets residing on the edges of a hexagon shape of etched GaN platform, and (**b**) magnified picture showing GaN cavity bounded by another set of $\{10\bar{1}0\}$ facets

self-assembled photo-mask made of platinum (Pt)/titanium (Ti) triangular pads on the $+c$ face of unintentionally n^- doped GaN. Such a 2D pattern was formed by using metal lift-off from a monolayer of self-assembled polystyrene spheres [80]. The 2D Pt/Ti metallic pads serve as the local cathode and the unmasked region of GaN as the local anode in the PEC processing.

We continued the fabrication process by taking the GaN samples into a process of 15 min of PEC etching in KOH electrolyte of pH $=$ 11.7, followed by 20 sec of molten KOH treatment to remove the material's defects. Prism-like cavities, residing on the corners of a close-pack hexagon with a side length $\sim$0.7 µm and sharp/straight boundaries, are clearly discerned in Fig. 2.15. Close examination of the magnified SEM micrographs indicated that these cavities are bounded by the optically smooth $\{11\bar{2}0\}_{GaN}$ or $\{10\bar{1}0\}_{GaN}$ facets. These observations signify that in a reaction-rate limited PEC etching process, one can preserve the non-polar $\{11\bar{2}0\}$ or $\{10\bar{1}0\}$ GaN facets to form prism-like cavities due to a vanishing surface field effect on these planes [81].

The optical emission properties of these prism-like GaN were investigated by using micro photo-luminescence (µ-PL) spectroscopy and the data are illustrated in Fig. 2.16. We see a characteristic spectral change from (a) a broad band edge emission (at a full width at half maximum FWHM $\delta\lambda \sim 14$ nm) centered at 365 nm for planar GaN Sample A to a *sharp* ($\delta\lambda = 0.34/0.29$ nm for Samples B/C) single resonant peak located either at the (b) *shorter* (364 nm for Sample B) or (c) *longer* (370 nm for Sample C) wavelength side of the emission spectrum. Here the Samples B/C had a nominal side length/cavity height of 0.75/0.5 µm but a different GaN thickness of 4 and 0.8 µm. A cavity Q factor exceeding 10^3 can be clearly resolved from the prism-like GaN Sample B and C. Indeed the emission from the prism-like cavity was found to polarize along the crystal C-axis (i.e., TM-polarized), whereas only broad emission from bulk GaN was detected when measured from the sample normal.

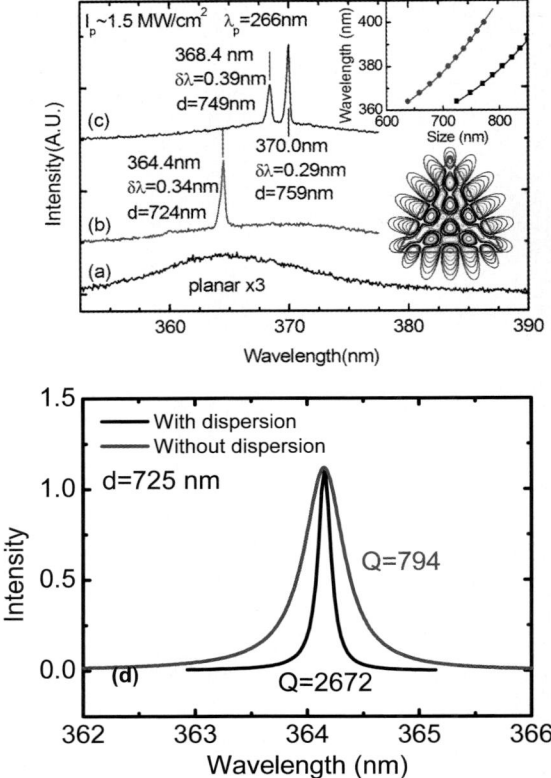

Fig. 2.16. Micro-photoluminescence spectra taken with a Nd:YAG 266 nm pulsed laser on (**a**) planar GaN sample, prism-like GaN cavities with a dimension in side/height/thickness as (**b**) 0.75/0.5/4 μm and (**c**) 0.75/0.5/0.8 μm. (**d**) Calculated cavity Q factor illustrating the effect by considering the material dispersion

The origin of such a high-Q cavity can be understood from considering the material's dispersion effect [82] using finite-difference time-domain (FDTD) analysis [83]. Our calculation, as shown in the inset of Fig. 2.16, indicates that the cavity emission wavelength λ sensitively depends on a phase-matching condition related to the prism side length d. The calculated Q factor in Fig. 2.16(d) exhibits a critical dependence on the material's dispersion. This can also be understood from a semi-classical analysis using $Q = Q_0 \cdot [1 - (\lambda/n)\partial n/\partial \lambda]$, where the material dispersion [84] near the band edge of GaN with $\partial n/\partial \lambda < 0$ increases the cavity Q value by a factor of 3.

The large free spectral range (≥ 20 nm) between two consecutive modes of the prism-like GaN cavity allows us to align one such cavity mode with the optical gain of GaN and to accumulate optical feedback for laser action. This implies that by properly designing the cavity size with a sufficiently high-Q factor, one can

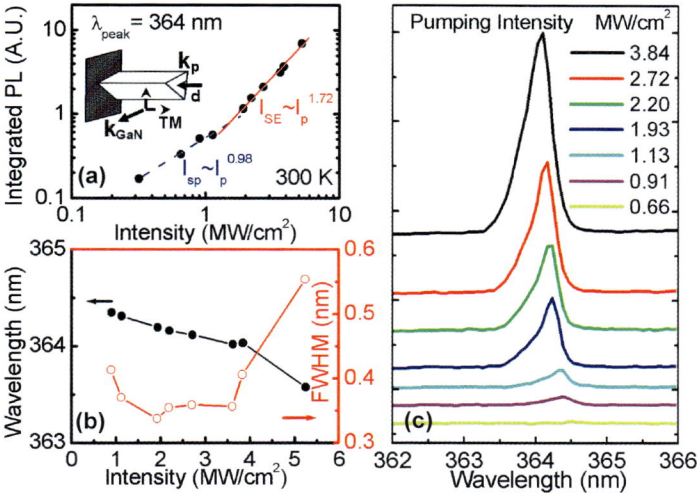

Fig. 2.17. Characteristics of single mode stimulated emission realized by the high-Q prism-like GaN cavity shown in Fig. 2.15

achieve amplification of the selected optical mode emission. Emission data shown in Fig. 2.17 confirm the following characteristics related to the stimulated emission of a single mode laser [85]:

(a) Linear to superlinear slope change at a threshold of $I_p \sim 2\,\mathrm{MW/cm^2}$.
(b) Slight spectral blue shift with wavelength clamped to the cavity mode.
(c) Little variation in the linewidth with pump intensity up to $I_p = 4\,\mathrm{MW/cm^2}$.

2.4 Conclusion

In Sect. 2.3 we demonstrated methods of fabricating micron and submicron domains on ferroelectric nonlinear crystals of $LiNbO_3$, $LiTaO_3$, and on ferroelectric semiconductor GaN. Various geometric structures and optical characterization in the QPM-SHG process and single mode laser action have also been analyzed. It would be interesting to compare with other contemporary poling techniques that have been developed to purse the goals of submicron to nano-domain engineering.

One such approach is to seek individual domain poling by applying a voltage over a conductive cantilever tip typically equipped in a scanning force microscope. Reference [87] provides a recent review of the instrumentation of this technique. If the conductive tip is modeled as a charged sphere, a highly anisotropic poling field $\mathbf{E}_z(\rho)$ having a spatial variation proportional to $(R+\delta)/[(R+\delta)^2+\rho^2]^{3/2}$ can develop, where R is the radius of the sphere, δ the distance from the tip apex to the ferroelectric surface, and ρ the horizontal surface distance measured from the projection point of the tip apex [88]. $\mathbf{E}_z$ can dramatically decrease from its peak value near the tip apex at the ferroelectric surface ($\rho = 0$) to a value that is five-order of magnitude lower at a distance 10 μm from the surface [89].

Such an inhomogeneous field can create a unique equilibrium state for reversing domains in the thin film or bulk ferroelectrics. Examples of engineered nanodomains have been found on the ferroelectric systems: thin film lead zirconate titanate (Pb(Zr,Ti)O$_3$) [90], thin film LiTaO$_3$ [91], thin film and bulk crystal LiNbO$_3$ [92]. It is worth noting that ultra-high density ferroelectric data recording/storage system with 10 Tbits/in^2 has recently been realized using thin (35 nm-thick) CLT with a sub-ns domain switching speed [91]. The corresponding domain shape of a circularly cylinder can also differ from its bulk form which obeys the crystal's point group symmetry [93]. For tip-field poling of a thick ferroelectric crystal, theoretical modeling confirms a saturation behavior of domain radius with respect to the applied voltage and field duration, whereas for tip-field poling of a ultra-thin ferroelectric film, the domain size increases with the applied voltage and pulse duration [92, 93].

An obstacle associated with the tip-field poling is its slow writing speed for making practical optoelectronic devices. To fabricate a 10 mm^2 area of periodically poled QPM structure would take more than 100 h considering a typical writing speed of 100 μm/s. The environmental and the machine stability issues can have vulnerable side-effects on the repeated writing process. The single-tip writing speed can be theoretically increased by using a multiple tip-array [94]. A recent study of poling with positively charged Ga^{++} focused ion beam (FIB) has revealed an increase in writing speed by two-order of magnitude compared with the tip writing technique [95]. The latter method also resembles our proposed method of positively charged potential discussed in Sect. 2.3.1. However, the high operating cost of FIB is probably another issue to be faced as the practical application is concerned.

Acknowledgments

The authors wish to acknowledge supporting from the National Science Council, Grant Nos. NSC 95-2221-E-002-377, NSC 95-2112-M-130-001, NSC 95-2120-M-01-06, and NSC 95-2120-M-001-009. We also acknowledge our MS students' participation in this research: C.-D. Lin, Y.-H. Chen, L.F. Lin, Y.-P. Tseng, C.-S. Lu, S.-H. Yang, Z.-X. Huang, K.-L. Lin, Y.-N. Hu for the fabrication and SHG work, and C.-T. Huang for the electrostatic modeling and coding with the Poisson equation solver.

References

1. J. Camlibel, J. Appl. Phys. **40**, 1690 (1969)
2. K. Nassau, H.J. Levinstein, G.M. Loiacono, J. Phys. Chem. Solids **27**, 989 (1966)
3. G.D. Miller, Periodically poled lithium niobate: modeling, fabrication, and nonlinear-optical performance. Ph.D. thesis, Stanford University, 1998

4. H.S. Nalwa (ed.), *Ferroelectric and Dielectric Thin Films*. Handbook of Thin Film Materials, vol. 3 (Academic Press, San Diego, 2002)
5. O. Auciello, J.F. Scott, R. Ramesh, Phys. Today **51**, 22 (1998)
6. J.V. Mantese, N.W. Schubring, A.L. Micheli, Appl. Phys. Lett. **80**, 1430 (2002)
7. Y.-Q. Lu, Y.-Y. Zhu, Y.-F. Chen, S.-N. Zhu, N.-B. Ming, Y.-J. Feng, Science **284**, 1822 (1999)
8. J.A. Armstrong, N. Bloembergen, J. Ducuing, P.S. Pershan, Phys. Rev. **127**, 1918 (1963)
9. M. Yamada, N. Nada, M. Saitoh, K. Watanabe, Appl. Phys. Lett. **62**, 435 (1993)
10. L.E. Myers, R.G. Eckardt, M.M. Fejer, R.L. Byer, W.R. Bosenberg, J.W. Pierce, J. Opt. Soc. Am. B **12**, 2102 (1995)
11. M.M. Fejer, G.A. Magel, D.H. Jundt, R.L. Byer, IEEE J. Quantum Electron. **28**, 2631 (1992)
12. Z.D. Gao, S.N. Zhu, S.Y. Tu, A.H. Kung, Appl. Phys. Lett. **89**, 181101 (2006)
13. V. Berger, Phys. Rev. Lett. **81**, 4136 (1998)
14. N.G.R. Broderick, G.W. Ross, H.L. Offerhaus, D.J. Richardson, D.C. Hanna, Phys. Rev. Lett. **84**, 4136 (2000)
15. L.-H. Peng, C.-C. Hsu, A.H. Kung, IEEE J. Sel. Top. Quantum Electron. **10**, 1142 (2004)
16. P. Xu, S.N. Zhu, X.Q. Yu, S.H. Ji, Z.D. Gao, G. Zhao, Y.Y. Zhu, N.B. Ming, Phys. Rev. B **72**, 064307 (2005)
17. M.-P. Bernal, N. Courjal, J. Amet, M. Roussey, C.H. Hou, Opt. Commun. **265**, 180 (2006)
18. Y.J. Ding, J.B. Khurgin, Opt. Lett. **21**, 1445 (1996)
19. M. Matsumoto, K. Tanaka, IEEE J. Quantum Electron. **31**, 700 (1995)
20. J.F. Scott, *Ferroelectric Memories* (Springer, Berlin, 2000)
21. A. Gruverman, A. Kholkin, Rep. Prog. Phys. **69**, 2443 (2006)
22. R.C. Miller, G. Weinreich, Phys. Rev. **117**, 1460 (1960)
23. E. Fatuzzo, W.J. Merz, Phys. Rev. **116**, 61 (1959)
24. R. Landauer, J. Appl. Phys. **28**, 227 (1957)
25. W.J. Merz, Phys. Rev. **95**, 690 (1954)
26. G. Rosenman, Kh. Garb, A. Skliar, M. Oron, D. Eger, M. Katz, Appl. Phys. Lett. **73**, 865 (1998)
27. L.-H. Peng, Y.-C. Fang, Y.-C. Lin, Appl. Phys. Lett. **74**, 2070 (1999)
28. V. Gopalan, T.E. Mitchell, K.E. Sicakfus, Solid State Commun. **109**, 111 (1999)
29. R.G. Batchko, V.Y. Shur, M.M. Fejer, R.L. Byer, Appl. Phys. Lett. **75**, 1673 (1999)
30. L.-H. Peng, Y.-C. Zhang, Y.-C. Lin, Appl. Phys. Lett. **4** (2001)
31. V.Y. Shur, E.L. Rumyantsev, E.V. Nikolaeva, E.I. Shishkin, Appl. Phys. Lett. **77**, 3636 (2000)
32. P. Urenski, M. Molotskii, G. Rosenman, Appl. Phys. Lett. **79**, 2964 (2001)
33. K. Mizuuchi, K. Yamamoto, M. Kato, Appl. Phys. Lett. **70**, 1201 (1997)
34. G.D. Miller, R.G. Batchko, W.M. Tulloch, D.R. Weise, M.M. Fejer, R.L. Byer, Opt. Lett. **22**, 1834 (1997)
35. L. Baudry, J. Tournier, J. Appl. Phys. **90**, 1442 (2001)
36. M. Asobe, O. Tadanaga, T. Yanagawa, H. Itoh, H. Suzuki, Appl. Phys. Lett. **78**, 3163 (2001)
37. A. Kuroda, S. Kurimura, Y. Uesu, Appl. Phys. **69**, 1565 (1996)
38. F. Abdi, M. Aillerie, M. Fontana, P. Bourson, T. Volk, B. Maximov, S. Sulyanov, N. Rubinina, M. Wöhlecke, Appl. Phys. B: Lasers Opt. **68**, 795 (1999)
39. N. Iyi, K. Kitamura, Y. Yajima, S. Kimura, Y. Furukawa, M. Sato, J. Solid State Chem. **118**, 148 (1995)

40. V. Gopalan, T.E. Mitchell, Y. Furukawa, K. Kitamura, Appl. Phys. Lett. **72**, 1981 (1998)
41. S. Ganesamoorthy, M. Nakamura, S. Takekawa, S. Kumaragurubaran, K. Terabe, K. Kitamura, Mater. Sci. Eng. B **120**, 125 (2005)
42. Y. Furukawa, K. Kitamura, S. Takekawa, K. Niwa, Y. Yajima, N. Iyi, I. Mnushkina, P. Guggenheim, J.M. Martin, J. Cryst. Growth **211**, 230 (2000)
43. N.E. Yu, S. Kurimura, Y. Nomura, K. Kitamura, Jpn. J. Appl. Phys. **43**, L1265 (2004)
44. K. Mizuuchi, A. Morikawa, T. Sugita, K. Yamamoto, J. Appl. Phys. **96**, 6585 (2004)
45. K. Nakamura, J. Kurz, K. Parameswaran, M.M. Fejer, J. Appl. Phys. **91**, 4528 (2002)
46. K. Mizuuchi, A. Morikawa, T. Sugita, K. Yamamoto, N. Pavel, T. Taira, Jpn. J. Appl. Phys. **43**, L1293 (2004)
47. H. Ishizuk, T. Taira, S. Kurimura, J.H. Ro, M. Cha, Jpn. J. Appl. **42**, L108 (2003)
48. K. Mizuuchi, A. Morikawa, T. Sugita, K. Yamamoto, N. Pavel, I. Shoji, T. Taira, Jpn. J. Appl. Phys. **42**, L1296 (2003)
49. L.-H. Peng, Y.-J. Shih, Y.-C. Zhang, Appl. Phys. Lett. **81**, 1666 (2002)
50. K. Nakamura, H. Ando, H. Shimizu, Appl. Phys. Lett. **50**, 1413 (1987)
51. Y.N. Korkishko, V.A. Fedorov, *Ion Exchange in Single Crystals for Integrated Optics and Optoelectronics* (Cambridge International Science, Cambridge, 1999)
52. M. Fujimura, T. Suhara, H. Nishihara, Electron. Lett. **27**, 1207 (1991)
53. J. Webjörn, F. Laurell, G. Arvidsson, IEEE Photonics Technol. Lett. **1**, 316 (1989)
54. V.D. Kugel, G. Rosenman, Appl. Phys. Lett. **62**, 2902 (1993)
55. L. Huang, N.A.F. Jaeger, Appl. Phys. Lett. **65**, 1763 (1994)
56. K. Mizuuchi, K. Yamamoto, Appl. Phys. Lett. **66**, 2943 (1995)
57. S. Grill, C. Canalias, F. Laurell, P. Ferraro, P. De Natale, Appl. Phys. Lett. **89**, 032902 (2006)
58. L.-H. Peng, Y.-C. Shih, S.-M. Tsan, C.-C. Hsu, Appl. Phys. Lett. **81**, 5210 (2002)
59. W.-L. Chen, D.-J. Chen, W.-S. Wang, IEEE Photonics Technol. Lett. **7**, 203 (1995)
60. A. Chernykh, V. Shur, E. Nikolaeva, E. Shishkin, A. Shur, K. Terabe, S. Kurimura, K. Kitamura, K. Gallo, Mater. Sci. Eng. B **120**, 109 (2005)
61. L.-H. Peng, C.-C. Hsu, Y.-C. Shih, Appl. Phys. Lett. **83**, 3447 (2003)
62. D.H. Jundt, Opt. Lett. **22**, 1553 (1997)
63. L.-H. Peng, C.-C. Hsu, J. Ng, A.H. Kung, Appl. Phys. Lett. **84**, 3250 (2004)
64. C.-S. Yu, A.H. Kung, J. Opt. Soc. Am. B **16**, 2233 (1999)
65. L.-H. Peng, C.-C. Hsu, A.H. Kung, IEEE J. Sel. Top. Quantum Electron. **10**, 1142 (2004)
66. L.-H. Peng, Y.-H. Chen, C.-D. Lin, L.-F. Lin, A.-H. Kung, J. Cryst. Growth **292**, 328 (2006)
67. P. Ferraro, S. Grilii, Appl. Phys. Lett. **89**, 133111 (2006)
68. S. Pezzagna, P. Vennéguès, N. Grandjean, A.D. Wieck, J. Massies, Appl. Phys. Lett. **87**, 062106 (2005)
69. D. Coquillat, G. Vecchi, C. Comaschi, A.M. Malvezzi, J. Torres, M. Le Vassor d'Yerville, Appl. Phys. Lett. **87**, 101106 (2005)
70. C. Meier, K. Hennessy, E.D. Haberer, R. Sharma, Y.-S. Choi, K. McGroddy, S. Keller, S.P. DenBaars, S. Nakamura, E.L. Hu, Appl. Phys. Lett. **88**, 031111 (2006)
71. L.-M. Chang, C.-H. Hou, Y.-C. Ting, C.-C. Chen, C.-L. Hsu, J.-Y. Chang, C.-C. Lee, G.-T. Chen, J.-I. Chyi, Appl. Phys. Lett. **89**, 071116 (2006)
72. S. Chang, N.B. Rex, R.K. Chang, G. Chong, L.J. Guido, Appl. Phys. Lett. **75**, 166 (1999)
73. K.C. Zeng, L. Dai, J.Y. Lin, H.X. Jiang, Appl. Phys. Lett. **75**, 2563 (1999)
74. K.J. Choi, H.W. Jang, J.-L. Lee, Appl. Phys. Lett. **82**, 1233 (2003)
75. S.V. Boriskina, T.M. Benson, P. Sewell, A.I. Nosich, IEEE J. Quantum Electron. **41**, 857 (2005)

76. L.-H. Peng, C.-W. Shih, C.-M. Lai, C.-C. Chuo, J.-I. Chyi, Appl. Phys. Lett. **82**, 4268 (2003)
77. C.-M. Lai, H.-M. Wu, P.-C. Huang, S.-L. Wang, L.-H. Peng, Appl. Phys. Lett. **90**, 141106 (2007)
78. L.-H. Peng, C.-H. Liao, Y.-C. Hsu, C.-S. Jong, C.-N. Huang, J.-K. Ho, C.-C. Chiu, C.-Y. Chen, Appl. Phys. Lett. **76**, 511 (2000)
79. L.-H. Peng, C.-W. Chuang, J.-K. Ho, C.-N. Huang, C.-Y. Chen, Appl. Phys. Lett. **72**, 939 (1998)
80. J.C. Hulteen, D.A. Treichel, M.T. Smith, M.L. Duval, T.R. Jensen, R.P. Van Duyne, J. Phys. Chem. B **103**, 3854 (1999)
81. L.H. Peng, C.-Y. Lu, W.-H. Wu, S.-L. Wang, Appl. Phys. Lett. **87**, 161902 (2005)
82. N. Antoine-Vincent, F. Natali, M. Mihailovic, A. Vasson, J. Leymarie, P. Disseix, D. Byrne, F. Semond, J. Massies, J. Appl. Phys. **93**, 5222 (2003)
83. S. Dey, R. Mittra, IEEE Microwave Guid. Lett. **8**, 415 (1998)
84. N. Antoine-Vincent, F. Natali, M. Mihailovic, A. Vasson, J. Leymarie, P. Disseix, D. Byrne, F. Semond, J. Massies, J. Appl. Phys. **93**, 5222 (2003)
85. H. Kressel, J.K. Butler, in *Semiconductor Lasers and Heterojunction LEDs* (Academic Press, San Diego, 1987), p. 557
86. H. Haken, in *Light: Laser Light Dynamics* (Elsevier, Amsterdam, 1985), p. 127
87. A. Gruverman, S.V. Kalinin, J. Mater. Sci. **41**, 107 (2006)
88. E.J. Mele, Am. J. Phys. **69**, 557 (2001)
89. G. Rosenman, P. Urenski, A. Agronin, Y. Rosenwaks, M. Molotskii, Appl. Phys. Lett. **82**, 103 (2003)
90. B.J. Rodriguez, A. Gruverman, A.I. Kingon, R.J. Nemanich, J.S. Cross, J. Appl. Phys. **95**, 1958 (2004)
91. Y. Cho, S. Hashimoto, N. Odagawa, K. Tanaka, Y. Hiranaga, Appl. Phys. Lett. **87**, 232907 (2005)
92. A. Agronin, M. Molotskii, Y. Rosenwaks, G. Rosenman, B.J. Rodriguez, A.I. Kingon, A. Gruverman, J. Appl. Phys. **99**, 104102 (2006)
93. A.N. Morozovaka, E.A. Eliseev, Phys. B **73**, 104440 (2006)
94. Y. Rosenwaks, D. Dahan, M. Molotskii, G. Rosenman, Appl. Phys. Lett. **86**, 012909 (2005)
95. X. Li, K. Terabe, H. Hatano, H. Zeng, K. Kitamura, Appl. Phys. Lett. **100**, 106103 (2006)

3 Sub-Micron Structuring of LiNbO$_3$ Crystals with Multi-Period and Complex Geometries

S. Grilli and P. Ferraro

3.1 Introduction

Lithium niobate (LN) is a ferroelectric material which has attracted a considerable interest in different fields, such as the optical and the laser and communications industry, due to its excellent nonlinear optical, electro-optic, piezoelectric and acousto-optical coefficients [1]. LN is widely used in the laser area, where fabrication of periodically poled materials has achieved high efficiencies in quasi-phased matched nonlinear interactions [2], but also in the field of microwave communications for surface acoustic wave delay lines [3]. Recently, the possibility to microstructure LN crystals has been attracted great interest for the useful applications foreseen in the fields of optics and optoelectronics. However, those applications require anisotropic etching techniques and different methods have been reported for machining and microstructuring in LN, as described in Sect. 3.2. Section 3.3 describes the electric field overpoling technique as a useful tool for sub-micron structuring by subsequent differential wet etching, while Sect. 3.4 describes the holographic lithography (HL) process used for submicron periodic poling. Sections 3.5 and 3.6 present the wide variety of surface structures obtained by overpoling and etching. Conclusions and discussion are summarized in Sect. 3.7.

3.2 Overview of the Etching Techniques Applied to Lithium Niobate

The desirability of etching LN to form surface structures is widely recognized and different techniques have been tried on this material in the past, such as mechanical grinding [4], ion milling [5], sputter etching [6], plasma etching [7]. None of these methods provided entirely satisfactory results. In fact, mechanical grinding can produce only rather large structures with a limited range of geometries, while ion milling and sputter etching can result in substrate faceting and redeposition of sputtered material on the substrate. Furthermore, these etching techniques do not differentiate strongly between substrate and mask so that poor etching selectivity is provided. Among those the plasma etching, which is basically a dry-chemical etch process, was the only technique able to give high selectivity but its first applications to LN substrates produced vary slow etch rates [7]. The wet etching is widely used in the field of material processing but is rarely applied to LN because of the well known material's strong etch resistance which results in low etch rates and sometimes non-uniform etching. On the other hand the wet etching of LN has been typically used to reveal domain polarity and defect topology [8–10] by immersion in a HF/HNO$_3$ acid mixture. In fact, this mixture exhibits the property to etch the negative Z face of the crystal while the positive is left essentially untouched. Nevertheless great effort has been spent during the last decade in order to improve the reliability of the wet etching in case of LN material. The introduction of structural defects accelerates the etching rate so that the combination of a defectiveness agent, such as ion bombardment or laser ablation, with subsequent wet etching has been demonstrated to be a useful method for etching of LN.

In this section a brief overview is given of the most important techniques, presented in literature during the last years, in order to get reliable microstructuring of LN crystal substrates. These include reactive ion etching (RIE) [11, 12], laser ablation below and over the band-gap of the material [13, 14], wet etching of proton exchanged material [15–17], wet etching of periodically poled crystals [18], electron beam bombardment [19], laser frustrated wet etching [20], ion implantation [21], focused ion beam bombardment [22], and more recently plasma etching of proton-exchanged samples [23], improved wet etching [24]. Another technique was proposed by the authors in [25] where periodic structures at sub-micron scale were fabricated by wet etching samples after specific ferroelectric domain manipulation.

The RIE presented in [11, 12] was a method combining some advantages of both plasma and sputter etching. The etching agent was represented by a low-pressure plasma of chemically active gases formed in a parallel plate radiofrequency sputtering station. The combination of a long mean free path and an electric field at the sample yielded anisotropic etching, thus minimizing the undercut and leading to high resolution. Ideally the reaction creates volatile products so that, unlike the case for sputter etching, the redeposition was strongly reduced. The laser ablation technique was presented in literature under different operating conditions. The method in [13] involved spatially localized melting of LN by high power density laser pulses with photon energies in excess of the band-gap of LN. Powdered or small crystals of KF were applied directly to the surface prior to irradiate the substrates with laser pulses. While molten, LN undergoes reaction with KF to form complex niobium oxyfluoride anions by fusion of salts. The resulting solid was highly water soluble so that the insolubility of LN permitted subsequent removal of only the irradiated area by rinsing in water. The technique presented in [14] made use of a laser irradiation process below the band-gap of LN by using a cw Argon laser emitting at 351 nm with a maximum power of 2.5 W. The process was basically a reduction process utilizing photon energies of about 3.5 eV, thus enabling selective removal of the laser-processed substrate by subsequent wet etching of the Y-cut LN substrates. The possibility to fabricate groove-like structures for precise fibre positioning was demonstrated and Fig. 3.1 shows the corresponding SEM images.

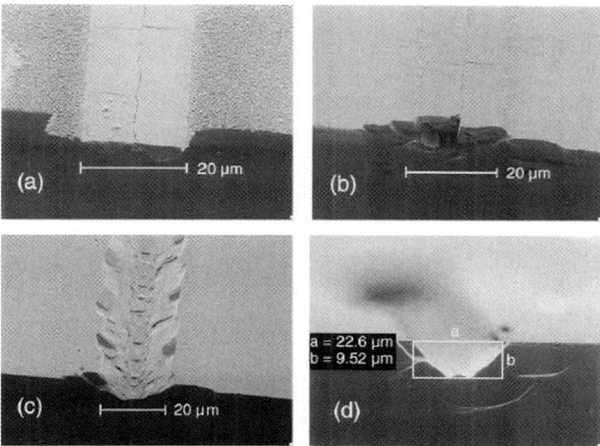

Fig. 3.1. LN sample after laser processing. (**a**) Before any cleaning; (**b**) after cleaning with ethanol; (**c**), (**d**) and after hot HF etch (from [14])

The wet etching process proposed in [15] consisted in selective etching of proton exchanged regions by using a mixture of HF and HNO_3. Since the first experiments in [26], proton exchange (PE) and annealed proton exchange (APE), by using benzoic acid as the proton source, are still widely used techniques to fabricate waveguides of relatively good quality in LN [27] and the results in [15] show the possibility to use this technique for selective surface etching of LN samples by a relatively simple process, consisting basically in the immersion of the LN sample in an acid, often a benzoic acid melt, at around 200°C for minutes up to hours.

LN has two stable domain orientations and microstructuring through domain manipulation followed by differential etching was developed in [18]. The process consisted in domain patterning by electric field poling followed by etching in a mixture of HF and HNO_3. A variety of structures at micron and sub-micron scale was produced by this technique and the surface quality of the structures was relatively smooth and straight as shown in Fig. 3.2.

The same group proposed later a laser frustrated etching process as an advanced technique for the fabrication of self-ordered sub-micron structures in iron-doped LN substrates [20], and Fig. 3.3 shows a SEM image of the fabricated structures as example.

One- and two-dimensional structures at sub-micron scale were fabricated by electron beam bombardment associated to chemical etching as presented in [19]. The electron beam was used to reverse the ferroelectric domains locally and subsequent etching revealed the surface structure. Figure 3.4 shows the SEM image of a two-dimensional structure obtained by this technique.

In [21] the possibility to fabricate high-quality ridge waveguides in Z-cut LN by oxygen ion implantation associated to wet etching was proposed. The technique allowed to obtain ridge structures up to 3.5 μm deep and thus suitable for ridged

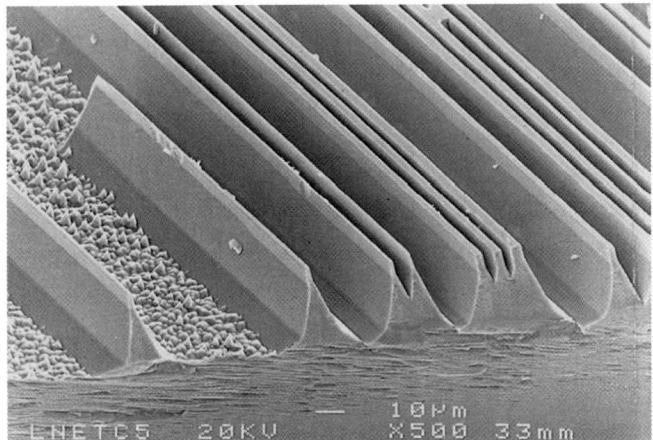

Fig. 3.2. SEM image of the structures obtained by domain manipulation and subsequent etching (from [18])

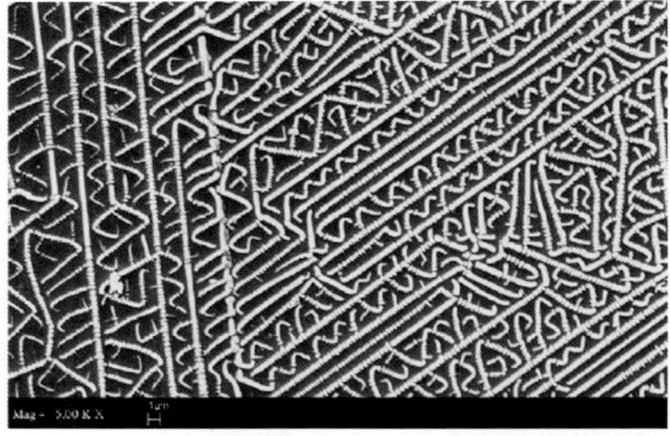

Fig. 3.3. SEM image of self-ordered structures, as example (from [20])

modulator applications. Figure 3.5 shows the cross sectional profile of the ridge waveguide fabricated by this technique.

Recently, the focused ion-beam bombardment was proposed as an alternative technique for surface etching of LN substrates [22]. Photonic band gap structures with a spatial resolution of 70 nm were obtained. The method provided high resolution and the ability to drill holes directly from the sample surface. The only constraint was that the sample surface had to be metallized and grounded to avoid charge accumulation. Figure 3.6 shows the focused-ion-beam image of the cross-section of the fabricated cavities. The array exhibited well defined circular holes with etching depth of approximately 2 μm.

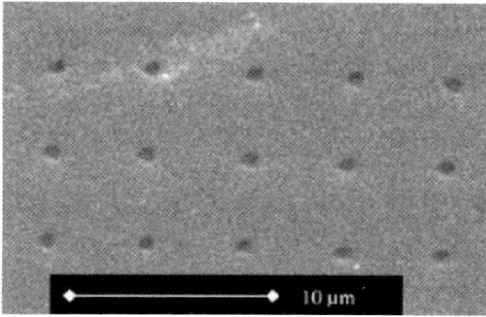

Fig. 3.4. SEM image of a two-dimensional structure obtained by electron beam bombardment and subsequent etching (from [19])

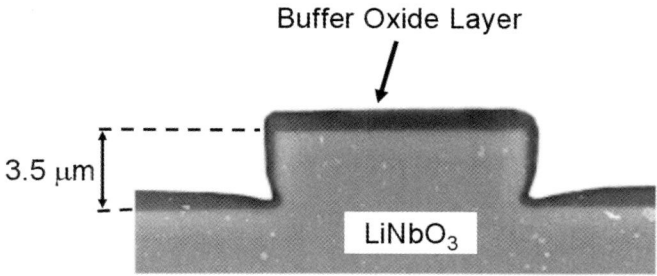

Fig. 3.5. Cross sectional profile of the ridge waveguide (from [21])

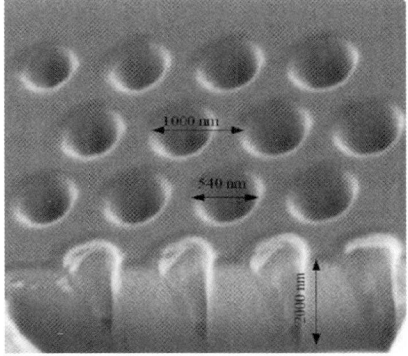

Fig. 3.6. Focused-ion-beam image of the fabricated structures (from [22])

The plasma etching of LN substrates was proposed very recently [23]. The novelty of the technique was based on the use of fluorine gases on proton-exchanged substrates in order to prevent redeposition of LiF and thus enabling the fabrication of deeper structures. Figure 3.7 shows the SEM image of the end face of the ridge structure obtained by this technique.

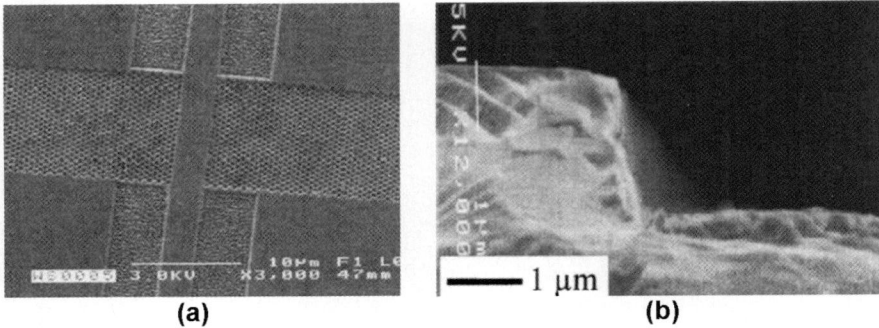

Fig. 3.7. (a) SEM image of the photonic crystal waveguide with a periodic structure at 500 nm distance between holes; (b) SEM image of the end face of the ridge structure (from [23])

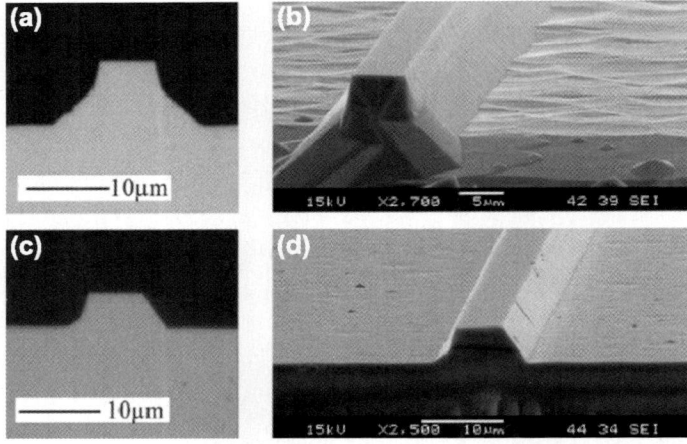

Fig. 3.8. (a) optical microscope image of the end face of a Y propagating ridge waveguide; (b) corresponding SEM image; (c) optical microscope image of the end face of a X propagating ridge structure; (d) corresponding SEM image (from [24])

The wet etching process is in continuous evolution and very recently new encouraging results were presented in case of LN substrates [24]. Ridge waveguides were fabricated by a mixture of HF and HNO_3 acids by using chromium stripes as masks. The results showed that smooth etched surfaces were obtained by adding some ethanol into the etching solution. The structures were also tested as low-loss mono-mode waveguides with height up to 8 μm and width from 4.5 to 7.0 μm. Figure 3.8 shows the optical microscope images and the corresponding SEM images of a couple of structures obtained by this technique.

Another surface structuring technique to be mentioned was proposed recently by the authors [25]. The method is based on the periodic domain reversal by electric field overpoling and subsequent wet etching in HF solution. The overpoling process

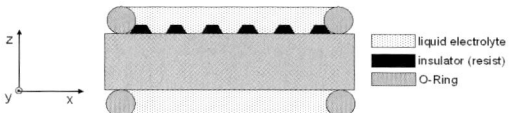

Fig. 3.9. Schematic view of the sample cross section after lithographic patterning

allowed the fabrication of sub-micron periodic structures while the conventional wet etching was used to replicate the domain pattern onto the Z cut LN crystal sample as a surface structure. The technique is relatively simple and the selective domain inversion at sub-micron scale is obtained by using fine pitch resist patterns realized by HL, as described in the following sections.

3.3 Electric Field Poling and Overpoling

Starting from a single-domain crystal, periodic domain structures can be achieved by applying external electric fields at room temperature [2, 28]. The challenge of fabricating high quality quasi-phase-matching structures by electric field poling (EFP) lies in achieving few micrometers wide domains in crystals of several millimetres in thickness. This put high demands on the poling process. Most of the last papers on periodically poled LN (PPLN), obtained by EFP, report interaction lengths of few centimeters in commercial samples of 0.5 mm thickness. Thicker samples cannot be used because of the dielectric breakdown appearing before domain inversion. Recently it has been reported the decrease of the ferroelectric coercive field by one order of magnitude between congruent and stoichiometric crystals. This makes possible the electric field fabrication of periodic structures at room temperature in few millimetres thick samples [29].

The periodic EFP (PEFP) consists of using lithographic techniques to produce a photoresist grating of the desired period to be used as a mask for applying the external electric field with a liquid electrolyte [28]. A positive voltage pulse slightly exceeding the coercive field of the material (around 21 kV/mm in LN and 2 kV/mm in KTP) is applied to the patterned crystal face by using a liquid electrolyte. The liquid electrode configuration has two electrolyte containing chambers which squeeze the sample between two O-ring gaskets, as shown schematically in Fig. 3.9.

Figure 3.10 illustrates the schematic view of the typical external electrical circuit. A conventional Signal Generator (SG) drives an High Voltage Amplifier (HVA-2000x) with a series current limiting resistor R_s in order to get a 12 kV positive voltage. A diode rectifier D is connected to the output of the HVA to prevent flowing of backswitch current in the circuit.

In case of LN the reversed domains typically grow beyond the width of the electrodes as result of the remaining fringing fields along the edges of the lithographic grating strips [28]. For example, in PPLN processed for infrared applications (periods >10 μm), the inverted domain width will typically be ∼3–4 μm wider than that of the electrode. To obtain the desired domain size, insulating strips wider than

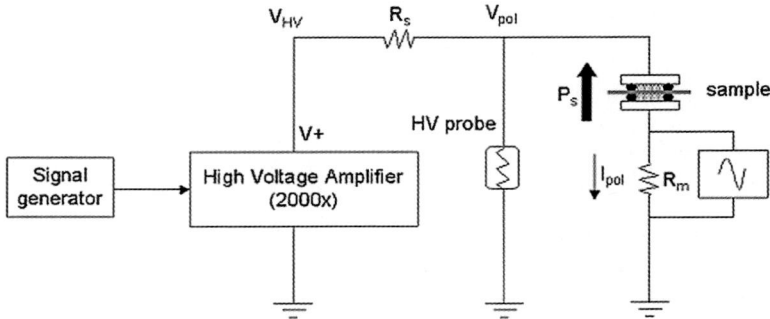

Fig. 3.10. Schematic view of the external circuit for EFP. SG signal generator; HVA high voltage amplifier; D diode rectifier; R_s series resistor; V_{pol} poling voltage; HVP high voltage probe; R_m monitoring resistor; OSC oscilloscope; I_{pol} poling current

the electrodes must be fabricated. The strategy for optimal domain patterning is to stop the voltage pulse before poling progresses under the photoresist layer. This is usually accomplished by delivering the *apriori* known amount of charge $Q = 2P_s A$ required for polarization reversal in the electrode regions [2]. An *in-situ* stopping criterion consisting in watching for a drop in the poling current I_{pol} and a corresponding rise in the poling voltage V_{pol}, both effects indicating that the sample has completely poled under the electrodes and that the domains are now laterally spreading under the insulating layer, is used [2, 28]. In fact, the conductivity of LN at room temperature is low enough that the poling current can be monitored readily by measuring the voltage drop across the R_m resistor (usually 10 kΩ) while a conventional High Voltage Probe (HVP) is used to measure the poling voltage V_{pol} across the sample. Both current and voltage waveforms are visualized on the oscilloscope OSC during the poling process. As mentioned previously, reversed ferroelectric domain patterns are usually inspected unambiguously by a well established technique based on selective wet etching [8]. Alternative approaches include the non-invasive domain visualization by crossed polarizers [30] and more recently the observation of domain structures by the EO effect [31].

The EFP is applied here by following a non-conventional procedure consisting in the application of additional electric field pulses slightly above the coercive field of the material till a drop to zero of the poling current is detected, thus ignoring both the total amount of charge delivered to the sample and the usual crucial stopping criterion. Figure 3.11(a), (b) show the typical waveforms of the poling voltage and of the poling current acquired in case of the first and the last electric field pulse, respectively. The technique is called "electric field overpoling" (EFO) and allows to fabricate periodic surface domain patterns at sub-micron scale, otherwise difficult to achieve by the conventional EFP described above.

The selective poling is achieved by using insulating resist gratings, as in case of conventional EFP, but obtained by the HL process, which allows to get finer pitch structures compared to mask lithography as described in the following section.

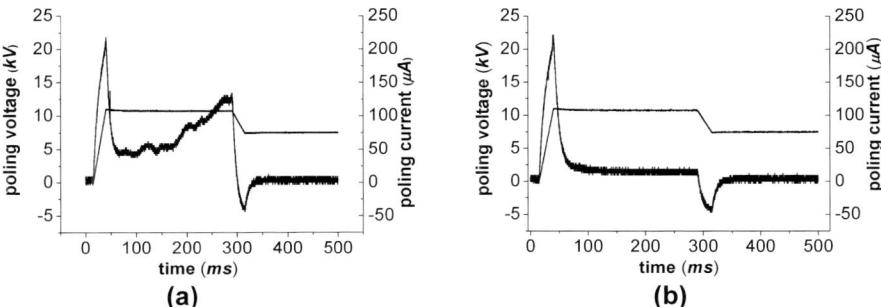

Fig. 3.11. Typical voltage and current waveforms acquired during electric field overpoling in correspondence of (**a**) the first and (**b**) the last electric field external pulse. The *thicker curve* corresponds to the current waveform while the *thinner* refers to the poling voltage

3.4 Holographic Lithography

HL is carried out by combining two coherent wavefields to form a sinusoidal intensity light pattern in space. The interferogram is generated within the volume of space defined by the overlapping beams. Since HL does not need any photolithographic mask, the field size depends only on that of the two beams, whereas in conventional mask lithography the imaging system generally limits the workable field size. A simple 1D grating can be obtained by exposing a layer of photoresist to the interference pattern. Furthermore, more complex 2D patterns, including square or hexagonal arrays of dots or holes, can be obtained by overlaying multiple exposures or combining more than two beams. HL has some unique advantages over conventional optical lithography. In particular, the spatial resolution of HL can easily exceed the resolution limits of today's optical steppers when comparable wavelengths are used. For example, structures as small as 100 nm are readily patterned by HL using a source at 351 nm [32, 33] with the additional advantage of a faster process, if compared to Electron Beam Lithography, especially when large exposed areas are required. Compared to mask optical lithography, HL is not diffraction limited and the depth of focus is effectively infinite on the scale of planar devices. This makes HL well suited for applications where substrate flatness and topography are critical issues. One more attractive feature of HL is its implementation with relatively simple optical components so that effects due to lens aberrations are dramatically reduced compared to the case of conventional lithography.

Three different set-ups have been used here to generate the interference pattern: Michelson (M); two-beams (TB); Loyd's mirror (LM). Figure 3.12 shows the schematic views of these interferometric configurations. The period of the fringe pattern depends on the overlapping angle 2θ between the two interfering beams according to the formula $p = \lambda/2\sin\theta$. The source used for these interferometers is a coherent He–Cd cw laser delivering a power of about 65 mW at 441.6 nm. Different fringe periods and exposed areas are provided by these interferometers, depending on the overlapping geometry of the two interfering wavefields. The M set-up pro-

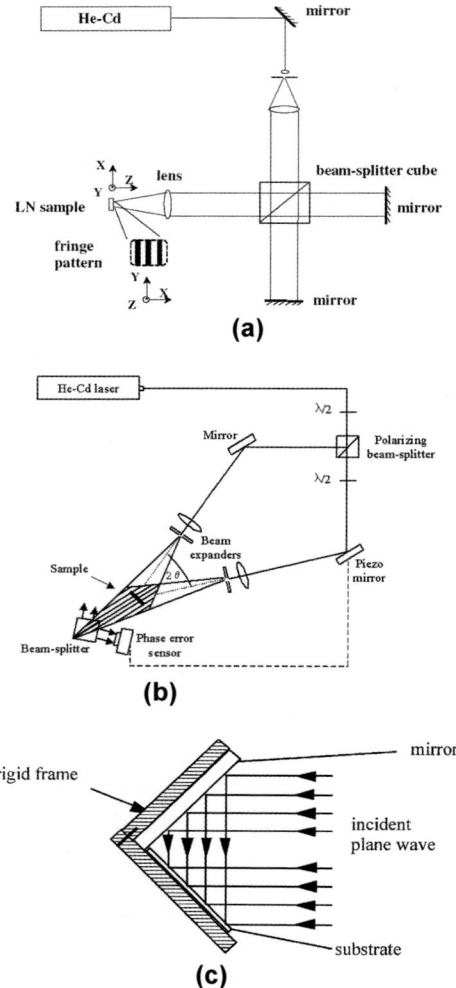

Fig. 3.12. Interferometric set-ups used for HL process: (**a**) Michelson type used for periods down to about 15 μm; (**b**) two-beams type used for periods down to the diffraction limit (about 220 nm) over relatively large areas (about 2 cm diameter); (**c**) Lloyd's mirror type used for periods down to about 220 nm over relatively small areas (about 5 mm large)

vides an interferogram covering a circular region of about 25 mm in diameter with a fringe period ranging from the zero fringe condition down to about 15 μm. Shorter periods are not possible due to the intrinsic limitations of the configuration which does not allow to enlarge the recombination angle of the two beams.

The TB set-up used in this work is suitable for shorter periods ranging from about 2 μm down to about 220 nm ($\lambda/2$ diffraction limit). Larger periods would be possible at narrower angles and thus at longer distances from the dividing beam-

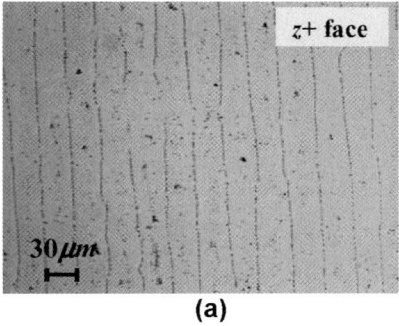 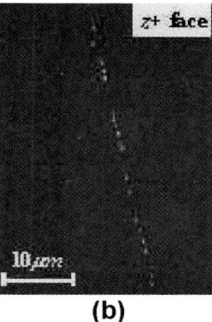

Fig. 3.13. (a) Optical microscope image of the aligned dot domains revealed by wet etching; (b) SEM image of the dot structures

splitter. The LM set-up allows to produce relatively short fringe periods like in case of the TB interferometer. In this case the coherence of the fringe gratings depends dramatically on the quality of the mirror. Dust particles on the mirror and any sharp edges give rise to scattered radiation which contributes to a coherent noise. The LM configuration is hence relatively simple and cheap, but does not produce the highest quality gratings and grids.

3.5 Periodic Sub-Micron Structuring

The EFO was applied to LN samples resist patterned with different geometries by HL as described in the following sections.

3.5.1 Overpoling Applied to One-Dimensional Michelson Resist Gratings

The EFO was applied to LN crystal samples patterned with a one-dimensional resist grating at 30 µm period obtained by the M set-up. The process was about 3 s long and the resulting ferroelectric domain pattern, revealed by a wet etching process of 60 minutes in HF:HNO$_3$ = 1:2 acid mixture, is shown in Fig. 3.13.

Aligned dot-like structures with sub-micron size are visible, corresponding to un-reversed regions under the photoresist strips. This effect is due to an incomplete merging of the adjacent reversed domains under the photoresist. The isolated dots aligned along the photoresist fringes correspond to the regions excluded by the merging of the hexagonal-type counter propagating domain walls, originating from two adjacent electrodes during poling and joining under the resist strips. The propagating domain walls are probably not straight because the sidewalls of the resist strips are affected by relatively high corrugation and low steepness. In fact, interference fringes with pitch values over the micron scale present quite large speckles such that the propagating domain walls find ways along which the velocity of motion varies. Consequently, the merging is not successful everywhere. The optical microscope image taken in a peripheral region of the pattern and presented in Fig. 3.14 clearly supports this interpretation. It illustrates the domain merging occurring in different regions of the pattern such that the merging process is frozen in the various stages of its evolution.

The mechanism of merging of adjacent domains leads to the formation of the aligned dot-like domains under the resist strips. Microscope inspection of the etched sample reveals that opposite crystal faces exhibit different structure morphologies, as shown by the magnified optical microscope images in Fig. 3.15. A line-shape structure on $Z-$ face corresponds to a dot-shape on the $Z+$. One of the interesting features of these dot-like domains is the sub-micron size, as shown by the SEM image in Fig. 3.13(b). In fact, despite the advantages achieved nowadays by the conventional EFP, such as repeatability, scalability and applicability over a wide range of materials, fabrication of periodically poled materials with arbitrarily small values of period, particularly at sub-micron scales, remains an elusive goal.

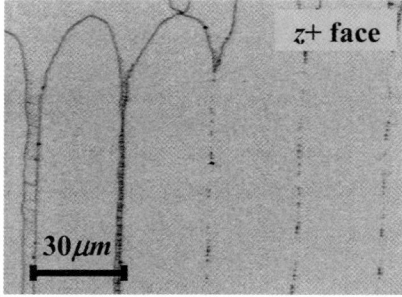

Fig. 3.14. Optical microscope image of the dot-like domains in a peripheral region of the pattern

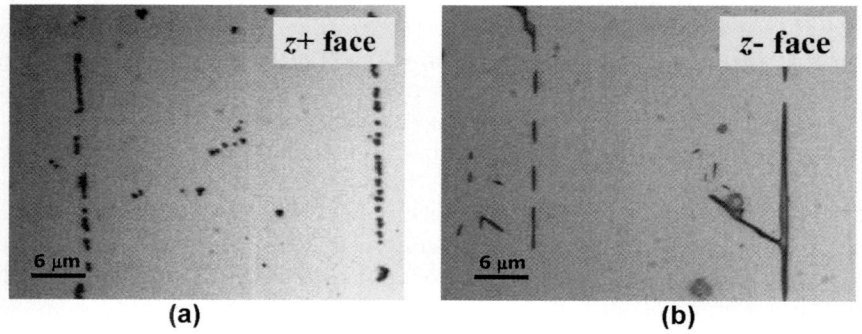

Fig. 3.15. (a) Magnified optical microscope image of the dot-like domains on $Z+$ and (b) the corresponding line-like structure on $Z-$

The EFO technique presented here can be considered as an effective and relatively simple solution for achieving high density and sub-micron ferroelectric reversed domain structures.

3.5.2 Overpoling Applied to Two-Dimensional Michelson Resist Gratings

The EFO technique was applied to LN crystal samples patterned with a two-dimensional resist grating obtained by two 90° crossed exposures by the M set-up. The resulting grating on Z+ face consists in a square array of photoresist dots with a period of about 23 μm along both the X and Y crystal axis direction, as shown by the optical microscope image in Fig. 3.16(a). The poling process was about 3.5 s long. Figure 3.16(b) shows the optical microscope image of the resulting square array of dot-like domains revealed by 30 minutes wet etching at room temperature. The two-dimensional geometry of the resist grating has clearly improved the period regularity of the dot-like domains along the Y crystal axis direction. In fact, these dot-like domains are equally spaced by about 20 μm along both the X and Y direction, in agreement with the pitch size of the resist grating.

As explained in the previous Section, the dot-like structures correspond to the regions left un-reversed after the domain spreading under the resist dots. The SEM

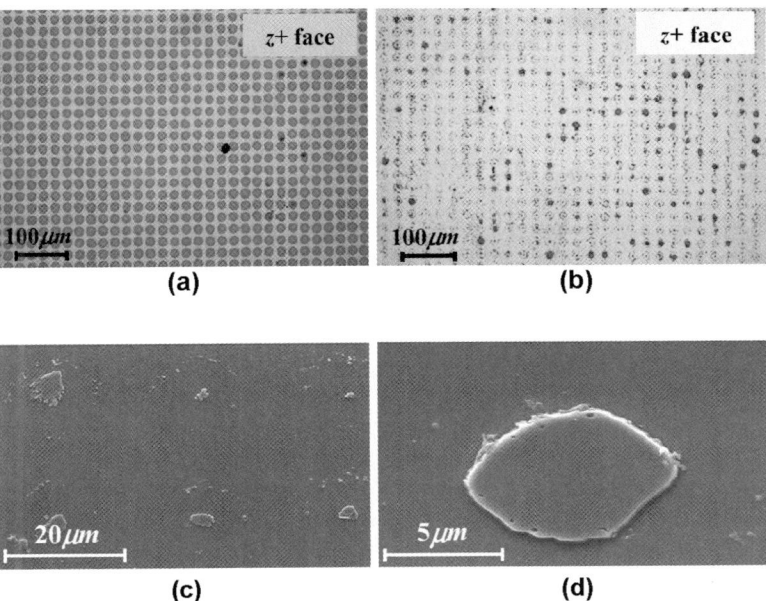

Fig. 3.16. (a) Optical microscope image of the square array of resist dots; (b) optical microscope image of the corresponding dot-like domains after wet etching; (c), (d) two different SEM magnified views of the dot-like domains. The period is around 23 μm along both X and Y crystal axis direction

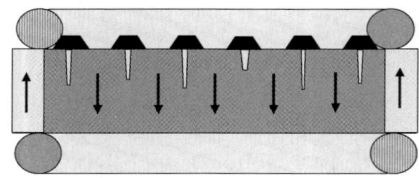

Fig. 3.17. Schematic cross section view of the un-reversed domains (*white regions*) formed under the resist dots after EFO of two-dimensional HL patterned LN samples

image in Fig. 3.16(c) shows that each dot-like domain is surrounded by smaller dot structures similar in nature to those obtained in case of one-dimensional resist pattern (see Fig. 3.13(a)) and forming a circle-like structure. The etching process reveals that the surrounding dots have the same polarization direction as the central ones. This means that the domain spreading under the resist dots evolves leaving such smaller dot-like regions un-reversed. In other words, the central dots correspond to the regions not reached at all by the domain spreading, whereas the surrounding ones result from non-homogeneous advancing of the domain wall towards the central part of the resist dots. The diameter of the circle-like structures is everywhere shorter than that of the printed resist dots by about 2 μm. This means that they originate under the resist dots but not in correspondence of their edges. A quantitative characterization of these surrounding dots is carried out by comparing the mean value of the ratio p/d estimated for the resist array (Fig. 3.16(a)) and for the etched domain structure (Fig. 3.16(b)), p being the pitch of the periodic structure and d the diameter of the resist dots and of the circle-like structures, respectively. The measurements give 1.6 in case of the resist array and 1.9 in the other. This confirms the geometrical non correspondence between the surrounding dots and the resist edges. Moreover, the $Z-$ face of the sample in Fig. 3.16(b) is perfectly flat, meaning that any structure is revealed by the etching process and thus a surface reversed domain structure has been obtained. Therefore the EFO clearly evolves from the $Z-$ to the $Z+$ face leaving shallow dot-like un-reversed domains under the resist dots, as shown by the schematic cross section view in Fig. 3.17.

3.5.3 Overpoling Applied to Two-Beams Resist Gratings at Sub-Micron Scale

The application of the EFO process to samples resist patterned by the TB interferometer allowed the fabrication of periodic surface structures at sub-micron scale with different geometries [25, 34]. Four samples with different kinds of topography and periods were investigated. The samples A and B have 2D periodic structure consisting of a square array of pillar-like structures (PLS) and hole-like structures (HLS), respectively, with 2 μm period; the sample C has a 2D periodic structure consisting of a square array of HLS with 530 nm period; the sample D has a 1D periodic structure with 600 nm period. All of these samples are first domain reversed by EFO and then etched in HF solution at room temperature for 5 minutes. The PLS or the HLS are obtained by applying the EFO to samples resist patterned with a

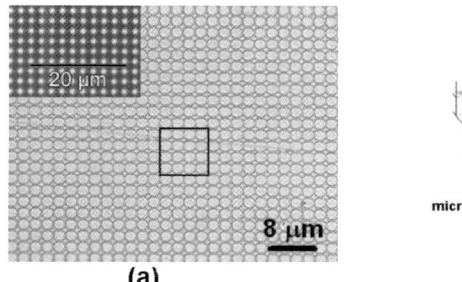
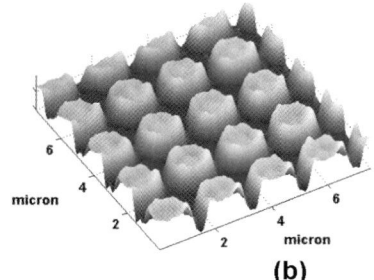

Fig. 3.18. (a) Optical microscope image of sample A and (b) corresponding surface representation of the region in the square frame. The *inset* shows the optical microscope image of the corresponding photoresist grid-like geometry

grid-like geometry (GLG) or a dot-like geometry (DLG), respectively. In fact, the EFO allows to reverse the ferroelectric polarization everywhere in the bulk crystal apart from shallow un-reversed regions under the photoresist layer which faithfully reproduce the resist pattern geometry. Figure 3.18(a) shows the optical microscope image of the $Z-$ face of sample A, after etching, and that of the corresponding GLG resist pattern as example (see the inset).

The PLS, surrounded by canyon-like structures, obtained in sample A is clearly visible in the surface representation image in Fig. 3.18(b), corresponding to the region highlighted by the square frame in Fig. 3.18(a). The AFM topography image taken on the $Z-$ face of the sample B is presented in Fig. 3.19(a). The HLS obtained in sample B is clearly visible in the surface view shown in Fig. 3.19(b).

The etch depth profile is around 220 nm as shown by the AFM profile measurement in Fig. 3.19(c). Figure 3.20(a) shows the AFM topography image taken on $Z-$ face of sample C which has been overpoled by using the DLG resist pattern shown in the inset as example. A magnified view of the structure is presented in Fig. 3.20(b) while the surface view is shown in Fig. 3.20(c). The etch depth is around 1 μm as shown by the AFM profile measurement in Fig. 3.20(d). Figure 3.21(a), (b) show the AFM topography image and the corresponding profile measurement taken on the $Z-$ face of sample D. The etch depth is around 200 nm.

Table 3.1 summarizes the main measurement results.

3.5.4 Complex Surface Structures by Moiré HL

Moiré effect [35] was used in the HL process by the TB set-up to fabricate periodic resist gratings with complex geometries to be transferred into 500 μm thick LN substrates as surface structures by EFO and subsequent wet etching. In particular, three kinds of structures are produced corresponding to three different beating geometries.

Sample M1 is resist coated on $Z-$ face and HL patterned with 2 μm period 2D square grating by 90° crossed exposures. Afterwards, $Z+$ face is coated and HL

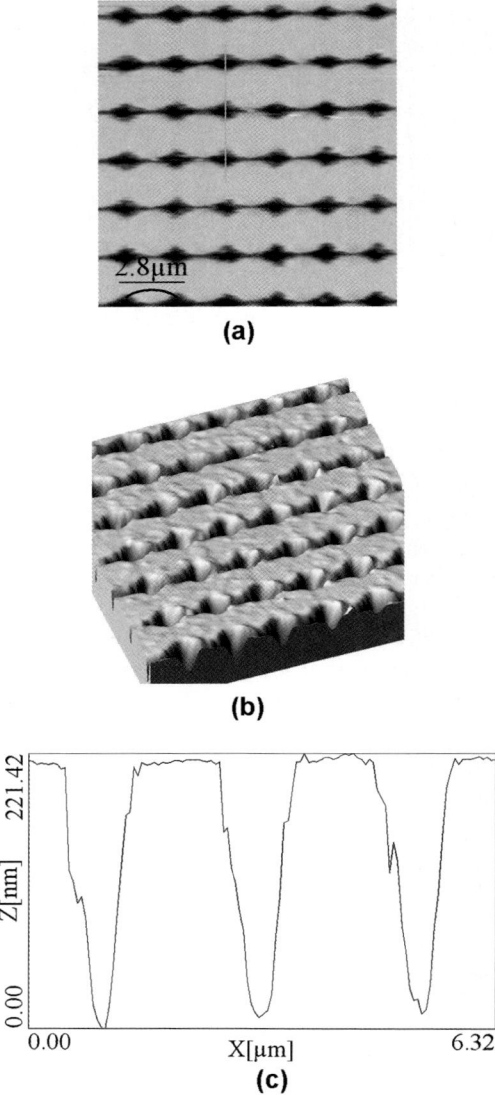

Fig. 3.19. (a) AFM topography image of sample B; (b) corresponding 3D representation; (c) depth profile along the line indicated in (a)

patterned by the same 2D pattern. Figure 3.22(a) shows the typical square array resist grating obtained on the $Z+$ face, after patterning $Z-$ face. The fringe period is slightly changed before exposing $Z+$ face, resulting in a typical moiré fringe pattern obtained on the $Z-$ face (see Fig. 3.22(c)). In fact, since the $Z-$ face is

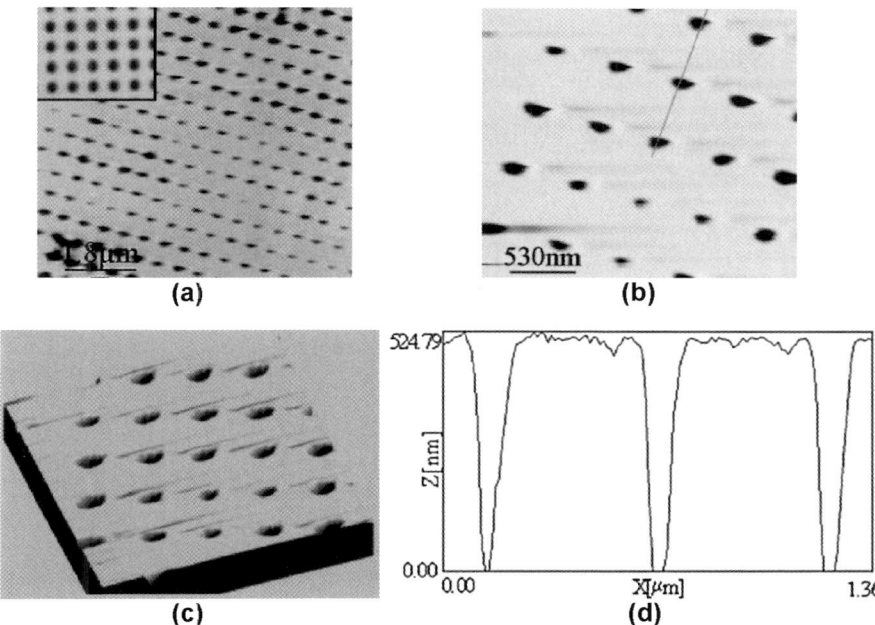

Fig. 3.20. (a) AFM topography image of sample C; (b) magnified topography view; (c) corresponding 3D image; (d) depth profile along the line indicated in (b). The *inset* shows the optical microscope image of the corresponding dot-like geometry resist pattern

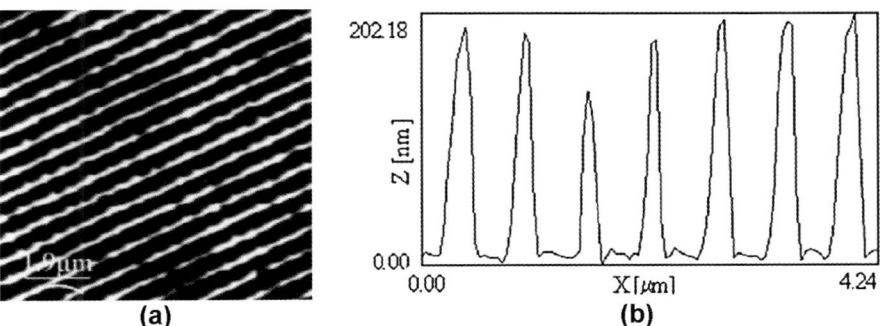

Fig. 3.21. (a) AFM topography image of sample D; (b) corresponding depth profile along a line perpendicular to the domain direction

patterned first, that effect is due to the residual resist photosensitivity on the $Z-$ face, during the exposure of $Z+$.

The interference fringes, transmitted by the resist on $Z+$ face and by the crystal itself, overlap with the resist pattern previously printed on the $Z-$ face, causing the well known moiré effect [35]. If necessary, this effect can be easily avoided by

Table 3.1. List of the main measurements obtained for each sample. In case of PLS the etch depth refers to the surrounding canyon-like structures

	Structure type	Period (nm)	Feature size (nm)	Etch depth (nm)	Area (mm × mm)
Sample A	PLS	2000	1500	1000	(4 × 4)
Sample B	HLS	2000	900	220	(2 × 2)
Sample C	HLS	530	200	530	(2 × 2)
Sample D	lines	600	300	200	(1 × 1)

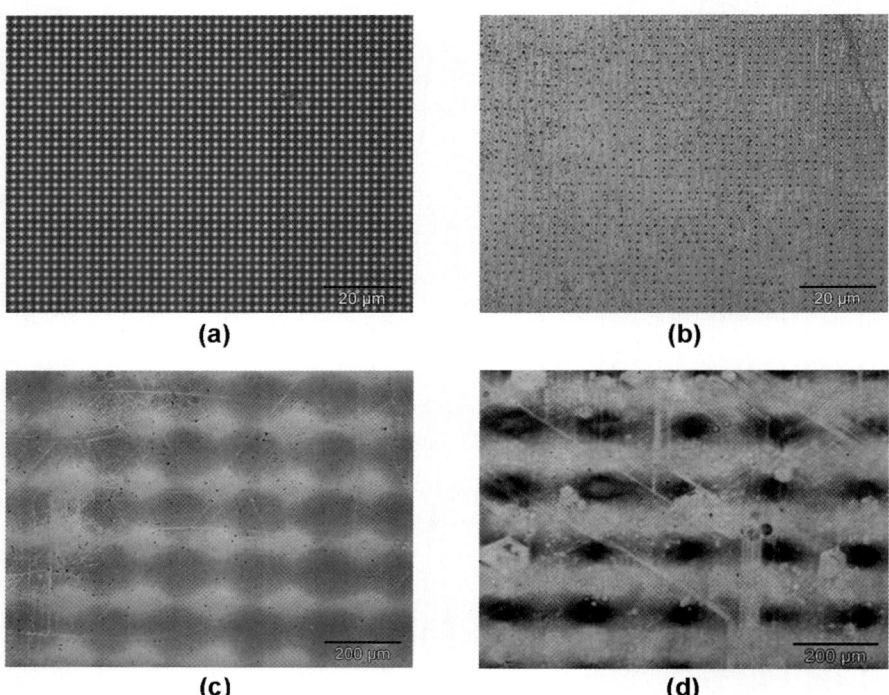

Fig. 3.22. Optical microscope image of (**a**) the resist grating printed on the $Z+$ face of sample M1; (**b**) the corresponding surface structure on $Z+$ after poling and etching; (**c**) the resist grating printed on the $Z-$ face; (**d**) the corresponding surface structure on $Z-$

post-baking the sample after developing the $Z-$ face and before spin-coating the $Z+$ face.

The moiré effect produces on $Z-$ face a double period resist grating with periods 2 μm and 200 μm (see Fig. 3.22(c)). The sample is then subject to EFO and Fig. 3.22(b)–(d) show the reversed domain pattern revealed by HF etching on $Z+$ and $Z-$ face, respectively. Referring to the wide field image in Fig. 3.22(d) a magnified optical microscope image is acquired in a dark region (see Fig. 3.23(a), and a SEM image is taken in one of the bright regions (see Fig. 3.23(b)). The SEM

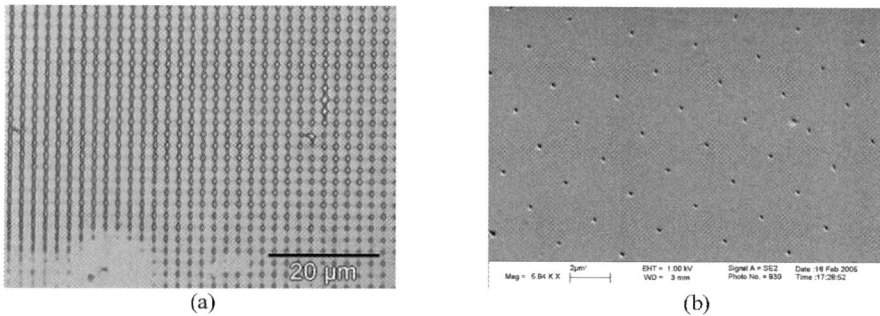

Fig. 3.23. (**a**) Magnified optical microscope image of sample M1 in correspondence of the regions appearing dark in wide field optical microscope image; (**b**) SEM image of the bright regions

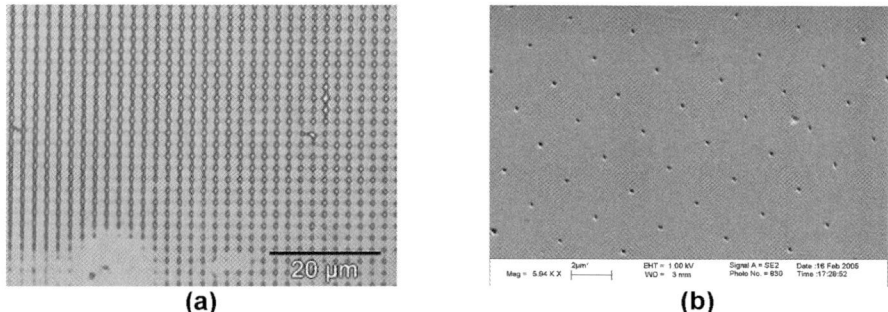

Fig. 3.24. Optical microscope image of (**a**) the resist grating printed on the $Z-$ of sample M2; (**b**) the corresponding surface structure

image clearly shows that periodic hole-like structures around 200 nm sized have been obtained.

Sample M2 is resist coated on $Z-$ face and subject to four HL exposures: two 90° crossed exposures at 2.4 µm period and two 90° crossed exposures at 2.5 µm period. A unique development process is performed after the four exposures. A 2D double period (2.4 µm and 2.5 µm) resist grating is obtained on $Z-$ face as result of the moiré beating between the two fringe gratings, as shown by the optical microscope image in Fig. 3.24(a). Subsequent EFO and HF etching allows to transfer such resist geometry into the LN substrate as surface structure shown by the optical microscope image in Fig. 3.24(b).

Sample M3 is resist coated on $Z-$ face and subject to three HL exposures: two crossed exposures at 2.4 µm period and one at 2.4 µm period. A unique development process is performed here too, after the four exposures. A 2D grating at 2.4 µm with 1D channels at 2.5 µm period is obtained as shown by the optical microscope image in Fig. 3.25(a). The surface structure obtained on $Z-$ face by EFO and subsequent

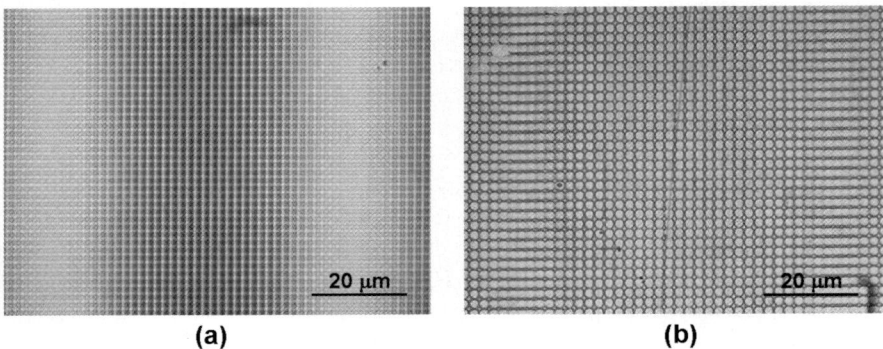

Fig. 3.25. Optical microscope image of (**a**) the resist grating printed on the $Z-$ face of sample M3 and (**b**) the corresponding surface structure

Table 3.2. Main features of resist gratings and corresponding surface domain structures for the three kinds of samples fabricated here

	Sample M1	Sample M2	Sample M3
Moiré process	$(Z+)$ 2D at 2 µm $(Z-)$ 2D at ∼2 µm	$(Z-)$ dual 2D (2.4 µm; 2.5 µm)	$(Z-)$ 2D at 2.4 µm $(Z-)$ 1D at 2.5 µm
Resist grating	$(Z+)$ 2D at 2 µm $(Z-)$ dual 2D (2 µm; 200 µm)	$(Z-)$ dual 2D (2.4 µm; 2.5 µm)	$(Z-)$ 2D at 2.4 µm + 1D at 2.5 µm (channels period 80 µm)
Domain structure	$(Z+)$ 2D pillar at 2 µm $(Z-)$ dual 2D holes (2 µm; 200 µm)	$(Z-)$ dual 2D holes (2.4 µm; 2.5 µm)	$(Z-)$ 2D holes at 2.4 µm + 1D PPLN at 2.5 µm (channels period 80 µm)

HF etching is shown by the optical microscope image in Fig. 3.25(b). The 1D channels are periodic with 80 µm period.

The moiré beating geometries and the corresponding structures fabricated here are listed in Table 3.2. It is important to note that, in every case, the domain pattern obtained by the EFO process on $Z-$ face faithfully reproduces the resist geometry.

3.6 Double-Face Sub-Micron Surface Structures

This section presents a technique for achieving double-face one-dimensional and two-dimensional sub-micron periodic surface structures over areas of about (5×5) mm^2 [36]. The period of the structures is 2 µm. The $Z-$ face was resist patterned and subject to post-bake to remove the residual photosensitivity and after that the $Z+$ face was resist patterned. The EFO was performed as in case of the previously described samples. Different domain pattern geometries were fabricated on $Z+$ and $Z-$ crystal faces to show the reliability of the technique. Sample M1 in the previous section is an example of double-face patterned sample where the bake process at 200°C of $Z-$ face has not been performed, thus allowing the moiré effect on $Z-$ face by using the residual resist photosensitivity. In this section three different kinds of double-face patterned samples are shown: D1, D2 and D3.

Sample D1 was patterned with one-dimensional resist grating on both faces. Figure 3.26(a), (b) show the optical microscope images of the $Z-$ and $Z+$ face of the sample D1, respectively, after EFO and etching. It is worth noting that several samples have been processed like sample D1 and the 1D surface structure generated on $Z-$ face always faithfully reproduces the resist grating in each sample. In contrast, the surface structure fabricated on $Z+$ face, even though patterned with the same quality resist grating as $Z-$, exhibits a corrupted geometry (see Fig. 3.26(b)). This demonstrates that the patterning of $Z-$ face provides better quality domain gratings.

Sample D2 was patterned with a two-dimensional array of resist dots on both faces. Figure 3.27(a), (b) show the optical microscope images of the $Z-$ and $Z+$ faces of sample D2, respectively, after EFO and etching. The EFO and etching applied to the array of resist dots clearly gives an array of pillars on $Z+$ face and an array of holes on $Z-$ face. Figure 3.26(c) shows the optical microscope image of the polished and etched Y face of sample D1. This picture clearly shows that the EFO applied to double-face resist patterned samples generates surface un-reversed domains on both faces.

Sample D3 was fabricated with completely different domain pattern geometries on the two faces. The $Z-$ face was patterned with a two-dimensional array of resist dots and $Z+$ face was patterned with a one-dimensional resist grating. Figure 3.28(a), (b) show the optical microscope images of the $Z-$ and $Z+$ face of sample D3, respectively, after EFO and etching.

The inspection of the polished and etched Y face of these three samples reveals that the EFO generates un-reversed regions under the resist coated areas on both faces, with average depths around 10 µm, largely compatible with waveguide fabrication. Figure 3.29 shows the optical diffraction pattern of the etched sample D3, taken with a He–Ne laser aligned nearly perpendicularly to the $Z+$ face.

The resultant diffraction pattern represents the reciprocal lattice of the periodically poled pattern and gives information about the quality of the realized grating. In fact, by scanning the laser beam over the whole structure area of about (5×5) mm^2, the diffraction orders are clear and exhibit constant diffraction angles, thus demonstrating the uniformity of the structures over the whole engineered region. Moreover, the measurement of the diffraction angles is in good agreement with the space periodicity, measured by standard optical microscope.

3.7 Possible Applications for Novel Photonic Crystal Devices

This section is aimed at proposing the LN structures described in this chapter for possible applications in the field of photonic devices. In fact, such structures exhibit specific features which allow to foresee the possibility to fabricate innovative

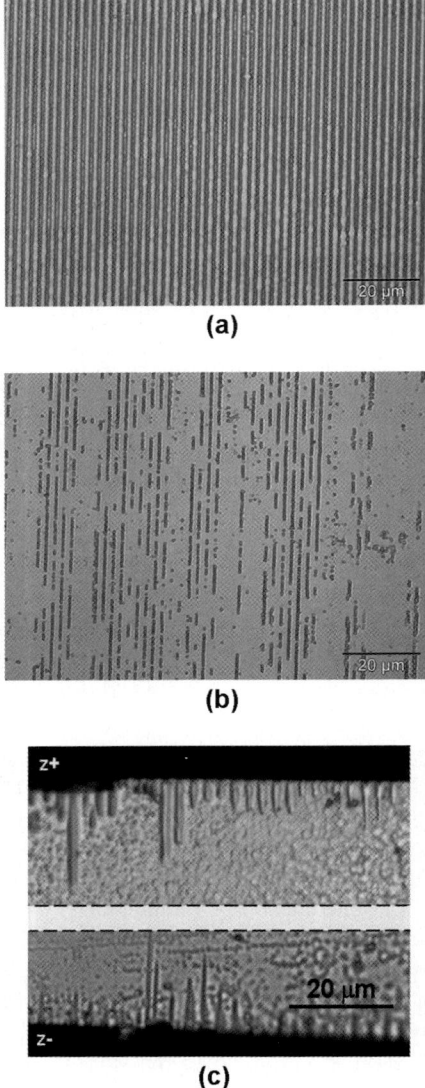

Fig. 3.26. Optical microscope image of (**a**) the etched $Z-$ face of sample D1; (**b**) the corresponding etched $Z+$ face; (**c**) the corresponding polished and etched Y face

photonic devices exploiting the specific properties of the LN material (EO, PZ, non-linearity, etc.). Further investigations are currently under consideration in order to demonstrate the actual feasibility of such photonic devices.

The one-dimensional sub-micron period surface reversed domain gratings (see Fig. 3.21) could be used for different purposes in etched as well as un-etched

Fig. 3.27. Optical microscope image of (**a**) the etched $Z-$ face of sample D2 and (**b**) the corresponding etched $Z+$ face. The *inset* shows a magnified surface view of the sample D2

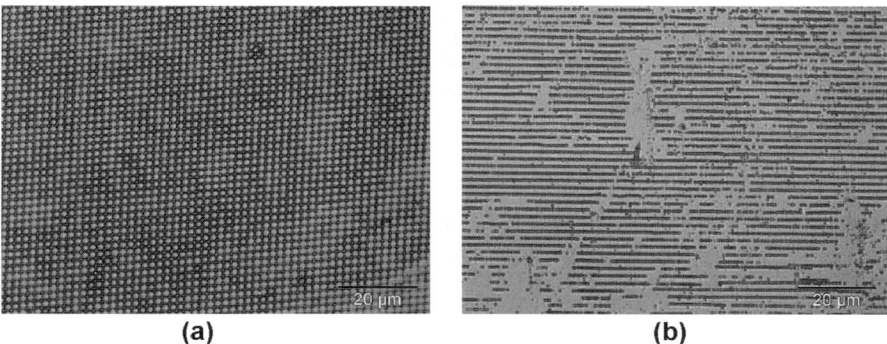

Fig. 3.28. Optical microscope image of (**a**) the etched $Z-$ face of sample D3 and (**b**) the corresponding etched $Z+$ face

Fig. 3.29. Optical diffraction pattern of the sample D3

version. In the last case the domain grating could be implemented in waveguide configuration for the fabrication of novel nonlinear devices providing short-wavelength conversions by QPM interaction. Moreover periods down to 300 nm would allow backward nonlinear wavelength interactions. The sub-micron period of these gratings allows to foresee also their implementation in the fabrication of different innovative Bragg reflectors, depending on the etched or un-etched nature of the structure, operating in the infrared wavelength region with subsequent interesting applications in the field of telecommunications. In case of un-etched gratings a tunable Bragg reflector device would be possible by exploiting the refractive index step induced electro-optically by appropriate electrodes. In case of etched gratings a different tunable Bragg reflector would be possible by using the refractive index step between air and LN where the tunability could be induced by the EO effect as before or for example by thermal expansion with further applications in the field of sensors.

The two-dimensional short period surface domain gratings after etching could be implemented in planar waveguides for the fabrication of innovative tunable photonic band-gap devices exploiting the refractive index contrast between air and LN and the unique properties of LN (nonlinearity; electro-optics; piezoelectricity; pyroelectricity; etc.).

The moiré method allows to introduce artificial defects in a host photonic bandgap structure, thus providing the possibility to manipulate photons by localizing the electromagnetic states and "trap the light" [37]. For example, photons can propagate through a linear defect within a two-dimensional pattern. This phenomenon may be used in ultra-small optical devices for optical communications.

An interesting advantage of the double-face patterning relies on the possibility to shrink into the same chip size a number of devices twice as many and with completely different geometries on the two faces according to the requirements.

References

1. A.M. Prokhorov, Yu.S. Kuz'minov, *Physics and Chemistry of Crystalline Lithium Niobate* (Adam Hilger, Bristol, 1990)
2. L.E. Myers, R.C. Eckardt, M.M. Fejer, R.L. Byer, W.R. Bosenberg, J.W. Pierce, J. Opt. Soc. Am. B **12**, 2102 (1995)
3. A.A. Oliver (ed.), *Acoustic Surface Waves*. Topics in Applied Physics, vol. 24 (Springer, Berlin, 1978)
4. B.U. Chen, E. Marom, A. Lee, Appl. Phys. Lett. **31**, 263 (1977)
5. H.L. Garvin, E. Garmire, S. Somekh, H. Stoll, A. Yariv, Appl. Opt. **12**, 455 (1973)
6. B.L. Sopori, C.M. Phillips, W.S.C. Chang, Appl. Opt. **19**, 790 (1980)
7. C.L. Lee, C.L. Lu, Appl. Phys. Lett. **35**, 756 (1979)
8. K. Nassau, H.J. Levinstein, G.M. Loiacono, Appl. Phys. Lett. **6**, 228 (1965)
9. N. Niizeki, T. Yamada, H. Toyoda, Jpn. J. Appl. Phys. **6**, 318 (1967)
10. J. Webjörn, F. Laurell, J. Arvidsson, J. Light. Technol. **7**, 1597 (1989)
11. J.L. Jackel, R.E. Howard, E.L. Hu, S.P. Lyman, Appl. Phys. Lett. **38**, 907 (1981)
12. C. Ren, J. Yang, Y. Zheng, L. Chen, G. Chen, S. Tsou, Nucl. Instrum. Methods Phys. Res. B **19**, 1018 (1987)

13. C.I.H. Ashby, P.J. Brannon, Appl. Phys. Lett. **49**, 475 (1986)
14. K. Christensen, M. Müllenborn, Appl. Phys. Lett. **66**, 2772 (1995)
15. F. Laurell, J. Webjörn, G. Arvidsson, J. Holmberg, J. Light. Technol. **10**, 1606 (1992)
16. T.-J. Wang, C.-F. Huang, W.-S. Wang, P.-K. Wei, J. Light. Technol. **22**, 1764 (2004)
17. T.-L. Ting, L.-Y. Chen, W.-S. Wang, Photonics Technol. Lett. **18**, 568 (2006)
18. I.E. Barry, G.W. Ross, P.G.R. Smith, R.W. Eason, G. Cook, Mater. Lett. **37**, 246 (1998)
19. C. Restoin, S. Massy, C. Darraud-Taupiac, A. Barthelemy, Opt. Mater. **22**, 193 (2003)
20. J.G. Scott, A.J. Boyland, S. Mailis, C. Grivas, O. Wagner, S. Lagoutte, R.W. Eason, Appl. Surf. Sci. **230**, 138 (2004)
21. D.M. Gill, D. Jacobson, C.A. White, C.D.W. Jones, Y. Shi, W.J. Minford, A. Harris, J. Light. Technol. **22**, 887 (2004)
22. F. Lacour, N. Courjal, M.-P. Bernal, A. Sabac, C. Bainier, M. Spajer, Opt. Mater. **27**, 1421 (2005)
23. H. Hu, A.P. Milenin, R.B. Wehrspohn, H. Hermann, W. Sohler, J. Vac. Sci. Technol. A **24**, 1012 (2006)
24. H. Hu, R. Ricken, W. Sohler, R.B. Wehrspohn, Photonics Technol. Lett. **19**, 417 (2007)
25. S. Grilli, P. Ferraro, P. De Natale, B. Tiribilli, M. Vassalli, Appl. Phys. Lett. **87**, 233106 (2005)
26. J.L. Jackel, C.E. Rice, J.J. Veselka, Appl. Phys. Lett. **41**, 607 (1982)
27. J.M.M.M. de Almeida, Opt. Eng. **46**, 064601 (2007)
28. M. Yamada, N. Nada, M. Saitoh, K. Watanabe, Appl. Phys. Lett. **62**, 435 (1993)
29. V. Bermúdez, L. Huang, D. Hui, S. Field, E. Diéguez, Appl. Phys. A **70**, 591 (2000)
30. V. Pruneri, J. Webjörn, P.St.J. Russell, J.R.M. Barr, D.C. Hanna, Opt. Commun. **116**, 159 (1995)
31. V. Gopalan, Q.X. Jia, T.E. Mitchell, Appl. Phys. Lett. **75**, 2482 (1999)
32. J.P. Spallas, A.M. Hawryluk, D.R. Kania, J. Vac. Sci. Technol. B **13**, 1973 (1995)
33. M.L. Schattenburg, R.J. Aucoin, R.C. Fleming, J. Vac. Sci. Technol. B **13**, 3007 (1995)
34. P. Ferraro, S. Grilli, Appl. Phys. Lett. **89**, 133111 (2006)
35. O. Kafri, A. Livuat, Opt. Lett. **4**, 314 (1979)
36. S. Grilli, P. Ferraro, M. Paturzo, L. Sansone, S. De Nicola, G. Pierattini, P. De Natale, Photonics Technol. Lett. **18**, 541 (2006)
37. J.D. Joannopoulos, P.R. Villeneuve, S. Fan, Nature **386**, 143 (1997)

4 Nonlinear Optical Waveguides in Stoichiometric Lithium Tantalate

M. Marangoni and R. Ramponi

Lithium tantalate (LT) is a ferroelectric uniaxial crystal characterized by a wide transparency range and a high optical nonlinearity, which make it particularly attracting for several frequency conversion processes from the ultraviolet to the mid-infrared region. Its nonlinear susceptibility tensor exhibits a very high diagonal coefficient (d_{33}) and very poor non-diagonal elements, which implies that only quasi-phase-matched (QPM) interactions [1, 2] between fields polarized parallel to the optical axis direction allow for high conversion efficiencies. Periods of the QPM gratings as short as 1.7 µm have been demonstrated, leading to second harmonic generation processes down to the UV spectral region [3, 4].

In the early 90's LT waveguides fabricated by proton-exchange [5] were considered as the most attracting choice for the realization of coherent compact blue sources to be applied in ultra-high density optical data storage devices, the blue light being produced by second harmonic generation of infrared diode lasers [6, 7]. The main advantages of LT waveguides with respect to the most widely employed lithium-niobate (LN) waveguides are the wider range of transparency towards the ultraviolet region, the higher optical damage threshold, and, most of all, the possibility of achieving QPM by selective proton-exchange followed by rapid thermal annealing [8]. This technique, not applicable to LN due to its much higher Curie temperature, allowed in fact for the periodic reversal of ferroelectric domains typical of QPM. By this method impressive normalized conversion efficiencies as high as 1500% W^{-1} cm^{-2} were demonstrated [7].

With the advent of the electric-field poling technique for ferroelectric domain reversal [9] at the middle of the 90's, technique applicable both to LT and LN, the performances of LN waveguides started growing very rapidly [10], because of the higher nonlinearity of the material, the availability of larger and cheaper substrates, and also of well consolidated techniques for fabricating highly homogenous waveguides, either by proton-exchange [11] or by Ti: indiffusion [12]. LN waveguides became thus the leading choice for nonlinear applications [13–15], most of all in the field of telecom all-optical devices [16, 17], in particular after the invention of the diode lasers in the blue region [18].

A resurgence of interest for LT, both in bulk and guided configuration, has arisen in the past few years thanks to the assessment of two techniques for synthesizing

LT in the so-called stoichiometric composition (SLT), one technique based on a double Czochralsky growth [19], the other one based on a vapour transport equilibrium method [20]. Indeed, although commonly referred to as LiTaO$_3$, conventional LT exhibits a sub-stoichiometric Li composition, the so-called congruent composition, given by a Li/(Li+Ta) atomic ratio approximately equal to 0.485 [21]. The Li deficiency, as widely demonstrated in the literature, strongly affects a number of properties, namely lattice parameters [22], Curie temperature [22, 23], absorption edge [22, 23], coercive fields [19], refractive indices [23] and photorefractive optical damage resistance [24, 25]. With respect to many of these properties, stoichiometric LT (SLT) is a much more performing material. In particular it exhibits an optical damage resistance much higher than periodically-poled LT and LN (PPLT and PPLN), which makes PPSLT one of the most interesting choices for several nonlinear applications involving non-critical phase matching and extremely high power densities. Some significant results achieved in bulk configuration are the following: (i) second harmonic (SH) generation in the green spectral region with SH power levels as high as 7 W cw [26, 27] and energy levels as high as 21 mJ in the pulsed regime [28], (ii) spectral compression and shift of near-infrared femtosecond pulses by cascaded SH generation processes [29–31], (iii) optical parametric amplification in noncollinear geometry leading to 16 fs long pulses in the near-infrared [32], (iv) RGB multicolour generation in aperiodic QPM gratings [33], (v) optical parametric generation in the near-infrared both at kHz [34] and MHz [35] repetition rate, (vi) optical parametric oscillation in the near-IR with multi-KHz nanosecond green pumping [36].

In guided configuration some relevant results have been recently obtained with the assessment of the reverse-proton-exchange (RPE) technique for the fabrication of high-quality optical waveguides [37]. Such technique, developed by Korkishko [38] and optimized by Fejer [10] for LN, was applied by the authors for realizing buried waveguides in SLT. The burial of the waveguide allows a reduction of attenuation and insertion losses, and an increase of the strength of the nonlinear interactions due to an improved spatial overlap between the field profiles of the interacting modes. The first nonlinear experiments in RPE waveguides fabricated in 1% MgO-doped periodically-poled SLT (PPSLT) allowed room-temperature second-harmonic-generation from telecom wavelengths with normalized efficiencies as high as 30% W^{-1} cm^{-2} and second harmonic power levels up to 41 mW without any evidence of optical damage [39, 40]. This result proves the RPE-PPSLT combination as one of the most innovative and promising solution for the realization of efficient nonlinear frequency converters and all-optical devices operating at room-temperature.

The chapter is organized as follows. The first section introduces the physical and optical properties of SLT and PPSLT that are relevant to the design and fabrication of optical waveguides for photonic applications. The second section illustrates the general characteristics and the details of the fabrication procedure of RPE, and analyses the modelling of the fabrication process in the case of SLT. The third paragraph summarizes the main results achieved on PPSLT RPE waveguides both in the linear and nonlinear regime.

4.1 Material Properties

Several material properties come into play in nonlinear frequency conversion processes: (i) linear properties, such as transparency, birefringence and dependence of refractive indices on temperature and wavelength, (ii) nonlinear properties, such as second and third order susceptibilities, (iii) applicability of periodic-poling techniques, and (iv) optical damage threshold. Since these properties are related to from the crystal structure, the first section of this paragraph provides an overview of the main physical properties of SLT, while the second section is devoted to summarize the most relevant optical properties.

4.1.1 Physical Properties

At low temperature Lithium Tantalate is a ferroelectric (FE) crystal belonging to the trigonal system and exhibiting a net spontaneous polarization. As sketched in Fig. 4.1(a) the crystal structure [42] consists of planar sheets of oxygen atoms in a distorted hexagonal close packed configuration. The axis perpendicular to the oxygen planes is commonly identified as the c-axis. With respect to it the structure exhibits the threefold rotation symmetry typical of the trigonal system. The interstices between oxygen octahedra are alternatively filled by Li atoms, Ta atoms and vacancies. This ordered sequence is responsible for the spontaneous polarization of the crystal: in fact, as represented in Fig. 4.1(b), the Li–Ta pairs oriented in a specific sense along the c direction can be thought of as dipoles that can be aligned either "up" or "down", giving rise to two possible domain polarizations and thus to two different signs for the c-axis. When the temperature is raised above the Curie temperature, ∼600°C for CLT and ∼700°C for SLT, the Ta atoms move towards the symmetry positions in the centres of oxygen octahedra, while Li ions align to the oxygen planes, thus cancelling the spontaneous polarization (see Fig. 4.1(c)).

The FE domain configuration can be modified either during growth or after growth by applying an electric coercive field to force the Li ions to take position on a specific side of the oxygen plane. If the poling is carried out at temperature close to the Curie value as in the case of poling during growth, a small field is sufficient to influence the position of the Li ions. However, if the crystal orientation is to be changed at room temperature, which occurs for fabricating periodically-poled crystals, fields close to the value of crystal breakdown are needed to force the Li ions through the oxygen planes. The value of the coercive field necessary to invert domains orientation depends strongly on the composition of the crystal. In Table 4.1 the coercive field and other significant physical properties are reported for three different compositions [42], as obtained by three different fabrication processes, namely a Czochralsky growth from a congruent melt, leading to CLT, a double crucible Czochralsky growth from a Li rich melt [19], leading to a near-stoichiometric

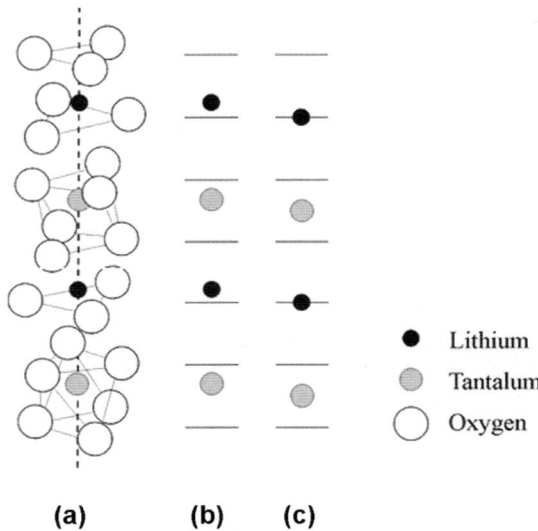

Fig. 4.1. Three-dimensional view of the position of Li and Ta atoms with respect to the oxygen octahedra in the FE phase (**a**); projection of the atom positions on a plane perpendicular to the oxygen octahedra (indicated as *solid lines*) in the FE (**b**); and paraelectric phase (**c**)

Table 4.1. Main physical and optical parameters for three differently grown LT substrates: congruent LT (CLT), stoichiometric LT grown by double-Czochralsky method (SLT-CZ), stoichiometric LT grown with VTE treatment (SLT-VTE). Superscripts indicate the references providing the data

Material	CLT	SLT-CZ	SLT-VTE
Composition, Li/(Li+Ta)	0.485[43]	0.498[19]	~0.5[42]
Curie Temperature [°C]	601[19]	685[19]	701[42]
Coercive fields [kV/cm] (forward)	211[44]	17[19]	1.61[42]
(backward)	126[44]	15[19]	1.48[42]
Domains stabilization time [s]	100–300[43]	~100[43]	<1[42]
UV absorption edge [μm]	0.275[21]	0.260[21]	0.256[42]

LT, and a vapor-transport-equilibrium (VTE) technique [20], which involves a post-growth treatment, leading to an almost pure SLT. The change in the atomic ratio Li/(Li+Ta) from ~0.485 for CLT to ~0.5 for SLT involves the Curie temperature to raise by ~100°C and the coercive field to decrease by more than two orders of magnitude. Furthermore, in SLT no significant difference exists between forward and reverse poling fields. The stabilization time of domains upon application of electric-field steps is strongly reduced for SLT, which helps in preventing domains back-switching. These properties enable the fabrication of SLT periodically-poled devices with much higher thickness with respect to CLT.

It is worth noting that the lack of a centre of symmetry of the crystal structure determines a number of interesting nonlinear properties, optical, electro-optical, pyroelectric and piezoelectric. The optical properties (linear and nonlinear), which are of main concern in this chapter, are summarized in the subsequent section, while for a review of the other ones, see, e.g., [5].

4.1.2 Optical Properties

Stoichiometric lithium tantalate has a wide transparency range from the UV, ~0.26 µm, to the mid-IR, ~5.5 µm, which makes it particularly attractive for both up-conversion and down-conversion nonlinear processes. Differently from CLT, which is a positive uniaxial crystal, SLT is a negative uniaxial crystal with modest birefringence, the extraordinary refractive index n_e being lower than the ordinary one n_o by about 0.003. This value is not sufficiently high to achieve birefringence phase-matching in nonlinear processes in bulk crystals. This fact, combined with the very low non-diagonal elements of the nonlinear susceptibility tensor, implies that the only efficient way to achieve nonlinear interactions is through QPM, coupling waves polarized along the optical axis direction (extraordinary direction). This is the reason why poor attention has been paid in the literature to the ordinary index. The wavelength and temperature dependence of the extraordinary refractive index is given by a Sellmeier curve [45]:

$$n_e = \sqrt{A + \frac{B + b(T)}{\lambda^2 - (C - c(T))^2} + \frac{E}{\lambda^2 - F^2} + \frac{G}{\lambda^2 - H^2} + D \cdot \lambda^2} \quad (4.1)$$

where the wavelength λ is expressed in µm and the temperature dependent parameters $b(T)$ and $c(T)$ are given by:

$$b = 3.48933 \times 10^{-8}(T + 273.15),$$
$$c = 1.607839 \times 10^{-8}(T + 273.15) \quad (4.2)$$

with the temperature T expressed in °C. Curve (4.1), with the coefficients reported in the first column of Table 4.2, holds in the range $\lambda = 0.39$–4.1 µm, while for longer wavelengths, up to 5.5 µm, the corrected coefficients reported in the second column can be adopted [46]. For the ordinary refractive index, less sensitive to temperature, the flowing simpler expression can be used:

$$n_o = \sqrt{A + \frac{B}{\lambda^2 - C^2} + D \cdot \lambda^2} \quad (4.3)$$

with the coefficients reported in the third column of Table 4.2. Such curve, as determined with an accuracy of 0.0002 through evanescent coupling of radiation modes [47], is valid at room temperature for the 0.5–1.6 µm range (for a temperature dependent curve of n_o, but in a narrower wavelength range 0.4–1 µm, see [48]).

Table 4.2. Coefficients of the Sellmeier equations (4.1) and (4.3) for SLT

Indices	n_e [45]	n_e [46]	n_o
A	4.502483	4.528254	4.533446
B	0.007294	0.012962	0.076772
C	0.185087	0.242783	0.226878
D	−0.02357	−0.022880	−0.031814
E	0.073423	0.068131	
F	0.199595	0.177370	
G	0.001	1.307470	
H	7.99724	7.061878	

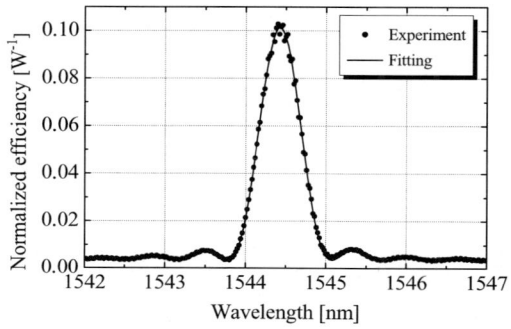

Fig. 4.2. Bulk second harmonic generation efficiency versus wavelength [experimental conditions: 25 mm long sample, 21 µm poling period, 107 µm beam radius]

Due to the lack of a centre of symmetry in the crystalline structure, SLT presents a second order nonlinear susceptibility (and consequently also electro-optical properties). Since the most significant element of the nonlinear tensor, d_{33}, couples equally polarized fields, efficient interactions can be obtained only by QPM techniques that allow for the compensation of material dispersion.

We have recently determined the d_{33} value of SLT at $\lambda = 1.55$ µm in a cw SHG experiment realized by focusing a Gaussian beam with a 107 µm radius into a 25 mm long PPSLT crystal with a 21 µm period [40]. From the fitting of the experimental efficiency curve reported in Fig. 4.2 a 10.6 pm/V value was retrieved, in excellent agreement with that obtained by fitting the intensity threshold of an optical parametric generation process from a 1.03 µm pump source [49]. This value is slightly higher than that reported in the literature for congruent LT [50] and about one half that of LN at the same wavelength, as confirmed by comparing in the same experimental conditions the SHG efficiency of the PPSLT sample with that of a PPLN sample having the same interaction length.

When dealing with FE crystals, the presence of light induced refractive index changes, also known as photorefractive effect, can provide a strong limitation to the maximum power density that can be used and thus to the conversion efficiency of the nonlinear processes. The photorefractive effect is initiated by space charge fields created by ionization of impurities, typically constituted by Fe ions, and ultimately leads to optical damage of the crystal [51]. The presence of impurities is however intrinsically much lower in stoichiometric crystals, where Li vacancies and Ta_{Li} antisite defects (Ta on Li site) are strongly reduced. Stoichiometric LT, in particular, was shown to exhibit index changes two orders of magnitude lower than CLT for a given irradiation, as a result of a much higher photoconductivity [52]. This implies the possibility of using the crystal at very high power densities without the need to increase its temperature for improving photoconductivity. It has been recently shown that at KHz repetition rate, a PPSLT crystal with 1% MgO doping from Oxide corp. could be pumped at $\lambda = 0.8\,\mu m$ with a peak intensity up to $300\,GW/cm^2$ without appreciable onset of optical damage [32].

4.2 Waveguide Fabrication through Reverse-Proton-Exchange

Reverse-Proton-Exchange (RPE) has become in the past few years one of the most attracting techniques for fabricating nonlinear waveguides in LN and LT. It allows the realization of buried waveguides with rather high index changes, which are favourable conditions both for minimizing fiber-waveguide insertion losses, due to the symmetry of the mode field profiles, and for maximizing the overlap between the field profiles of the interacting modes in the nonlinear processes. As a result of this latter circumstance, RPE waveguides have exhibited the highest ever observed normalized conversion efficiencies in second harmonic generation processes from telecom wavelength, both for LN [10] and SLT [40]. RPE fabrication technique consists of three steps, namely proton-exchange (PE), annealing (ANN) and reverse-exchange (RE). All of them are low temperature processes, i.e. below the Curie temperature of the material, and can thus be applied on periodically-poled substrates (in other cases such as for Titanium in-diffused LN waveguides, the periodic poling must be applied after waveguide fabrication [53]).

Before analysing the modelling of RPE waveguides in the specific case of SLT, let us briefly recall the main general properties of the three processes composing RPE (for a more detailed analysis see [5]).

The PE process, proposed for the first time by Jackel et al. in 1982 for LN [54], consists of an ion exchange process where lithium ions are substituted by hydrogen ions, i.e. protons. The latter ones are provided by appropriate melts, typically benzoic acid, at temperatures between 160 and 300°C. Unless using strongly diluted acids, the PE region presents very high hydrogen concentration leading to a strongly modified crystallographic phase (with respect to the un-doped crystal). The optical quality is rather modest, in terms both of attenuation (generally higher than 1–2 dB/cm), and of reduction of the nonlinearity (by more than 90%). Another optical peculiarity is that PE increases the extraordinary index of the material while decreasing the ordinary one. Only extraordinary modes can thus be supported, with TM polarization in the Z-cut samples used for periodic-poling. The solid curve in Fig. 4.3 shows a typical refractive index profile of a PE layer in SLT, exhibiting a

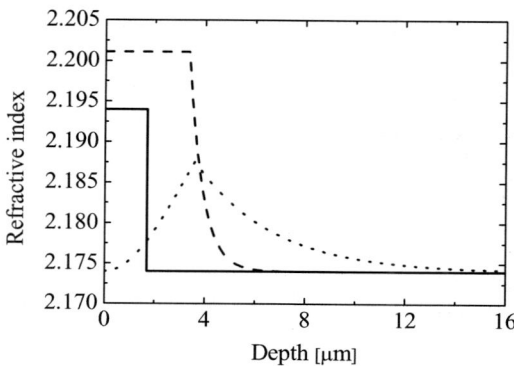

Fig. 4.3. Typical refractive index profiles of a waveguide after proton-exchange (*solid*), annealing (*dash*) and reverse-exchange (*dot*)

nearly step-like shape as a result of a strong nonlinear diffusion of protons inside the crystal.

The ANN process applied after PE gives rise to the so-called annealed-proton-exchanged (APE) waveguides. During ANN the sample is heated at temperatures between 300 and 450°C allowing hydrogen ions to diffuse into the substrate. The hydrogen concentration thus decreases while the waveguide depth increases. The ANN process strongly improves the optical quality of the waveguide: it favours a recovering of the optical nonlinearity of the exchanged layer, it produces a deeper waveguide whose modes propagate mostly in a region unaffected by PE (which, al already pointed out, is a rather invasive process), it reduces the internal stresses of the material by making the waveguide region to evolve towards crystallographic phases closer to that of the un-doped material. A typical refractive index profile as obtained in an APE SLT waveguide is represented by the dashed curve in Fig. 4.3 (note that the edges in the curves are due to the use of different fitting functions in different depth regions; real profiles will have smoothed edges). Unexpectedly, despite the lower local concentration of protons, the index change is increased with respect to the PE case. This is a peculiar property of LT (stoichiometric and congruent), which can be traced back to the crystallographic-phases diagram shown in Fig. 4.4. The figure reports the index change Δn_e as a function of the normal deformation ε_{33} of the cell, which is in turn proportional to the hydrogen concentration. The curve is not continuous and also not-monotonic because of the presence of a discrete number of crystallographic phases, indicated as $\beta_3, \beta_2, \beta_1, \kappa, \alpha$, each presenting a linear Δn_e vs ε_{33} relationship, but with different slope and intercept. After PE in a non-diluted acid, the hydrogen concentration is high and the crystallographic phase is typically a β_3 one; further ANN makes the hydrogen concentration to lower and the exchanged region to evolve towards the α phase (almost identical to that of the un-exchanged substrate) by passing through the intermediate ones. The index change thus firstly increases (phase β_2 and β_1) and subsequently decreases (phase κ, α).

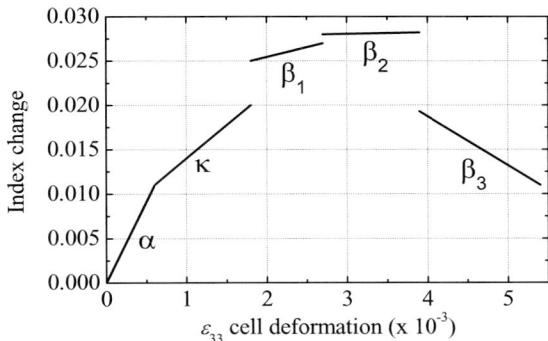

Fig. 4.4. Index change as a function of the normal cell deformation for the different crystallographic phases that can be obtained by PE or by PE followed by ANN in lithium tantalate (data are from [5])

The RE process is applied after ANN in order to obtain buried waveguides with lower attenuation and insertion losses and higher frequency conversion efficiencies in the nonlinear processes. The burial step is performed by dipping the APE waveguide in a Li rich melt, generally a salts mixture, where Li ions replace protons at the waveguide surface, so to recover the substrate composition. In this way the index profile, represented by the dotted curve in Fig. 4.3, results to be buried. The process requires quite high temperatures, of the order of 330–350°C, so that the Li-protons exchange at the surface is accompanied by further diffusion of protons on the substrate side. The modelling of the RE process can not thus be separated by that of ANN.

The following sections will analyse the techniques adopted for the fabrication and characterization of the waveguides and the modelling of the three processes in SLT.

4.2.1 Fabrication and Characterization Procedures

PE is realised in pure benzoic acid at 280°C by means of the sealed ampoule technique [55]. ANN and RE are both performed at $T = 350°C$, the former one in air, the latter one in an eutectic melt composed by $LiNO_3$, KNO_3 and $NaNO_3$ (molar ratio 37.5:44.5:18.0 [38]) again employing the sealed ampoule technique. The use of the same temperature for the two latter processes gives very good results in terms of waveguide quality and provides a simplification of the modelling due to the fact that the diffusion of protons towards the substrate during RE occurs in the same conditions as during ANN.

Empirical relationships have been established on the basis of the fabrication of several planar samples. Four different PE depths d_e, reported in Table 4.3, have been taken as a starting point. All these samples have been annealed for a constant time $t_a = 3\,h$ and then reverse-exchanged in three intermediate steps, $t_{r1} = 4\,h$, $t_{r2} = 7\,h$, $t_{r3} = 16\,h$, for an overall RE time $t_r = t_{r1} + t_{r2} + t_{r3} = 27\,h$. Since ANN

Table 4.3. PE depths d_e of the samples used for the modelling

Sample	#1	#2	#3	#4
d_p [μm]	3.07	2.38	2.06	1.71

prosecutes during RE, we monitored the annealing evolution for $t_a > 3$ h by dividing the samples into two parts, the one being reverse-exchanged for t_{r1}, t_{r2}, t_{r3}, the other one being merely annealed for the same time durations. After each intermediate step both parts were characterized, so as to recover the refractive index profile resulting from ANN only, and the index profile as modified at the surface by RE.

As a characterization technique, we used m-lines spectroscopy at $\lambda = 0.633$ μm for determining the effective indices n_{eff} of the guided modes [56], and then the inverse-WKB method [57] for reconstructing the refractive index profiles. The measurement of propagation losses was performed in channel waveguides supporting a single mode at $\lambda = 1.55$ μm, this wavelength being used in the nonlinear experiments as a pump wavelength. To this purpose the intensity profile of the fundamental mode was measured by a vidicon camera and, by comparison with the intensity profile of a conventional single-mode polarization-maintaining telecom fiber, the upper value for butt-coupling efficiency was calculated [58]. By comparing the waveguide output power with the power exiting the fiber, and by taking into account both Fresnel losses and the estimated butt-coupling efficiency, a slightly value for propagation losses was found. It is worth noting that, in order to avoid power fluctuations induced by parasitic Fabry-Perot effect in the waveguide, the amplified spontaneously emitted light (ASE) from an optical amplifier was used in the experiments.

4.2.2 Modelling

The above described modelling refers to waveguides realized in 0.5% MgO doped SLT, the doping contributing to a further increase of the optical damage threshold. We experimentally verified this modelling to be adequate also for slightly more doped materials, in particular for 1% MgO doping.

Proton-Exchange

PE was found to give rise, as expected, to a step-like refractive index profile with an index-change Δn_e (e standing for PE) slightly higher than that achievable under the same fabrication conditions in CLT. It ranges from 0.0190 to 0.0200 depending mainly on the aging of the exchanged layer, which takes a few days to stabilize after PE. In Fig. 4.5 we report the depth d_e of the profiles as a function of the proton-exchange duration t_e. The experimental values follow the typical Arrhenius law

$$d_e = 2\sqrt{D_e t_e} \tag{4.4}$$

with a diffusion coefficient $D_e = 0.41$ μm^2/h at $T = 280°$C, which is about 1.2 times lower than that reported for the congruent crystal [59].

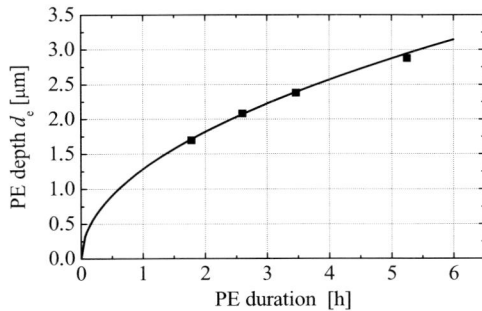

Fig. 4.5. Waveguides depth after PE as a function of the exchange time t_e. Experimental values: *filled squares*. *Solid line*: fitting curve according to Arrhenius law

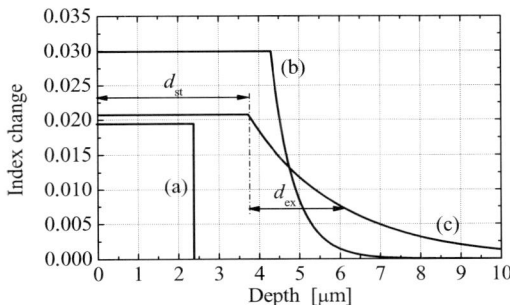

Fig. 4.6. Refractive index profiles of sample #2 after PE (**a**) and after 3 h (**b**) and 30 h (**c**) of ANN

Annealing

The ANN process was found to give rise to an index profile that remains flat at the surface and decays exponentially towards the substrate (see Fig. 4.3). The overall profile can thus be described by three parameters, namely the index change of the flat part Δn_a, its depth d_{st}, and the 1/e depth of the exponentially decreasing part d_{ex}. When comparing the index profile before and after ANN, the area subtended by the profile increases abruptly during the first annealing step ($t_a = 3$ h), while it remains constant during the subsequent steps. For all the samples, i.e. independently on d_e, the area of the annealed profiles, equal to $\Delta n_a \cdot d_a$, with $d_a = d_{st} + d_{ex}$, is 3.05 ($\pm 5\%$) times the area of the starting step-profile of PE. In Fig. 4.6 we report as an example the index profiles that were obtained from sample #2. After the first 3 h of annealing Δn_a was $\sim 1.5 \Delta n_e$ and d_a was $\sim 2 d_e$, which gives a threefold increase in the area. In the following 27 h of ANN the area does not change appreciably, and only a slight increase in d_a and a monotonic reduction of the ratio d_{st}/d_{ex} occurs. The first effect is illustrated in Fig. 4.7, which reports for each sample the experimental depth values d_a as a function of the annealing times t_a, together with the corresponding

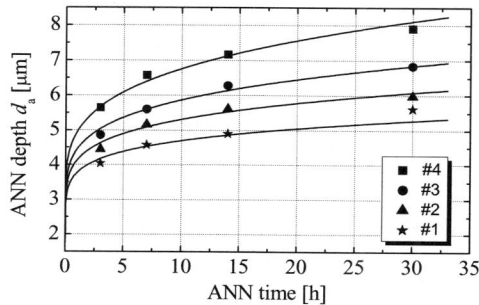

Fig. 4.7. Index profile depths for different ANN times as determined for the various samples. The fitting curves are calculated with (4.5)

fitting curves. All the curves satisfy the following empirical law:

$$d_a = d_e + D_a \cdot t^{\gamma d_e} \quad (4.5)$$

with $D_a = 2.059\,\mu\text{m/h}$ and $\gamma = 0.0967\,\mu\text{m}^{-1}$. The standard deviation is 0.13 µm, corresponding to a 2% average spread of the experimental results. In (4.5) the saturation effect of the depth with the annealing time is parameterised by the exponent γd_e, which ranges from 0.16 to 0.28 depending on the sample. To complete the study of the ANN process we determined an empirical law fitting the monotonic reduction of the ratio d_{st}/d_{ex} as a function of t_a. Such reduction is lower for deeper PE layers, according to the formula:

$$\frac{d_{st}}{d_{ex}} = \frac{1}{d_e}(\alpha t_a + \beta) \quad (4.6)$$

with $\alpha = -0.823\,\mu\text{m/h}$ and $\beta = 3.279\,\mu\text{m}$, valid for $t_a > 3\,\text{h}$. Equation (4.6) gives a normalized standard deviation of the order of 20%, which is a rather high value. However, good predictions of the optical properties of a waveguide can still be achieved by employing (4.6) since the effective indices of the modes do not depend critically on the ratio d_{st}/d_{ex}, but mainly on the profile area and on the total depth d_a. Thus, by combining (4.6) with (4.5), and by taking into account the constant area, one obtains a powerful tool for designing the modal structure of an annealed SLT waveguide.

The analysis of the ANN process clearly shows that after a first, very rapid phase, where the annealing strongly modifies the structural and thus the optical properties of the guiding layer, a second phase occurs, with very smooth changes even if the annealing is performed for many hours. Such a high nonlinear behaviour is very well suited for the realization of buried waveguides through RPE. Indeed, by starting RE after the first ANN step, one obtains at the same time a fast diffusion of Li ions at the surface and a slow diffusion of protons towards the substrate. Under these conditions a very fine control on the optical properties of the buried waveguides can be easily achieved, since the deepest part of the index profile is determined

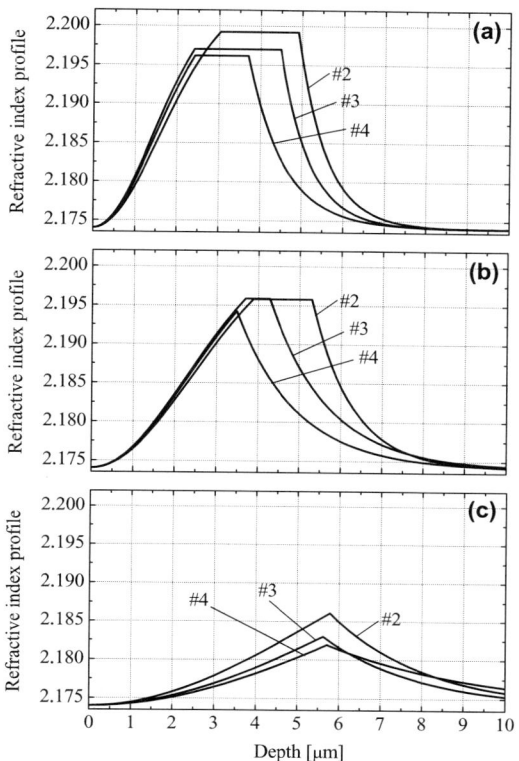

Fig. 4.8. RPE index profiles of samples #2, #3, #4 after 4 h (**a**), 11 h (**b**) and 27 h (**c**) of RE process

essentially by the annealing step, and the surface part by the subsequent reverse-exchange step. Actually, in our case, starting with PE waveguides having depth from 1.69 μm to 2.87 μm, 3 h of annealing were sufficient as a first annealing step, this being the reason why we considered the time of the ANN process as fixed.

Reverse Exchange

Figure 4.8 reports the refractive index profiles after RE of three samples, namely #2, #3, #4, at $t_r = t_{r1} = 4\,\text{h}$ (a), $t_r = t_{r1} + t_{r2} = 11\,\text{h}$ (b), $t_r = t_{r1} + t_{r2} + t_{r3} = 27\,\text{h}$ (c).

The profiles can be fitted with a combination of the step-exponential profile due to ANN, with a second profile due to RE. The mathematical expression that describes this latter profile was found to be a Gaussian function with complete recovery of the substrate index at the surface:

$$n(x) = n_s + \Delta n_m \left(1 - \exp(-y^2/d_r^2)\right) \tag{4.7}$$

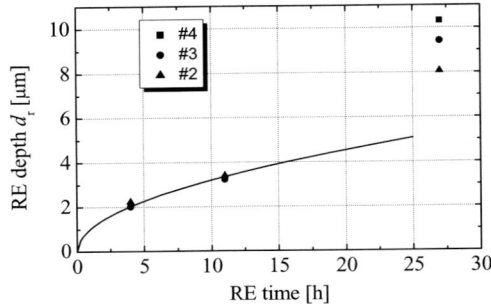

Fig. 4.9. Burial depth of samples #2, #3, #4 for different RE times. *Solid line*: fitting curve

where y is the depth spatial coordinate, n_s is the extraordinary substrate index, Δn_m is the maximum index change that we determined on annealed samples, equal to 0.03, and d_r is the 1/e depth of the Gaussian curve, as calculated for each sample by a least square fitting of the measured n_{eff}. It is worth pointing out that (4.7) best predicts the modal structure of the samples reverse-exchanged for longer times (11 h and 27 h). Indeed, these waveguides are the most useful for applications since they have a burial depth higher than the PE depth, and thus unspoiled nonlinear coefficients and low scattering losses.

In order to study the speed of the RE process we refer to Fig. 4.9, which reports the burial depths d_r as a function of the RE time t_r for the different samples. After 4 h and 11 h the d_r values are the same for all the samples, at least within the experimental error, and the dependence of d_r on t_r approximately follows a $\sqrt{t_r}$ law (more precisely, $d_r = 2\sqrt{D_r t_r}$ with $D_r = 0.257\,\mu\text{m}^2/\text{h}$). However, for $t_r = 27$ h the d_r values are different from one another and far from the $\sqrt{t_r}$ law. Indeed, in all the samples the RE was performed when the crystallographic phase was already completely stabilized after the first annealing step (as discussed with reference to Fig. 4.7). In such conditions no difference arises from sample to sample as long as the flat parts of the annealed index profiles are excavated, and a diffusion-like behaviour is found. However, once the RE process reaches the exponential part of the profile, an acceleration of the process occurs, which is the higher the lower is the depth of the waveguide, i.e. the fewer are the protons to be substituted. This is the reason why the burial depths of the samples for $t_r = 27$ h are inversely ordered with respect to their PE depths. Actually, the most interesting waveguides for nonlinear interactions are those obtained with RE times around 11 h, which can be easily modelled with the $\sqrt{t_r}$ law. In fact, they exhibit at the same time high optical quality, as a result of a burial depth being higher than the PE depth, and tight field confinement, which is essential for maximizing the overlap integral between the interacting modes in nonlinear processes.

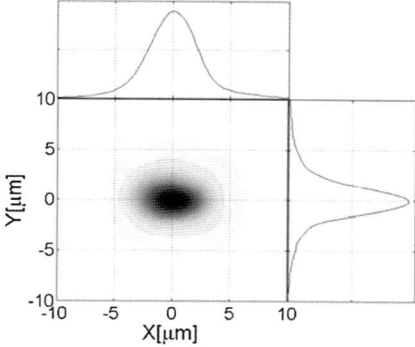

Fig. 4.10. Intensity profile of the fundamental mode of the waveguide at 1.55 μm. *Top* and *right* boxes report respectively the horizontal and vertical cross-sections of the waveguide mode

4.3 Second-Harmonic Generation in RPE-PPSLT Waveguides

Nonlinear characterization of RPE waveguides was performed through second harmonic generation from $\lambda = 1.55$ μm in cw regime. This process is in fact at the basis of most nonlinear processes, either cascaded or not, allowing for the implementation of all-optical functionalities [60–64]. We exploited the above described modelling to select fabrication parameters giving rise at the same time to single-mode propagation at the pump wavelength and to a burial depth d_r higher than the PE depth d_e, so as to fully exploit the nonlinearity of the material. Two index profile configurations were tested: a highly confining one, allowing for the best conversion efficiency, and a weakly confining one, allowing for less critical phase-matching conditions [6] and optimized waveguide-fiber mode matching. The waveguides were fabricated in a 25 mm long 1% MgO doped PPSLT sample provided with 5 poling periods from 17.7 to 18.5 μm. On each poled region a group of 5 channels having widths between 6 and 10 μm was defined with a Ti mask, which was removed after the PE process.

4.3.1 Highly Confining Waveguides

The fabrication parameters chosen to obtain highly confining single-mode waveguides were the following: 1 h 45 m for PE, giving rise to a 1.7 μm thick step-index film, 3 h for the annealing, changing the refractive-index-profile into a step-exponential one with a 4.9 μm 1/e depth, 11 h for the RE, producing a buried index profile with 3.7 μm transverse dimension (these parameters reproduce indeed those of #4). Such parameters result in single-mode waveguides at 1.55 μm with almost circular mode profile, as shown in Fig. 4.10 for a 10 μm-wide channel. The upper and right-hand side of the figure report the vertical and horizontal cross-sections of the mode profile, presenting a FWHM of 4.8 μm and 3.2 μm, respectively. With such dimensions the fibre-waveguide coupling losses due to mode spatial mismatch were

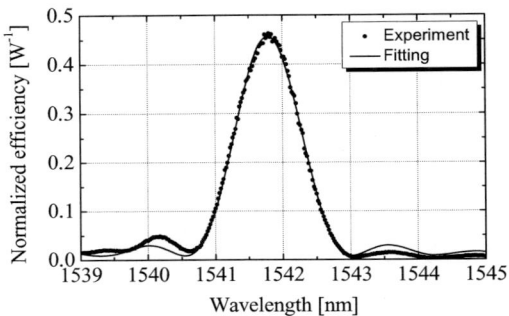

Fig. 4.11. Waveguide second harmonic generation efficiency versus wavelength

about 22%, and the propagation losses (measured as described in Sect. 4.2.1) were between 0.3 dB/cm and 0.5 dB/cm depending on the channel, which is a satisfactory result.

As a pump laser source for the nonlinear characterization we employed a cw extended-cavity laser-diode with broad wavelength tunability and output power as high as 2.5 mW in the wavelength range used in the experiments (Agilent 81600 B). The nonlinear characterization of the waveguides gave best results for the 9 and 10 μm wide channels, whose phase-matching curves are less-critical in the presence of local in-homogeneities of the refractive index profile. The tuning curve of the 10 μm-wide channel of the 18.1 μm periodically-poled region is reported in Fig. 4.11. Its FWHM corresponds to an interaction length of ~1.3 cm, and its peak to a normalized efficiency of 30% W^{-1} cm^{-2}. The interaction length is quite low with respect to the sample length, most likely due to waveguide in-homogeneities related to undesired thermal gradients during the fabrication process. On the other hand, the normalized conversion efficiency is lower but still consistent with the highest values reported for PPLN waveguides (i.e. 40–50% W^{-1} cm^{-2} with APE [60] and 150% W^{-1} cm^{-2} with RPE [10]), due to a 4 times reduction of the nonlinear coefficient in waveguides having comparable effective interaction area. Such area in our case is to 19.4 μm^2, indicating an excellent spatial overlap between fundamental (FF) and second harmonic (SH) fields by virtue of the buried refractive index profile

For checking the optical damage resistance of the waveguides, we measured the SH power for increasing values of the coupled pump power up to a level of 313 mW corresponding to a FF peak intensity of 1.8 MW/cm^2. The measurement was performed in cw regime by keeping the waveguide at 25°C, so to verify the possibility of room temperature high power density operation. For each pump power the waveguide was exposed for 20 minutes and the SH power sampled every minute, in order to appreciate SH power changes due to possible optical damage [16]. Extremely stable values were obtained independently of the pump power level. These values are reported in Fig. 4.12 (dots) together with the theoretical curve (solid line) as expected for a 30% normalized efficiency. A very good agreement is found, indicating that no degradation of the waveguide performances occur up to a SH power

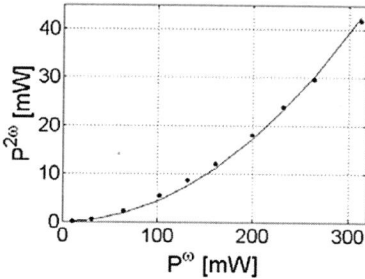

Fig. 4.12. *Dots*: second harmonic output power vs input pump power as measured in room temperature operating conditions. *Solid line*: expected curve from low pump-depletion characterization

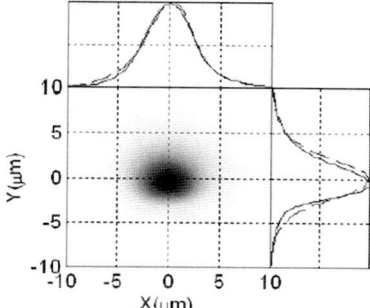

Fig. 4.13. Intensity profile of the fundamental mode at $\lambda = 1.55\,\mu\text{m}$. *Top* and *right boxes* report respectively the horizontal and vertical cross-sections of the waveguide mode (*solid lines*) as compared to that of a fiber (*dashed line*)

level of $\sim$41 mW (corresponding to a peak intensity of 0.56 MW/cm^2 if taking into account the SH intensity profile). The optical damage threshold is thus particularly high if compared to waveguides in congruent PPLN.

4.3.2 Weakly-Confining Waveguides

These waveguides were fabricated with the same PE and ANN times of the former ones, but with an extended RE duration of 25 h. In such conditions all channel waveguides were found to be single-mode at $\lambda = 1.55\,\mu\text{m}$. The intensity profile of the mode of the 10 μm wide channel is reported in Fig. 4.13, together with its horizontal and vertical cross-sections (solid lines in the up and right boxes). Both sections are highly symmetric and very well matched to those of a standard telecom fiber mode (dashed lines), so that coupling losses of only 0.08 dB are estimated. Propagation losses ranging from 0.2 to 0.3 dB/cm were measured, thus very close to the state of the art.

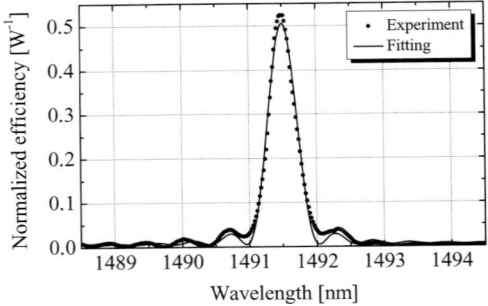

Fig. 4.14. Second harmonic generation efficiency normalized to the input pump power as a function of wavelength. The curve refers to a 10 μm wide channel and to a $\Lambda = 18.3$ μm poling period

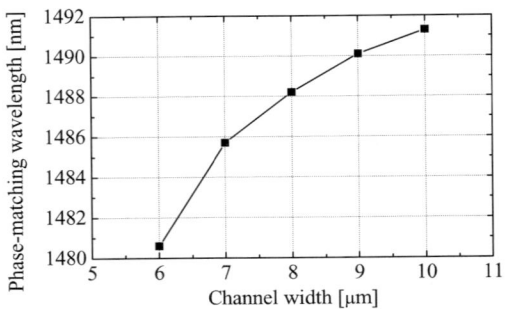

Fig. 4.15. Phase matching wavelength versus channel width

In nonlinear regime tuning curves like that reported in Fig. 4.14 were obtained by measuring the second harmonic generation efficiency as a function of the fundamental wavelength. The curve, which refers to the 10 μm wide channel of the 18.3 μm poling period, presents a ∼440 pm FWHM. By considering both material and waveguide dispersion (the former one being calculated from Sellmeier curves reported in [45], the second one from the above reported refractive index profile) this value corresponds to an interaction length of ∼25 mm, thus equal to the sample length. Slightly larger curves, up to 1000 pm in the worst cases, were obtained for smaller channels, indicating more critical phase-matching conditions. This was confirmed by the measurement of phase-matching wavelength versus channel width, reported in Fig. 4.15. For narrow channels the phase-matching wavelength rapidly increases with channel-width, which implies a strong dependence of the phase-matching condition on waveguide geometry, and thus on waveguide homogeneity. For wider channels the homogeneity constraints are reduced, and a rather satisfactory non-critical phase-matching condition is achieved for the 10 μm width.

In correspondence of the peak of the tuning curve a normalized second harmonic generation efficiency of 8.4% W^{-1} cm^{-2} was calculated. By taking into account

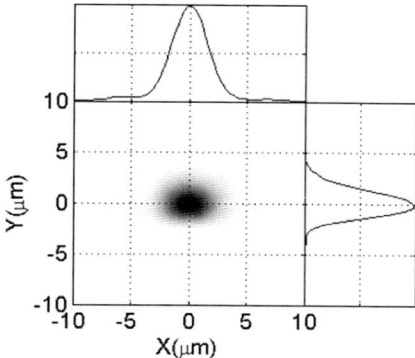

Fig. 4.16. Intensity profile of the second harmonic mode together with its horizontal (*top box*) and vertical (*right box*) cross-sections

SLT nonlinear coefficient, $d_{33} = 10.6\,\text{pm/V}$, this efficiency corresponds to a rather large interaction area, equal to $70\,\mu\text{m}^2$. The consistency of such result was verified by a direct measurement of the intensity profile of the second harmonic mode, which is reported in Fig. 4.16.

The profile clearly corresponds to the TM_{00} mode and exhibits, as well as the fundamental mode, a highly symmetric behaviour. The numerical superposition of the profiles of Figs. 4.13 and 4.16 gives an effective interaction area of $80\,\mu\text{m}^2$, in good agreement with the expected value. This result attests that the waveguiding region exhibits full nonlinearity, which is not trivial if detrimental effects due to PE are considered.

By comparing the results as obtained on the two waveguide configurations, the main advantage of the weakly confining one concerns insertion losses. As to the overall conversion efficiency the two configuration are almost equivalent. This is however valid in our fabrication conditions. With improved thermal uniformity of the oven, much longer interaction lengths, limited only by the poling uniformity, can be foreseen.

The above reported results indicate that MgO-doped PPSLT is a promising alternative to PPLN for the realization of all-optical photonic devices in guided geometry, mainly designed for applications where room-temperature operation and resistance to optical damage are key issues.

References

1. J.A. Armstrong, N. Bloembergen, J. Ducuing et al., Phys. Rev. **127**, 1918 (1962)
2. S. Somekh, A. Yariv, Opt. Commun. **6**, 301 (1972)
3. K. Mizuuchi, K. Yamamoto, Opt. Lett. **21**, 107 (1996)
4. J.P. Meyn, M.M. Fejer, Opt. Lett. **22**, 1214 (1997)

5. Y.N. Korkishko, V.A. Fedorov, *Ion Exchange in Single Crystals for Integrated Optics and Optoelectronics* (Cambridge International Science Publishing, Cambridge, 1999), pp. 270–341
6. M.L. Bortz, S.J. Field, M.M. Fejer et al., IEEE J. Quantum Electron. **30**, 2953 (1994)
7. S. Yi, S. Shin, Y. Jin, Y. Son, Appl. Phys. Lett. **68**, 2493 (1996)
8. K. Yamamoto, K. Mizuuchi, K. Takeshige et al., J. Appl. Phys. **70**, 1947 (1991)
9. M. Houé, P.D. Towsend, J. Phys. D **28**, 1747 (1995)
10. K.R. Parameswaran, R.K. Route, J.R. Kurz et al., Opt. Lett. **27**, 179 (2002)
11. M.L. Bortz, M.M. Fejer, Opt. Lett. **16**, 1844 (1991)
12. S. Fouchet, A. Carenco, C. Daguet et al., IEEE J. Light. Technol. **5**, 700708 (1987)
13. G.I. Stegeman, Introduction to nonlinear guided wave optics, in *Guided Wave Nonlinear Optics*, ed. by D.B. Ostrowsky, R. Reinisch (Kluwer Academic, Norwell, 1992), pp. 11–27
14. G. Assanto, G.I. Stegeman, M. Sheik-Bahae et al., IEEE J. Quantum Electron. **31**, 673 (1995)
15. G.I. Stegeman, D.J. Hagan, L. Torner, Opt. Quantum Electron. **28**, 1691 (1996)
16. M.H. Chou, I. Brener, M.M. Fejer et al., IEEE Photonics Technol. Lett. **11**, 653 (1999)
17. T. Pertsch, R. Iwanow, R. Schiek et al., Opt. Lett. **30**, 177179 (2005)
18. G. Fasol, Science **272**, 1751 (1996)
19. K. Kitamura, Y. Furukawa, K. Niwa et al., Appl. Phys. Lett. **73**, 3073 (1998)
20. P.F. Burdui, R.G. Norwood, D.H. Jundt et al., J. Appl. Phys. **71**, 875 (1992)
21. Y. Furukawa, K. Kitamura, E. Suzuki et al., J. Cryst. Growth **197**, 889 (1999)
22. C. Baumer, C. David, A. Tunyagi et al., J. Appl. Phys. **93**, 3102 (2003)
23. I.G. Kim, S. Takekawa, Y. Furukawa, J. Cryst. Growth **229**, 243 (2001)
24. Y. Liu, K. Kitamura, S. Takekawa et al., J. Appl. Phys. **95**, 7637 (2004)
25. F. Holtmann, J. Imbrock, C. Bäumer et al., J. Appl. Phys. **96**, 7455 (2004)
26. A.G. Getman, S.V. Popov, J.R. Taylor, Appl. Phys. Lett. **85**, 3026 (2004)
27. S.V. Tovstonog, S. Kurimura, K. Kitamura, Appl. Phys. Lett. **90**, 051115 (2007)
28. D.S. Hum, R.K. Route, M.M. Fejer, Opt. Lett. **32**, 961 (2007)
29. M. Marangoni, C. Manzoni, R. Ramponi et al., Opt. Lett. **31**, 534 (2006)
30. F. Baronio, C. De Angelis, M. Marangoni et al., Opt. Express **14**, 4774 (2006)
31. M. Marangoni, D. Brida, M. Quintavalle et al., Opt. Express **15**, 8884 (2007)
32. G. Cirmi, D. Brida, C. Manzoni et al., Opt. Lett. **32**, 2396 (2007)
33. J. Liao, J.L. He, H. Liu et al., Appl. Phys. Lett. **82**, 3159 (2003)
34. G. Marcus, A. Zigler, D. Eger et al., J. Opt. Soc. Am. B **22**, 620 (2005)
35. F. Brunner, E. Innerhofer, S.V. Marchese et al., Opt. Lett. **29**, 1921 (2004)
36. S.Y. Tu, A.H. Kung, Z.D. Gao et al., Opt. Lett. **31**, 3632 (2006)
37. M. Marangoni, R. Osellame, R. Ramponi et al., Opt. Express **12**, 2754 (2004)
38. Y.N. Korkishko, V.A. Fedorov, T.M. Morozova et al., J. Opt. Soc. Am. A **15**, 1838 (1998)
39. M. Marangoni, M. Lobino, R. Ramponi et al., Opt. Express **14**, 248 (2006)
40. M. Lobino, M. Marangoni, R. Ramponi et al., Opt. Lett. **31**, 83 (2006)
41. S.C. Abrahams, J.M. Reddy, J.L. Bernstein, J. Phys. Chem. Solids **27**, 997 (1966)
42. L. Tian, V. Gopalan, L. Galambos, Appl. Phys. Lett. **85**, 4445 (2004)
43. S. Kim, V. Gopalan, K. Kitamura et al., J. Appl. Phys. **90**, 2949 (2001)
44. V. Gopalan, M. Gupta, Appl. Phys. Lett. **68**, 888 (1996)
45. A. Bruner, D. Eger, M.B. Oron et al., Opt. Lett. **28**, 194 (2003)
46. V. Kolev, M. Duering, B. Luther-Davies, LEOS 2005 Meeting, Sydney, Australia, MC4 paper, 23–27 October 2005
47. R. Ramponi, M. Marangoni, R. Osellame et al., Appl. Opt. **39**, 1531 (2000)

48. M. Nakamura, S. Higuchi, S. Takekawa et al., Jpn. J. Appl. Phys. **41**, L465 (2002)
49. S.V. Marchese, E. Innerhofer, R. Paschotta et al., Appl. Phys. B **81**, 1049 (2005)
50. I. Shoji, T. Kondo, A. Kitamoto et al., J. Opt. Soc. Am. B **14**, 2268 (1997)
51. Y. Liu, K. Kitamura, S. Takekawa et al., J. Appl. Phys. **95**, 7637 (2004)
52. F. Holtmann, J. Imbrock, C. Bäumer et al., J. Appl. Phys. **96**, 7455 (2004)
53. Y.L. Lee, H. Suche, G. Schreiber et al., Electron. Lett. **38**, 812 (2002)
54. J.L. Jackel, C.E. Rice, J.J. Veselka, Appl. Phys. Lett. **41**, 607 (1982)
55. M.J. Li, M.P. De Micheli, D.B. Ostrowsky et al., Opt. Commun. **62**, 17 (1987)
56. P.K. Tien, R. Ulrich, J. Opt. Soc. Am. **60**, 1325 (1970)
57. G.B. Hocker, W.K. Burns, IEEE J. Quantum Electron. **11**, 270 (1975)
58. R.G. Hunsperger, *Integrated Optics: Theory an Technology* (Springer, Berlin, 1982), pp. 88–90
59. P.J. Matthews, A.R. Mickelson, S.W. Novak, J. Appl. Phys. **72**, 2562 (1992)
60. M.H. Chou, I. Brener, M.M. Fejer et al., IEEE Photonics Technol. Lett. **11**, 653 (1999)
61. M.H. Chou, I. Brener, G. Lenz et al., IEEE Photonics Technol. Lett. **12**, 82 (2000)
62. K.R. Parameswaran, M. Fujimura, M.H. Chou et al., IEEE Photonics Technol. Lett. **12**, 654 (2000)
63. K. Gallo, G. Assanto, K.R. Parameswaran et al., Appl. Phys. Lett. **79**, 314 (2001)
64. T. Ohara, H. Takara, I. Shake et al., IEEE Photonics Technol. Lett. **15**, 302 (2003)

5 3-D Integrated Optical Microcircuits in Lithium Niobate Written by Spatial Solitons

E. Fazio, M. Chauvet, V.I. Vlad, A. Petris, F. Pettazzi, V. Coda, and M. Alonzo

5.1 Review of Waveguide Fabrication Techniques

Integrated optical microcircuits are structures based on optical waveguides to confine light and consequently to make it realize operations and functionalities. Consequently, the basic element of any integrated circuit is the waveguide. It is constituted of 3 main spatial regions: the propagating core and the surrounding media, the upper and lower ones. In order to confine light, the propagating core must show a higher refractive index than the surrounding media in order to ensure a total reflection regime of the optical rays trapped inside the core area. Consequently, the propagating light proceeds inside the waveguide with a particular configuration called mode, which means the overall interference between all the trapped waves. According to the refractive index contrast between core and surrounding media and according to the transverse dimension of the waveguide, one or many modes can propagate: however, if there is just one mode or many, the characteristic of every mode is to keep its transverse profile constant along the whole propagation. Thus, the mode is a steady-state solution of the light propagation equation inside the confining structure called eigenstate. Many techniques [1] have been developed in the past to realize such waveguide structures. However, they are mainly based on two principles:

1. deposit the propagating layer on a substrate;
2. modify the substrate surface in order to obtain the propagating layer.

The deposition of a layer on a substrate can be realized by many different techniques: evaporation, radio-frequency sputtering, spin- dip-coating, chemical vapour deposition and epitaxial growths by melting, by liquid-phase, by vapour-phase, by metal-organic vapour phase and by molecular beams. Among all these techniques, epitaxial growths by melting [2], by liquid-phase [3] and by molecular beams [4] were efficiently applied to grow lithium niobate films on substrates. However, the most efficient methods for realising waveguides in such material are those based on material modification, which means out-diffusion of atoms from the substrate, in-diffusion of metallic impurities inside the material, ion exchange (Li–Na, Cs–Na, Ag–Na, K–Na, Ti–Na/K) and high-energy ion implantation. Almost all of them have been efficiently applied in Lithium Niobate: however, the most efficient and used one is for sure Ti in-diffusion. Ti ions can strongly increase the refractive index of the substrate (up to 10^{-2}–10^{-3}), penetrating for microns with concentrations up to 0.8% and with semi-Gaussian profiles. Consequently, single-mode waveguides can be obtained with a technique whose propagation losses might be quite low (0.1 dB/cm). However, every technique here described is only able to realize either planar or channelled waveguides on the top of substrates, i.e. surface structures. Historically this is also the reason for the name "substrate," which clearly describes the material under the structure. Integrated optical circuits are consequently only bi-dimensional on the external surface of substrates. Some efforts have been done to bury waveguides inside the substrates either growing on the top of the realized other layers or pushing the in-diffusion for long time in order to force the highest doping concentration deep below the external surface. All these efforts can bury the realized structure few microns or tens of microns inside, anyhow far from a real three-dimensional circuit. A promising technique for real 3-dimensional circuits is represented by soliton waveguides. A soliton beam is an non-diffracting wave propagating in a material. The diffraction compensation is ensured by a local modification of the material refractive index induced by a light-excited electric-charge distribution. Such a charge distribution via the electro-optic effect writes a light confining structure. With respect to the previous techniques, many advantages can be identified in this case: the guiding structures can be written everywhere in the volume of the host material (we do not speak anymore of substrates but host materials) giving a perfect 3-D structure; the associated waveguide is perfectly symmetrical, single-mode, with very low propagation losses, being the refractive index profile self-written by a non-diffracting light beam. an lithium niobate, due to the negative origin of the electro-optic effect, the first soliton waveguides were associated to dark soliton beams [5, 6], which means a dark shadow propagating without diffraction in a lighten background. Recently [7] bright solitons have been observed in lithium niobate, thanks to the application of intense static electric biases in order to decrease the influence of the photovoltaic electro-motive force. The associated soliton waveguides are indeed very attractive for applications: in fact, they can live for long time after the writing procedure even without any fixing treatment; thus, the writing technology is very simple and low costly (the writing beam should have powers of order of μW–mW), the final structure is perfectly round and single-mode, with low propagation losses (<0.04 dB/cm).

5.2 Theory of Photorefractive–Photovoltaic Spatial Solitons in Biased LiNbO$_3$

5.2.1 Photorefractive Model

Theory of 1-D spatial solitons in photovoltaic photorefractive materials with an external applied field was developed by Keqing et al. [8] in steady-state regime, however, a time dependent model is useful to bring some insight on the soliton dynamical formation. In order to derive such a 1-D time dependent model, we assume a beam uniform along the y-axis that propagates along the z-axis and is described by the x-dependent intensity function I_{em}. To solve the photorefractive space charge field formed by this inhomogeneous beam, we use a one carrier one-dimensional time-dependent band-transport model. In a photovoltaic medium, the set of equations which describe the photorefractive effect [9] consist of

$$\frac{\partial N_d^+}{\partial t} = (\beta + SI_{em})(N_d - N_d^+) - \xi n N_d^+, \quad (5.1a)$$

$$\frac{\partial (\epsilon_0 \epsilon_r E)}{\partial x} = \rho, \quad (5.1b)$$

$$\frac{\partial J}{\partial x} + \frac{\partial \rho}{\partial t} = 0, \quad (5.1c)$$

$$J = e\mu n E + \mu k_B T \frac{\partial n}{\partial x} + \beta_{ph}\left((N_d - N_d^+)\right) I_{em}. \quad (5.1d)$$

Here the space and time dependent variables are the density of ionized donors N_d^+, the density of electrons n, the electric field E due to both the PR space charge field and possible presence of an external applied field, the charge density ρ and the current density J. The six variables depend on the x transverse coordinate. N_d is the total donor density, N_a is the density of ionized acceptors, β and S are the thermal and photoexcitation coefficients, ξ is the recombination coefficient, μ is the electron mobility, ϵ_r is the relative dielectric constant at low-frequency, β_{ph} is the photovoltaic constant and T is the temperature. ϵ_0 is the vacuum permeability, $-e$ is the electron charge and k_B is the Boltzmann constant.

5.2.2 Time Dependent Electric Field Distribution

In order to establish a time dependent relation between the electric field and the intensity, we use the same method originally developed by Fressengeas et al. [10] for bright screening or photovoltaic solitons. We recall that in photorefractive crystals at moderate intensity the electron density n is negligible compare to the density of ionized donors or acceptors. As a consequence, we can express the density of ionized donors with (5.1b–5.1d):

$$N_d^+ = N_a\left(1 + \frac{\epsilon_0 \epsilon_r}{eN_a}\frac{\partial E}{\partial x}\right). \quad (5.2)$$

To obtain the density of free electrons n, we make the hypothesis that steady-state regime is reached for (5.1a), which is valid for slowly varying intensity. This previous assumption is valid because the recombination time $\frac{1}{\xi N_a}$ that characterize the response time for the free electron build-up process is very short compare to the dielectric response time $\frac{\epsilon_0 \epsilon_r}{e\mu n}$ that is associated with the dynamics of the ionic charges build-up process. The space charge field being essentially due to ionic charges accumulation, it is justified to neglect the time necessary to generate the electrons.

Consequently, the density of free electrons is then given by:

$$n = \frac{(\beta + SI_{em})\left(N_d - N_a\left(1 + \frac{\epsilon_0 \epsilon_r}{eN_a}\frac{\partial E}{\partial x}\right)\right)}{\xi N_a\left(1 + \frac{\epsilon_0 \epsilon_r}{eN_a}\frac{\partial E}{\partial x}\right)}. \tag{5.3}$$

The spatial derivative of the current density J calculated from (5.1e) and the temporal derivative of the charge density ρ from (5.1b) are opposite according to (5.1c). We consequently deduce the following relation between the electric field E, the intensity I_{em}, the electron density n and the density of the ionized donors N_d^+:

$$-\epsilon_0\epsilon_r\frac{\partial^2 E}{\partial x \partial t} = e\mu\frac{\partial(nE)}{\partial x} + \mu k_B T\frac{\partial^2 n}{\partial x^2} + \frac{\partial}{\partial x}\left(\beta_{ph}\left(N_d - N_d^+\right)I_{em}\right). \tag{5.4}$$

After one integration of (5.4) relative to the parameter x, we obtain

$$-\epsilon_0\epsilon_r\frac{\partial E}{\partial t} = e\mu n E + \mu k_B T\frac{\partial n}{\partial x} + \beta_{ph}\left(N_d - N_d^+\right)I_{em} + J_0 \tag{5.5}$$

where the x-independent J_0 value is determined by experimental conditions. Substitution of N_d^+ and n by their expressions, respectively given by (5.2) and (5.3), leads to a time dependent differential equation involving the intensity distribution and the electric field E. Such an equation can be numerically solved to give the formation dynamic of the electric field distribution E for any arbitrary intensity distribution I_{em}. In the experimental conditions usually encountered to form PR spatial soliton, the width of the trapped beam (for bright solitons) or the width of the dark notch (for dark solitons) is usually very small compare to the crystal width l. As a consequence, far from the beam center, the intensity I_∞ is homogeneous, and all spatial derivatives vanish. Furthermore, when the crystal is electrically biased with an applied voltage V across the crystal width l, we can consider that, far from the beam center, $E = V/l = E_0$ at all time. The value of J_0 is then given by:

$$J_0 = -\beta_{ph}(N_d - N_a)I_\infty - e\mu\frac{(\beta + SI_\infty)(N_d - N_a)}{\xi N_a}E_0. \tag{5.6}$$

We note that J_0 corresponds to the x-independent component of the current density flowing through the crystal. For open circuit condition (no applied voltage), we then have to set $J_0 = 0$. At this stage only numerical methods can be employed to solve (5.5).

5.2.3 PR Space Charge Field

While (5.5) does not lead in general to an analytical solution for the photorefractive electric field, it is however possible to simplify it under additional assumptions. Particularly, if we assume that the spatial modulation of the charge density is weak, the normalized term $\frac{\epsilon_0 \epsilon_r}{eN_a}\frac{\partial E}{\partial x}$ can be neglected. This hypothesis is valid when the charges modulation is small. From (5.2) and (5.3) we can therefore conclude that $N_d^+ \simeq N_a$, and the density of free electron n is then given by:

$$n = \frac{(\beta + SI_{em})(N_d - N_a)}{\xi N_a}. \tag{5.7}$$

Equation (5.5) then yields the differential equation governing the time dependent electric field formation under a given intensity I_{em}:

$$T_d I_d \frac{\partial E}{\partial t} + (I_d + I_{em})E + \frac{k_B T}{e}\frac{\partial I_{em}}{\partial x} = E_{ph}(I_\infty - I_{em}) + (I_d + I_\infty)E_0 \quad (5.8)$$

where $T_d = \frac{\epsilon_0 \epsilon_r}{e\mu} \frac{\xi N_a}{\beta(N_d - N_a)}$ is the dielectric relaxation time in absence of illumination, $I_d = \frac{\beta}{S}$ is the equivalent dark irradiance and $E_{ph} = \frac{\beta_{ph} \xi N_a}{e\mu S}$ is the photovoltaic field. With the initial condition $E = E_0$ and assuming a time independent light intensity distribution I_{em}, (5.8) leads to the following time dependent solution for the electric field distribution E due to both the applied field E_0 and the photorefractive space charge field:

$$E = \left(\frac{(E_{ph} + E_0)(I_{em} - I_\infty)}{I_{em} + I_d} - \frac{k_B T}{e}\frac{1}{I_{em} + I_d}\frac{\partial I_{em}}{\partial x}\right)$$
$$\times \left(\exp\left(-\frac{I_{em} + I_d}{T_d I_d}t\right) - 1\right) + E_0. \quad (5.9)$$

5.2.4 Soliton Solutions

The optical beam propagates as a spatial soliton if it keeps a constant profile as a function of propagation distance. In the case of the photorefractive effect, a soliton is formed when the natural light diffraction is exactly compensated by the focusing effect due to the space charge field refractive index change. The soliton optical field amplitude can be written in the form [11]:

$$A(x, z) = \sqrt{r I_d} u(x) \exp(i\Gamma z). \quad (5.10)$$

Here, r is the ratio of the soliton peak intensity I_{max} to the dark irradiance I_d, u is the normalized soliton profile, which is only x dependent and Γ is the soliton propagation constant. Propagation of an optical field in a weakly perturbed medium is theoretically described by the scalar wave propagation equation:

$$\left(\frac{\partial}{\partial z} - \frac{i}{2k}\frac{\partial^2}{\partial x^2}\right)A(x, z) = -i\frac{k}{n_o} 0.5 n_o^3 r_{eff} E A(x, z) \quad (5.11)$$

where $k = \frac{2\pi n_o}{\lambda}$ is the wavenumber in the medium whose average refractive index is n_o (λ is the wavelength in vacuum). The refractive index variation due to the space charge field E is $\Delta n = -0.5 n_o^3 r_{eff} E$ with r_{eff} the effective electro-optic coefficient. We then introduce the dimensionless parameter $X = x/d$ with $d = (|k^2 n_o^2 r_{eff} E_{ph}|)^{-1/2}$. To find pure solitonic solutions, we consider that charge diffusion is negligible in (5.9) thus leading to

$$E = \frac{(E_{ph} + E_0)(I_{em} - I_\infty)}{I_{em} + I_d}\left(\exp\left(-\frac{I_{em} + I_d}{T_d I_d}t\right) - 1\right) + E_0. \quad (5.12)$$

Note that charges diffusion is known to induce a slight beam bending [12]. The simplified space charge field E expression (5.12) is then inserted into the propagation equation (5.11), which leads to the equation

$$u''(X) = 2kd^2 \Gamma u(X) - \alpha u(X)$$
$$- (1+\alpha) \frac{r(u^2(X) - u_\infty^2)}{1 + ru^2(X)} \left(\exp\left(-\frac{1 + r|u(X)|^2}{T_d} t \right) - 1 \right) u(X) \quad (5.13)$$

with $\alpha = E_0/E_{\text{ph}}$. The electric field E_0 is considered positive when applied in the direction of the crystal c-axis. It is well known that without external applied field ($E_0 = 0$), the refractive index of LiNbO$_3$ decreases in the illuminated region due to the photorefractive effect. In the same condition our model establishes in accordance with Valley et al. [13] that the steady-state refractive index change is given by:

$$\Delta n = 0.5 n_o^3 r_{\text{eff}} E_{\text{ph}} \frac{I_{\text{em}} - I_\infty}{I_{\text{em}} + I_d}. \quad (5.14)$$

In LiNbO$_3$, r_{eff} is positive and equal to r_{13} or r_{33} for respectively ordinary or extraordinary polarized light. As a consequence, from (5.14) we conclude that E_{ph} has to be negative in order to give a light-induced reduction of the refractive index in concordance with the experimental observations. To obtain solitonic profile solutions $u(X)$, (5.13) is first integrated in space using quadrature, taking into account boundary conditions:

$$u'^2(X) = (2kd^2\Gamma)(u^2(X) - u_\infty^2) - (1+\alpha)\left(\frac{1}{r} + u_\infty^2\right) \text{Ln} \frac{1 + r|u^2(X)|^2}{1 + ru_\infty^2}$$
$$+ \frac{T_d(1+\alpha)}{rt} \left[\exp\left(-\frac{t}{T_d}(1 + r|u(X)|^2)\right) - \exp\left(\frac{t}{T_d}(1 + ru_\infty^2)\right) \right]$$
$$+ \left(\frac{1}{r} + u_\infty^2\right)(1+\alpha)\left[E_i\left(-\frac{t}{T_d}(1 + r|u(X)|^2)\right) \right.$$
$$\left. - E_i\left(\frac{t}{T_d}(1 + ru_\infty^2)\right) \right] \quad (5.15)$$

where $\int_{-\theta}^{+\infty} \frac{\exp(-y)}{y} dy$ is the exponential integral function.

5.2.5 Dark Solitons

A 1-D dark soliton is an extended wave with an abrupt π phase change in its center forming a 1-D dark notch that propagates unchanged. We can apply the dark soliton boundary conditions $u''(\infty) = 0$ and $u(\infty) = u_\infty = 1$ in (5.13) to deduce its propagation constant Γ:

$$\Gamma = \frac{\alpha}{2kd^2}. \quad (5.16)$$

We observe that the propagation constant is time independent. This is explained by the fact that the dark soliton can be considered as a spatially extended wave

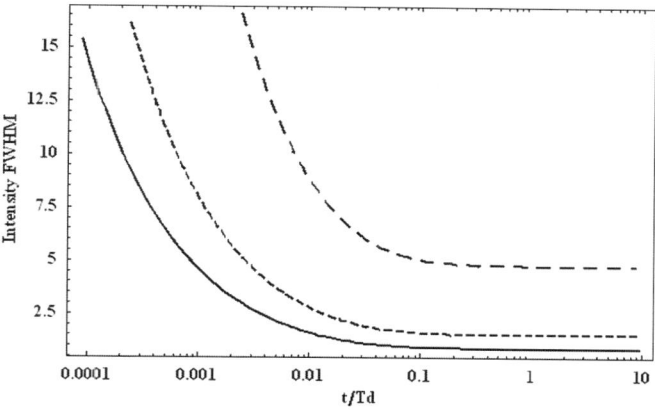

Fig. 5.1. Dark soliton normalized FWHM versus time for $\alpha = 2$ (*continuous line*), 0 (*dotted line*), -0.9 (*dashed line*), $r = 100$

traveling in a medium under constant applied field E_0. The central part of the soliton where the electric field is inhomogeneous has a negligible impact on the value of the propagation constant. The normalized propagation equation for dark solitons is then

$$u'^2(X) = (\alpha + 1)(u^2(X) - 1) - (1 + \alpha)\left(\frac{1}{r} + 1\right)\operatorname{Ln}\frac{1 + r|u^2(X)|^2}{1 + r}$$
$$+ \frac{T_d(1 + \alpha)}{rt}\left[\exp\left(-\frac{t}{T_d}(1 + r|u(X)|^2)\right) - \exp\left(\frac{t}{T_d}(1 + r)\right)\right]$$
$$+ \left(\frac{1}{r} + 1\right)(1 + \alpha)\left[E_i\left(-\frac{t}{T_d}(1 + r|u(X)|^2)\right) - E_i\left(\frac{t}{T_d}(1 + r)\right)\right].$$
(5.17)

Note that a dark soliton can be solution of (5.13) only if $\alpha > -1$ that is $E_0 > -E_{\text{ph}}$. In such conditions and for a given time t, (5.17) can be numerically integrated to deduce the dark soliton profile $u(X)$. For given values of the intensity (fixed r) and applied field (fixed α), we first examine the soliton FWHM that can be formed as a function of time. Soliton FWHM in normalized unit d are presented in Fig. 5.1 for $r = 100$ and for three different values of α. Note that to strictly reproduce these curves experimentally it would necessitate adjusting the dark notch size present at the entrance of the medium over time in order to stay in soliton regime at any instant, which is experimentally difficult to perform. However, the predicted dynamic depicted in Fig. 5.1 can also be seen as a fair description of the evolution of the dark notch at the exit face for a fixed width at the entrance face. We observe that the dark notch is expected to gradually focus over time to reach a minimum width at steady-state. This behavior differs from the open circuit regime [14], where a transient focusing effect was predicted. Moreover the soliton width can be controlled by varying the amplitude of the applied field. The narrowest width is obtained when E_0

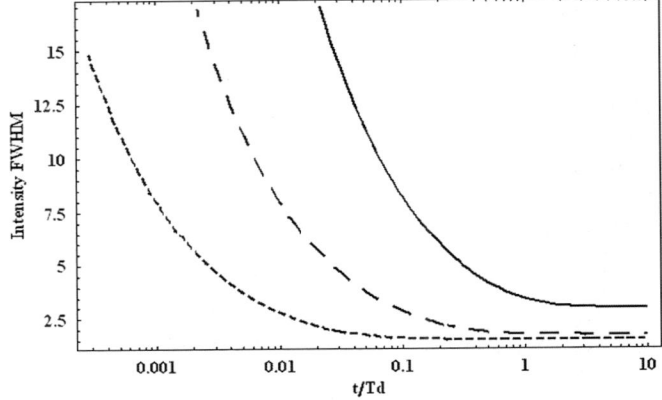

Fig. 5.2. Dark soliton normalized FWHM versus time for $r = 1$ (*continuous line*), 10 (*dotted line*), 100 (*dashed line*), $\alpha = 100$

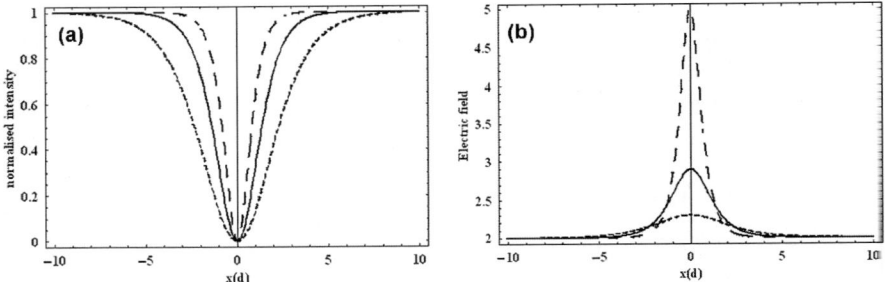

Fig. 5.3. Dark soliton normalized intensity profile (**a**) and electric field profile in unit E_{ph} (**b**) at $t = 0.001T_d$ (*dotted line*), $t = 0.003T_d$ (*continuous line*), $t = 0.01T_d$ (*dashed line*), $\alpha = 2, r = 100$

is applied in the same direction as E_{ph}. The absence of applied field ($\alpha = 0$) gives rise to photovoltaic dark soliton in closed circuit condition and when E_0 is opposite to E_{ph}, dark solitons exist up to the limit $E_0 = -E_{ph}$ where the nonlinearity vanishes. For a fixed value of E_0, we observe that the dark soliton width tends to decrease as the intensity is raised (Fig. 5.2). In addition, self-focusing occurs more rapidly for higher intensity. As usual for PR effect, the response time is inversely proportional to the intensity. To have a better insight of the physics responsible for this beam self-trapping, it is wise to look at the electric field distribution and the associated soliton profile. In Fig. 5.3 the electric field distribution and the soliton profile are plotted at three successive instants of the self-focusing process. We observe that in the dark region the electric field amplitude is at all time larger than the external applied field. Analysis at steady-state reveals that the electric field amplitude reaches $r(\alpha + 1)E_{ph}$. As a consequence, in steady-state regime the peak refractive index is proportional to intensity and applied electric field. It thus gives a

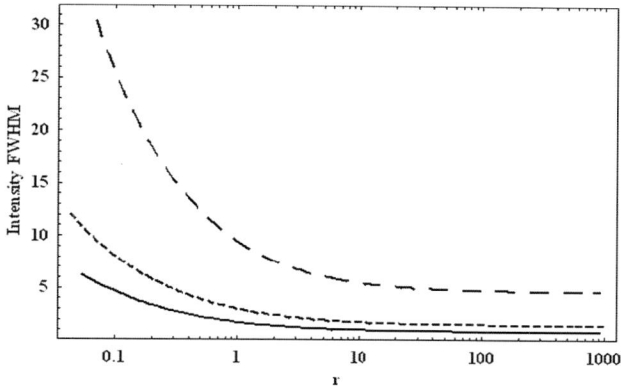

Fig. 5.4. Dark soliton normalized FWHM at steady state regime versus intensity to dark irradiance ratio for three different values of the applied field. $\alpha = 2$ (*continuous line*), $\alpha = 0$ (*dotted line*), $\alpha = -0.9$

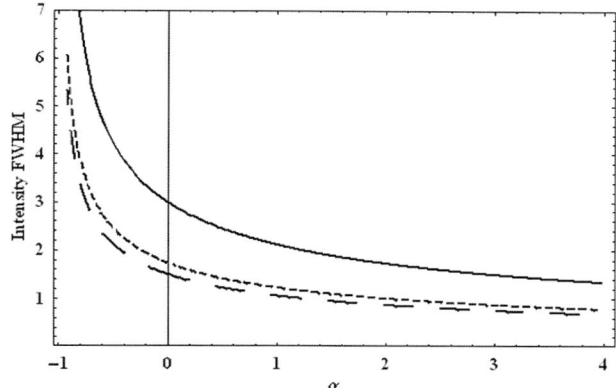

Fig. 5.5. Dark soliton FWHM at steady state regime versus applied field for three different intensity to dark ittadiance ratio r. $r = 1$ (*continuous line*), 10 (*dotted line*), 100 (*dashed line*)

steady-state soliton width that monotonically decreases as a function of intensity as shown in Fig. 5.4. This behavior is consistent with the theory from Segev et al. [15] for short-circuited dark PV soliton in steady-state regime. Similarly soliton width versus applied field is presented in Fig. 5.5. We would like to point out that the large values of the electric field may not be attainable for different physical limits not included in the model. For instance, space charge field may be limited by available charge density, or charge diffusion may not be negligible when dealing with narrow dark notch width. Experimental observation is not straightforward since it implies illumination of the LiNbO$_3$ crystal with a beam that extends up to the electrodes. If in addition to the dark soliton, dark regions are present in between electrodes, it would strongly limit the electric current and would thus modify considerably the behavior.

5.2.6 Bright Solitons

A 1-D bright soliton is a localized beam shaped as a stripe with a constant phase. Its time dependent propagation constant Γ can be deduced from the boundary conditions $u(\infty) = u_\infty = 0$, $u'(0) = 0$ and $u(0) = 1$ using (5.15):

$$\Gamma = \frac{1}{2kd^2}\left(-1 + \frac{1+\alpha}{r}\left(\text{Ln}(1+r) - \frac{T_d}{t}\exp\left(-\frac{t}{T_d}(1+r)\right)\right.\right.$$
$$\left.\left. + \frac{T_d}{t}\exp\left(-\frac{t}{T_d}\right) - E_i\left(-\frac{t}{T_d}(1+r)\right) + E_i\left(-\frac{t}{T_d}\right)\right)\right), \quad (5.18)$$

$$u'^2(X) = \frac{1+\alpha}{r}\left(u^2(X)\text{Ln}(1+r|u(X)|^2) - \text{Ln}(1+r)\right.$$
$$+ \frac{T_d}{t}\left((u^2(X)-1)\exp\left(-\frac{t}{T_d}\right) - u^2(X)\exp\left(-\frac{t}{T_d}(1+r)\right)\right.$$
$$\left.+ \exp\left(-\frac{t}{T_d}(1+r|u(X)|^2)\right)\right) + (u^2(X)-1)E_i\left(-\frac{t}{T_d}\right)$$
$$\left. - u^2(X)E_i\left(-\frac{t}{T_d}(1+r)\right) + E_i\left(-\frac{t}{T_d}(1+r|u(X)|^2)\right)\right). \quad (5.19)$$

Bright solitons are solutions of (5.19) for $\alpha < -1$ that is $E_0 < -E_{\text{ph}}$. In such conditions and for a given time t, (5.19) can be numerically integrated to deduce the dark soliton profile $u(X)$. For given values of the intensity (fixed r) and applied field (fixed α), we first calculate the soliton FWHM that can be formed as a function of time. Soliton FWHM in normalized unit d are presented in Fig. 5.6 for $r = 100$ and for three different α values. We observe that unlike the behavior of dark solitons, a transient self-focusing occurs. This phenomena was originally reported for bright solitons under external applied field [16] and is consequently not a characteristic of the simultaneous presence of the photovoltaic effect and the applied field. The soliton formed in this transient regime is called "quasi-steady-state" soliton. For high r values, the beam tends to enlarge as it reaches steady-state. Note that the induction time to obtain a steady-state regime is close to T_d. Stronger values of the applied field attenuate the difference between steady-state and quasi-steady-state widths. Moreover the transient regime gradually disappears when r is lowered as shown in Fig. 5.7. The presence of transient self-focusing followed by beam relaxation for high r values can be understood thanks to the soliton profile and corresponding electric field distribution shown in Fig. 5.8. Using the theory of guided-waves, the evolution of the soliton width can be linked to the variations of the photo-induced refractive index which is proportional to the electric field distribution. Early in the process ($t/T_d = 0.002$), a wide spatial soliton is formed due to a low magnitude space charge field. As time evolves, the amplitude of the photo-induced refractive index increases dramatically as the photorefractive space charge field builds up. For

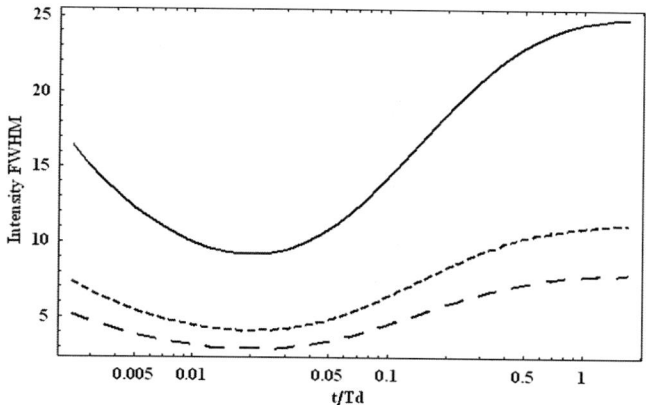

Fig. 5.6. Bright soliton normalized FWHM versus time for $\alpha = -1.1$ (*dotted line*), -1.5 (*continuous line*), -2 (*dashed line*), $r = 100$

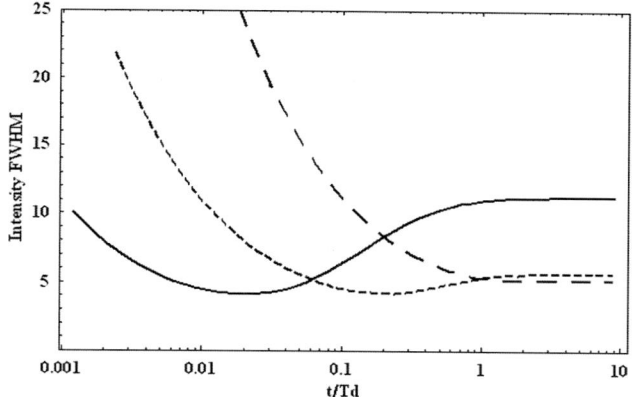

Fig. 5.7. Bright soliton normalized FWHM versus time for $r = 1$ (*dotted line*), 10 (*continuous line*), 100 (*dashed line*), $\alpha = -1.5$

this reason, the soliton is more and more efficiently confined. An optimally confined soliton is reached when the peak value of the space charge approaches the saturation value given by the photovoltaic field $E_{p}h$ ($t/T_d = 0.02$). As time further evolves, the refractive index distribution associated with the soliton keeps a saturated amplitude but the width of the waveguide enlarges ($t/T_d = 0.2$). As a consequence, the bright soliton which is the first mode of the waveguide broadens until steady-state regime is reached. as depicted in Fig. 5.9, narrowest beams are obtained in quasi-steady-state regime for high r values and large applied field. However, narrow quasi-steady-state solitons can also be formed for intensity to dark irradiance ratio close to unity as previously observed for screening soliton. When the applied field E_0 approaches E_{ph} ($\alpha = -1$), self-confinement is lost as shown in Fig. 5.10.

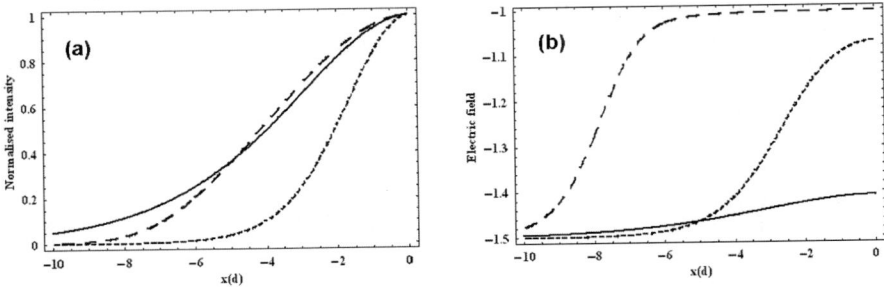

Fig. 5.8. Bright soliton normalized intensity profile (**a**) and electric field profile in unit E_{ph} (**b**) at $t = 0.002T_d$ (continuous line), $t = 0.02T_d$ (dotted line), $t = 0.2T_d$ (dashed line), $\alpha = -1.5, r = 100$

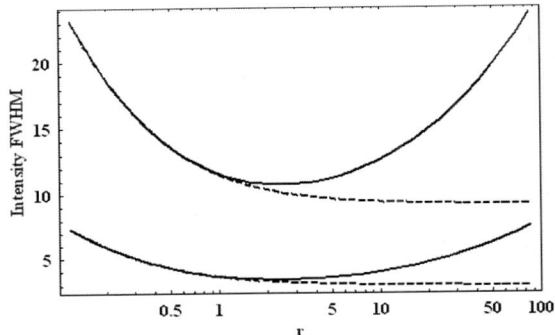

Fig. 5.9. Bright soliton normalized FWHM as a function of r in quasi-steady state regime (*dotted line*) and steady state regime (*continuous line*) for $\alpha = -1.1$ (*upper curves*), and $\alpha = -1.5$ (*lower curves*)

We can also clearly observe that the soliton width in transient regime is very similar to the quasi-steady-state soliton for r close to 1.

5.3 Photorefractive Bright Soliton Observation

As previously introduced, bright soliton formation can be observed in congruent lithium niobate crystals when biased by static electric fields. The experimental set-up of such experiment is very simple (Fig. 5.11(a)): a CW laser beam (in the visible spectrum) is focused on the input face of a lithium niobate crystal in order to reach the theoretical soliton waist (typically around 8–9 m). The input power is ranging up to 1–2 mW. The crystal is electrically biased along the extraordinary axis, and so is the direction of the light polarisation (Fig. 5.11). The output face plane of the lithium niobate crystal is imaged on a CCD camera, whose imaging optics introduces magnification in order to record a clean light profile. Orthogonally to the propagating direction, a background beam can be used to reduce the ratio between the light beam

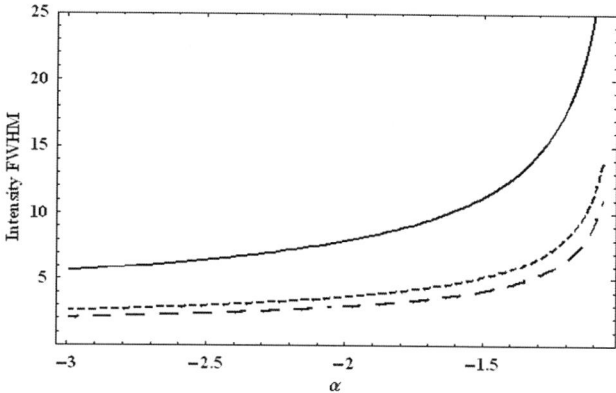

Fig. 5.10. Bright soliton normalized FWHM as a function of α. For $r = 100$, in steady state regime (*continuous line*) and in quasi-steady-state regime (*dashed line*), for $r = 1$ in steady state regime (*dotted curve*)

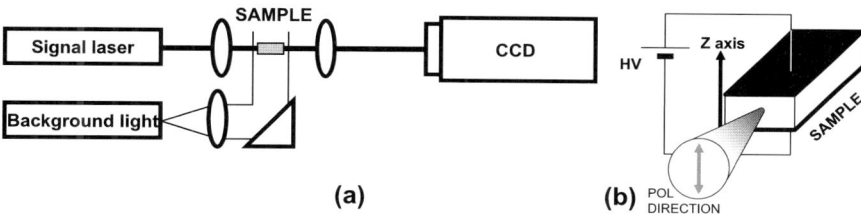

Fig. 5.11. The simple experimental set-up (**a**) for bright solitons and consequently for writing soliton waveguides. A laser beam from a visible source is focused onto the input face of the sample (**b**), which is biased by a high voltage along the z-crystalloghaphic axis. The light polarisation is along z as well. The output face is imaged on a CCD camera

and the dark irradiance. The sample is typically several light-diffraction lengths in order to easily follow and analyse the soliton formation: the one used by [7] was about 6 diffraction lengths. The images of the output beam with soliton formation are shown in Fig. 5.12. The laser beam experiences a shrinkage firstly along the z direction, i.e. in the direction of the crystal extraordinary axis, and only later along the other direction. At the end it is perfectly round, with an overall dimension similar to the input one. These features are pointed out in Fig. 5.13, where the output waists, normalised to the input ones, are reported as function of time, as function of the trasverse directions (open circles: z-direction) and on the applied bias (15, 20 and 25 kV/cm). The formation speeds are different for the transverse directions parallel (open circles) and orthogonal (closed squares) to the optical axis. Such formation dynamics follow negative exponential trends; whole time constants are shown in Fig. 5.14 for the transverse direction parallel to the optical axis (the fastest one). Peculiarity of solitons in lithium niobate is the long-living modification of the re-

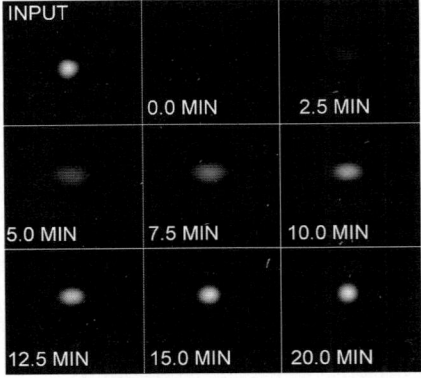

Fig. 5.12. Images of soliton formation

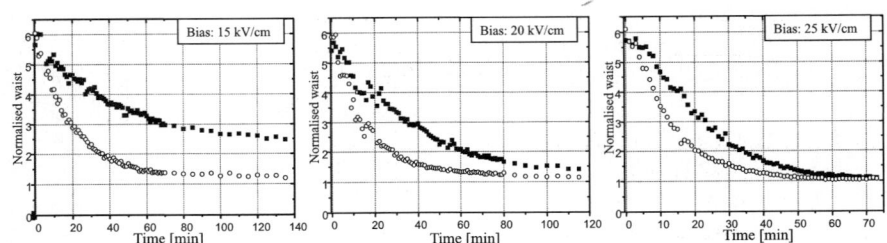

Fig. 5.13. Dynamics of the beam waist during self-focusing and soliton formation for different biases. The open circles refer the waists along the vertical direction of Fig. 5.12, which corresponds to the crystallographic z-axis; closed squares describe the waist evolution along the other direction

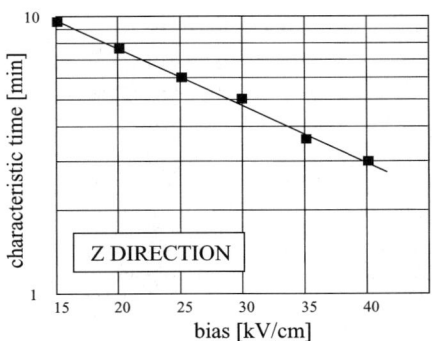

Fig. 5.14. Characteristic formation time of solitons along the crystallographic optical axis as function of the applied bias

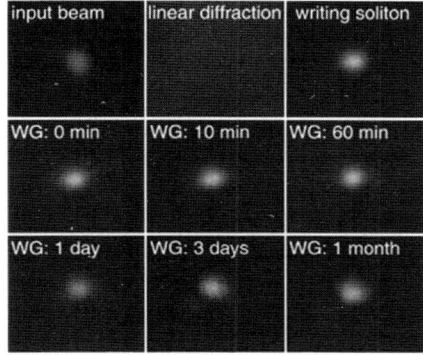

Fig. 5.15. Long term stability of the SWGs. The propagating mode has the same cross-section of the writing soliton even after 1 month since the waveguide is written

fractive index induced by light. In fact it was experimentally demonstrated that the soliton channel will act as a waveguide when the whole writing process is switched off, which means as soon as applied bias, background illumination and soliton light are turned off.

5.4 Waveguiding in Soliton Channels/Strips

5.4.1 Experimental Observation

It was demonstrated in Sect. 5.3 that optical waveguides can be produced in $LiNbO_3$ by recording spatial solitons inside the bulk of the material [17]. This procedure is really attractive because soliton waveguides (SWGs) are by definition single mode structures [18]. Dark photovoltaic (PV) solitons have been observed in $LiNbO_3$ [5, 6] due to the defocusing nature of the photovoltaic effect, even if bright screening-PV solitons [19] have been assumed by the application of static biases stronger than the photovoltaic field. Bright solitons are much more stable than dark solitons and offer the simplest way to photo-induce slab and channel waveguides. The application of intense biases has been already tested [20, 21] with other photorefractive materials. SWGs were obtained in a bulk pure Lithium Niobate crystal by adopting the setup discussed in Sect. 5.3. The temporal stability of the induced SWGs was checked for times up to 1 month by monitoring their guiding properties when coupling to waveguide, for short times, the light with the same wavelength as used for writing (Fig. 5.15). No effective change of the output profile of the beam transmitted through waveguide was observed.

In addition to testing the guiding properties of the SWGs with cw beams, we tested them by guiding femtosecond pulses generated by a mode-locked Ti:Sapphire laser, at 800 nm. Figure 5.16 shows the corresponding beam profiles fitted with

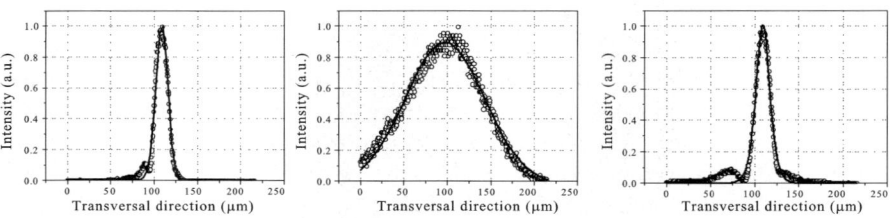

Fig. 5.16. Long term stability of the SWGs. The propagating mode has the same cross-section of the writing soliton even after 1 month since the waveguide is written

Gaussian curves (the input beam and output beam, passing through the crystal, outside SWGs) and hyperbolic secant curve (the output beam passing through a SWG) respectively.

5.4.2 Fixing Soliton Waveguides and Circuits in Lithium Niobate Crystals

We have previously presented that soliton waveguides remain active for long time after the writing procedure is switched off, mainly storing the SWG's in the dark. On the contrary, when needed, they can be easily erased by uniform illumination (without any external field), which leads to an uniform re-distribution of charges and of the internal electric field.

In many cases, optical information processing asks for permanently written components, which means for SWGs, the possibility of fixing. Since early moments of dynamic holography in lithium niobate crystals [22, 23], at least two methods for "freezing" a spatial structured electronic space charged field into an ionic permanent pattern were developed.

In the first method, one uses the storage in deformed ion lattices by a high photovoltaic electric field (the case of lithium niobate crystals), which can induce permanent optical "damage." Itoh et al. [24] have produced single-mode waveguides by very precise scanning of single tightly focused Ar-ion laser beam. Malis et al. [25] have reported direct writing of waveguides in lithium niobate crystals by very precise scanning the sample under a focused cw UV beam, probably due to a local Li-ion out-diffusion or inter-diffusion. Matoba et al. [26] have used arrays of non-diffracting beams (yielded from an Ar-ion laser) to generate narrow waveguide arrays.

In the second method, Bekker et al. [27] and Klotz et al. [28] have used the ferroelectric spatial structured poling, which allows fixing by domain reversal within a special hysteresis cycle of the signal beam self-trapping and the high external electric field (electric polarization). These SWGs can be erased at demand, at a later moment, by application of an external electric field, which is larger than the coercive field of the crystal.

5.4.3 Waveguide Characteristics

With respect to traditional waveguides, the soliton ones present many advantages: first off all, they are written in the volume of the host material, thus consequently they are perfect 3-dimensional structures. They are also perfectly symmetric waveguides, allowing always at least one mode to propagate inside, without cut-off. According to the initial focusing of the writing light, the associated waveguide can be planar, channelled or conical: in fact, also not-completely formed solitons are able to modify the material refractive index, realising conical stigmatic structures. The waveguide characteristics depend on the writing procedure: according to the writing wavelength, different dimensions can be obtained, ranging from 18–20 μm FWHM for red light, down to 6–8 μm FWHM for blue light. The refractive index profile might be either Gaussian (for short time formation) or hyperbolic secant (for long time formation) with a total variation ranging between 10^{-4} to 10^{-6}. Such a low refractive contrast guarantees that also large structures might behave as single-mode ones. One of the most important and evident advantages of soliton waveguides depend on the self-writing process of the light. Being the refractive profile not depending on any physical-chemical process instead being self-written by undiffracting light beams, it is optimised for both single-mode propagation (clearly using it at the same writing wavelength), low dispersion and low losses. Single mode operation and low dispersion were controlled and measured propagating ultra-short pulses at 800 nm inside a soliton waveguide written by a CW green beam at 532 nm. In such a case, it was measured that the material dispersion of the material outside the waveguide was 10 fs/mm, while the dispersion introduced by the waveguiding was below 0.5 fs/mm. Propagation losses due to scattering have been measured to be below 0.04 dB/cm, a very low value for common waveguides, whose losses in the best case are able to arrive as low as 0.1 dB/cm.

5.5 Optical Microcircuits with Soliton Waveguides

5.5.1 Passive

Types of Waveguides: Planar, Circular, Elliptical, Tapered

Research on spatial optical solitons and soliton waveguides (SWGs), as building blocks in passive and active optical microcircuits, is stimulated by their potential in future all-optical signal processing devices [29, 30]. Spatial solitons offer a simple and cheap solution for the recording of efficient single mode waveguides in the bulk of the material, without any change of its crystalline structure [18]. This opens the way to realize 3D passive and active integrated optical circuits, with simple or complex functionalities. Although a SWG can be recorded in a photorefractive material, particularly in lithium niobate (LN), by a very low-power (from nW to mW) soliton beam, in the sensitivity range of the material, it can guide much more powerful (W) cw or pulsed beam of a longer wavelength (including the telecom windows), at which the medium is non-photosensitive, due to the fact that the nonlinear index change is only weakly wavelength dependent [31, 32]. This property can be used to realize waveguides close to surface or deep in the bulk of materials, using an all-optical method and allowing integration into the volume of the material of waveguides and waveguide based devices without costly equipment and crystal-specific fabrication techniques [32]. The shape of the photorefractive SWGs can be controlled by the transversal shape of the incident beam at the entrance face of the photorefractive crystal and by the specific recording procedure (recording time, external field applied on the crystal [33]). Different incident shapes and writing protocols give planar or either cylindrical or conical (tapered) waveguides (optical funnels [31]), with circular or elliptical cross-sections, which are optical elements important for integrated optics applications [34]. One important parameter influencing shape and guiding properties of SWGs in LN is the recording time, due to the faster light confinement along the c-axis direction in respect with the transversal orthogonal direction. If cutting the external field on the photorefractive crystal when the confinement is reached on the c-axis direction mainly, the soliton channel has stronger guiding properties on this direction and a shape close to a planar or elliptical waveguide. Tapered SWGs with transverse shape circular at one face of the crystal and elliptical at the other one could be used, for example, as tapered couplers, in order to match the astigmatic elliptic beam of a laser diode with single mode fibres and other integrated circular channel waveguides. Recording of waveguides with different transversal shapes at the output face of the crystal, varying only the writing time, for the same circular shape of the incident beam at the input (Fig. 5.17) were reported in [35–37]. In Fig. 5.17, there are presented the guiding properties of three SWGs, recorded in $LiNbO_3$ with low-power (10 µW) cw Ar ion laser (514 nm) at different exposure times (WG1 – 15 min, WG2 – 30 min and WG6 – 90 min) for pulsed femtosecond laser beams (800 nm).

Spatial solitons induced in $LiNbO_3$ allow not only the recording of SWGs with different transversal shapes but even recording of waveguides with complex trajectories, different from the linear one. Curved SWGs, based on the self-bending of spatial solitons, [7, 38], dependent on the recording parameters and material, and

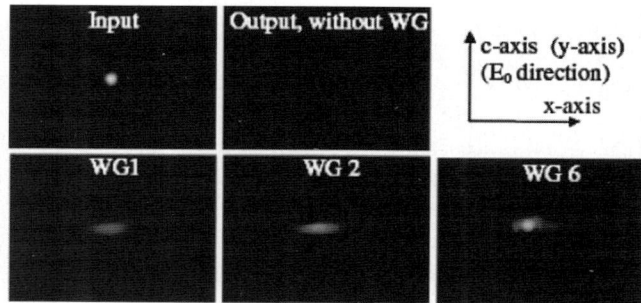

Fig. 5.17. Guiding fs laser beams through tapered SWGs created by a low-power cw Ar ion laser

even SWGs with zero-radius (90°) turns [39] were demonstrated. The guiding properties of the SWGs with sharp bends for light with longer wavelength than that used for recording and the robustness of these waveguides against total internal reflection were proved. The recording of SWGs with controlled trajectories is very important for optical interconnects and high-density integrated optics [39].

5.6 Optical Microcircuits with Solitons Waveguides

5.6.1 Passive

In Sect. 5.4, straight photo-induced waveguides have been demonstrated, but complex trajectories can also be given to these waveguides. Examples of curved waveguides and waveguides with sharp bends are presented in the following paragraphs.

Large Self-Deflection

In [38], curved photo-induced waveguides are realized with an optical set-up similar to Fig. 5.11. The light beam from a HeNe laser at 633 nm, linearly polarized along the crystal c-axis, is focused at the entrance face of a z-cut photonic grade congruent $LiNbO_3$ crystal to a 10 μm FWHM spot.

When a positive 30 kV/cm electric field is applied along the crystal c-axis, the beam diameter at the exit face gradually focuses down to 10 μm, similarly to the report in Sect. 5.3, and a spatial soliton is formed. However, when the same experiment is replicated using a higher applied field, intriguing beam changes are observed at the exit face. A typical observation is depicted in Fig. 5.18 for a 100 μW incident beam and a 50 kV/cm external applied field E_0. In the initial stage, the beam at the exit face gradually focuses (Fig. 5.18(a)–(b)) and a 10 μm diameter circular spot is reached after about an hour. During this time, a small beam shift (15 μm) opposite to the c-axis is also observed. Then in a second stage, the beam is fairly stable in position and width for a few hours (Fig. 5.18(c)). Thorough examination of the beam diameter however reveals that the beam continues to slowly self-focus during this stage. Then the entire beam shifts and diffracts (Fig. 5.18(d)–(e)). Subsequently, a large beam shift takes place accompanied with a refocusing effect (Fig. 5.18(f)–(j)). Finally at steady-state, a tightly focused beam of about 7 μm FWHM displaced by over 200 μm in the direction opposite to the crystal c-axis is formed. Moreover, it has been observed that the sudden defocusing will occur sooner in time and the total shift will be larger as we increase E_0 [38]. As a result, a 300 μm maximum shift has been measured experimentally for a 60 kV/cm applied field.

Although beam bending has already been reported in the literature essentially attributed to charge diffusion [12], it typically identifies a weak beam deflection of about 10–12 μm for a typical 1 cm long crystal [40]. To understand the origin of the giant deflection, we have analyzed theoretically the formation of the internal field E which is due to both the applied field E_0 and the photorefractive space charge field. The field E, as a function of time t and spatial coordinate x, satisfies the differential

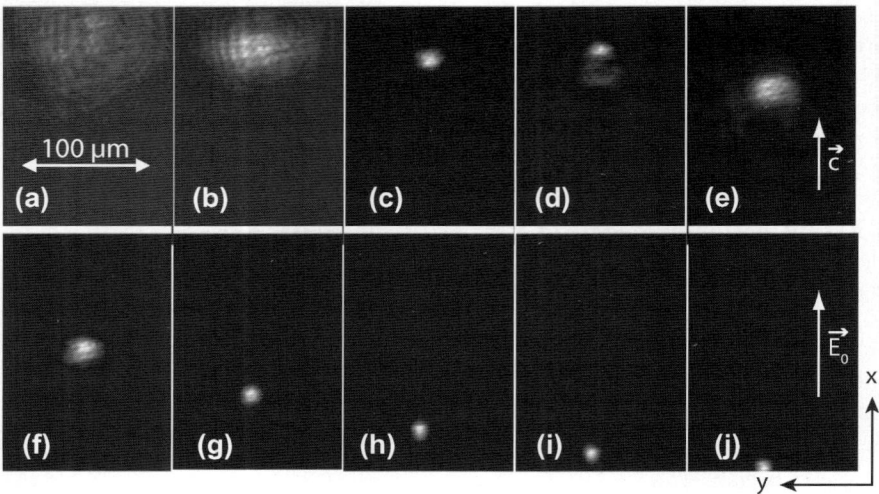

Fig. 5.18. Beam evolution at the exit face of a 7 mm long LiNbO$_3$ sample. Experimental parameters: $E_0 = 50$ kV/cm, $P = 100$ µW. Images from a to j have been taken at time respectively equal to 0 h, 0.5 h, 2.1 h, 4 h, 4.6 h, 6.3 h, 9.6 h, 13 h, 16.3 h and 19.5 h (reproduced from [38])

equation (5.5) obtained in Sect. 5.2. It is essential to note that in unintentionally doped crystal, such as the one used in theses experiments, the concentration of acceptors N_a is low, and as a consequence the term $(\epsilon_0 \epsilon_r / e N_a)(\partial E / \partial x)$ cannot be neglected anymore as it was done in Sect. 5.2. The complete expressions of N_d^+ (5.2) and n (5.3) must now be used in (5.5). The differential equation (5.5) now requires a numerical resolution to obtain the evolution of E. Then we used an iterative method considering a 10 µm FWHM Gaussian beam centered at the origin to solve (5.5) and a *Beam Propagation Method* to solve the propagation equation (5.11). A comparison of experimental and numerical results is given by Fig. 5.19, where we plotted the evolution of the beam width and the position of the beam maximum. Parameters used in the numerical calculations are realistic for unintentionally doped samples [41, 42]. The calculation reveals the same behavior as experiments. In particular, both show the same dependence of the bending with the applied field. Moreover, our model shows that N_a is the key parameter to explain the large self-deflection, because the lateral shift disappears with a ten times greater value of N_a.

After formation of a bent soliton, the applied field and the beam can be switched off leaving a curved waveguide in the medium like for experiments described in Sect. 5.4. These low loss bent waveguides open up new possibilities to create complex optical circuits inside LiNbO$_3$. This possibility is strengthened by another experiment described in next section.

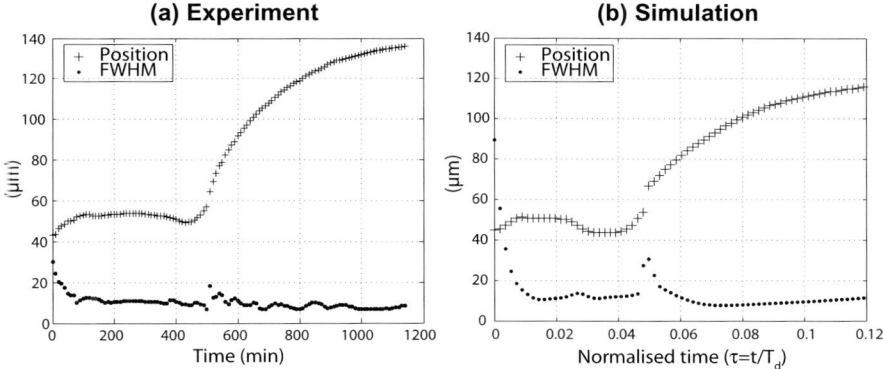

Fig. 5.19. Beam FWHM (*dots*) and position of beam maximum (*crosses*) versus time. a) Experimental results with $E_0 = 50\,\text{kV/cm}$ and $P = 100\,\mu\text{W}$. (**b**) Numerical results $E_0 = 50\,\text{kV/cm}$, $E_{\text{ph}} = -15\,\text{kV/cm}$, $r = 100$ and $N_a = 3 \times 10^{15}\,\text{cm}^{-3}$)

Total Reflection

Development of applications such as optical interconnects necessitates optical circuits with very sharp bends. Realization of such waveguides based on standard integrated optics [43] or photonic crystals [44] involves costly technology and is still a challenging problem. As they provide specific features and much simpler implementation, soliton-induced waveguides have thus the potential to contribute to this field. It has been demonstrated than solitons are robust against total multiple internal reflections and can consequently self-generate waveguides with zero-radius bends [39].

To reach this goal, a prism-shaped LiNbO$_3$ sample had been realized with a right angle as sketched in Fig. 5.20(a). The sample is cut from a 0.5 mm thick photonic-grade z-cut LiNbO$_3$ wafer. Electrodes are deposited on z-faces, and other faces are optically polished. The prism base is about 1 cm wide. A continuous-wave 532 nm beam is delivered from a frequency-doubled Nd:YAG laser. The input beam, with a linear polarization parallel to the c-axis, is focused on with a normal incidence on the LiNbO$_3$ prism to a 15 μm FWHM waist (Fig. 5.20(b)). It then propagates perpendicularly to the input face and experiences two total internal reflections at a right angle before exiting the sample. In the linear regime, we first observe that the beam strongly diffracts during the 1 cm-long trip inside the crystal (Fig. 5.20(c)). Once the 35 kV/cm bias electric field is applied, the light gradually focuses to finally give rise to a solitonic beam after about 20 minutes (Fig. 5.20(d)).

The ability of the soliton-induced waveguide to guide light at a different wavelength is then tested with a 633 nm beam coming from a cw He:Ne laser in the complex waveguide previously inscribed inside the prism. Note that both the 532 nm soliton beam and the applied potential difference are switched off during the guiding test. Efficient light transmission in the waveguide has been observed. We measured an overall light transmission in the waveguide of about 96% taking into ac-

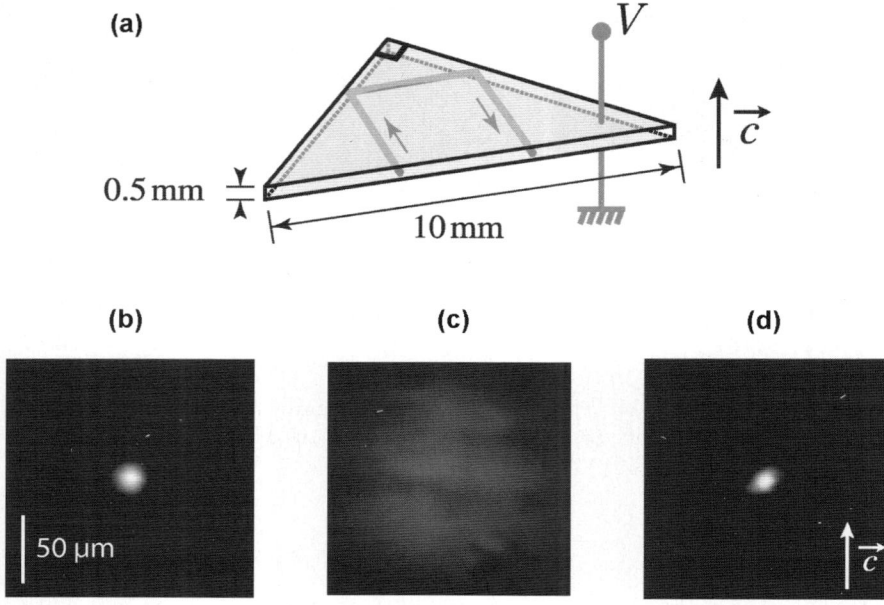

Fig. 5.20. (a) Scheme of the prism shaped sample for demonstration of two successive internal reflections of a spatial soliton. (b) Image of the launched beam, image of output (c) in linear regime and (d) with $E_0 = 35\,\text{kV/cm}$ after 20 min exposure time (reproduced from [39])

count Fresnel reflections, which gives an absorption coefficient of about $0.04\,\text{cm}^{-1}$ ($0.17\,\text{dB/cm}$). This result establishes that photorefractive solitons with complex trajectories offer an unique way to fabricate efficient waveguides with sharp direction changes, like zero-radius bends.

5.6.2 Active

Switching and Routing with Self-Focused Beam

This section is devoted to an integrated routing device fabricated inside a photonic grade undoped $LiNbO_3$ crystal. A 3-D component based on multiple light-induced single-mode waveguides generated by self-focused beams has been realized and is used to test a 1×4 interconnect [45].

The device is composed of four independent waveguides inscribed inside a $LiNbO_3$ crystal. All four waveguides start from the same entrance place but travel in slightly different directions, as depicted in Fig. 5.21(a). In order to induce the complex guiding structure, the setup presented in Fig. 5.21(b) is realized. A 400 µW He–Ne laser beam at 633 nm, linearly polarized along z, is focused down to about 25 µm (FWHM) at the input face of a 7 mm long $LiNbO_3$ crystal. The beam propagation direction can be adjusted with a glass plate that can be tilted along both y and z axis. This simple beam steering device maintains the same beam position at

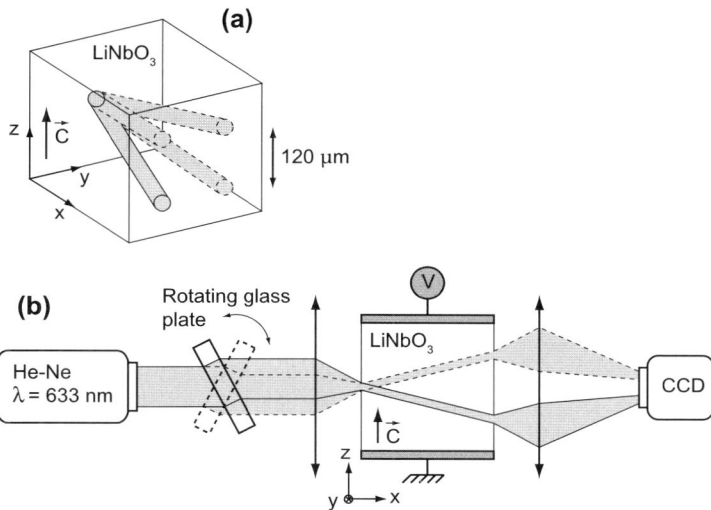

Fig. 5.21. Scheme of the experimental set-up (**a**) and sketch of 3-D 14 router (**b**) (reproduced form [45])

the entrance face of the crystal while offering the possibility to accurately choose the location of the beam at the output face. For this demonstration, we selected four propagation directions so that the beam can exit the crystal to any corner of a square. Once the LiNbO$_3$ crystal is electrically biased along its optical c-axis with a 40 kV/cm D.C. field, it initiates a photorefractive self-focusing effect as depicted in previous sections. The confinement along both transverse directions is complete after a 35 min exposure time per induced waveguide. At this point we have created a 3-D integrated structure composed of four efficient waveguides induced by self-focused beams. The next step is to test the ability of this device to route an optical signal. The router reconfiguration time is given by the response time of the phase modulator and is not related to the photorefractive response time. As a consequence, if fast switching is required, a modulator such as an acousto- or electro-optic device should be used instead of the mobile glass plate. However, in the following experiments the glass plate is sufficient to characterize the router potential. To analyze the interconnect, the applied D.C. field is turned off and similarly as for the fabrication method, a 632 nm signal beam extraordinary polarized is focused at the input face of the index structure, and a phase tilt is introduced across the beam in order to direct it. Four privileged directions for which the signal is efficiently guided to the output face of the device are clearly obtained as shown in Fig. 5.22(a)–(d). Transmission losses have been measured to be lower than 0.6 dB and are mainly due to LiNbO$_3$ absorption. In addition, crosstalk between ports is evaluated to be better than 24 dB. Intermediate position can also be found for which the signal is split between multiple ports as shown for instance in Fig. 5.22(e), where even distributions between two adjacent waveguides are obtained. This latter configuration shows

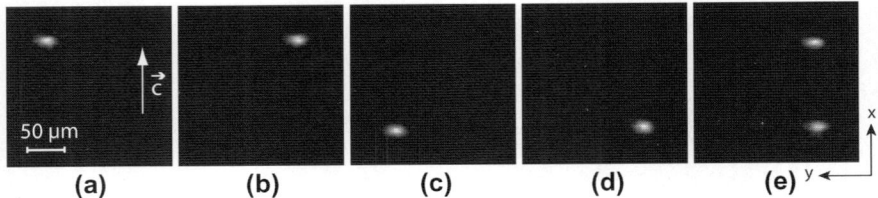

Fig. 5.22. Image of the output face of the photo-induced 1 × 4 router for 5 different routing configurations (reproduced form [45])

that y-junctions can be formed. Thanks to the very long dielectric response time of $LiNbO_3$, the associated photo-induced waveguides are memorized for months, even if as developed in Sect. 5.4.2 methods to create permanent structures should be investigated. This integrated component constitutes a basic building block to conceive more complex N × M optical crossconnects. The demonstration reveals the potential of $LiNbO_3$ crystal to host complex 3-D optical circuits.

Parametric Amplification Generation

Second order phenomena such as harmonic generation and parametric amplification arise when high intensity laser beams, propagating in a nonlinear material, produce polarization fields proportional to the product of two or more interacting waves: these polarization fields radiate electric fields at a different frequency, resulting in a conversion process that is strongly dependent on the intensity of the beam. For this reason, high frequency conversion is expected in optical waveguides, where the electromagnetic field is confined by a transversal variation of the refractive index in the propagation direction. Many experiments in the past years have shown efficient frequency generation and amplification in optical waveguides obtained with different methods, such as titanium indiffusion and proton exchange, and different configuration, as for example channel or planar waveguides. A review of the major results in this field is contained in [46]. Many of these results were obtained in ferroelectric crystals, which usually have high nonlinear coefficients and permit high conversion efficiency when the phase matching (PM) condition is fulfilled, such as when the interacting waves propagate with the same phase velocity. This condition can be achieved in many ferroelectric crystals by varying the temperature of the sample or by applying an external field and using the index modification induced by the electro-optic effect.

Self-induced waveguides obtained by spatial PR solitons are an interesting alternative to perform nonlinear parametric interactions in a ferroelectric crystal: this fascinating issue was explored in the past ten years by several authors [47, 48], because self-induced waveguides can be written in the bulk material thus accessing the third spatial dimension and offer reconfigurability, giving the possibility to design new schemes to efficiently perform parametric processes. Moreover, self-induction technique based on photorefractive effect does not modify the lattice of the crystal, as happens for example with titanium indiffusion: for this reason, nonlinear susceptivity and electro-optic coefficient in the waveguide are as high as in the bulk sample, resulting in large conversion efficiency.

Many straightforward experiments were performed, in the past, in $KNbO_3$ by Lan et al. [47, 49]; in their analysis, those authors employed two different techniques based on the fact that almost all the photorefractive crystals are sensitive to visible light but are insensitive to infrared illumination. In the first scheme, called active configuration, the authors launch an infrared fundamental frequency (FF) beam in a PR waveguide, previously induced in the crystal using a visible laser source and applying an external field. Using this approach, they demonstrate improved conversion efficiency and tunability. In the second scheme, the second harmonic (SH) generated when focusing the FF beam at the input face of the sample is used to write the self-induced waveguide via the PR effect under a strong external bias. This latter configuration, the passive one, is advantageous, since it is a self-aligning scheme, in which the interacting beams are always overlapped in the sample.

Up to date, however, few SH experiment involving PR effect were performed in $LiNbO_3$. The main reason of that is related to the defocusing nonlinearity associated with the PV effect that causes a strong distortion of the beam, thus affecting the transversal superimposition between waves and consequently the conversion efficiency of the nonlinear parametric process. However, LN possesses high nonlinear coefficients, and strong harmonic generation seems possible in such crystals. The first attempt to characterize the role of PR-PV effect in SHG process was made by Orlov et al. [50], who studied the effect of a strong pulsed infrared illumination on a LN sample in different PM conditions.

In their paper, they launch an o-polarized FF beam in a direction orthogonal to the optical axis at a wavelength of 1064 nm, obtaining an e-polarized SH beam propagating in the same direction at 532 nm. They can tune the PM condition by varying the incident angle at the input of the sample. In this scheme, known as noncritical type-I configuration, the parametric interaction is maximized because spatial walk-off typical of waves propagating in birefringent crystals is absent. In their analysis, the authors show that SH generated light experience strong defocusing associated to the PV nonlinearity. At the same time, they record the SH output power as a function of the time of illumination, a parameter that describes how the conversion efficiency changes during the exposure time. Their results show that the PV field arising in the crystal in the vicinity of the beam strongly modifies the PM condition, causing an increase of the SH output power as a function of the exposure time, if properly initial PM condition are selected. This self induced modification of the PM condition is called Nonlinear Self-Phase Matching (NLSPM) and can be used to take advantage of the PR effect and increase harmonic generation.

More recently, Lou et al. [51] have investigated SHG in planar waveguides obtained inducing a 1D dark spatial soliton in LN, which can be efficiently created exploiting the defocusing nonlinearity associated with the PV effect [5]. In their work, the authors write a 1D waveguide by launching at the input face of an iron doped LN sample a dark notch beam from a Nd:YAG doubled laser at 532 nm. After forming the dark 1D soliton, they use the *e*-polarized pulsed light from a mode-locked Ti:Sapphire laser at 800 nm in order to study the SHG process. In this scheme, PM is not possible with the standard techniques (temperature [52], angle [50] or external bias tuning [53]), and very low conversion efficiency is expected in such configuration. Anyway, the strong confinement induced by the waveguide causes an increase of the generation as high as 60%, testifying that PR waveguides are a good mean to increase the conversion efficiency even in a clearly unmatched configuration.

Finally, in their recent paper, Pettazzi et al. [54] have performed SHG in self-induced waveguides in LN employing the passive method, as described by Lan et al. [47] The aim of this study was to experimentally verify if the SH generated visible light can be self-confined in a waveguide thanks to the PR effect and to clarify if NLSPM plays a role in the self-focusing process. To demonstrate this assumption in LN, the authors used a setup similar to the one in [7], employing the noncritical type-I SHG configuration. The light source used in this experiment is a Nd:YAG laser, with a wavelength of 1064 nm, a pulse width of 1.2 ns and a repetition rate of 12 kHz. Light ordinary polarized from the Nd:YAG laser is focused on the input face of a 2 cm long congruent LN sample; the temperature of the crystal is controlled in order to vary the PM condition at the beginning of the experiment. Electrodes are deposited on both the *c* faces in order to apply static fields as high as 40 kV/cm. Light emerging from the sample is then imaged with a lens and splitted in two parts with a glass plate: one part is used to image both FF and SH light at the output of the sample, while in the other part color filters and a photodetector are employed to measure the SHG power as a function of time, as in [50]. The authors previously performed a temperature scan in order to derive the composition of the sample and obtain the temperature and composition dependent Sellmaier equations following [55]. From these equations, they calculate the PM condition as a function of the applied field and temperature finding that the perfect PM condition, in a crystal uniformly biased with 40 kV/cm field, can be reached at a temperature of 6°C.

However, when PR-PV field take place in the crystal, static fields arise in the region of the material shined with visible light; these static fields are algebraically added to the applied external bias. If self-focusing occurs, it is accompanied with a decrease of the total field that locally increases the refractive indices, modifying also the PM conditions via the NLSPM process. The results regarding the self-focusing and waveguide formation in congruent LN are reported in Fig. 5.23, in which the evolution of FF and SH beam at the output face of the sample during the measurement is depicted. This experiment is performed at a temperature of 5°C, with an external bias of 40 kV/cm. According to the calculations, these temperature and electric fields values correspond to a condition in which the ordinary index n_o is slightly larger than the extraordinary one n_e; this initial negative mismatch between FF and SH waves can be compensated by the local increase of the indices obtained by the self-focusing effect thanks to the NLSPM.

The 760 µW infrared *o*-polarized beam, focused down at the input face of the sample to 18 µm (FWHM), experiences linear diffraction at the beginning of the experiment ($t = 0$ sec). As a consequence, both the FF and the SH beams are large as

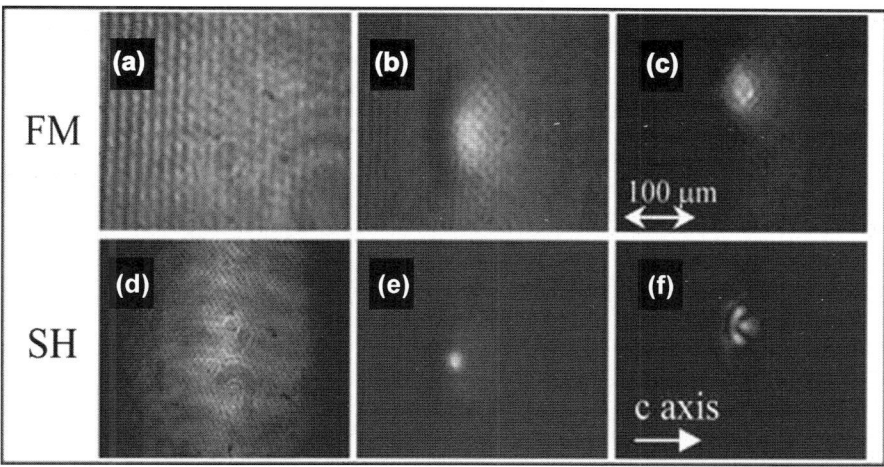

Fig. 5.23. Evolution of the FF (*upper pictures*) and of the SH beam (*lower pictures*) during the experiment. (**a**)–(**b**) ($t = 0$ sec), diffraction at the beginning of the experiment. (**b**), (**c**), FF and SH output when the SH beam attains its minimum waist ($t = 165$ sec). (**e**)–(**f**), FF and SH output beam at the end of the experiment ($t = 324$ sec)

shown in Fig. 5.23(a)–(b). As time is progressing, a strong focusing of both FF and SH beam is observed: Figs. 5.23(c) and (d) are taken when the SH beam attains its minimum dimension ($t = 165$ s). In this case, the index change efficiently traps also the FF beam, as can be seen in Fig. 5.23(c). By comparing Figs. 5.23(c) and (d) it is also possible to notice that the SH beam dimension (17 μm) is smaller compared to the FF dimension (65 μm). This difference can be explained taking into account the different wavelengths involved and that waveguide deepness, as derived by (5.14), on depends on the value of the electro-optic tensor element. In the noncritical type-I configuration, the electro-optic coefficient seen by the SH (r_{33}) is five times greater than the one seen by the FF beams (r_{13}), justifying such a large difference between the two output spots seen in Fig. 5.23(c)–(d).

At the end of the experiment ($t = 324$ sec), the FF beam is still well focused (Fig. 5.23(e)), while a multispot pattern appears for the SH beam (Fig. 5.23(f)). This multispot pattern can be explained considering that, during the experiment, the deepness, as well as the transversal dimension of the waveguide, is increasing [18] and finally saturates, according to the theory of PR soliton formation. In this condition, the waveguide becomes multimode for SH extraordinary light but remains monomode for FF light. This waveguide build-up evolution also reveals that the best focusing condition for SH and for FF appears at different instant, such as for different waveguides parameters, as is confirmed by waveguide theory [56]. The final evolution of the SH and FF beams in the self-focusing experiment reported in Fig. 5.23(e)–(f) can be then attributed to the occurrence of a saturation in the waveguide formation, which is always happening when the intensity of the beam

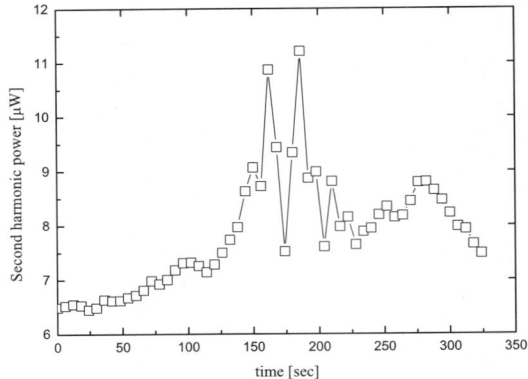

Fig. 5.24. Temporal evolution of the SH output power for the experiment with $P_{FF} = 760\,\mu W$, $T = 5°C$, $E = 40\,kV/cm$

that experience the PR effect is much greater than the dark irradiance of the crystal [18].

The effect of the PR-PV nonlinearity on PM condition is instead depicted in Fig. 5.24, where the SH output power as a function of the exposure time is reported for the experiment in Fig. 5.23. This plot shows a strong increase in the SH output during the exposure time, suggesting that PR effect acts on the conversion efficiency of the parametric process. The strong increase reported in Fig. 5.24 is due to two different processes: from one side, the confinement effect obtained in the induced waveguide increases the local intensity and then the conversion efficiency. From the other side, the local increase of the refractive index is also responsible of a local rectification of the initial negative mismatch. As a result, the crystal approaches locally the perfect phase matching condition, and then the output SH power is maximized. In this condition the authors obtained a maximum conversion efficiency of $19.03\,W^{-1}$. After the first maximum is reached, oscillations in the SH output power appears in the plot, with a large variation of the conversion efficiency during the exposure time. This phenomenon is probably related to the appearance of different higher orders SH modes in the self-induced waveguides that beat together giving an oscillating behavior at the output of the waveguide. The oscillations obtained in Fig. 5.24 are then consistent with the output pattern in Fig. 5.23(f), in which the interference between several modes propagating in the waveguide appears in the near field pattern.

Despite this nonlinear dynamic that led to beam break-up, experiments performed in [54] testify that it is possible to induce PR waveguides in congruent LN using parametric interactions. The efficiency of the nonlinear process can then be tuned using the electro-optics effect, taking advantage of the previously used electrodes to apply an external bias to the sample. To prove efficient tunability of PR waveguides, the authors write another waveguide by repeating the experiment

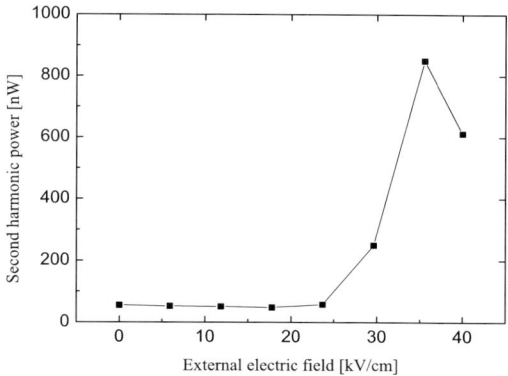

Fig. 5.25. Second harmonic optical power in the self-induced waveguide as a function of the applied field for a temperature of $T = 5°C$ and with a FF power of 230 µWC

above but interrupting the induction process before the appearance of saturation (corresponding to Fig. 5.23(d)).

The tuning experiment is then performed varying the external field in the range 0–40 kV/cm at a fixed temperature of 5°C, and with a FF power of 230 µW, to prevent the sample from any further optical PR damage. In Fig. 5.25, the tuning curve is reported, showing that induced waveguide permits to efficiently vary the PM condition by using applied field of the same order of magnitude of the ones required to obtain self-focusing. This configuration, together with a properly designed electronic circuit, can be used to obtain an optically tunable green source with a response time in the nanosecond range.

5.7 Three-Dimensional Optical Micro-Circuits with SWGs

Modern technologies lead to a strong development of both one-dimensional (1D) and two-dimensional (2D) optical WGs [57, 58]. Three-dimensional (3D) optical WG technology is still under investigation.

The progress of technology is pushing optical circuitry towards the third dimension, into the substrate depth. Recently, direct laser writing of buried WGs was demonstrated using a transversal configuration, whereby ultrafast laser pulses were focused through the lateral crystal face inside the material. At the focal point, these pulses have such a high intensity that they can cause controlled damage to the material, which can guide the light [59, 60]. Femtosecond laser writing produced: volume Bragg gratings [61] micro-optical elements, 3D memories [62] and planar devices as couplers [63–66], interferometers [67, 68], active WGs [69, 70]. This technique can produce a multilevel array of vertically stacked WG structures, spaced over several tens of micrometers in depth [66]. Recently, by the use of an amplified laser, 3D devices were written: 1-to-3 couplers [71], directional couplers and 3D microring resonators, as well as WGs over a depth range of 1 mm [72]. The direct writing of WGs by use of femtosecond laser oscillators is attractive because these laser systems are less complex and costly than amplified systems. In addition, because the pulse period is much shorter than the thermal diffusion time (1 ms), the material modification is a multi-pulse cumulative heating. This allows increased device fabrication speeds, approximately 3 orders of magnitude faster than possible with amplified lasers (millimeters per second versus micrometers per second). However, the relatively small pulse energy available from laser oscillators imposed focusing with very high numerical aperture objectives in order to reach sufficient intensities for the nonlinear interaction and, consequently, strong limitations in the working distance and in the waveguide depths inside materials. Recently, Kowalevicz et al. [73] described waveguide fabrication with an extended cavity Ti:sapphire system. The use of higher-energy pulses permits less restrictive focusing requirements and greater flexibility in the fabrication process.

Direct laser writing of waveguides is clearly a more versatile technique, even though it is still affected by some limitations, mainly due to the focusing procedure and by the refractive-index profile that can be obtained by this laser treatment. All techniques such as layer growing, in-diffusion and direct laser writing create waveguides with refractive-index profiles, which are determined by the physical fabrication process and are not yet optimized for the best propagation performances (best modal distribution, losses and dispersion).

In photonics, lithium niobate is a widely used material for WGs [74, 75]. Among all of the effects that this material offers, we shall focus our attention on the electro-optic one. It yields optical nonlinearities in $LiNbO_3$, which are based either on the generation of a local photovoltaic electric field, or on the generation of bound carrier populations that can produce the screening of a uniform externally applied bias or on a combination of the two [8, 76]. These processes can be used for spatial soliton propagation. Dark solitons have been observed without any external bias in $LiNbO_3$, in the open-circuit configuration [14]. The transition from defocusing to focusing behavior of photovoltaic nonlinearities has been predicted and experimentally observed in the open-circuit configuration [77], opening the possibility of bright photovoltaic soliton generation in $LiNbO_3$. Screening-photovoltaic bright solitons were indeed observed and experimentally characterized by cw green laser beams [7, 33]. Pulsed lasers were used to test the WGs written with screening-photovoltaic solitons.

The photorefractive spatial solitons (PRSS) show important advantages over other types of solitons:

- PRSS are stable when trapped in either one or two transverse dimensions;
- the required optical power for formation of PRSS is very low (as low as $10\,\mu W$);
- PRSS enables the steering and controlling of intense beams (of a wavelength at which the material is less photosensitive) by use of weak (soliton) beams.

Therefore, photorefractive spatial solitons induced WGs seem very promising for various applications, such as light-induced Y and X couplers, beam splitters, directional couplers, waveguides, all-optical control of beam steering, adaptive multichannel interconnects, frequency conversion, etc. [78].

The number of possible guided modes in a BSS-WG depends on the intensity ratio of the soliton, which is the ratio between the soliton peak intensity and the sum of the background illumination and dark irradiance. The number of guided modes increases monotonically with increasing intensity ratio starting from a single mode. When the intensity ratio is much smaller than the Kerr limit, the waveguide can support only one mode for guided beams of wavelengths equal to or longer than that of the soliton. On the other hand, DSS-WG can support only a single guided mode for all intensity ratios.

Spatial solitons with two-dimensional light confinement open the possibility of writing 3D optical WGs and WG arrays, which have an optimum graded refractive-index profile matched to the fundamental laser mode profile and can be temporarily stored, erased, rewritten or can even be permanently fixed [17, 28, 37, 79, 80].

Petter et al. [79] demonstrated the creation of an array of coherent solitons propagating through a ferroelectric crystal in parallel. In the first step, the creation of the soliton array and its waveguiding properties were examined. A regular pattern of 81 spots, with a diameter of 15 mm and a signal intensity of $20\,\text{mW}/\text{cm}^2$ for each spot, was imaged onto the front face of the crystal and wrote the WG.array. Furthermore, the waveguide properties of this array were tested with a probe laser beam, in red, finding that each channel guides properly this probe beam. In the second step, using a separate beam located between two channels of the array, they exploited the mutually attractive force between coherent solitons to let them fuse into a single output at the back face of the crystal. This ability demonstrates the potential for all-optical control of single channels in a large photorefractive SWG array. Furthermore, Chen and Martin [80] have shown the possibility to write SWG arrays even with partially incoherent light.

In Sect. 5.4, we have shown good waveguiding of ultrafast laser beams in SWGs with low pulse dispersion. SWGs are written with low-power cw laser beams or using high-repetition-rate femtosecond laser pulses [35, 37]. In the second case, an efficient writing of SWGs is possible, after accumulating a large number of pulses, because the characteristic photorefractive build-up time is much longer than the pulse period, and the efficient two-photon absorption may contribute to the solitonic confinement. Several SWGs were induced in the volume of the same LN crystal (3D), in different positions. The SWGs show reproducible features with no perturbations of previously recorded SWGs produced by the multiple writing procedures in the same crystal. They show a controllable transversal profile in terms of their writing time, as well as a long lifetime and a low dispersion for ultrashort laser pulse propagation.

Spatial photorefractive SWGs are natural candidates for 3D large arrays, for self-adjustable waveguiding and for versatile interaction capabilities. The geometries with many solitons propagating in parallel, so-called soliton pixels, arrays, or lattices have been suggested for applications in information processing and image reconstruction. In a group of recent papers, from which we mention that of Terhalle et al. [81], the properties of spatial PR solitons were controlled as to form pixel-like lattices and to investigate experimentally and numerically the generation and interactions in large arrays of spatial solitons. The creation of solitonic lattices requires the stable noninteracting propagation of arrays of self-focusing beams, which was ensured by the control of their anisotropic mutual interactions.

Recently, discrete solitons (DS) in nonlinear lattices have also received considerable attention, as the nonlinear waveguide arrays provide an excellent system, where these entities can be experimentally studied and possibly used for all-optical applications. Eugenieva et al. [82] have shown that the corresponding DS-WG can be employed to realize intelligent functional operations such as routing, blocking, logic functions, and time gating. DSs can be routed at array intersections by use of vector/incoherent interactions with other discrete solitons. These intersections behave as DS switching junctions and by appropriate engineering of the intersection site, the switching efficiency of these junctions can be fairly high.

References

1. B. Chen, T. Findakly (eds.), *Optical Waveguide Fabrication in Integrated Optical Circuits and Components* (Marcel Dekker, New York, 1987)
2. S. Miyazawa, Appl. Phys. Lett. **23**, 198 (1974)
3. S. Miyazawa, S. Fushimi, S. Kondo, Appl. Phys. Lett. **23**, 198 (1974)
4. R.A. Betts, C.W. Pitts, Electron. Lett. **21**, 980 (1985)
5. M. Taya, M.C. Bashaw, M.M. Fejer, M. Segev, G.C. Valley, Phys. Rev. A **52**, 3095 (1995)
6. Z. Chen, M. Segev, D.W. Wilson, R. Muller, P.D. Maker, Phys. Rev. Lett. **78**, 2948 (1997)
7. E. Fazio, F. Renzi, R. Rinaldi, M. Bertolotti, M. Chauvet, W. Ramadan, A. Petris, V.I. Vlad, Appl. Phys. Lett. **85**(12), 2193–2195 (2004)
8. L. Keqing, Z. Yanpeng, Y. Tiantong, H. Xun, J. Opt. A Pure Appl. Opt. **3**, 262 (2001)
9. P. Yeh (ed.), *Introduction to Photorefractive Nonlinear Optics* (Wiley, New York, 1987)
10. N. Fressengeas, J. Maufoy, G. Kugel, Phys. Rev. E **54**, 6866 (1996)
11. M. Segev, G.C. Valley, B. Crosignani, P. Di Porto, A. Yariv, Phys. Rev. Lett. **73**, 3211 (1994)
12. M.I. Carvalho, S.R. Singh, D.N. Christodoulides, Self-deflection of steady-state bright spatial solitons in biased photorefractive crystals. Opt. Commun. **120**, 311 (1995)
13. G.C. Valley, M. Segev, B. Crosignani, A. Yariv, M.M. Fejer, M.C. Banshaw, Phys. Rev. Lett. **73**, 3211 (1994)
14. M. Chauvet, J. Opt. Soc. Am. B **20**, 2515 (2003)
15. M. Segev, G.C. Valley, M.C. Banshaw, M. Taya, M.M. Fejer, J. Opt. Soc. Am. B **14**, 1772 (1997)
16. G. Duree, J.L. Shultz, G. Salamo, M. Segev, A. Yariv, B. Crosignani, P. Di Porto, E. Sharp, R.R. Neurgaonkarm, Phys. Rev. Lett. **71**, 533–536 (1993)
17. G. Couton, H. Maillotte, R. Giust, M. Chauvet, Electron. Lett. **39**, 286 (2003)
18. M.F. Shih, Z. Chen, M. Mitchell, M. Segev, H. Lee, R.S. Feigelson, J.P. Wilde, J. Opt. Soc. Am. B **14**, 3091 (1997)
19. L. Jinsong, L. Keqing, J. Opt. Soc. Am. B **16**, 550 (1999)
20. E. Fazio, F. Mariani, M. Bertolotti, V. Babin, V. Vlad, J. Opt. A Pure Appl. Opt. **3**, 466 (2001)
21. E. Fazio, W. Ramadan, A. Belardini, A. Bosco, M. Bertolotti, A. Petris, V.I. Vlad, Phys. Rev. E **67**, 026611 (2003)
22. J. Amodei, D. Staebler, Appl. Phys. Lett. **18**, 540 (1971)

23. F. Micheron, G. Bismuth, Appl. Phys. Lett. **20**, 79 (1972)
24. K. Itoh, O. Matoba, Y. Ichioka, Opt. Lett. **19**, 652 (1994)
25. S. Malis, C. Riziotis, I.T. Wellington, P.G.R. Smith, C.B.E. Gawith, R.W. Eason, Opt. Lett. **28**, 1433 (2003)
26. O. Matoba, K. Kuroda, K. Itoh, Opt. Lett. **145**, 150 (1998)
27. A. Bekker, A. Peda'el, N.K. Berger, M. Horowitz, B. Fischer, Appl. Phys. Lett. **72**, 3121 (1998)
28. M. Klotz, H.-X. Meng, G. Salamo, M. Segev, S.R. Montgomery, Opt. Lett. **24**, 77 (1999)
29. A.W. Snyder, F. Ladouceur, Opt. Photonics News **10**, 35 (1999)
30. Y.S. Kivshar, G.I. Stegeman, Opt. Photonics News **13**, 59 (2002)
31. M. Morin, G. Duree, G. Salamo, M. Segev, Opt. Lett. **20**, 2066 (1995)
32. P. Gunter, J.P. Huignard (eds.), *Photorefractive Materials and Their Applications: 1 Basic Effects* (Springer, Berlin, 2006)
33. E. Fazio, W. Ramadan, A. Petris, M. Chauvet, A. Bosco, V.I. Vlad, M. Bertolotti, Appl. Surf. Sci. **248**, 97–102 (2005)
34. I. Duport, P. Benech, D. Khalil, R. Rimet, J. Phys. D Appl. Phys. **25**, 913–918 (1992)
35. V.I. Vlad, E. Fazio, M. Bertolotti, A. Bosco, A. Petris, Appl. Surf. Sci. **248**, 484–491 (2005)
36. A. Petris, A. Bosco, V.I. Vlad, E. Fazio, M. Bertolotti, J. Optoelectron. Adv. Mater. **7**, 2133–2140 (2005)
37. V.I. Vlad, A. Petris, A. Bosco, E. Fazio, M. Bertolotti, J. Opt. A Pure Appl. Opt. **8**, S477–S482 (2006)
38. M. Chauvet, V. Coda, H. Maillotte, E. Fazio, G. Salamo, Opt. Lett. **30**(15), 1977–1979 (2005)
39. R. Jager, S.P. Gorza, C. Cambournac, M. Haelterman, M. Chauvet, Appl. Phys. Lett. **88**, 061117 (2006)
40. M. Shih, P. Leach, M. Segev, M. Garret, G. Salamo, G.C. Valley, Opt. Lett. **21**, 324 (1996)
41. L. Ren, L. Liu, D. Liu, J. Zu, Z. Luan, J. Opt. Soc. Am. B **20**, 2162 (2003)
42. W. Yan, Y. Kong, L. Shi, L. Sun, H. Liu, X.N Li, D. Zhao, J. Xu, S. Chen, L. Zhang, Z. Huang, S. Liu, G. Zhang, Appl. Opt. **45**, 2453 (2006)
43. E. Cassan, S. Laval, S. Lardenois, A. Koster, IEEE J. Sel. Top. Quantum Electron. **9**, 460 (2003)
44. S. Fan, S.G. Johnson, J.D. Joannopoulos, C. Manolatou, H.A. Haus, J. Opt. Soc. Am. B **18**, 162 (2001)
45. V. Coda, M. Chauvet, F. Pettazzi, E. Fazio, Electron. Lett. **42**, 463 (2006)
46. G.I. Stegeman, C.T. Seaton, J. Appl. Phys. **58**, R57 (1985)
47. S. Lan, M. Shih, G. Mizell, J.A. Giordmaine, Z. Chen, C. Anastassiou, M. Segev, Opt. Lett. **24**, 1145 (1999)
48. A.D. Boardman, W. Ilecki, Y. Liu, J. Opt. Soc. Am. B **19**, 832 (2002)
49. S. Lan, M. Shih, G. Mizell, J.A. Giordmaine, Z. Chen, C. Anastassiou, M. Segev, Appl. Phys. Lett. **77**, 2101 (2000)
50. S. Orlov, A. Yariv, M. Segev, Appl. Phys. Lett. **68**, 1610 (1996)
51. C. Lou, J. Xu, H. Qiao, X. Zhang, Y. Chen, Z. Chen, Opt. Lett. **29**, 953 (2004)
52. R.L. Byer, Y.K. Park, R.S. Feigelson, W.L. Kway, Appl. Phys. Lett. **39**, 17 (1981)
53. N. Uesugi, K. Daikoku, K. Kubota, Appl. Phys. Lett. **34**, 60 (1979)
54. F. Pettazzi, V. Coda, M. Chauvet, E. Fazio, Opt. Commun. **272**, 238 (2007)
55. U. Schlarb, K. Betzler, Phys. Rev. B **48**, 15613 (1993)
56. T. Tamir, W. Tomlinson, Appl. Opt. **28**, 2262 (1989)

57. A.M. Prokhorov, Y.S. Kuz'minov, O.A. Khachaturyan, *Ferroelectrics Thin-Film Waveguides in Integrated Optics and Optoelectronic* (Cambridge International Science Publishing, Cambridge, 1997)
58. S.K. Korotky, R.C. Alferness, in *Integrated Optical Circuits and Components, Chap. 6.*, ed. by L.D. Hutcheson (Dekker, New York, 1987), p. 169
59. K.M. Davis, K. Miura, N. Sugimoto, K. Hirao, Opt. Lett. **21**, 1729 (1996)
60. H. Varel, D. Ashkenasi, A. Rosenfeld, M. Waehmer, E.E.B. Campbell, Appl. Phys. A Mater. Sci. Process. **65**, 367 (1997)
61. L. Sudrie, M. Franco, B. Prade, A. Mysyrowicz, Opt. Commun. **171**, 279 (1999)
62. E.N. Glezer, M. Milosavljevic, L. Huang, R.J. Finlay, T.-H. Her, J.P. Callan, E. Mazur, Opt. Lett. **21**, 2023 (1996)
63. D. Homoelle, S. Wielandy, A.L. Gaeta, N.F. Borrelli, C. Smith, Opt. Lett. **24**, 1311 (1999)
64. C.B. Schaffer, A. Brodeur, J.F. Garcia, E. Mazur, Opt. Lett. **26**, 93 (2001)
65. A.M. Streltsov, N.F. Borrellim, Opt. Lett. **26**, 42 (2001)
66. K. Minoshima, A.M. Kowalevicz, I. Hartl, E.P. Ippen, J.G. Fujimoto, Opt. Lett. **26**, 1516 (2001)
67. K. Minoshima, A.M. Kowalevicz, E.P. Ippen, J.G. Fujimoto, Opt. Express **10**, 645 (2002)
68. C. Florea, K.A. Winick, J. Light. Technol. **21**, 246 (2003)
69. R. Osellame, S. Taccheo, M. Marangoni, R. Ramponi, P. Laporta, D. Polli, S. Silvestri, G. Cerullo, J. Opt. Soc. Am. B **20**, 1559 (2003)
70. Y. Sikorski, A.A. Said, P. Bado, R. Maynard, C. Florea, K.A. Winick, Electron. Lett. **36**, 226 (2000)
71. S. Nolte, M. Will, J. Burghoff, A. Tuennermann, Appl. Phys. A **77**, 109–111 (2003)
72. S. Nolte, M. Will, J. Burghoff, A. Tuennermann, in *CLEO Conf. Digest Paper CWI4*, Baltimore (2003)
73. A.M. Kowalevicz, V. Sharma, E.P. Ippen, J.G. Fujimoto, K. Minoshima, Opt. Lett. **30**, 1060 (2005)
74. K.K. Wong, *Properties of Lithium Niobate. Lithium Tantalate and Potassium Tytanil Phosphate*, 30, 1060 (IEEE, London, 2005)
75. L. Arizmendi, Phys. Status Solidi A **201**, 253 (2004)
76. A.M. Glass, Opt. Eng. **17**, 11 (1978)
77. C. Anastassiou, M. Shih, M. Mitchell, Z. Chen, M. Segev, Opt. Lett. **23**, 924 (1998)
78. M. Segev, M. Shih, G.C. Valley, J. Opt. Soc. Am. B **13**, 706 (1996)
79. J. Petter, J. Schroeder, D. Traeger, C. Denz, Opt. Lett. **28**, 438 (2001)
80. Z. Chen, H. Martin, Opt. Mater. **23**, 235 (2003)
81. B. Terhalle, D. Traeger, L. Tang, J. Imbrock, C. Denz, Phys. Rev. E **74**, 57601 (2006)
82. E.D. Eugenieva, N.K. Efremides, D.N. Christodoulides, Opt. Lett. **26**, 1978 (2001)

Part II

Characterization

6 Light Aided Domain Patterning and Rare Earth Emission Based Imaging of Ferroelectric Domains

V. Dierolf and C. Sandmann

6.1 Introduction and Background

6.1.1 Overview

Trivalent rare earth and transition metal ions in insulating materials have been the topic of extensive investigations over the years due to their application as active ions in solid-state lasers. In lithium niobate ($LiNbO_3$) with its favorable nonlinear and electro-optical properties, Er ions and other rare earth ions have been utilized in integrated optical lasers and optical amplifiers (see, e.g., [1–4]).

In this chapter, we want to summarize a shift of focus from studying the properties of the defect ions for their own sake to applying the acquired knowledge and using the ions as probes to study the properties of the host material. In particular, we will highlight the use of erbium (Er^{3+}) and other rare earth ions as probes for local electrical fields and defect configurations that are found in integrated optical devices such as periodically poled wave-guide devices. In the latter, the introduction of, e.g., Ti ions for waveguide production and the domain inversion process induces changes in the local structure of the material that can be probed using emission spectra of the Er ions and exploited for imaging purposes. With this nondestructive imaging technique, we can demonstrate that the observed changes are not independent of each other and that this interaction needs to be considered for precise device fabrication. We will use Er ions as the prime example throughout this chapter because this ion is the one most commonly used in devices such that suitable samples were more easily available to the authors.

The unprecedented degree of control of the local electric fields obtained with this probe further allowed the development of a laser-aided domain inversion writing process, which offers the promise of sub-micron structure with precise dimensions as needed in next-generation nonlinear and electro-optical devices.

6.1.2 Rare Earth Ions in $LiNbO_3$

Erbium belongs to the group of rare earth ions for which the 4f shell is continuously filled. For more details, see, e.g., [5]. The 4f electrons are well shielded from external fields by the completely filled $5s^2$ and $5p^6$ shells which have a larger radial extension than the 4f shell. The neutral rare earth atoms have, in addition to the filled shells and the 4f shell, two or three electrons in the 5d and 6s shells. These electrons are removed first during ionization. In the case of a trivalent ion, which is the most common valence state of the rare earth ions, only the partially filled 4f shell and the [Xe]-configuration are remaining. For the energy levels of the ions, the Coulomb interaction between the 4f electrons and the spin-orbit interaction has to be taken into account. Both interactions are of about equal size such that neither the LS coupling nor the jj coupling schemes of the angular momenta is strictly applicable. Nevertheless, the LS-scheme (e.g., $^{2S+1}L_J$) is used for naming the various levels. For the 11 electrons of the Er^{3+} ion and the coupling of their individual orbital momentum, the total orbital momenta can be $L = 0, \ldots, 8$, and for the total spin, the two values $S = 1/2, 3/2$ are possible. This results in 41 multipletts. The lowest multipletts are shown in Fig. 6.1.

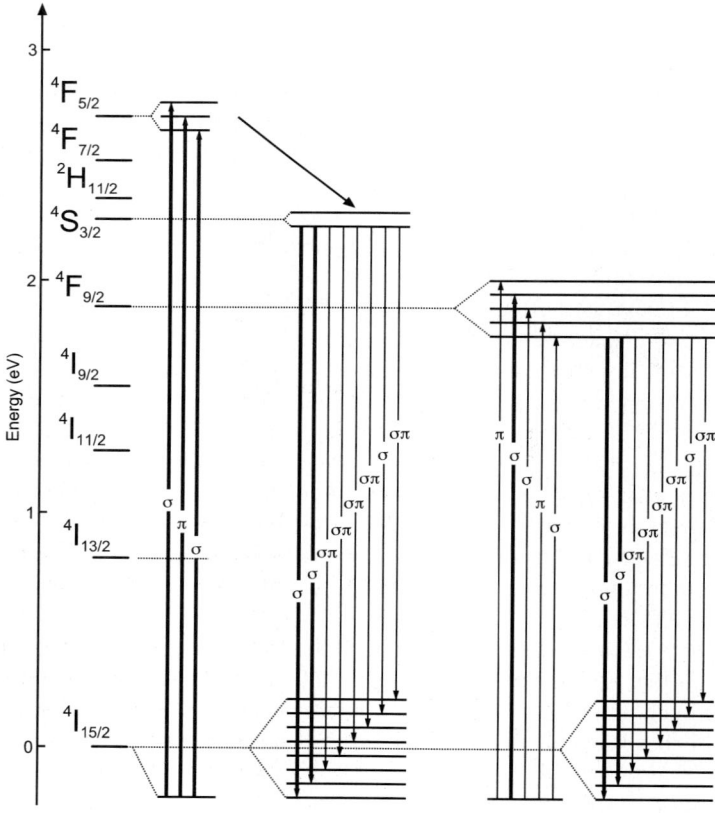

Fig. 6.1. Energy levels and optical transitions of Er in LiNbO$_3$ used in this chapter. The expected polarization for C_{3v} symmetry is indicated

When the ion is introduced into a crystal, the ion experiences the crystal field. However, due to the effective shielding of the outer shells, the 4f electrons are only weakly influenced. This is demonstrated by the fairly sharp absorption and emission lines of the rare earth ions in crystals. The crystal field interaction however is sufficient to have these ions act as probes for their local environment. From an experimental point of view, the small interaction allows one to work in a narrow spectral range for which lasers and spectrometers can be optimized. Due to the narrow lines, even small spectral changes can be resolved.

For concentrations below 1 mol%, rare earth ions in $LiNbO_3$ occupy Li sites [6–9] and require two singly-charged compensators leading to multiple incorporation environments ("sites") and consequently to "multi-site" spectra that are quite complicated. Moreover, the number and relative abundances of sites depend on the stoichiometry of the crystal, the thermal history, and the presence of other co-dopants. Nevertheless, a quite good understanding has been achieved [10–14].

The observed incorporation sites can be divided into three classes (see also Fig. 6.4: (1) A site with C_{3v} local symmetry for which no local charge compensation is present (A1). (2) Sites with charge compensations along the c-axis of the $LiNbO_3$ crystal in which the C_{3v} symmetry is maintained (B1, C1). (3) Site with off-axis charge compensators that lower the symmetry (A2, A3). Combinations of (2) and (3) to achieve a complete local charge compensation leads to a large number of observed sites. In our imaging work, sites of class (1) and (2) will play a role as probes for defect dipoles that are independent (i.e., A1) or related to the probe ion (i.e., for example, B1), respectively. The electric dipole transitions for the trivalent rare earth ions are parity forbidden and appreciable absorption and emission cross sections are due to the admixture of states with different parity (mostly 5d). For this reason, emission strengths are also very sensitive to the local environment and cannot be used directly for the determination of relative abundances of different incorporation sites. In fact, we will see below that the emission strength for different sites changes under domain inversion in $LiNbO_3$.

6.1.3 Combined Excitation Emission Spectroscopy

Principle and Experimental Realization

Excitation and emission spectroscopy are classical tools to investigate defects and dopants in solids, such as rare earth ions in $LiNbO_3$. Often both techniques are applied in sequence. In such studies [11], an emission spectrum is recorded for a particular excitation wavelength followed by an excitation spectrum for which a particular emission line is chosen and the excitation wavelength is varied. Through careful evaluation and multiple repetition of the sequence with different emission and excitation wavelengths, it is possible to assign emission and excitation peaks to a certain defect. However, this approach is often tedious and time consuming. Moreover, small variations and changes are usually missed. With the availability of CCD and other area detectors, it becomes possible to collect whole emission spectra within a very short time. This can be utilized to simplify excitation emission spectroscopy and record in a rapid sequence a large number of emission spectra while changing the excitation wavelength in steps comparable to the resolution in emission. This way a 2D-data set of emission intensity as a function of excitation and emission wavelength or energy is obtained, which can be inspected most conveniently by plotting the data as contour or image plots. In this data representation, emission peaks show up as mountains similar as in a topographical map. This technique has first been applied to $LiNbO_3$ by Gill et al. [10] and named by the authors' group "Combined Excitation Emission Spectroscopy." In their set-up, shown in Fig. 6.2, they use a whole array of tunable lasers (dye-lasers, Ti: Sapphire, external cavity semiconductor laser) that are coupled in a single mode fiber allowing a reliable exchange of light sources with minimal readjustment efforts.

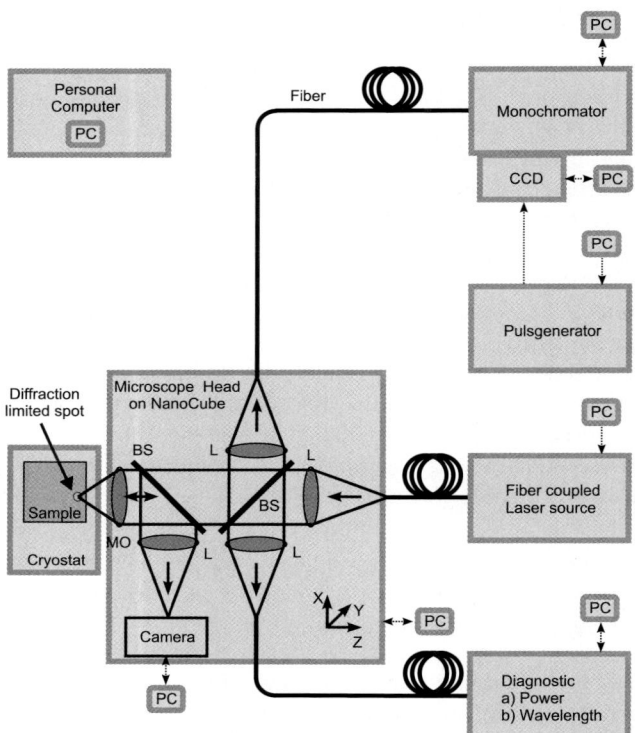

Fig. 6.2. Schematic of the confocal luminescence microscope and combined excitation-emission spectroscopy set-up. Depending on mode of operation, spectra are recorded either as a function of location or as function of excitation wavelength produced by a fiber-coupled tunable laser source. BS: Beamsplitter, L: lens, MO: microscope objective, NA = 0.7. A sample holder that allows application of electric field can easily be put in place of the cryostat

6.1.4 Confocal Microscopy and Spectroscopy

In this chapter, we discuss the use of scanning confocal optical microscopy for the imaging of ferroelectric domain walls. Scanning confocal microscopy is now widely used in biological research [15] and differs from regular wide-field microscopy by recording data for one point at a time and is hence – in combination with fast CCD detection – very amenable to spectral imaging. In this mode, an emission or Raman spectra is collected for each spatial point, which is often referred to as hyperspectral imaging. In a perfect microscope system without aberration, the image size of a point source is diffraction limited and determined by the numerical aperture NA (NA $= n \cdot \sin \alpha$; n, refractive index; α, semi-aperture angle) of the system.

The 3D intensity distribution of the diffraction limited spot for a circular lens system can be found in [16]. The optical system possesses cylindrical symmetry and is best described in cylindrical coordinates (r, z). The resolution of a confocal microscope according to the Rayleigh criterion can be found to be:

$$r_0 = \frac{0.61 \cdot \lambda}{\text{NA}}, \tag{6.1}$$

$$z_0 = \frac{2 \cdot n \cdot \lambda}{\text{NA}^2}. \tag{6.2}$$

An alternative way to define the resolution of a system is to use the full width half maximum of the intensity distribution Γ (FWHM). While the two definitions give almost identical results for a regular microscope system, they differ for the confocal luminescence microscope (CLM). Only in Γ, the CLM offers an advantage over a regular microscope. Using the Gaussian approximation for the intensity distribution and assuming that the wavelength for excitation and emission are very similar, the FWHM Γ^r_{conf} and Γ^z_{conf} are improved by a factor of $\sqrt{2}$. In this case, the intensity distributions for excitation and collection light are simply squared leading to a narrower distribution, which still has the zero-points at the same position. This explains why the numerical manifestation of the improved resolution depends on the definition of the resolution. In practice, the theoretical diffraction limits are hard to achieve in particular in the presence of a sample possessing a high refractive index. The schematic of our home-built confocal microscope is shown in Fig. 6.2, and picture is shown in Fig. 6.3. In this fiber-based approach, single mode fibers act as excitation and emission pinholes of the microscope. With this approach, the whole microscope head is moved for scanning purposes, which is the most suitable mode of operation for the used sample holders in our experiments, a helium cryostat or high voltage sample holder. We were able to achieve a resolution of 0.5 µm and 2 µm in lateral and axial direction respectively.

Data Evaluation

For both the CEES and the CLM techniques, a large number of spectra are recorded, and it is often convenient to convert these spectra into single numbers. For individ-

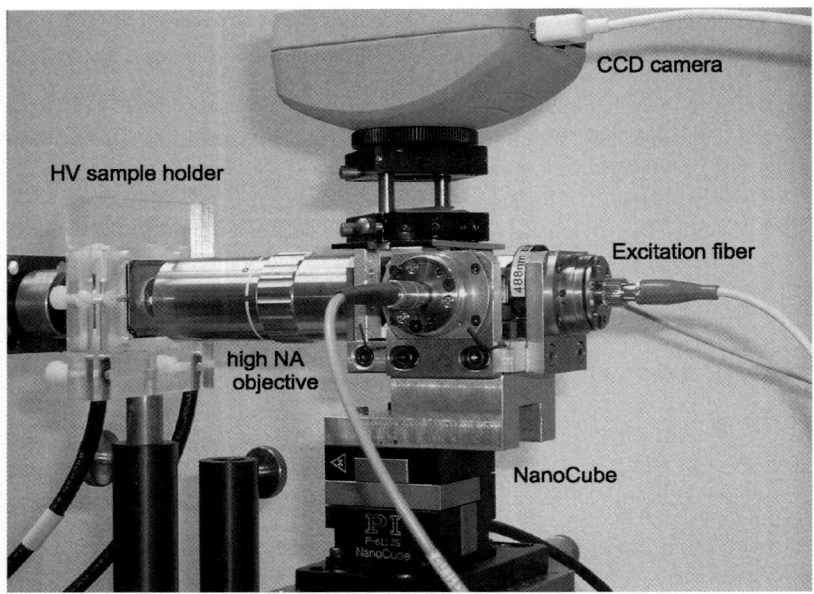

Fig. 6.3. Image of the confocal luminescence microscope head

ual emission peaks, curve fitting is often used to extract such numbers. However, this requires the choice of the expected waveform to give reasonable results. Alternatively, moments of the spectra can be calculated. We define the zeroth (A), first (M_1), and second (M_2) moments for a spectrum $I(E)$ as follows:

$$A = \int_{-\infty}^{\infty} I(E)\,dE, \tag{6.3}$$

$$M_1 = \frac{1}{A} \int_{-\infty}^{\infty} I(E)E\,dE, \tag{6.4}$$

$$M_n = \frac{1}{A} \int_{-\infty}^{\infty} I(E)E^2\,dE. \tag{6.5}$$

For an individual emission line, the moments correspond to the total intensity, the average emission energy, and the spectral width of the emission peak, respectively. Changes in these moments can be calculated very accurately when difference spectra $\Delta I(E)$ are calculated first. For instance, a spectral shift can be determined as:

$$\Delta M_1 = \frac{1}{A} \int_{-\infty}^{\infty} \Delta I(E)(E - M_1)\,dE. \tag{6.6}$$

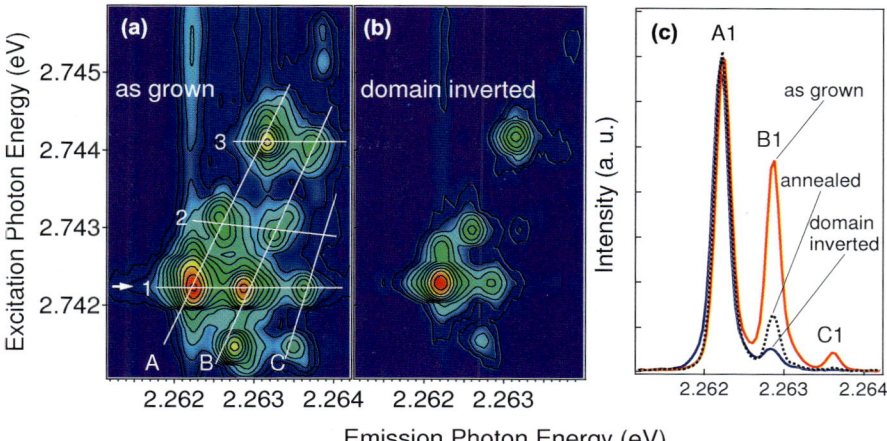

Fig. 6.4. (a) Image and contour plot of CEES data taken at 4 K for single excitation and emission transitions of Er^{3+} in stoichiometric $LiNbO_3$. (b) Same after domain inversion. (c) Individual spectra taken at the excitation energy indicated by *white arrow* in (a). In (a), the *white lines* illustrate the hierarchy of sites

6.2 Application of RE Spectroscopy to the Imaging of Integrated Optical Devices in Lithium Niobate

Using our CEES technique, we can obtain a detailed picture of the different incorporation sites (see Fig. 6.4). The different categories of sites mentioned above can easily be seen in the obtained data for excitation in the blue ($^4I_{15/2}$ to $^4F_{5/2}$) and emission in the green ($^4S_{3/2}$ to $^4I_{15/2}$) spectral region. Several of the spectral features are labeled A1...A3, B1...B3, and C1...C3. This yields a solid basis for the use of the ions as a probe.

Several groups [10, 12] have shown that the relative emission intensity and site distribution is modified by increasing the number of intrinsic defects that are associated with the lack of Li ions in congruent $LiNbO_3$. For instance, while site A1 is most abundant in stoichiometric material, site B1 dominates the emission in as grown congruent crystals. Besides this change in abundance, an increase in inhomogeneous spectral line width is observed that reflects the increase of disorder. The relative abundance of the rare earth ion sites is further dependent on the dopant concentration. In stoichiometric $LiNbO_3$, one finds that site A1 is dominant for low concentrations of Er^{3+}, while other sites become more abundant as the concentration is increased. Combined these trends indicate that the Er^{3+} ions compete for charge compensating defects and that this competition is influenced by the abundance of both intrinsic defects and Er^{3+} ions.

6.2.1 Rare Earth Ions as Probes

In order to use the ions as quantitative probes, we need to perform a calibration in regards to their sensitivity to certain perturbations. The ability to obtain detailed fingerprints that can be achieved with CEES can be uniquely utilized for site selective studies under application of external perturbations. We have performed studies under application of magnetic fields (yielding site-selective g-factors) [17], electric fields, and hydrostatic pressure [13, 14]. We will focus here on the latter two, as they are relevant for the imaging of waveguides and domain patterns.

Under application of hydrostatic pressure (produced inside a diamond anvil cell), both excitation and emission energies change, and hence the peaks are shifting in the contour and image plots. With the CEES technique, it is easily possible to follow the peaks of distinct sites as indicated for two emission transitions in Fig. 6.5. From these data we can extract site-selective pressure-induced energy shifts [14, 18].

The changes that can be produced by application of electric fields experimentally (10 kV/mm) are quite small and are best seen by taking the differences between spectra taken with and without field or, even better, opposite field direction. The directions of the shifts are indicated in Fig. 6.5. In comparing these shifts with the ones observed under application of hydrostatic pressure, we see that they are not identical. On the contrary, they are almost perpendicular to each other. It should also be noted that the emission transition (not shown here) which goes to the 5th level of the ground state multiplet is virtually insensitive to the application of an electric field but shows maximum sensitivity under application of hydrostatic pressure. As we will see below, this is one of our main indicators to decide whether certain spectral changes are due to changes in local E-field or not. The shifts observed under hydrostatic pressure and electric field can be used for comparison with shifts observed under variation of the material. Calibration values for these shifts can be found in [18].

6.2.2 Imaging of Waveguides

As a first application of the imaging concept, we discuss the imaging of the Ti-profile in a waveguide produced by Ti-indiffusion.

As shown in Fig. 6.6, a shift of the emission energies and a broadening of the lines can be observed within such waveguides. The shifts reflect a change in the internal fields due to the Ti concentration. The most reliable change is found for the width of the fifth peak (labelled E). This broadening effect can be used for a quantitative imaging purposes of the Ti-profile. For example, the Er ions can be excited at 488 nm with a standard Argon laser and the emission from the $^4S_{3/2}$ state can be observed in the green spectral region. Other combinations of excitation and emission transitions work as well. The spectra obtained at low temperatures for different regions of the waveguides are depicted in Fig. 6.6 showing the mentioned shift and broadening. In order to quantify, the moments of the emission peaks can be calculated. Alternatively, we can fit part of peak E to a Lorentzian line shape.

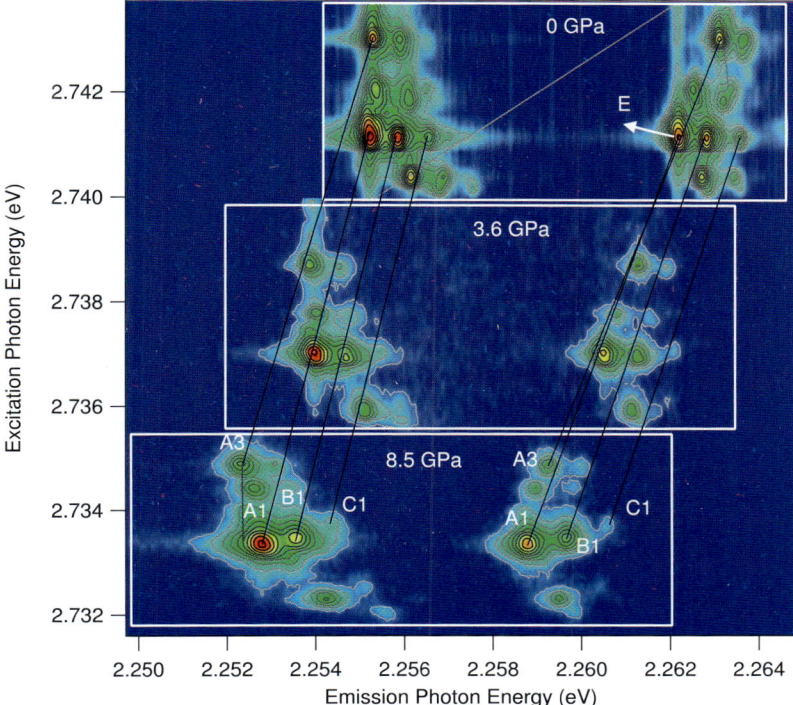

Fig. 6.5. Image and contour plots of CEES data from Er^{3+} in stoichiometric $LiNbO_3$ for three different values of hydrostatic pressure. The site-specific shifts are indicated by *lines* for four different sites (A1, A3, B1, C1). Note that the direction of the shift corresponds quite well to the shift from A3 to A1. The direction of spectral shifts that are observed under application of hydrostatic pressure have been indicated for four different sites. For site A1, the direction of the shift under application of an electrical field is shown by a *white arrow* in the *upper right corner*

When we compare the spectral width Γ (FWHM) of the spectra measured in a line scan with the calculated extraordinary refractive index profile of the waveguide an almost linear relation is obtained. Since this index profile itself is almost linearly related to the Ti^{4+}-concentration, one finds that the spectral width can be translated into Ti^{4+}-concentration by a linear relation (as shown in Fig. 6.7). For calibration, the following relation is obtained:

$$\Gamma_E = 1.41 + 1.2 \times 10^{-21} [Ti^{4+}] \, \text{meV}. \tag{6.7}$$

This way a good relative accuracy can be achieved ($\approx 5\%$). The absolute concentration values that are obtained with this method carry errors of about 10%. An alternative calibration can be found in [19].

With such a calibration in hand, a quantitative imaging of the complete Ti^{4+} profiles across the waveguide is possible. An example is shown in Fig. 6.8. It de-

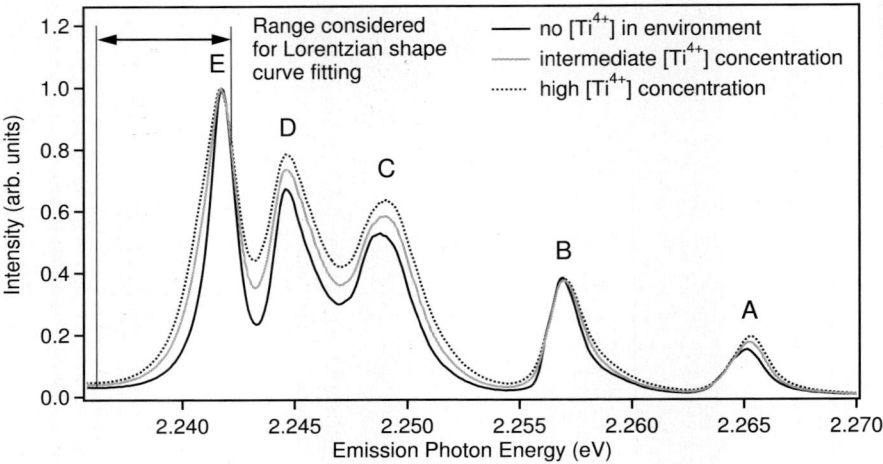

Fig. 6.6. Emission spectra of sample regions containing no Ti^{4+} ions, an intermediate concentration of Ti^{4+} ions, and a high concentration of Ti^{4+} ions measured at low temperature $T = 4$ K. The 488 nm emission line of an argon ion laser served as excitation source. In the *left upper corner* the range is indicated, which was used for a Lorentzian shape curve fitting

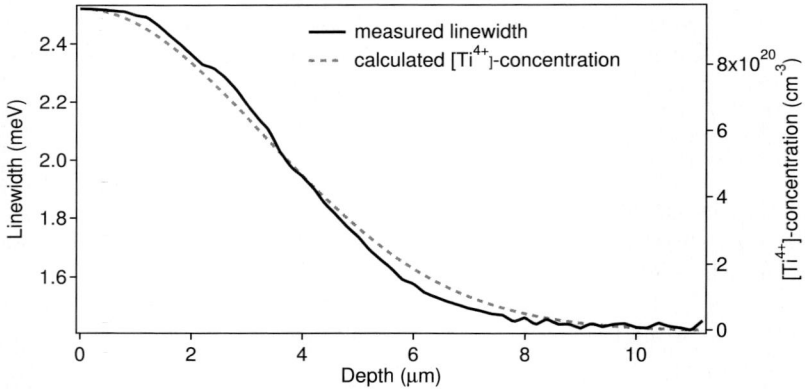

Fig. 6.7. Comparison between measured linewidth in dependence on depth and calculated $[Ti^{4+}]$-concentration profile

picts the profile for a cross section parallel to the front surface located about 5 μm inside of the sample. We can see considerable details of the Ti-distribution and apparent deviation from the ideal smooth patterns – caused most like by crystal imperfection – become apparent. Although the front surface had a scratch, it did not show up in the image indicating that the method is insensitive to topographical artifacts.

Fig. 6.8. Image of the Ti^{4+} profile in a waveguide using the width of the emission spectra (Peak 5 in Fig. 6.6) obtained in the different regions of the waveguide. The waveguide was designed to be single mode at a wavelength of 1.5 mm

6.2.3 Imaging of Ferroelectric Domains and Domain Wall Regions

Let us next turn our attention to the changes that occur under domain inversion [20].

Figure 6.4 shows CEES data before and after a domain inversion performed at room temperature in nearly stoichiometric $LiNbO_3$ material. We can see that the emission strength of the center type A remains virtually unchanged while the strength of emission from center of type B, and C is reduced. All emission lines are shifted slightly to lower energies. After a subsequent domain inversion back to the initial direction of the ferroelectric axis, these changes are almost completely reversed. Annealing the samples ($T = 250°C$, 5 h) at temperatures for which the Li-ion is mobile also yields changes in the relative emission intensities of the sites of type B and C. Annealing of as-grown and domain inverted samples results in the same emission spectra, which are, however, different from the ones seen in as-grown samples. These observations suggest that the thermal equilibrium of the defect configuration is perturbed by the domain inversion but can be re-established by annealing. The thermal equilibrium obtained at the conditions used in our experiments is different from the one that is obtained during crystal growth.

These results make it apparent that the contrast in imaging the domain patterns in real devices will be limited since most devices undergo an annealing procedure as part of the patterning process. However, as we will see, such procedures are often incomplete in terms of defect realignment such that some contrast is maintained. We will further see that the level to which thermal equilibrium is achieved depends on the presence of other dopants and of the defect type that we are considering.

In order to understand the origin of the observed changes, we need to recall that the domain inversion process has been performed at temperatures at which the charge compensating defects cannot rearrange in their location relative to the ferroelectric axis and the Er^{3+} position. However, the Er^{3+} ion itself will rearrange similar to the Li ion it replaces. This leads to frustrated defect dipoles, which are not arranged according to the thermal equilibrium. This is true for both intrinsic defect dipoles and for those that include an Er ion. With the spectroscopy studies outlined above, one has the tools to measure the intrinsic electric field changes and frustrated defect configuration across a domain wall. For this matter, the earlier described con-

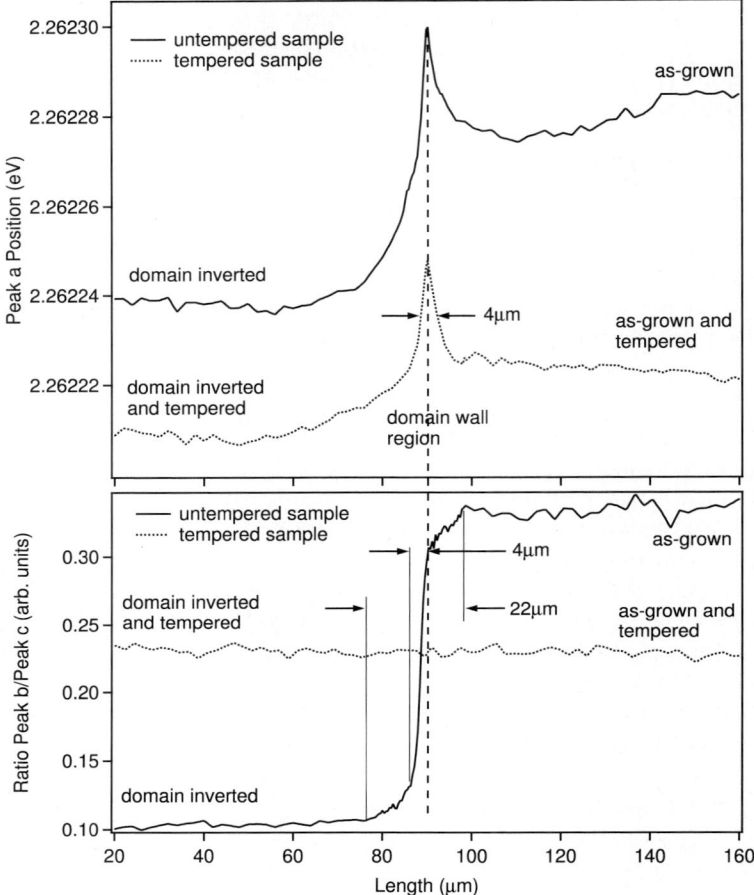

Fig. 6.9. Evaluated emission data for a scan over a domain wall region (see *inset*) for a nearly stoichiometric sample before and after annealing (5 h, 250°C). *Top*: Spectral position of the emission from site A1 (see Fig. 6.4). *Bottom*: Intensity ratio of main site A1 and satellite peak B1. The data have been collected at 4 K

focal luminescence microscope is used to measure spectra with a spatial resolution of about 500 nm. Results exist so far only for the near-stoichiometric LiNbO$_3$ samples (2 kV/mm coercive field) with Er concentrations of about 10^{-2} mol% (see Fig. 6.9).

For these experiments, freshly prepared inverted domain regions were used, and line scans across a domain wall were collected. The same experiment was repeated after the sample has been annealed for 5 hours at $T = 250$°C. The effect of intrinsic defect dipoles can be observed by the shift of peak A1 that originates from Er ions that have no local charge compensation and exhibit a spectral shift due to the intrinsic fields created by distant, frustrated intrinsic defects. In order to translate the spectral shifts, an electric field calibration is used. For the depicted transition the emission shift is 4×10^{-3} meV $\pm 1.5 \times 10^{-3}$ meV per kV/mm of applied field. We see the following easily resolved spatial features:

1. A 4 µm wide peak at the center of the domain wall region present before and after annealing. The peak represents a maximum shift of 0.02 meV that corresponds to a change in the electric field of 5 ± 2 kV/mm that is experienced by the Er ion probe. This peak at the wall may come from very local distortions at the wall location. Since the peak remains after annealing, this effect is most likely not due to non-stoichiometric defect clusters but rather due to the presence of Er in the wall region where a polarization gradient exists. This may lead to a difference in the off-center position of the Er ion away from the position determined for the bulk material [6–9].
   In this environment, the Er ion will measure a different field. Another possibility is that any local tilts in the wall will also induce polarization charges and fields that will be symmetric across the wall and give rise to a symmetric peak as seen for Er at the wall. A further possibility is the accumulation of (surface-) charges in the vicinity of a domain wall.
2. The step transition of the peak positions across the domain wall occurs within 22 µm. This effect is reduced after annealing; before annealing, a shift of 0.04 meV and a corresponding change in the local field of $E \sim 11.5 \pm 4.5$ kV/mm is observed. Though a similar measurement is not currently available for congruent composition, we assume that this is the lower limit for the domain wall fields. After annealing these values are 0.015 meV and $\Delta E \approx 4.35 \pm 1.65$ kV/mm, respectively, indicating that the sample has not reached a new thermal equilibrium for the annealing conditions used in this experiment. Such a step change in peak position across the wall cannot arise from local wall tilts, which will be expected to have symmetric response across the wall. It could arise however from an uneven charge accumulation across the domain wall.

In the ratio of the emission from A1 and B1, we can observe the behavior of the Er ions that participate in defect cluster formation. One finds a short-range component over ~4 µm and another one with an extent of 22 µm. Both of them disappear completely after annealing indicating that the rearrangement of defects clusters is different for the A and B type Er-sites considered here.

The domain wall imaging technique based on rare earth probes can be extended to allow in-situ observation of a moving domain wall. In this mode of operation, the point of observation is kept constant, and the growth of an existing domain nucleus is induced by the application of an electric field. Spectra were recorded in rapid fashion (100 spectra/s), while a domain wall was moving through the spot of observation. The velocity of domain wall motion can be controlled by the applied field and is determined independently using regular microscope imaging with a CCD camera. For these room temperature measurements, the spectra of individual sites are no longer resolved, but the ratio of site A and B is still reflected in a

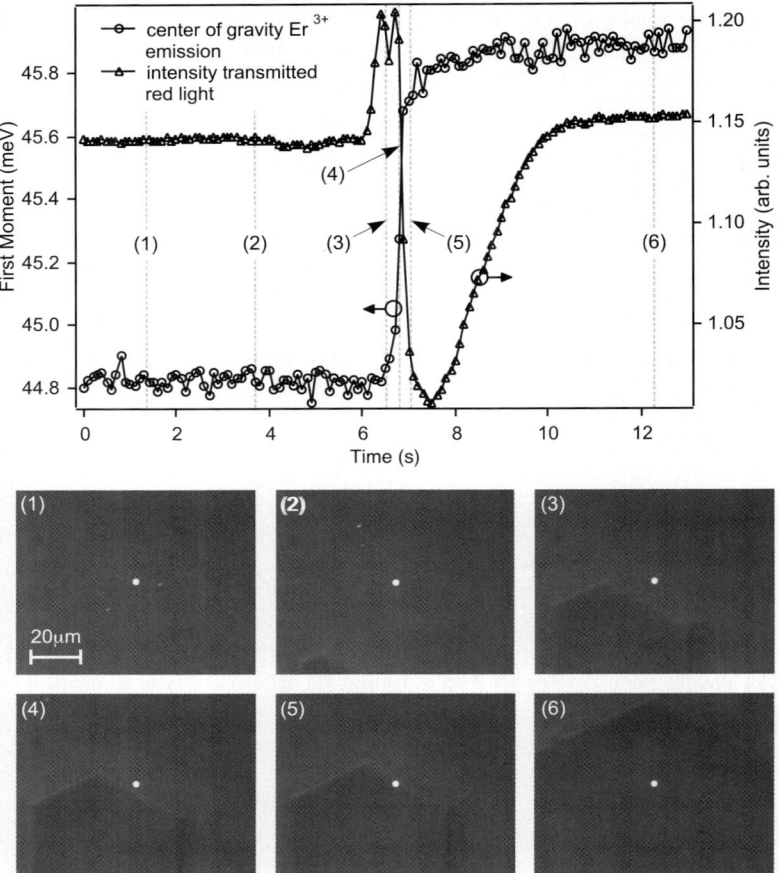

Fig. 6.10. Evaluated emission data recorded at the laser spot as a function of time, while a domain wall is passing it (velocity 25 m/s). *Top*: Relative emission peak position and intensity of transmitted light. *Bottom*: Time sequence from camera observing the moving domain wall. The *white dots* indicate the point of observation

shift of the unresolved emission band. Figure 6.10 shows the result for a growing domain wall for the sequence shown on the bottom of the figure. The measurements were performed for several of the sequences with different wall velocities. It was found that the domain wall regions are expanded and reach about 16 µm width for a speed of 25 µm/s, up from 4 µm that were determined in the static case. This is a significant result since it suggests that adjacent domain walls may interact over long distances.

The transmitted light that detects phase differences due to local strains in the sample was measured simultaneously with the Er emission. This direct comparison shows that the two methods probe changes on somewhat different length scales with clearly different spatial details. While the transmission shows increased and decreased signals indicating compressive and tensile regions, the Er emission shows as function of time the same type of dependence as in the static case.

6.2.4 Imaging of Periodically Poled Waveguide Structures

The methods introduced in the previous two subsections can be combined in order to study periodically poled waveguide devices. One possible application of such devices are self-frequency doubling waveguide lasers. They are suitable for our imaging technique since the rare earth doping is already present intentionally. In the waveguide environment in congruent $LiNbO_3$, the spectral responses are much wider compared to the stoichiometric material considered in the previous section. For that reason, we cannot expect the same details of information. Moreover, the devices have been annealed after the domain inversion process to relief strain from the structure. This further reduces the contrast that can be achieved. For this reason, imaging of such devices requires low temperatures (<100 K) to obtain good contrast.

Again, excitation with 488 nm light is used, and the emission spectra in the green are evaluated. It turns out that the different emission peaks of Fig. 6.6 show different sensitivity to the different perturbations (waveguides and domain pattern). We see good sensitivity for the Ti^{4+} concentration in almost all peaks (A...E), while the domain inversion is best visible in peak A and B. Furthermore, the presence of the Ti^{4+} ion influences the defect reconfiguration after annealing such that the contrast for the domain patterns depends on location (in or outside of the waveguide) [21].

Nevertheless, important information about the device and the interaction between waveguide and domain pattern can be obtained through the evaluation of the Er^{3+} spectra across the device. In Fig. 6.11, an overview of a PPLN device is shown. While in regular reflection confocal microscopy the domain pattern cannot be seen, it becomes clearly visible by evaluating the emission. The scans over a wider range can be used to determine the duty cycle of domain up vs down, inside, and outside the waveguide. It is found that this duty cycle is 48:52 and 53:47 along the white and black line in the figure; both deviated from the 50:50 duty cycle imposed by the patterned electrodes. This shows that the domain growth is hindered in the presence of the Ti^{4+} ions.

This behavior becomes even more evident in the detailed image shown in Fig. 6.12. Within the waveguide region, the dark domain inverted regions is clearly smaller. While the deviation found in the present device has little influence on the performance of the investigated device, the difference in domain growth behavior is expected to become more important when domain patterns with smaller periods are desired. In Fig. 6.12, it is also evident that the contrast for domain patterns is much stronger in the waveguide. This can be explained as follows. Domain inversion shifts peak B (that is used here) to lower energies [20]. The latter trend is strongly reduced when the sample is annealed, as it was the case for our sample. For a horizontal linescan in a region with the as-grown dipole direction, the contrast

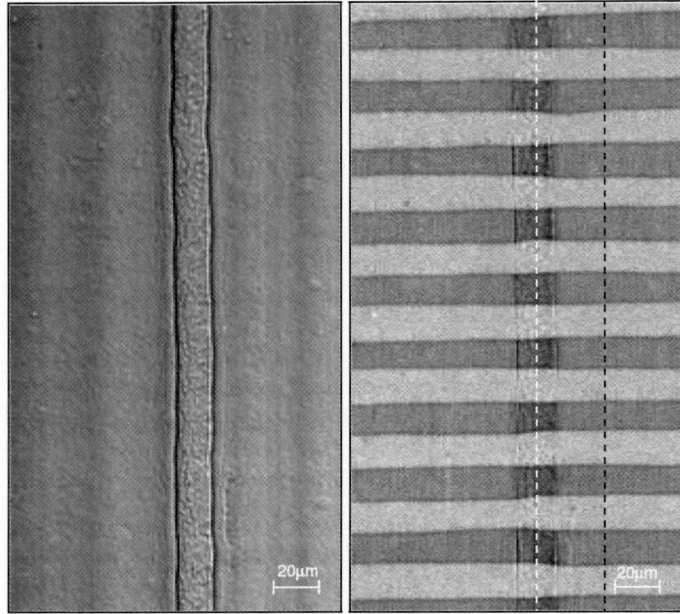

Fig. 6.11. Scan over a large area of a PPLN waveguide device. *Left*: Confocal microscope image created by detecting the reflected excitation laser light. *Right*: This image was created by calculating the first moment of peak A in Fig. 6.6

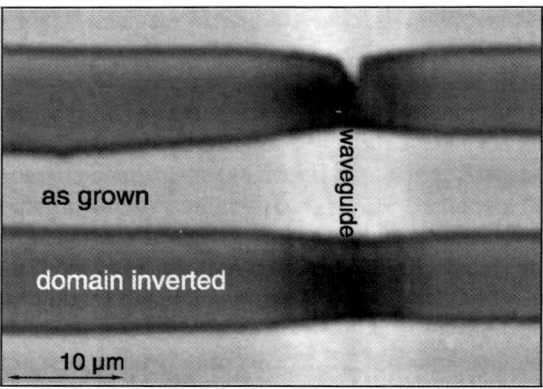

Fig. 6.12. Example of a domain which stopped growing inside the waveguide. The image was obtained by evaluating peak B. The tendency of domains to bend inside the waveguide is in this image also nicely visible. Light grays reflect higher emission energy

exhibits the same behavior seen in regular waveguide samples, i.e., the first moment resembles closely the Ti-concentration profile [19] as seen above. In the domain inverted region, the behavior is quite different. Instead of an increase, a reduction of emission energy is seen in the waveguide regions. Inside the waveguide, the Ti hinders the thermally induced center reconfiguration that occurs during the annealing. Therefore, the strong contrast seen before tempering is much better maintained inside the waveguide. As a result, we see a much better contrast for the domain structures within the waveguide. The contrast value is in fact similar to the one seen for domain patterns in samples which have not been tempered [20]. This indicates that little reconfiguration takes place within the waveguide during tempering. As disadvantage of this high contrast for domain structure, the overlapping contrast due to Ti concentration seen in the as-grown region is not observable in the domain inverted region anymore using lines A and B. However, using line D or E, the waveguide can be imaged reliably in all regions.

In the study of waveguide devices, the strength of the spectral imaging technique comes out very nicely. The possibility of evaluating different aspects of the spectrum can be utilized to untangle the mutual interaction effects between Ti^{4+} doping and domain inversion. The apparent down side of the presented imaging technique using rare earth doping is, of course, the need for an additional dopant. This makes the technique not suitable for routine inspection of devices that do not have the dopant already. To remedy this problem, Raman spectroscopy can be used as the spectroscopic tool. First success with this approach has been already reported [21–23]. However, due to the smaller Raman signal, intensities increase the measurement time significantly.

6.3 Light Induced Domain Inversion

The most common way to produce ferroelectric domain patterns is the application of electric fields greater than the coercive field through patterned electrodes. While this technique is quite mature and allows one to produce domain patterns with feature sizes above 5 μm routinely, it lacks flexibility and has severe limits for whenever smaller features sizes are desired. To overcome these limitation, several groups have been working on the use of light to define the inversion regions.

6.3.1 Methods

Different approaches have been used to exploit light for domain inversion. Several of them are covered in other chapters of this volume. Light can influence the domain inversion process mainly in three ways:

1. Increase of temperature leading to a reduction of the coercive field.
2. Photoconductivity can lead to inhomogeneous fields along the z-direction leading to domain inversion in the dark region.
3. Build up of local charges due to photo-ionization.

For the fabrication of domain engineered devices, it is desirable to have a parallel approach in which the elaborate patterns are written simultaneously. This can be

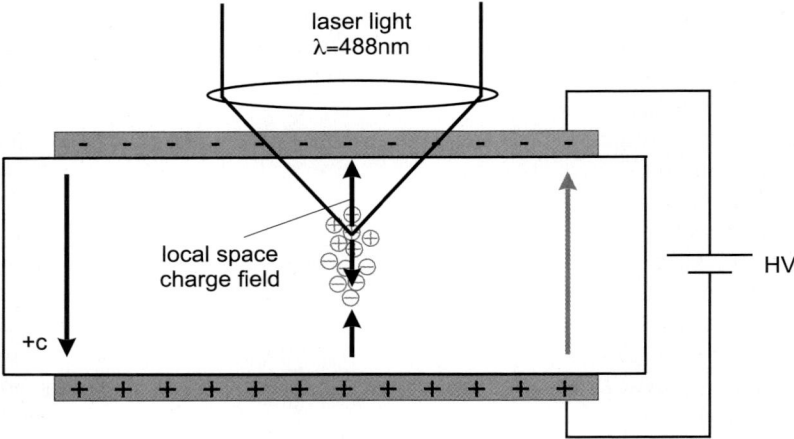

Fig. 6.13. Electric field due to space charges. As indicated, the field direction is above the laser focus in direction of the applied field necessary to perform a domain inversion

achieved through holographic exposure [24] or photomasks [25, 26]. However, both methods lack the detail control that is possible when the domain patterns are written in a serial process.

The group of Eason in Southampton [27] have demonstrated that under pulsed laser excitation domain patterns can be obtained in a self-organized way even in the absence of any applied electric field.

In this chapter, we focus on the effect of focussed visible laser light on local electric fields and how they can be used to write domain patterns.

6.3.2 Build-Up of Charge under Focussed Laser Irradiation

It is well known that congruent $LiNbO_3$ contains impurity defects that can be photo-ionized. In particular, Fe^{2+} defects can be photoionized by light in the blue/green spectral region to become Fe^{3+}. This will lead in a tight focus of a laser to a redistribution of charges. Within the focus, we will dominantly have Fe^{3+}, while the charges will be driven out of the focused area. These charges may accumulate on the surface. As a result, we obtain an electric field. This mechanism, shown in Fig. 6.13, can be best demonstrated at low temperatures using the ability to measure local electric fields of Er ions.

In a systematic study [28], laser powers between 3.6 mW and 60 mW were focused just below the $-c$ or $+c$ surface of the congruent sample, and emission spectra were recorded until no change in the emission spectrum could be identified. As an example, Fig. 6.14 shows the position of the emission peak A for laser powers of 11.7 mW, 36 mW, and 60 mW. The peak position was calculated by evaluating the first moment of the peak. The shift of the peak position versus time reflect the build-up of a local electric field. It does not follow a mono-exponential function, but rather

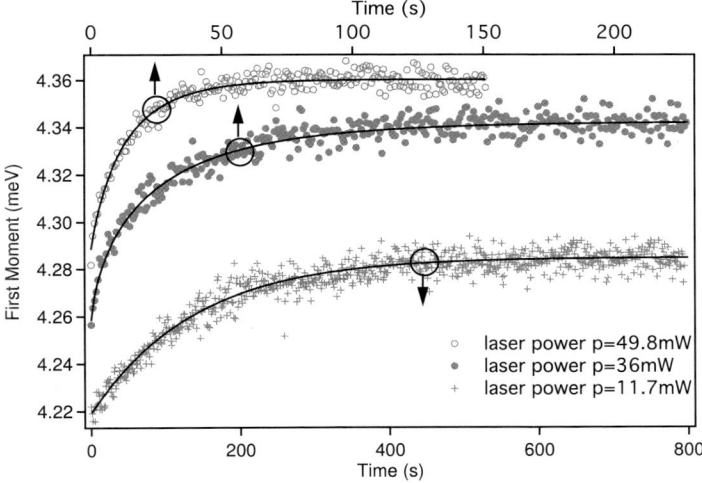

Fig. 6.14. Peak position of Er^{3+} emission peak A as a function of time after a laser focussed on the sample is switched on

a stretched exponential function. In Fig. 6.14, the initial peak position is shifted with increasing power to lower emission photon energies indicating a fast component of the charging process which is power-dependent. The slow component on the other hand depends in its built-up time on power, while the absolute value of the change is power independent. The underlying different charging mechanisms will result in different dynamics of the light induced domain inversion depending on laser intensities. We expect that intense pulsed lasers will exploit the fast first process, while cw-lasers will rely on the slower second charging mechanism.

The charging is most pronounced at low temperatures when the conductivity is low but is present at room temperature as well. This is demonstrated in Fig. 6.15, which shows a sequence of pictures taken from a movie of a PPLN device that is viewed in a regular microscope just after a focussed laser is switched on. As one can see, a contrast is slowly building up, due to the build up of the field which makes the domain patterns visible due to the phase contrast induced by the electro-optical effect. Over time the effect spreads over a fairly large region showing that the picture of Fig. 6.13 is too simplified and that timing will be critical in any domain patterning process that utilizes these light induced fields. In the next section, we will see that such fields strongly influence domain growth in a fairly complex way.

6.3.3 Influence of Light on Domain Inversion and Growth

As we have seen in the previous section, intense light irradiation is capable of producing substantial fields that can influence the domain inversion process. However, by itself the fields are not necessarily strong enough to trigger an inversion and

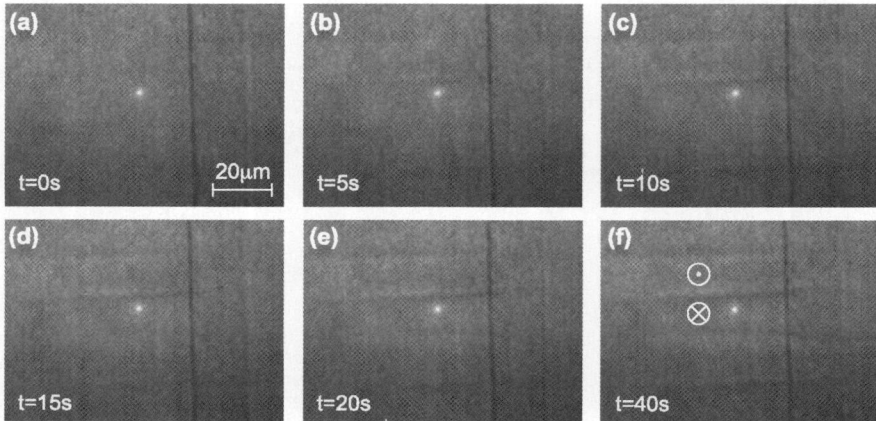

Fig. 6.15. PPLN device observed in a microscope under intense laser radiation: Image sequence taken from a movie recorded during illumination of the device. The *white spot* in the center of the images corresponds to the laser focus. At $t = 0$ s, the laser was switched on. The sample was kept at room temperature

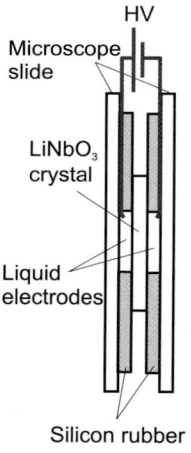

Fig. 6.16. High voltage holder used for application of electric fields at room temperature

hence, the application of a bias electric field is required. This applied field can be kept below the coercive field. In order to combine microscope observation, tight focus, and the application of fields in the order of 10 kV/mm, a special sample holder is required. A design used in the authors' work is shown in Fig. 6.16.

It is well known from LiNbO$_3$ that the coercive field required for domain inversion depends on the thermal history and the direction of domain inversion in respect to the polarization established at high temperature. This asymmetry in the hysteresis loops of electrical polarization is due to intrinsic defect dipoles, which are in

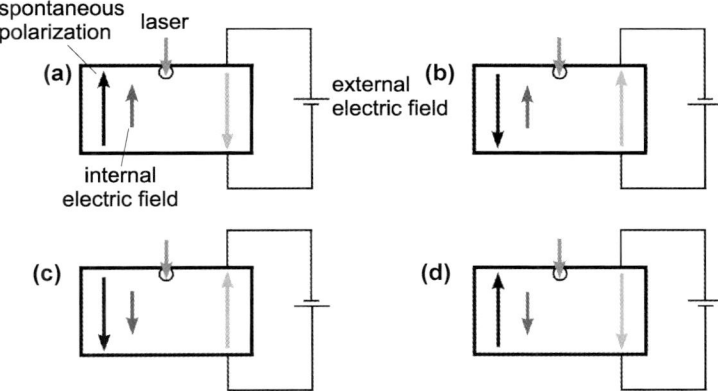

Fig. 6.17. Four different configurations of sample orientation and direction of incident laser light considered for the performed experiments to investigate the influence of light on the domain inversion process with electric field poling. In the configurations (**a**) and (**d**), the $+c$ side of the crystal is illuminated, and these two cases are different in the orientation of the internal electric field. For the configurations (**b**) and (**c**), the $-c$ side is illuminated, and these cases differ again only in the orientation of the internal electric field

the original state aligned along the polarization axis and after a "forward" domain inversion end up in a frustrated state opposing the polarization direction [29]. For that reason, domain inversion requires less voltage for the domain inversion back to the original state. Furthermore, the position of the laser spot relative to the surface and the polarization axis plays a role in the way the induced electric field will influence the domain inversion. This leads to four different configurations that have to be studied. They are shown in Fig. 6.17.

With these configurations, the depth of the focus, the light intensity, and the applied field can be modified. The outcome of each experiment critically depends on these parameters as well as the history of light illumination and surface structure of the sample leading to the impression of unpredictability of the outcome. Realizing and controlling these dependencies results in reproducible outcomes. All experiments were performed on congruent $LiNbO_3$ either undoped or doped with Er^{3+} ions. No significant deviations for the discussed behavior between undoped and low concentration Er^{3+} doped $LiNbO_3$ were found. This enables us to use the Er^{3+} ions to monitor the process. The observations can be summarized as follows: Several general observation can be made:

1. In all four cases, the irradiation with a focused laser with a wavelength of $\lambda = 488$ nm influenced the domain inversion process but 980 nm light had no influence. This supports the assumption that photo-ionization and note heating is governing the phenomena. An independent proof of the negligible heating effect was obtained by Raman measurements [30].
2. The observations for the configurations (b) and (c) are the same except that higher bias voltages have to be used for inversion in forward direction (c). Sim-

ilarly configurations (a) and (d) are equivalent. Observation however depend on whether the laser is focussed closed to the $-c$ or $+c$ surface.
3. For illumination of the $-c$ surface, domain inversion occurs first close to the laser spot or exactly at the laser spot. The required electric field for domain inversion depends on how deep the laser focus is below the surface. At a depth of about 130 µm, a minimum was found with about 70% of the usual required field. This value increases to about 80% just below the surface. An image sequence of such an experiment is shown in Fig. 6.18. It shows a case in which first nuclei of domains appear about 10 µm away from the laser spot. In other experiments, we find that when the focus is deeper into the samples, the domain inversion occurs close to the laser spot.
4. For illumination of the $+c$ side, i.e., configurations (a) and (d), nucleation appears laterally close to the focal spot but on the opposite side of the sample ($-c$ side again), starting at electric fields somewhat higher than in configurations (b) and (c). This behavior is shown in an example in Fig. 6.19. Nucleation of domain appears in (a) and (b) only on the backside (which is the $-c$ side) in this experiments. Only closer to the coercive field, some nucleation appears on the $+c$ side as well.
5. When the electric field was increased quickly, above the usually observed coercive field, domain inversion occurs over the whole area covered by the electrodes but not at the laser spot and at a certain area around it. In this case, illumination from the $+c$ side prevents domain inversion.

With this understanding of the process and the in-situ diagnostics using the rare earth probe ions, the authors' group was able to exploit the light induced field effects for the writing of domain patterns as will be described in the final section.

6.3.4 Direct Writing of Domain Patterns

We have seen in the previous subsection that light focussed laser light (488 nm) reduces the coercive field needed for domain inversion by up to 30%. This gives the bases for a direct write scheme of domain patterns in LiNbO$_3$ in which a constant electric field is applied that is somewhat lower than the needed coercive field for domain inversion, and at the same time moving a focused laser beam over the sample. It is also evident from the experiments shown in the previous subsection that only the experimental cases (b) and (c) shown in Fig. 6.17 are suitable. In the configuration case (b), a backward domain inversion is carried out, which takes place at a lower coercive field and is experimentally easier. Domains can be nucleated in a matter of seconds at the laser spot or at the vicinity in a radius of about 5 µm. A focal depth of around 130 µm was most favorable for this step. However, attempts of writing patterns by moving the laser laterally at this depth give only very poor results. Domains tend to grow very fast and arbitrarily around the laser. This observation made it clear that for a successful writing scheme the procedure has to be divided into two steps:

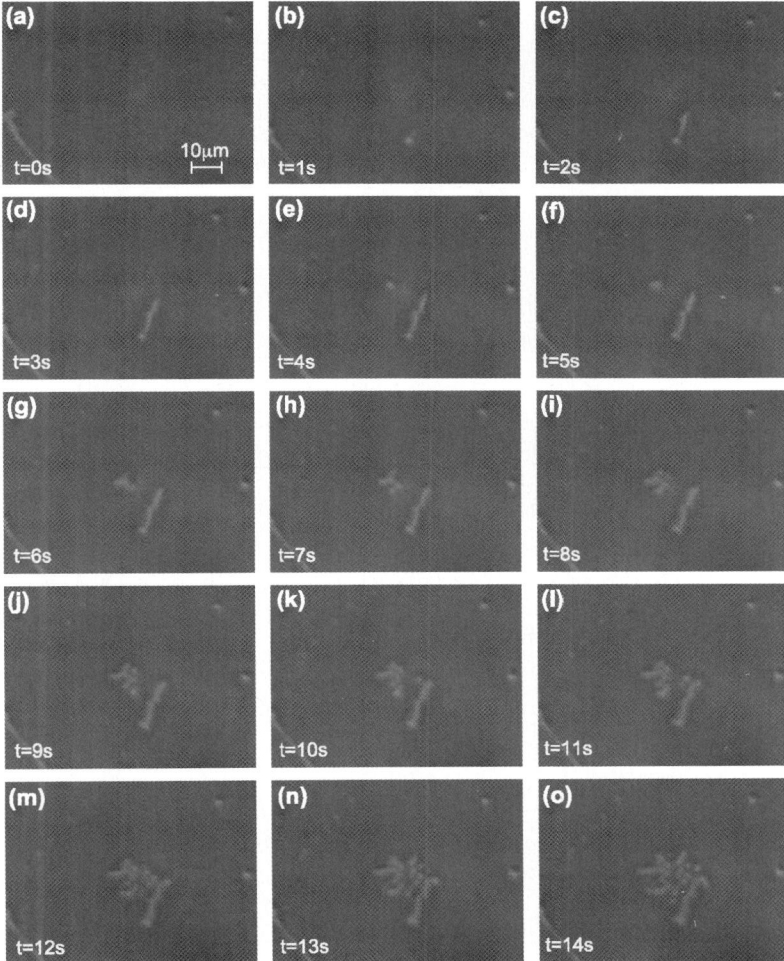

Fig. 6.18. Image sequence taken from a movie recorded during an experiment with configuration case (**b**). The laser focus is located in the center of the image and was focused about 50 μm below the sample surface and centered in the image. The laser power was about 140 mW

1. Domain nucleation, which happens on a time-scale of seconds and is more reliable when the laser focus was well below (≈ 100 μm) the sample surface.
2. Domain writing with a speed of about 50 μm/s and with the laser focus close to the sample surface (5 μm).

Phenomenologically, this division into two steps can be understood by the much higher energy that is required to overcome the electrostatic barrier to nucleate a region with opposing polarization direction. Once this nucleation is achieved, the

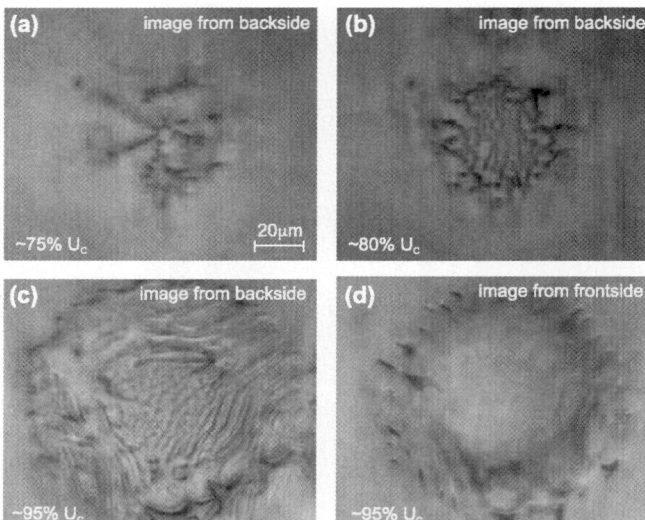

Fig. 6.19. Image sequence taken from a movie recorded during an experiment with configuration case (**d**). The laser focus is located in the center of the image and was about 5 μm below the sample surface and centered in the image. The laser power was about 140 mW. The applied electric field was slowly increased

writing of the domain pattern corresponds to a guided growth of the nucleated domain. The required electrostatic energy for this process is lower, and the availability of charges to neutralize the increasing domain surface area becomes the guiding principle.

On this basis, the domain pattern shown in Fig. 6.20 for a Er^{3+} doped congruent $LiNbO_3$ sample was created with the following sequence [30]: In order to start the writing process, the focused laser with a power of about 200 mW was positioned 130 μm below the sample surface, and an electric field 25% below the regular coercive field 20 kV/mm was applied to the 0.5 mm thick sample. Under these specific conditions, domain nucleation occurs within 2 s in the vicinity of the laser focus. Then the laser focus was moved upwards just below the sample surface (5 μm), followed by a fast movement with 50 μm/s parallel to the sample surface. During this lateral motion, the previously nucleated domain grows closely following the laser spot. After 100 μm the electric field and laser were switched off. The laser focus is then moved to a new location 15 μm away from the previously created domain. By repeating this sequence a periodic domain pattern can be quickly generated. In order to inspect the result, the RE-based imaging scheme was used. The image in Fig. 6.20(a) is a horizontal and in Fig. 6.20(b) is a vertical cross section image of the light-induced domain pattern. For the horizontal scan, the laser focus was 2 μm below the sample surface, and the vertical scan was recorded approximately in the center of the sample. The positions where the respective cross sections were taken are indicated by dotted lines in each image.

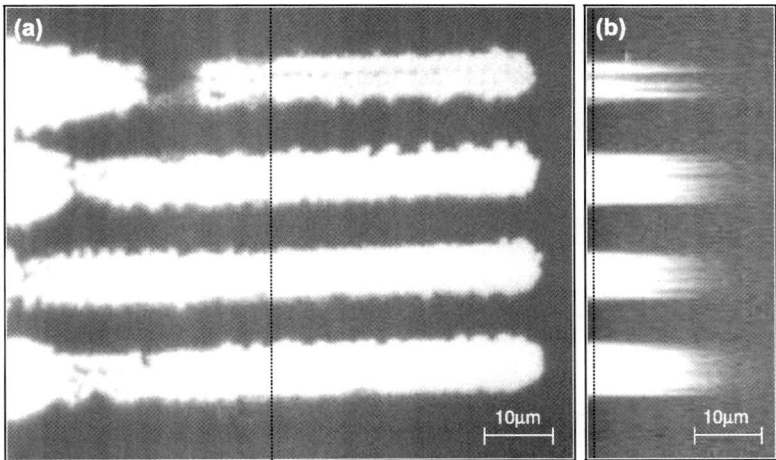

Fig. 6.20. (a) Horizontal (2 µm below surface) and (b) vertical cross-section image of a directly written periodically domain structure in a Er^{3+} doped $LiNbO_3$ crystal recorded with the confocal luminescence microscope

We see the following features. At the beginning of the structure, the domains are almost grown together and are irregular. This is the region where the laser focus was deep inside the sample and where the domain inversion was initialized. After the laser focus was positioned close to the surface and moved away from the starting point, the domains become parallel. In the case shown, the domain walls are a little frayed, but overall a very regular structure was achieved. The period of this structure (15 µm) was determined by the positioning of the laser focus. The width of each line is determined by the depth of focus and the speed of motion. It is interesting to see that in one instance the domain growth was interrupted due to some dust particle on the surface of the sample but could be continued of few µm later. The vertical cross section in Fig. 6.20(b) shows that the domains are in this case only about 20 µm deep and a characteristic sawtooth-like structure is visible. Apparently, the domains do not grow vertically much beyond the illuminated regions. This can be understood by the small growth velocity of domains in the absence of light. Deeper structures can be produced by reducing the speed of writing, by focusing deeper into the sample, or by increasing the applied electric field. Variation of these parameters, however, will also influence the width of the domains. The produced patterns are stable, and hence an inspection of the created structure after months revealed a practically unchanged pattern.

While the presence of Er^{3+} ions is useful for instant inspection of the pattern, they do not influence the writing process. To illustrate that point, a domain pattern that was written into an undoped congruent $LiNbO_3$ is shown in Fig. 6.21. Moreover, a smaller period of 8 µm was achieved by altering the parameters from the previous case (laser power $P = 220$ mW, electric field about 10% below coercive field, focused laser during writing 25 µm below the surface). Due to the lack of the

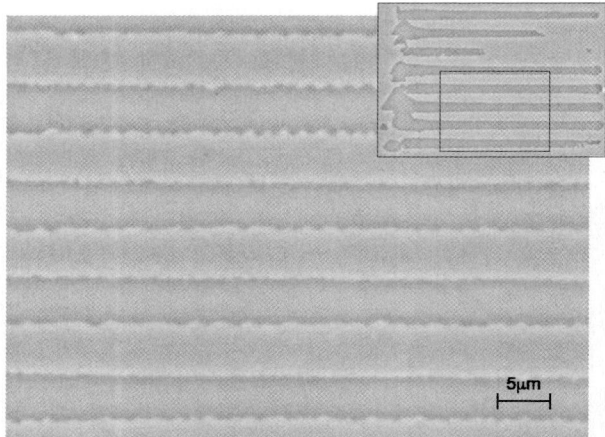

Fig. 6.21. Phase-contrast image of an undoped LiNbO$_3$ crystal with a directly written periodically domain structure that has been etched. The smaller image at the *upper right hand corner* is the whole created domain structure. The *black square* indicates the part which is shown with a higher magnification in the main image

Er^{3+} doping, the sample had to be examined according to the classical method of selective etching technique. The etching in a mixture of hydrofluoric acid and nitric acid transforms the periodic pattern into a topographic one, which can be inspected, for example, in a phase-contrast microscope. The depth of the created domain could not be determined with this procedure.

6.4 Summary and Conclusions

In this chapter, we illustrated how the detailed knowledge of dopants such as Er^{3+} ions can be utilized to probe modifications in LiNbO$_3$. In particular, this approach allows the detection of local internal fields, which will have a major impact on the understanding of materials. Since the method is linked to optical spectroscopy, it offers the possibility to explore changes in small dimensions down below 200 nm using a near field optical microscope and as a function in time. These capabilities have been demonstrated for the domain inversion process and the structure of domain walls in LiNbO$_3$. The knowledge of local electric fields lead to the development of domain writing processes in which the induced local electric fields are used to locally lower the required electric fields for domain inversion. Due to the maturity of the crystal growth and sample preparation of the LiNbO$_3$, this material is an ideal model case. The basic approaches are certainly not limited to LiNbO$_3$, and it is apparent and has been observed in the authors' lab already that other materials such as LiTaO$_3$ and KNbO$_3$ exhibit similar behavior. In judging if a material system is suitable for the imaging techniques, it is important to realize that the sensitivity to intrinsic electric fields is based on the linear Stark shift of spectral lines that only exist in materials that lack inversion symmetry. The laser-aided domain inversion process relies on defects that can be photo-ionized and on how the charge can propagate in the sample. Hence, the process depends even in $LiNbO_3$ strongly on details of the growth process and co-doping. For instance, while a light induced effect is present in stoichiometric $LiNbO_3$ and Mg-doped $LiNbO_3$, the parameters are drastically different. The parameter space for optimization of the writing process for particular domain patterns is very wide and optimization is tedious. However, the capability to use RE ions as probes may again be proof critical in such work as real-time diagnostics, and the realization of an active feedback is possible.

References

1. C. Becker, T. Oesselke, J. Pandavenes, R. Ricken, K. Rochhausen, G. Schreiber, W. Sohler, H. Suche, R. Wessel, S. Balsamo et al., IEEE J. Sel. Top. Quantum Electron. **6**, 101 (2000)
2. I. Baumann, R. Brinkmann, M. Dinand, W. Sohler, S. Westenhofer, IEEE J. Quantum Electron. **32**, 1695 (1996)
3. J. Amin, J. Aust, D. Veasey, N. Sanford, Electron. Lett. **34**, 456 (1998)
4. V. Voinot, R. Ferriere, J. Goedgebuer, Electron. Lett. **34**, 549 (1998)
5. S. Hüfner, *Optical Spectra of Transparent Rare Earth Compounds* (Academic Press, San Diego, 1970)
6. T. Gog, M. Griebenow, G. Materlik, Phys. Lett. A **181**, 417 (1993)
7. T. Gog, M. Griebenow, T. Harasimowicz, G. Materlik, Ferroelectrics **153**, 249 (1994)
8. L. Rebouta, M. da Silva, J. Soares, D. Serrano, E. Dieguez, F. Agullo-Lopez, J. Tornero, Appl. Phys. Lett. **70**, 1070 (1997)
9. L. Kovacs, L. Rebouta, J. Carvalho Soares, M. Fernanda da Silva, Radiat. Eff. Defects Solids **119–121**, 445 (1991)
10. D.M. Gill, L. McCaughan, J.C. Wright, Phys. Rev. B **53**, 2334 (1996)
11. O. Witte, H. Stolz, W. von der Osten, J. Phys. D (Appl. Phys.) **29**, 561 (1996)
12. V. Dierolf, M. Koerdt, Phys. Rev. B **61**, 8043 (2000)
13. V. Dierolf, C. Sandmann, A.B. Kutsenko, T. Tröster, Radiat. Eff. Defects Solids **155**, 253 (2001)
14. V. Dierolf, A.B. Kutsenko, C. Sandmann, T. Tröster, G. Corradi, J. Lumin. **87–89**, 989 (2000)
15. J.B. Pawley, *Handbook of Biological Confocal Microscopy* (Kluwer Academic/Plenum, Amsterdam/New York, 1995)
16. M. Born, E. Wolf, *Principles of Optics* (Pergamon Press, Oxford, 1980)
17. G. Corradi, T. Lingner, A.B. Kutsenko, V. Dierolf, K. Polgar, J.M. Spaeth, W. von der Osten, Radiat. Eff. Defects Solids **155**, 223 (2001)
18. V. Dierolf, C. Sandmann, J. Lumin. **125**, 67 (2007)
19. V. Dierolf, C. Sandmann, J. Lumin. **102–103**, 201 (2003)
20. V. Dierolf, C. Sandmann, S. Kim, V. Gopalan, K. Polgar, J. Appl. Phys. **93**, 2295 (2003)
21. V. Dierolf, C. Sandmann, Appl. Phys. B **78**, 363 (2004). ISSN 0946-2171
22. P. Capek, G. Stone, V. Dierolf, C. Althouse, V. Gopalan, Phys. Status Solidi C **3**, 830 (2007)
23. Y. Zhang, L. Guilbert, P. Bourson, Appl. Phys. B **78**, 355 (2004). ISSN 0946-2171

24. M. Wengler, U. Heinemeyer, E. Soergel, K. Buse, J. Appl. Phys. **98**, 64104 (2005) ISSN 0021-8979
25. M. Fujima, T. Sohmura, T. Suhara, Electron. Lett. **39**, 719 (2003)
26. C. Valdivia, C. Sones, S. Mailis, J. Mills, R. Eason, Ferroelectrics **340**, 75 (2006). ISSN 0015-0193
27. R. Eason, S. Mailis, C. Sones, A. Boyland, A. Muir, T. Sono, J. Scott, C. Valdivia, I. Wellington, Ceram. Trans. **196**, 93 (2006)
28. C. Sandmann, V. Dierolf, Phys. Status Solidi C **2**, 136 (2005). ISSN 1610-1634
29. V. Gopalan, T. Mitchell, Y. Furukawa, K. Kitamura, Appl. Phys. Lett. **72**, 1981 (1998)
30. V. Dierolf, C. Sandmann, Appl. Phys. Lett. **84**, 3987 (2004). ISSN 0003-6951

7 Visual and Quantitative Characterization of Ferroelectric Crystals and Related Domain Engineering Processes by Interferometric Techniques

P. Ferraro, S. Grilli, M. Paturzo, and S. De Nicola

7.1 Introduction

Nonlinear crystals are, nowadays, key devices to build coherent sources emitting radiation from the UV to the IR spectral range [1]. Applications of nonlinear optics are primarily based on frequency conversion, through harmonic generation or sum and difference frequency mixing. These nonlinear frequency conversion techniques make possible coherent light sources in spectral regions where laser sources are limited, or do not exist. Light sources based on nonlinear crystals, like optical parametric oscillators as well as harmonic and difference frequency generators, are finding increasing application in high sensitivity spectroscopy [2–4].

It has to be outlined that recent results obtained in micromachining and engineering at micrometric and nanometric level of ferroelectric materials and in particular of lithium niobate (LN or $LiNbO_3$) makes strategic the accurate knowledge and characterization of the optical ferroelectric materials with aim at realizing and fabricating micro-optical devices. On the other hand it is important to have reliable fabrication processes of micro-devices for non-linear or linear applications. Among various methods and techniques adopted to investigate and study the engineering processes a very important role has been played by interferometric techniques.

In this chapter we focus the attention on different interferometric approaches that have been adopted for characterizing the materials and engineering processes. In Sect. 7.3.1 are described method for measuring refractive index. Section 7.3.2 will be devoted to method allowing the visualization and study of the kinetics of domain formation in ferroelectric crystals during the electric filed poling (EFP). In Sect. 7.3.3 will be illustrated how through a digital holographic (DH) technique it is possible to obtain a 2D mapping of the electro-optic coefficient in standard wafers and how it i s possible investigate the internal electric field. Finally, Sect. 7.3.3 is instead dedicated to the investigation of the optical birefringence near the wall between two opposite domains in a z-cut congruent $LiNbO_3$ through a full-field polarimetric method.

7.2 Measuring the Refractive Indices and Thickness of Lithium Niobate Wafers

In electro-optic crystal materials, the index of refraction can be controlled by changing an electric field across the material. Many materials exhibit a small EO effect including quartz; however, the EO materials with the largest EO effect that are commonly used in laser based systems are GaAs, KH_2PO_4, $NH_4H_2PO_4$, CdTe, $LiNbO_3$, $LiTaO_3$, and $BaTiO_3$. $LiNbO_3$ has become the most common EO material widely used for optical telecommunication [5–7] such as amplitude modulating fiber optic communications systems operating at data rates exceeding 10 gigabits per second and nonlinear optics [8].

There are several applications in these fields where permanent modifications of the refractive index are desired, e.g., waveguides or Bragg reflectors require refractive-index modulations [9, 10]. An indispensable condition of some of these applications is a high degree of optical uniformity of LN crystals used for fabrication of active elements. As the specification of ready-made crystal slabs is often limited by manufacturing tolerances, accurate inspection after production is usually required. Knowledge of the refractive indices is fundamental to the characterization of crystals that are useful in optoelectronic devices. Properties of birefringent crystal materials are generally characterized by application of interferometric [11–14] or ellipsometric techniques [15–17]. Moiré deflectometry with circular gratings [17] or Talbot interferometry [18] have also been employed to characterize birefringent crystals.

All these techniques enables one to determine the axes of orientation or the refractive indices of the birefringent material but not its thickness. This is a long-standing problem, common to all interferometric methods applied to measuring the refractive indices: interferometers are sensitive to the optical path difference, i.e., to the product of the refractive index and thickness of the sample under test. Using the refractive indices of the crystal, chromatic polarization interferometry was employed for simultaneous determination of both the cutting angle and thickness of birefringent crystals, acting as a wave plate [19].

An interferometric technique based on the measurement of the rotation-dependent phase changes of the optical path length in crystal plates was employed for measuring both the ordinary and extraordinary refractive indices of a LN crystal plate of known thickness. The basic principle of this technique will be briefly discussed in the following in order to give an idea of the potentialities of the interferometric techniques for crystal characterization. The experimental set-up is a typical Mach–Zehnder interferometer shown in Fig. 7.1.

The experimental set-up of the Mach–Zehnder interferometer is shown in Fig. 7.1. The linearly polarized expanded and collimated beam from a He–Ne laser (wavelength $\lambda = 632$ nm) is used as the light source. The beam is divided by the beam splitter into two beams. One of these, the object beam, is incident on a Glan–Thompson polarizer (GT) and $k = 2$ plate in order to rotate the linearly polarized state of the object beam. The object beam which passed through the $\lambda/2$ plate is normally incident on a disk of congruent LN plane-parallel substrate with a diam-

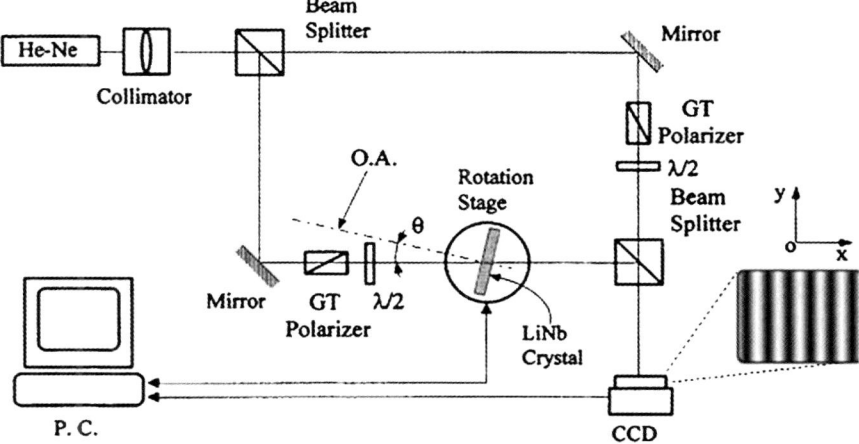

Fig. 7.1. Mach–Zehnder interferometer for measuring the refractive indices of lithium niobate crystals

eter of 76.2 mm and a nominal thickness of $d = 0.5$ mm. The second beam is the reference beam which passes through another combination of GT polarizer and $\lambda/2$ plate in order to rotate the linearly polarized state of the beam. The reference beam is combined at the output beam-splitter with the object beam. The fringe pattern resulting from the interference of these two beams is a family of equally spaced parallel straight fringes whose spatial frequency can be easily varied by tilting one of the two mirrors of the interferometer. The crystal has the optic axis perpendicular to its surfaces and it is mounted on a high resolution motorized precision rotation stage. The object beam incident on the LN crystal and the reference beam are polarized either s or p. Because of symmetry, the polarization of the incident wave is not affected by refraction through the uniaxial LN plate. Of course, the refracted waves experience different indices of refraction depending on their polarization, i.e. the ordinary index n_o for s polarization or the extraordinary index n_e for p polarization. When the object beam impinging on the sample is s polarized, the ordinary o-ray exiting the test object is refracted at an angle θ_o determined by the condition $\sin \theta = n_o \sin \theta_o$ where θ is the incidence angle. In order to obtain o-fringes, also the reference beam is s polarized. In this condition, a rotation of the LN crystal plate of an angle θ with respect to the direction of the s polarized incident light produces a variation in the optical path length of the o-beam exiting the test object that results in a phase change ϕ_o given by

$$\phi_o = kd\left[\left(n_o^2 - \sin^2 \theta\right)^{1/2} - \cos \theta - n_o + 1\right] \quad (7.1)$$

where $k = 2\pi/\lambda$ is the wavenumber. In order to produce e-fringes both the reference and object beams are p polarized. The corresponding phase change ϕ_e experienced by the e-ray exiting the test object is given by the following expression as a function of the incidence angle θ, namely

$$\phi_e = kd\left[\left(n_o^2 - \left(\frac{n_o \sin\theta}{n_e}\right)^2\right)^{1/2} - \cos\theta - n_o + 1\right]. \tag{7.2}$$

Equations (7.1)–(7.2) allow the determination of the refractive indices n_o and n_e by independent measurements of the phase changes ϕ_o and ϕ_e. When the incident angle θ of light is varied by rotating the crystal plate, the e or o interference patterns, consisting of vertical straight fringes, cross the field of view of the detector array.

In principle, different approaches can be adopted, depending on the homogeneity of the sample under test. In fact, it is possible to process the full field interference e- or o-fringe pattern imaged by the CCD array and digitized by the frame grabber or portions of it, in order to obtain a spatially resolved information of the refractive index or thickness. Figure 7.2(a)–(b) show a typical interference signals, recorded while the interference fringe pattern crosses the filed of view of CCD array (array size 768 × 581 pixel with size 11 μm × 11 μm) as a function of the rotation angle of the LN plate.

The s and p polarization interference signals were taken at rotation angle increment $\Delta\theta = 0.01°$ ranging from zero, at normal incidence, when the optical axis of the crystal plate is along the direction of the incidence beam, up to 30° for the s signal and 40° for the p signal, respectively. According to the well known Fourier transform method (FTM) for fringe pattern analysis [21], both signals can be Fourier transformed, in order to retrieve the rotation–dependent phase change wrapped in the range $-\pi, +\pi$. The phase changes of the ordinary and extraordinary waves ϕ_o and ϕ_e are determined by applying the unwrapping procedure to the previously determined wrapped phase data and both the ordinary and extraordinary indices of refraction of the LN crystal $n_o = 2.2772 \pm 0.0009$ and $n_e = 2.2035 \pm 0.0002$ are obtained by fitting the unwrapped phase data to (7.1)–(7.2), respectively.

In principle, a nonlinear fitting procedure would allow to determine the refractive index n and the thickness d of the sample. Unfortunately, the nonlinear least square fit of the phase change does not in general provide accurate results for both n and d. As remarked before, to employ interferometric systems for determining the refractive index of a thin plate requires accurate knowledge of the thickness and vice versa. Therefore, efforts have been made to find valid approaches for measuring n and d simultaneously or at least by using the same apparatus [22–25].

These methods require complex instrumentations, such as translation stages with high resolution. An alternative simpler method employed a lateral shear interferometer [26] in which the shear plate itself is the sample to be measured. In the following, we will briefly illustrate the principle of measurement of the technique and will describe a typical application of the method for measuring both the refractive indices and thickness of lithium niobate plates. The optical configuration is schematically shown in Fig. 7.3.

The laser source consists of a distributed–feedback (DFB) diode laser with a peak wavelength near 1435 nm and a time averaged spectral linewidth at -27 db

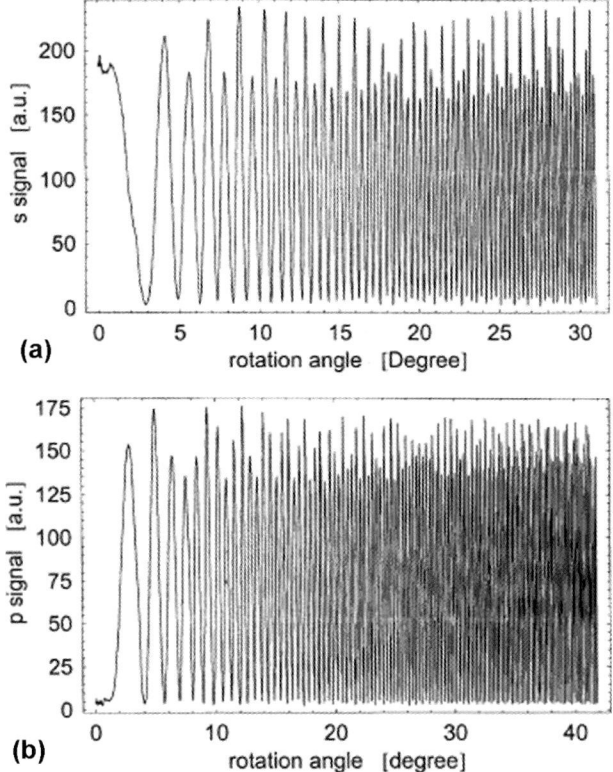

Fig. 7.2. Recorded signal intensity vs. rotation angle of the crystal plate: (**a**) s polarization signal; (**b**) p polarization signal

of 0.4 nm. Linear wavelength tuning is achieved by varying the laser temperature. A fiber optic collimator is mounted upon the fiber pigtail end of the DFB laser. An InGaAs photodiode with a 0.8 mm sensitive area and a 800–1800 nm spectral range was used as a detector along the path of the emerging beams. The sample, an uniaxial LiNbO$_3$ z-cut crystal, with nominal ordinary n_o and extraordinary n_e refractive index of 2.20 and 2.13, respectively and a thickness of 500 ± 50 µm, was mounted vertically upon a rotating dc motor, and a stable dc voltage applied to the motor guaranteed a constant rotation speed. The interferometric signal was digitized on an oscilloscope with a sampling frequency of 2.5×10^6 points/s. The detected signal was averaged over 50 periods of rotation, reducing the noise. The temperature induced laser wavelength variation was in the range 1535.28–1536.04 nm, with 150 steps of $\Delta\lambda = 0.005$ nm.

For each wavelength the signal is composed was composed of 5000 sampled points for an angular range of $-25°$ to $+25°$, corresponding to an angular resolution

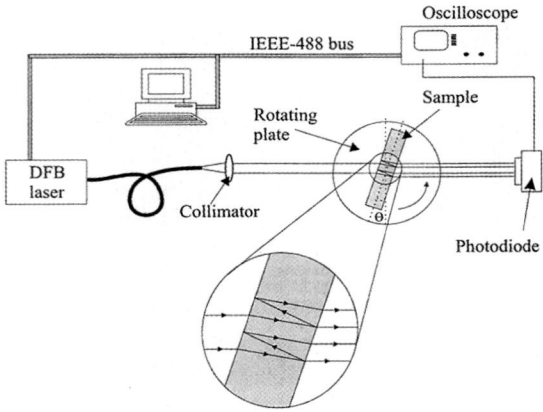

Fig. 7.3. Experimental setup employed for measuring both refractive indices and thickness of lithium niobate plates: DFB, distributed feedback laser

of 0.01° per point. The rotation period ~22.0 ms and the rate of change of the laser wavelength (<0.1 s) led to an estimate of the total measurement time of ~165 s.

A Glan-Thompson polarizer was inserted into the optical set-up between the collimator and the crystal sample to permit selection of the polarization state of the incident light. The emerging wavefronts are laterally sheared, with a shear that depends on the actual angle of incidence θ and interfere in the area where they are superimposed, producing an interference fringe pattern whose geometry depends on the phase front of the beams. If the sample is rotated, the angle of incidence will change and the interference fringe pattern will change its phase because the interfering wave fronts will have a different optical path difference at each angle of incidence. Considering the direct transmitted and the first internally reflected beam, the interference light intensity at the detector as a function of the wavelength λ, rotation angle θ is given by a set of two equations each for n_o and n_e, namely

$$I_{n_p}(\theta) = I_0 + \gamma \cos\left[\frac{4\pi n_o d}{\lambda}\left(1 - \frac{\sin^2\theta}{n_p^2}\right)^{1/2}\right], \quad p = o, e \quad (7.3)$$

where I_0 is a constant offset that is due to the intensities of the two interfering beams and γ is related to the degree of coherence of light and n_p is equal to n_o or n_e, depending on the king of measurement. When the wavelength of the laser is changed in steps $\Delta\lambda$ provided that $\Delta\lambda/\lambda \ll 1$, the θ dependent phase shift $\Delta\phi(\theta)$ is

$$\Delta\phi(\theta) = \frac{4\pi n_o d}{\lambda}\left(\frac{\Delta\lambda}{\lambda}\right)\left(1 - \frac{\sin^2\theta}{n_p^2}\right)^{1/2} \quad (7.4)$$

which becomes at normal incidence $\Delta\phi(\theta = 0) = 4\pi n_o d \Delta\lambda/\lambda^2$ which allows to obtain the optical thickness $n_o d$ through a nonlinear fit of the wavelength dependent interference intensity curve $I_{\theta=0}(\lambda)$ for normal incidence ($\theta = 0$) when the optic

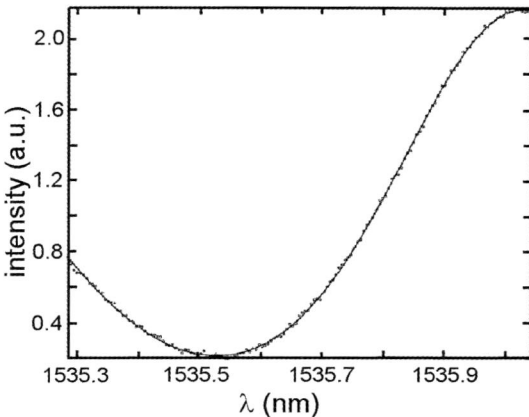

Fig. 7.4. Experimental (*circles*) and fitted (*solid curve*) wavelength dependent interference signal for normal incidence for a LiNbO$_3$ sample

axis of the crystal plate lies along the direction of the incident beam. Figure 7.4 shows the experimental and fitted intensity curve for normal incidence for a LiNbO$_3$ sample.

To take into account multiple reflections that occur at normal incidence, the experimental data were fitted by $I_{\theta=0}(\lambda) = A + \sum_{m=1}^{3} B_m \cos(4\pi md/\lambda + D)$, where A, B_m and D are parameters of the fit. From the knowledge of the optical thickness $n_o d$, one can obtain the refractive indices by retrieving the phase of the overall interference signal for each state of polarization of the incident light beam. The phase change of (7.2) can be determined by applying the FTM at fixed wavelength. Figure 7.5(a) and (b), respectively, show the typical experimental interference signal and corresponding cosine of the phase retrieved by the FTM method at a fixed wavelength $\lambda = 1535.68$ nm.

Applying the described method to an interference signal composed of 5000 sampled points it is obtained obtain $n_o = 2.2004 \pm 0.0001$, $n_e = 2.1360 \pm 0.0002$, and $d = 500.0 \pm 0.3$ μm. To determine by the optical thickness $n_o d$ by nonlinear fit, the incremental phase change at normal incidence needs to satisfy the condition, dictated by the Nyquist sampling theorem, $\Delta\phi(\theta = 0) \geq \pi$, which poses a limitation on the minimum thickness of the plate to be characterized as well as on the wavelength-scanning range $\Delta\lambda_{\max}$. Thickness d has to be grater than $d_{\min} = \lambda^2/(4n_o\Delta\lambda_{\max})$. In the example show $\Delta\lambda_{\max} = 0.76$ nm and $d_{\min} = 220$ μm and the scanning range is small enough that for LiNbO$_3$ the index change due to dispersion $dn_o/d\lambda = -3.4 \times 10^{-5}$ nm^{-1} can safely be assumed to be negligible.

7.3 Visualization and *In-Situ* Monitoring of Domains Formation

Ferroelectric crystals, such as lithium niobate (LN) and other ferroelectric crystals, have many applications in photonic related fields. In fact due to their special

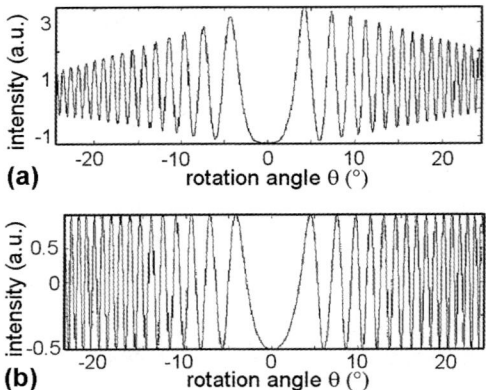

Fig. 7.5. (a) Experimental interference signal vs. rotation angle of the plate and (b) corresponding cosine of the retrieved phase at fixed wavelength $\lambda = 1535.68$ nm

nonlinear and electro-optic properties, fabrication of periodically poled crystals for nonlinear frequency generation by quasi-phase-matching (QPM) [27–30] is possible. The ability to micro-engineering ferroelectric domains is central to all of these applications, and techniques for visualizing domain structure and dynamics are important for the characterization of ferroelectric materials and devices. For example, the fundamental prerequisite for using the QPM technique, in nonlinear devices, is the ability to fabricate crystals with periodic domain structures of good quality, typically with periods in the 3 μm to 30 μm range by electric field poling [31].

Various approaches have been developed to micro-engineering ferroelectric domains. The most common is based on electric field poling. Electric field poling is based on the application of high external electric field pulses to periodically reverse the spontaneous polarization using the *in-situ* monitoring method which measures the displacement current flowing in the external electrical circuit [32]. The major limitation of this technique is the minimal achievable period length, limited by electric field inhomogeneities. In fact, it is well known [33] that domain growth under the electrodes can be simplified into four main stages: domain nucleation at the electrode edges, domain tip propagation towards the opposite crystal face, the merging of the domains under the electrodes and the domain walls propagation away from the electrodes. A 50% duty cycle between regions of opposite domain orientation is desirable for QPM applications and the last stage makes it difficult to obtain.

Other approaches have been developed to apply electric poling without patterned electrodes based on scanning probe devices. For example it has been investigated microscale to nanoscale ferroelectric domain and surface engineering of a near-stoichiometric $LiNbO_3$ crystal by using scanning force microscopy, in single crystals $LiNbO_3$ fixed on metal substrates and polished to a 5 μm thickness. Patterns of inverted-domain structures were fabricated by scanning the samples with a conductive cantilever while applying voltages [34]. Submicrometer ferroelectric domains in bulk ferroelectrics were tailored with 1D and 2D configuration in $LiNbO_3$,

RbTiOPO$_4$, and RbTiOAsO$_4$. By a high voltage atomic force microscope [35]. By using only the tip of the SFM probe a so named "Calligraphic poling" was developed to generate domains structures in crystals that are up to 250 μm thick [36].

In all the mentioned poling methods, visualization of the domain growth and its subsequent morphological development, through the domain wall motion, would be desirable to control in real-time and *in-situ* the domain pattern fabrication. Unambiguous *ex-post* identification of domain structures is usually achieved by an invasive technique based on selective chemical etching [37]. Domain polarity is determined by transforming the reversed domain pattern into a topographic structure on the crystal surface, observable by an optical microscope. This technique provides high resolution and comparatively ease of use but it is destructive and it is not able to give *in-situ* information about the structure of reversed domains.

Since one decade, investigation started to visualize and study the kinetics of domain formation in ferroelectric crystals. Imaging of the 180° domain walls in ferroelectric materials [38] at room temperature by ordinary light microscopes under polarized or unpolarized light has been reported. A variety of monitoring techniques have been developed. At the beginning much effort has been addressed for studying the domain kinetics by partly *ex-post* techniques. Constant pulses were applied to the crystal to nucleate and grow domains and the evolution process was separately monitored, after every pulse, by optical observation under a light microscope [39, 40]. The electric field application and the observation steps were separate. Afterwards, other monitoring imaging techniques have been developed, such as confocal frequency doubling [39], near-field optical microscopy [40], atomic force microscopy [41], interferometry [42] and electro-optic imaging microscopy [43–46]. Real-time observation of domain formation was also be developed [47]. These techniques provide amplitude-contrast images of the ferroelectric domain pattern during formation, giving information about nucleation and growth kinetics of the reversed domains. Recently, illumination with coherent light along the z axis of electro-optic crystals has been demonstrated to contain information about the domain structures. It comes from an analysis of the near field diffraction pattern generated by waves travelling through domains of opposite direction and phase shifted by the combined electro-optic and piezoelectric effects in lithium Niobate (LiNbO$_3$) and lithium tantalate (LiTaO$_3$) [48–50].

In this section will be described the interferometric methods that allows *in-situ* and quasi-realtime investigation of electric poling in LiNbO$_3$ and KTiOPO$_4$ crystals. Firstly, it will be introduced a description of a flexible imaging technique named Digital Holography that allows quantitative full filed analysis of the poling process. Then various experimental methods will be described and results illustrated to show how the interferometric visualization can aid to investigate kinetics of the poling process with various spatial and temporal resolutions. Results will be reported about the investigation of so named *free poling*, and poling with spatial constrain due to the presence of photoresist pattern on the surface of the crystals.

Moreover it will be shown approaches, even using interferometric techniques, to allow some kind of characterization of ferroelectric crystal such as: study of internal field, mapping of electro-optics properties of wafers. Furthermore investigation of the optical birefringence near the wall between two opposite domains by means of a full field polarimetric technique will be illustrated. Finally measurement of induced photorefractive grating in a Fe^+ doped $LiNbO_3$ will be reported.

7.3.1 Digital Holography and Experimental Configuration for *In-Situ* Investigation of Poling

We describe here digital holography (DH) technique [51–53] for *in-situ* interferometric analysis of domain reversal process in congruent LN providing quantitative spatially resolved phase shift data. The phase shift distribution is generated simultaneously by linear electro-optic and piezoelectric effect along the z crystal axis. Such phase shift distribution is numerically reconstructed by the DH method and it is denoted here as the phase-map of the sample. Differently from the conventional phase-contrast imaging performed with a microscope, in which an optical mechanism is used to translate phase variations into corresponding amplitude changes, the technique presented here provides direct measurement of the phase shift distribution of the object wavefield.

DH is a quantitative imaging method in which the hologram resulting from the interference between the reference and the object complex fields, $w_1(x, y)$ and $w_2(x, y)$ respectively, is recorded with a CCD camera and numerically reconstructed. The hologram is multiplied by the reference wavefield in the hologram plane, namely the CCD plane, to calculate the diffraction pattern in the image plane. The reconstructed field $\Gamma(\nu, \mu)$ in the image plane, namely the plane of the object, at a distance d from the CCD plane, is obtained by using the Fresnel approximation of the Rayleigh-Sommerfeld diffraction formula

$$\Gamma(\nu, \mu) \propto \iint h(\xi, \eta) r(\xi, \eta) \exp\left[\frac{i\pi}{\lambda d}(\xi^2 \cos\alpha^2 + \eta^2)\right]$$
$$\times \exp[-2i\pi(\xi\nu + \eta\mu)] \, d\xi \, d\eta \qquad (7.5)$$

where $r(\xi, \eta) \equiv w_1(\xi, \eta)$ is the reference wave which, in the case of a plane wave, is simply given by a constant value, $h(\xi, \eta) = |w_1(\xi, \eta) + w_2(\xi, \eta)|^2$ is the hologram function, λ is the laser source wavelength, the factor $\cos\alpha$, with $\alpha \cong 49°$, is introduced for correcting the anamorphism effect [52] and d is the reconstruction distance, namely the distance measured between the object and the CCD plane along the beam path. The coordinates (ν, μ) are related to the image plane coordinates (x', y') by $\nu = x'/\lambda d$ and $\mu = y'/\lambda d$. The reconstructed field $\Gamma(\nu, \mu)$ is obtained by the Fast Fourier Transform (FFT) algorithm applied to the hologram $h(\xi, \eta)$ multiplied by the reference wave $r(\xi, \eta)$ and the chirp function $\exp[(i\pi/\lambda d)(\xi^2 + \eta^2)]$. The discrete finite form of (7.2) is obtained through the pixel size $(\Delta x', \Delta y')$ of the CCD array [20], which is different from that $(\Delta \xi, \Delta \eta)$ in the image plane and related as follows:

$$\Delta x' = \frac{\lambda d}{N \Delta \xi}; \qquad \Delta y' = \frac{\lambda d}{N \Delta \eta} \qquad (7.6)$$

where N is the pixel number of the CCD array. The wavefields impinging on the CCD surface are digitized at 8-bit rate and stored as numerical array data corresponding to the hologram patterns which are processed by Matlab program according to the discrete finite form of (7.5). The great advantage of this technique is the possibility to numerically reconstruct the complex field of the object beam. The two-dimensional amplitude $A(x', y')$ and phase $\phi(x', y')$ distributions of the object wavefield can be re-imaged by using one hologram acquisition and performing simple calculations on the object wavefield $\Gamma(\nu, \mu)$ reconstructed from the numerical solution of diffraction equations:

$$A(x', y') = \text{abs}[\Gamma(x', y')]; \qquad \phi(x', y') = \arctan \frac{\text{Im}[\Gamma(x', y')]}{\text{Re}[\Gamma(x', y')]}. \qquad (7.7)$$

The DH technique is a useful tool for microscopic visualization and optical analysis of ferroelectric domain boundaries. In fact, conventional microscope imaging suffers from small depths of focus [25], because of the high numerical apertures of the lenses and the high magnification ratios, so that mechanical motion along the optical axis is required to check the focus of the image. In contrast, DH allows 3D simultaneous calculation of the complex wave front, in amplitude and phase, by numerically solving diffraction equations. In comparison with standard interference techniques, which also provide amplitude and phase map imaging, DH provides many advantages. Recording of only one hologram is necessary while recording of four or more interferograms is required in phase-shifting interferometry, such that the required stability for the acquisition of multiple images is very much relaxed and, in addition, dynamic events can be recorded. Numerical reconstruction of the back propagated beam does not suffer for spreading of diffraction since the amplitude and the phase of the object beam are reconstructed at right focusing distance. This allows a correct exact correlation between the spatial distribution of the phase-map and the area of the sample under investigation. Furthermore, DH provides the possibility to digitally correct the aberrations due to the objective lens.

Different optical configuration can be used for applying DH as non-destructive methods and monitoring technique. Essentially we describe here two different configurations: a reflective grating interferometer (RGI) and a Mach–Zehnder.

RGI Configuration for Relatively Low Spatial and Temporal Resolution

A RGI set-up generates an interference fringe pattern obtained by the recombination of a reference wavefield with a wavefield transmitted by the sample during the poling process [53]. While an external electric field that reverses the spontaneous polarization of the material is applied, the fringe pattern changes due to electro-optic and piezoelectric effects. A charge-coupled device (CCD) continuously records the evolving fringe pattern. The digital holograms acquired during the poling process are used to numerically reconstruct both the amplitude and the phase of the wavefield transmitted in quasi real-time by the sample. Sequences of amplitude- and phase-maps of the domain walls moving under the effect of the external electric

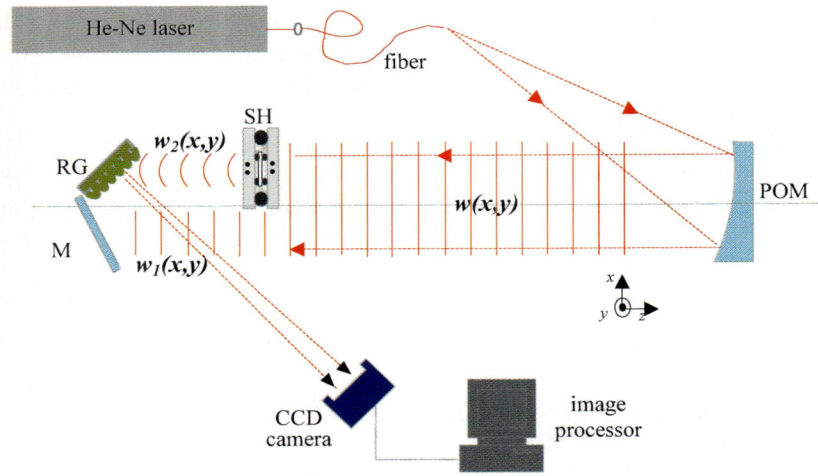

Fig. 7.6. Schematic view of the RGI set-up. The laser source is a He–Ne laser emitting at $\lambda = 632.8$ nm. POM – parabolic off-axis mirror; M – mirror; RG – reflective grating; SH – sample holder

field are obtained and collected into movies. Quasi real-time *in-situ* qualitative information about the spatial evolution of the reversed domain pattern is given by the reconstructed amplitude-map movie, while quantitative information is provided by the phase-map movie.

The poling system is positioned into the RGI set-up as shown in Fig. 7.6. It is made of two components, a mirror and a reflective grating (RG) with 1200 lines/mm and (44×44) mm sized. A He–Ne laser emitting at 632.8 nm is launched into a single mode optical fiber and expanded to a spherical wave which impinges onto the off-axis parabolic mirror (POM) to obtain a plane wavefront beam. The collimated wavefront $w(x, y)$ is spatially divided in two half wavefronts $w_1(x, y)$ and $w_2(x, y)$ by the mirror and the grating while the wavefront $w_1(x, y)$ is reflected onto the grating by the mirror. The angle of incidence of the two half wavefronts on the grating is such that both of them are diffracted along the normal to the reflective grating. The wavefront $w_1(x, y)$ is folded on the other wavefront $w_2(x, y)$ and interferes with it giving place to a non-localized fringe pattern in front of the grating. The fringe pattern is digitized by a CCD camera with 512×512 square pixels 11.7 μm sized. The field of view captured by the CCD array is around (5×5) mm sized.

The poling set-up is positioned in front of the reflective grating RG before the recombination occurs, so that the half wavefront $w_2(x, y)$ is that transmitted by the crystal. A careful alignment is necessary to position the wave vector of the incident light parallel to the z crystal axis and the laser beam is linearly polarized with the polarization direction along the x-axis of the crystal.

Congruently melting z-cut LN samples $(25 \times 25 \times 0.5)$ mm 3-*inch* diameter crystal wafers polished on both sides were used. Figure 7.7(a) shows the sample holder

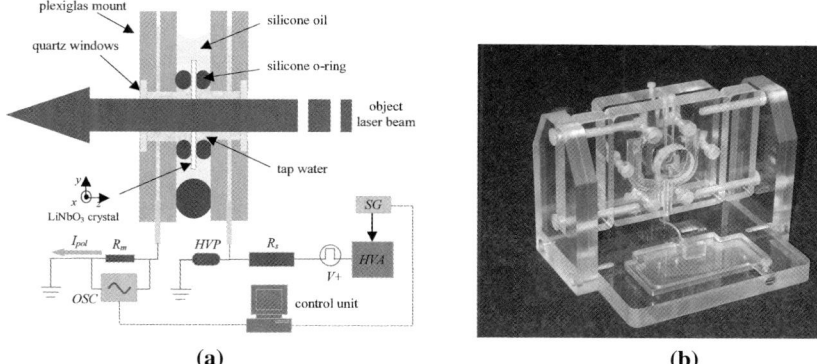

Fig. 7.7. (a) Sample holder and the electrical circuit used for domain inversion. SG – signal generator; HVA – high voltage amplifier (2000×); R_s – series resistor (100 MΩ) for current limitation; R_m – monitoring resistor (10 kΩ); HVP – high voltage probe; OSC – oscilloscope. (b) Picture of the sample holder. The electrical circuit is integrated into the Plexiglas mount

and the electric circuit used for the poling process. Figure 7.7(b) shows a picture of the sample holder. The structure of the sample holder is inspired by that used by Wengler et al. [42]. Domain inversion is achieved by applying one positive high voltage, slightly exceeding the coercive field of the material (~21 kV/mm for LN), to the $z+$ face of the crystal sample. Electrical contact on the sample surfaces is obtained by a liquid electrode configuration consisting of two electrolyte containing chambers which squeeze the sample between two O-ring gaskets. Tap water is used as liquid electrolyte. This configuration insures both the uniformity of the applied electric field within the sample and transparency along the z direction, which allows illumination of the sample through the quartz windows during the poling process.

A Signal Generator (SG) drives an High Voltage Amplifier (HVA – 2000×) to deliver +12 kV and a series resistor $R_s = 100$ MΩ is used to limit the current flowing in the circuit. Because the applied voltage exceeds the coercive field of LN, a displacement current I_{pol} flows in the external circuit due to the charge redistribution within the crystal. This current is measured through the voltage drop across the resistor R_m by the oscilloscope OSC. A computer-controlled technique is used [31] to apply the voltage for a duration T such that $T = Q/I_{pol} = 2P_s A/I_{pol}$ where Q is the delivered electric charge, P_s the spontaneous polarization of the crystal and A the area to be polarization reversed.

For congruent LN samples, spontaneous polarization is reversed by application of a single external electric field pulse exceeding the coercive field of the crystal. The interferometric technique provides a full-field simultaneous information about the optical amplitude and phase deformation induced by domain reversal process during poling. This technique is non-invasive tool for *in-situ* monitoring of the electric field periodic poling in ferroelectric materials. By means of the technique the behaviour of independently growing and merging domains can be studied in order to investigate interaction between domains which arise from independent nucleation events.

In view of the fact that LN is an electro-optic material, the refractive index n increases from n to $n + \Delta n$ under a uniform external electric field, in one domain, while in the oppositely oriented one it decreases from n to $n - \Delta n$, thus providing an refractive index contrast across the domain wall [55]. The refractive-index change, due to the linear electro-optic effect along the z crystal axis, depends on the domain orientation according to $\Delta n \propto r_{13} E_3$, where E_3 is the external electric field parallel to the z crystal axis. The index difference across a domain wall is equal to $2\Delta n$ and causes phase retardation of the transmitted beam in the crystal. This effect is widely used for *in-situ* electro-optic imaging of domain reversal in ferroelectric materials [55], through the crystal thickness, without any polarizers, thus avoiding the *ex-situ* invasive chemical etching process. The investigation of the optical birefringence near the wall between two opposite domains optical phase retardation of a plane wave passing through the domain is also affected by the piezoelectric effect, which induces a negative or positive sample thickness variation Δd in reversed domain regions. Therefore, during the electric field poling, an incident plane wave experiences a phase shift $\Delta \phi$, mainly due to the linear electro-optic and piezoelectric effects along the z crystal axis, according to

$$\Delta\phi = 2\left[\frac{2\pi}{\lambda}\Delta n d + \frac{2\pi}{\lambda}(n_0 - n_w)\Delta d\right] = \left[-r_{13}n_0^3 + 2(n_0 - n_w)k_3\right]U \quad (7.8)$$

where thickness change Δd due to the piezoelectric effect is dependent on k_3, the ratio between the linear piezoelectric and the stiffness tensor ($k_3 = 7.57 \times 10^{-12}$ m/V), $n_w = 1.33$ is the refractive index of water and U is the applied voltage.

In Fig. 7.8(a) is shown an example of the *in-situ* interferogram recorded by the CCD during the electric field poling of a LN z-cut sample. Video frames can be recorded with a temporal resolution of 10 frame/s. The $z+$ crystal face, has a total electrode area about (5 × 5) mm sized. The polarization axis is normal to the image and the field of view is slightly larger than the electrode area. The diffraction pattern of the photoresist square window is visible. The out of focus diffraction image of two principal domain walls which grows from the upper and lower regions is clearly superimposed on the interference fringe pattern. They move towards the center until the spontaneous polarization of the crystal under the electrode area is reversed. The visibility of the domain walls under transmitted light are determined by the opposite phase shift induced on the reversed domains by the linear electro-optic and piezoelectric effects along the z-axis.

As described before, DH technique is applied to all of the recorded interferograms for reconstructing the amplitude and the phase-maps of the domain pattern during the electric field poling process. The domain switching of the crystal took about 5 s. Figure 7.8(b) shows a frame from a movie obtained by collecting the two-dimensional distribution of the numerically reconstructed object wavefield amplitude from one interferogram recorded during the poling process.

7 Visual and Quantitative Characterization 179

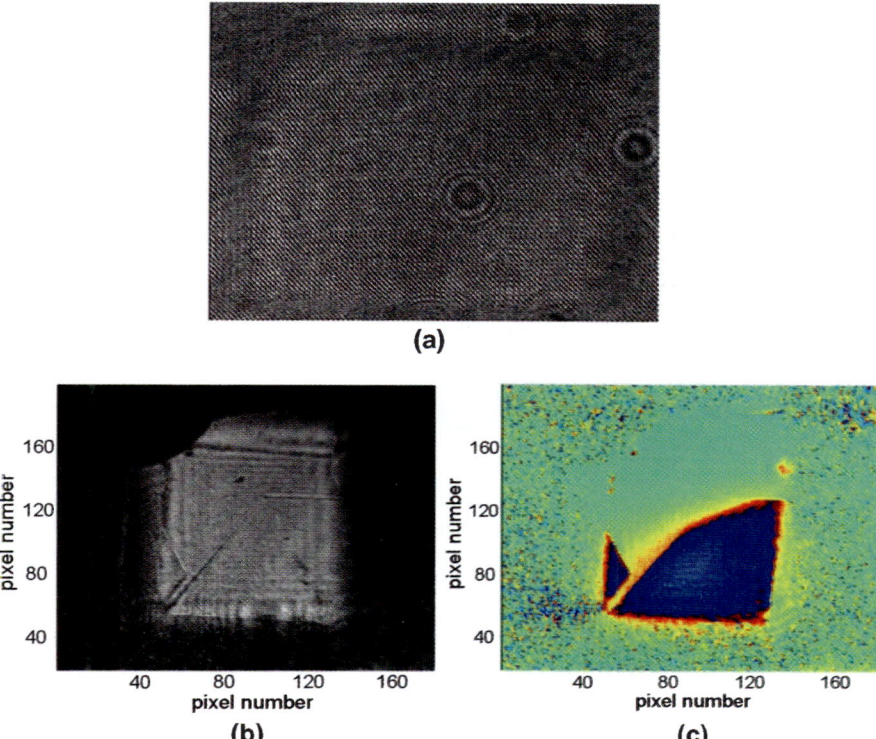

Fig. 7.8. (**a**) Interferograms recorded during the poling process. (**b**) Two-dimensional distribution of amplitude and (**b**) wrapped phase numerically reconstructed from the hologram in (**a**) recorded during the poling process. The reconstruction distance is $d = 540$ mm

It is clearly visible that domain nucleation preferentially starts at the edges of the electrode area. Then, domains merge to form a few dominant domain wall fronts. The lateral resolution achieved in these images is around 55 µm, according to (7.6). The resolution can be easily improved by using a smaller distance between the sample and the CCD array. In fact, the reconstruction pixel in the image plane is lower for smaller reconstruction distances. Without any magnification, about 11.7 µm (CCD pixel size) spatial resolution can be achieved by using the so-called convolution numerical procedure for resolving the diffraction problem. Otherwise, by using a DH set-up in a microscope configuration, it is possible in principle to obtain lateral resolutions down to 0.5 µm.

A reference interferogram of the sample at its starting virgin state can be recorded when no voltage is applied. Optical phase shift experienced by the object wavefield during poling can be calculated with respect to this by (7.7). In fact the holographic reconstruction can be performed for both the reference hologram and the nth hologram recorded during the domain switching, to obtain the corresponding

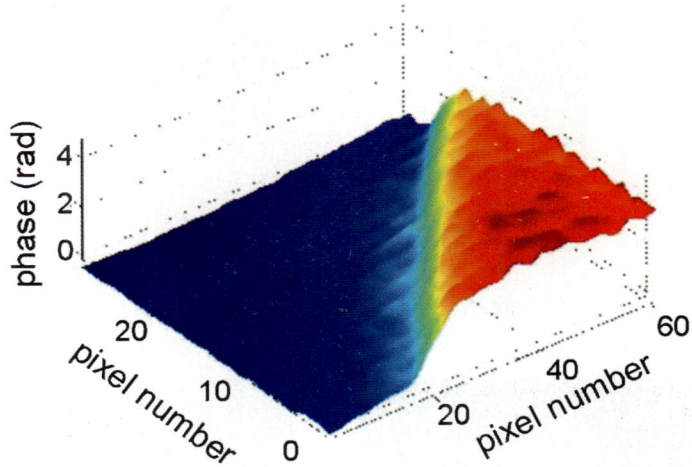

Fig. 7.9. Surface plot representations of the unwrapped phase shift distributions retrieved from the inner (60 × 60) pixel sized region of the phase map in Fig. 7.8(c)

object wavefield phase distributions $\phi_0(x', y')$ and $\phi_n(x', y')$. The two-dimensional map of the phase shift $\Delta\phi(x', y') = \phi_n(x', y') - \phi_0(x', y')$ is calculated for each hologram and the corresponding images are collected into the movie presented in Fig. 7.8(c). In Fig. 7.8(b) and (c) the reconstruction distance is $d = 540$ mm. Both the out of focus real image and the zero-order diffraction term, generated by the DH numerical procedure, are filtered out for clarity. Moreover, the anamorphism effect, visible in Fig. 7.8(a) by the deformed shape of the square resist window, was numerically corrected in the reconstruction process by inserting the $\cos\alpha$ factor in the diffraction integral of (7.5). A surface plot representation of the unwrapped phase-map is performed for the central region of the more representative frames, to reveal the profile of the phase shift occurring at domain boundaries. Figure 7.9 shows the resulting movie, which shows the evolution of the object wavefield phase-map profile during the poling process. The mean value of the phase-shift occurring across the domain boundary is calculated to be around 3.6 rad. Such value is in good agreement with the theoretical value calculated by (7.1) ($n_o = 2.29$; $n_w = 1.33$; $r_{13} = 10$ pm/V; $k_3 = 7.57$ pm/V) apart from a modulo 2π shift. Mean value and variance of the two-dimensional distribution of both the amplitude $A_{ref}(x', y')$ and the phase $\Delta\phi_{ref}(x', y')$ reconstructed for the reference hologram were calculated to estimate the accuracy of the method. The mean value of the reference amplitude is (4.7×10^{-4}) a.u. with a variance of 0.09 a.u., while the mean value of the reference phase is 0.82 rad with a variance of 0.11 rad corresponding to about $\lambda/60$.

The *in-situ* reconstruction of the object wavefield phase distribution is useful essentially for two reasons: it gives information about the local orientation of the ferroelectric axis by means of the phase shift profile and it provides quantitative

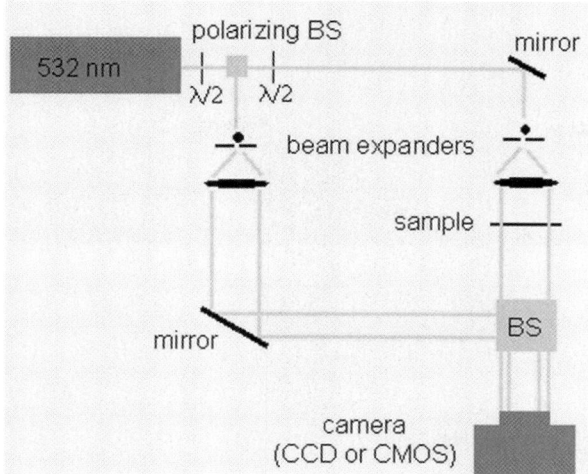

Fig. 7.10. Schematic view of the MZ set-up. The laser source is a frequency doubled Nd:YAG laser emitting at 532 nm. PBS – polarizing beam splitter; $\lambda/2$ half-wavelength plate; BE – beam expander; BS – beam splitter

quasi real-time evaluation of the amount of optical path difference in adjacent reversed domains.

M–Z Configuration for Higher Spatial and/or Temporal Resolution

A different optical configuration was adopted by using a M–Z interferometer to get higher spatial resolution. In fact by means of a MZ it is possible to record and reconstruct at distances much shorter in respect to the RGI arrangement, having consequently a lower reconstruction pixel for the numerical holographic reconstruction. Schematic drawing of the Mach–Zehnder configuration is shown in Fig. 7.10. Furthermore to address the need of having a better temporal resolution about the electric poling two diverse experimental conditions will be illustrated in following. In case denoted by letter "S" the LN sample is subjected to *slow poling* process by using high series resistor in the external circuit (100 MΩ). The whole area under investigation (diameter 5 mm) is reversed in less than 10 s and the interferograms are acquired by the CCD camera. In case indicated by the letter "F" another virgin LN crystal sample is reversed by *fast poling* (series resistor 5 MΩ) in order to reverse the whole crystal area in less than 1 s and the interferograms are recorded by a CMOS camera. Amplitude and phase maps of the object wavefield are numerically reconstructed by DH method as described in the previous section. The reconstruction distance is $d = 125$ mm in case S and 180 mm in case F, while the lateral resolution obtained in the reconstructed amplitude and phase images is 9.7 μm in case S and 16 μm in case F, according to (7.3). A reference interferogram of the sample at its initial virgin state is digitized before applying the external voltage

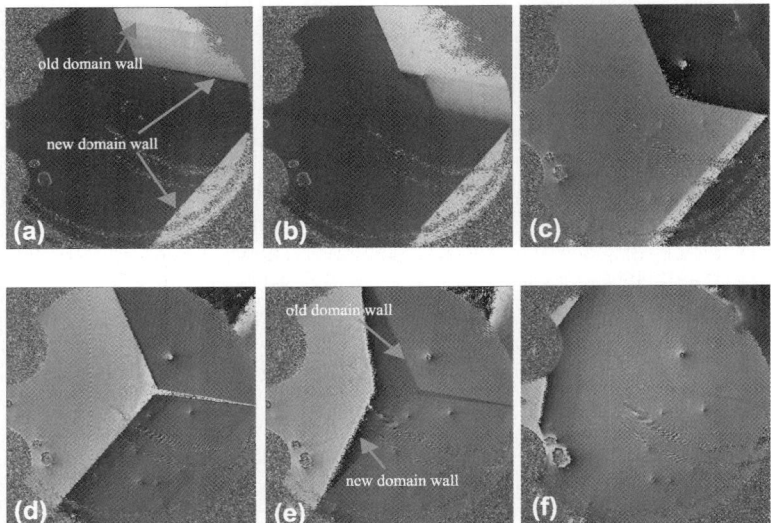

Fig. 7.11. Selected frames from the phase-map movie obtained in case con *slow poling* (S). The frame area is $(5 \times 5)\,\text{mm}^2$ and the time t (in seconds) corresponding to each frame is (**a**) 4.2, (**b**) 4.6, (**c**) 5.0, (**d**) 6.3, (**e**) 6.7, (**f**) 7.9. The polarization axis is normal to the image plane

and it is used to calculate the phase shift experienced by the object wavefield during poling. The DH reconstruction is performed for both the reference hologram and the nth hologram, recorded during the domain switching, to obtain the corresponding phase distributions $\phi_0(x', y')$ and $\phi_n(x', y')$. The 2D phase shift map $\Delta\phi(x', y') = \phi_n(x', y') - \phi_0(x', y')$ is calculated for each hologram and the corresponding images are collected into a movie. Figure 7.11–7.12 show some of the frames extracted from such movie in case S and F, respectively. The out of focus real image term, generated by the DH numerical procedure [53, 54] is filtered out for clarity. The in-focus image of the domain wall propagating during the application of the external voltage is clearly visible.

Switching process always starts with nucleation at the electrode edges. It is interesting to note that a residual phase shift gradient is present in correspondence of previously formed domain walls, as indicated in the Figs. 7.11–7.12. This is probably due to the decay effect of the internal field related to the polarization hysteresis in ferroelectric crystals. It is important to note that crystal defects and disuniformities are clearly visible and readily detectable by observing Figs. 7.11–7.12, due to their different EO behaviour. Moreover in Fig. 7.12 the high temporally resolved frames obtained by using the CMOS camera allow to notice that the evolution of the domain walls is clearly influenced by the crystal defects where the domain wall propagation appears to be partially blocked.

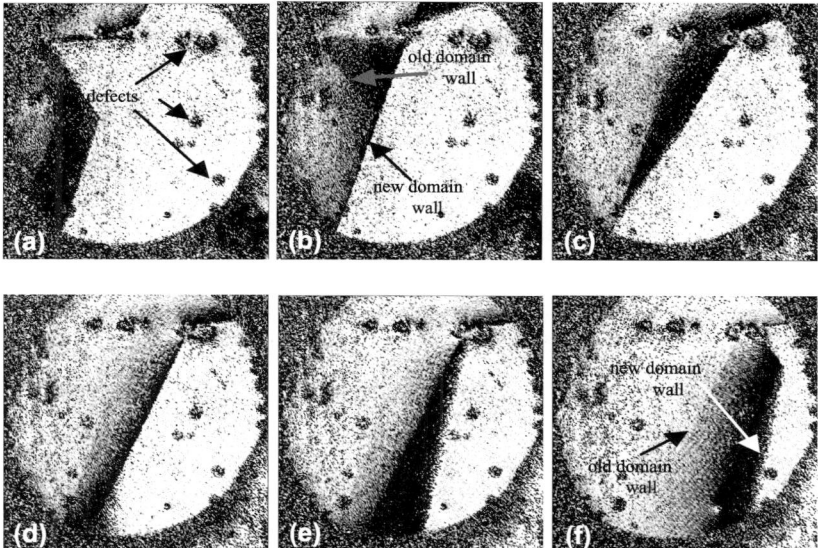

Fig. 7.12. Selected frames from the phase-map movie obtained in case of *fast poling* (F). The frame area is $(5 \times 5)\,\text{mm}^2$ and the time t (in milliseconds) corresponding to each frame is (**a**) 390, (**b**) 430, (**c**) 470, (**d**) 510, (**e**) 530, (**f**) 580. The polarization axis is normal to the image plane

In-situ Investigation of Periodic Poling Into Resist Patterned LiNbO$_3$ Samples

The dynamic evolution of the periodic poling into resist patterned LiNbO$_3$ samples is also very important to investigate. It is described here how proceed with this aim by measuring the variation of the optical phase shift distribution across domain walls. Online quantitative and high contrast imaging can be obtained through an interference microscope based on digital holography. Phase map movies of the evolving domains are presented and discussed. The method provides remarkable online information about the poling of ferroelectric crystals in case of insulating constraints, under different voltage and resist conditions, providing a deeper understanding of the electric field periodic poling.

Congruent z-cut LN samples $(20 \times 20 \times 0.5)\,\text{mm}^3$ polished on both sides were used. Samples were spin coated with 1.3 μm thick photoresist layer (Shipley Microposit S1813-J2) and then subject to standard mask photolithography to realize different periodic resist gratings: (a) square lattice of round resist openings (sample "A"); (b) square lattice of hexagonal resist openings (sample "B"). The samples were always baked after resist development to harden the resist before poling. Figure 7.13 shows the schematic view of the two different resist patterns.

The pitch is 200 μm along both x and y directions. The samples were positioned in the usual special Plexiglas holder previously described. The liquid electrolyte was tap water and external circuit was the same described in the previous sections. The

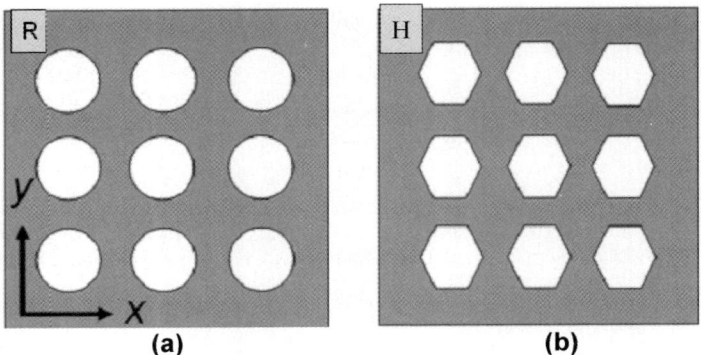

Fig. 7.13. Schematic views of the two resist patterns used in the experiment: (**a**) square array of round openings (R); (**b**) square array of hexagonal openings (E)

Table 7.1. List of the investigated LN samples

Sample	Voltage condition	Mask
A	Fast (forward)	R
B1	Slow (forward)	R
B2	Slow (backward)	R
C	Slow (forward)	H

EFP was performed under two different external voltage conditions corresponding to two poling regimes which are nemed here slow (S) and fast (F): (i) continuous voltage and 100 MΩ series resistor (F); (ii) train of 10 square pulses 300 ms long and 5 MΩ series resistor (F). The external voltage always exceeded the LN coercive field (21 kV/mm) in order to have ferroelectric domain switching in the areas without photoresist. In case of the train of square pulses, if necessary, more than one train was applied in order to complete the polarization switching in the electrode regions. Results of four experiments are summarized here, as reported in Table 7.1.

A MZ interferometer was adopted in this case with a frequency doubled Nd:YAG laser emitting at 532 nm. An optical microscope objective equivalent to 5X magnification was used to image the poling process onto the CCD camera. The reference beam and the spherical object beam were recombined by the beam splitter BS to obtain an interferogram pattern behind the BS. The fringe pattern was digitised by the CCD camera with (1024 × 1024) square pixels 7.4 μm sized and a frame rate of about 10 fps.

This work provides an important improvement compared to that reported in the previous section on RGI. The online imaging of the reversing domains is performed here in case of insulating constrains represented by a lithographic pattern, while in the previous work the domain evolution was investigated in case of crystal samples without any resist grating. Therefore, the experiments performed here provide new and remarkable information about the evolution of reversing domains in case of selective poling and thus useful for innovative visual and quantitative monitoring of periodic poling in ferroelectric crystals.

The phase shift across reversed regions is measured to be about 5 rad. This value, apart from the 2π ambiguity, is in accordance with the expectations. Two trains of square pulses were used to reverse the ferroelectric polarization in all of the electrode regions. The vertical image shift fro different frames (visible in the frame shown in Fig. 7.14(e)) is related to a slight sample displacement occurred between the two trains of voltage pulses. The frames in Fig. 7.14 clearly show the in-focus image of the domain walls evolving within the resist openings during the application of the external voltage. The reversing domains nucleate basically across the resist edges and subsequent domain walls are formed and evolve according to the hexagonal symmetry of the LN unit cell. The external voltage was removed when the domain walls did not move anymore according to the online diffraction image provided by the CCD camera. It is interesting to note that, without controlling the poling current as performed in usual periodic EFP, a uniform square array of hexagons was formed. The reversed domains are perfectly circumscribed around the resist openings, with sides parallel to the y crystal axis. Figure 7.15 shows the optical microscope image of sample A after poling and resist stripping.

Several more samples were poled under the very same conditions as for sample A, and a uniform square array of hexagons was always obtained, thus demonstrating the reliability of the fabrication process. The sample B was processed by slow poling, making use of continuous external voltage and high series resistor (100 MΩ). It was intentionally overpoled in order to investigate the phenomenon of domain wall merging. Moreover, the poling behaviour in case of backward poling was investigated online by preserving the resist grating. Figures 7.16 and 7.17 show some frames of the phase map movie reconstructed for the sample B, in case of the forward and backward poling, respectively. The periodic domain reversal was achieved in about 600 ms before the intentional overpoling was performed. Different information are provided by such online investigation. Firstly, a reliable fabrication of a square array of hexagons is achievable also in case of slow poling. Then, the resist defects act as nucleation sites and merge with the main domain walls only after they spread out of the electrode areas. The regions resulting from the merging of adjacent hexagonal domains appear as a sequence of dot-like un-reversed domains and are affected by remarkable stress; Finally, the backward poling allows to obtain the same array of hexagons, as in case of forward poling, with relatively good uniformity.

The results presented here provide information on the periodic poling behaviour in congruent LN crystal samples. The experiments show that spontaneous hexagonal domains are formed when using a square lattice of round resist openings by using train of pulses as well as by slow poling. Such resist geometry appears to favour the formation of domains with the hexagonal shape typical of the LN unit cell. The technique allows to visualize with high contrast the evolution of the reversing domains which appear to circumscribe perfectly the round openings. Moreover, the dynamic events related to the merging phenomenon of adjacent reversing domains was also

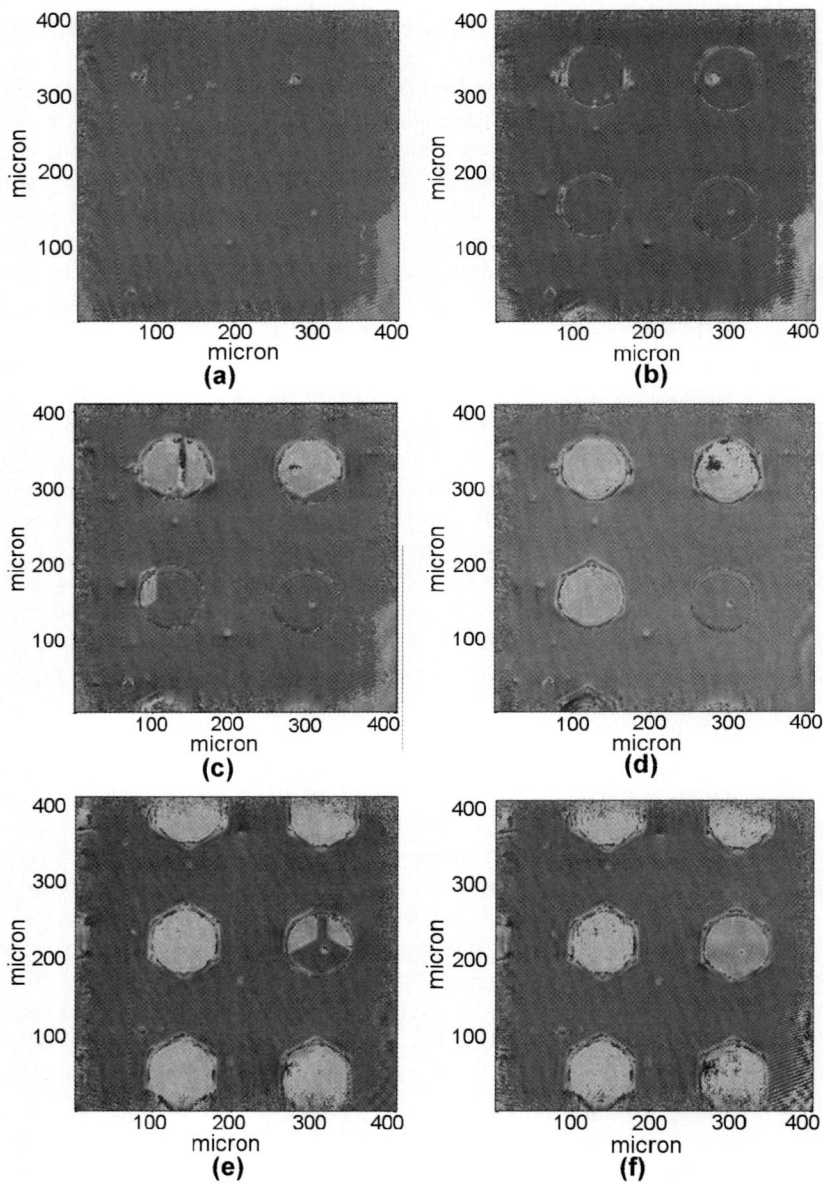

Fig. 7.14. Some frames of the phase map movie reconstructed for sample A during EFP

investigated showing the formation of remarkable stress across merging regions in addition to the dot-like regions with un-reversed polarization. The results also show the possibility to get uniform array of hexagonal domains even by backward poling.

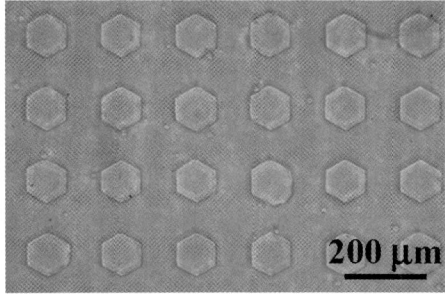

Fig. 7.15. Optical microscope image of sample A after EFP and resist stripping

It is important to note that investigation of the EFP in LN samples under insulating constraints provided by photoresist periodic patterns on the z-face of the crystal is possible. In fact, the wide field and relatively high spatial resolution of the phase shift movies allow the visualization and analysis of both the morphology and the kinetics of reversing domains under different poling regimes and photoresist conditions. The method allows to study the behaviour of the EFP in proximity of the photoresist boundaries, useful to understand deeply how to model the presence of the insulating layer and how to evaluate its effect on the EFP. The sample B was patterned with a square array of hexagonal resist openings (grating B) on the z-face and subject to slow poling. Since the samples A showed that the poling occurred preferentially according to hexagonal regions with sides parallel to the y crystal axis, the sides of the hexagons patterned in B were intentionally aligned to the x crystal axis (see Fig. 7.13) in order to investigate the poling behaviour in case of such geometry constraint. Figure 7.18 shows the phase map movie reconstructed for the forward poling of the sample H.

As for the other hard baked samples, the periodic domain reversal of the sample C occurred without appreciable spreading outside the electrode regions and took about 200 ms. Figure 7.19 shows the optical microscope image of the sample C after poling and resist stripping. It is important to note that the reversed domains exhibit a particular morphology resulting from the mismatch between the preferential hexagon orientation with sides parallel to the y axis and the resist openings geometry with sides parallel to the x axis.

It is noteworthy to observe the remarkable birefringence occurring all around the boundary of reversed domains. The regular, bright and fascinating patterns along the walls of reversed domains in Fig. 7.10(b) show the remarkable stress occurring in the crystal lattice as a consequence of the geometrical mismatch between the resist pattern and the spontaneous orientation of the LN hexagonal unit cell. The PPLN structures experience spatially periodic change in refractive index over extended areas that has never taken into account in non-linear QPM conversion processes. It is well known that such effect could be reduced by appropriate thermal annealing but not completely removed.

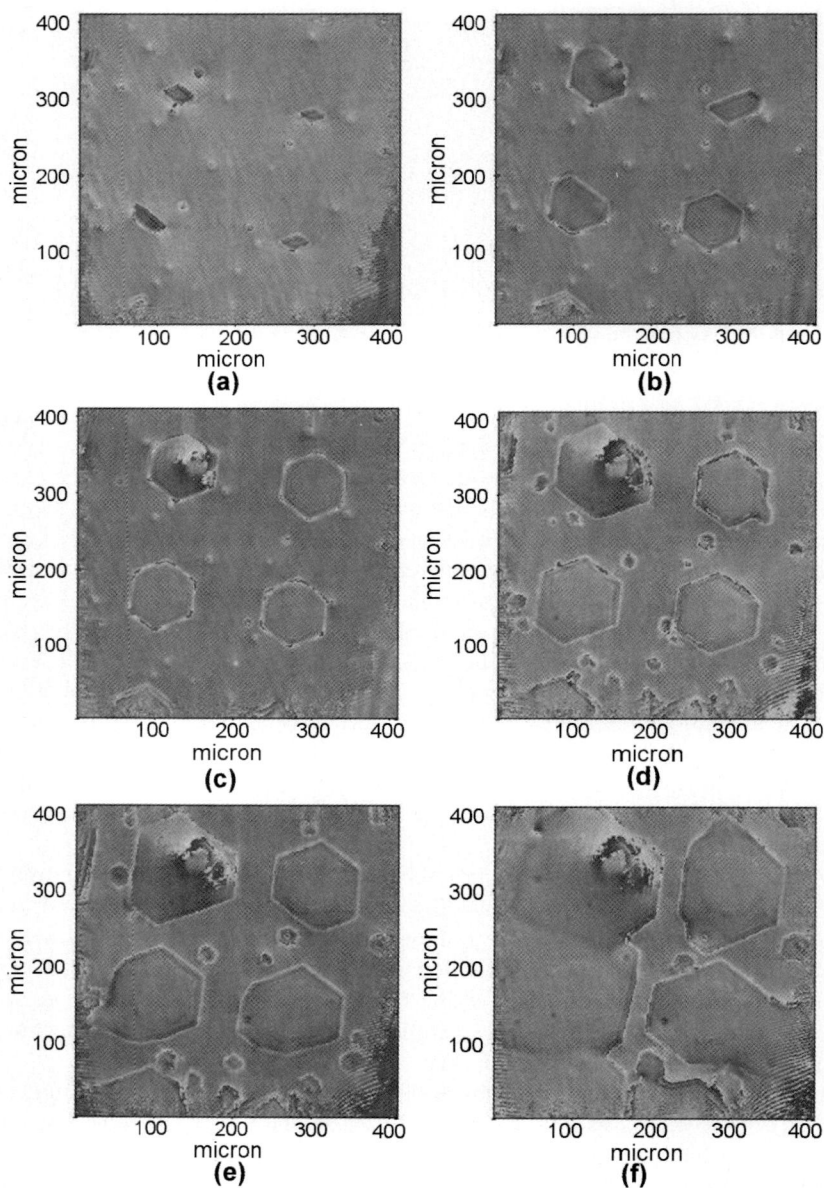

Fig. 7.16. Some frames of the phase map movie reconstructed for the sample B1 during forward EFP

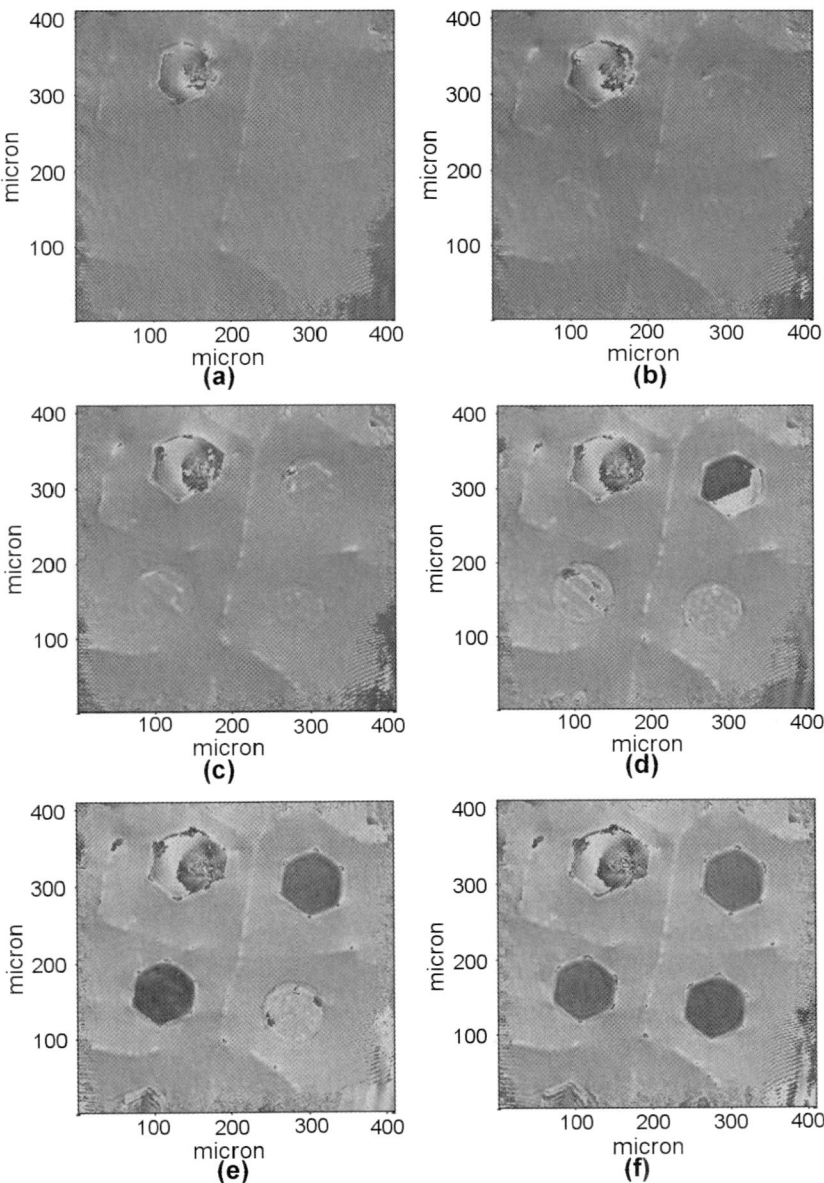

Fig. 7.17. Some frames of the phase map movie reconstructed for the sample B2 during backward EFP

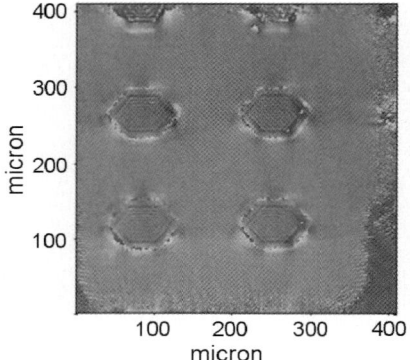

Fig. 7.18. Phase map for the sample C in forward EFP

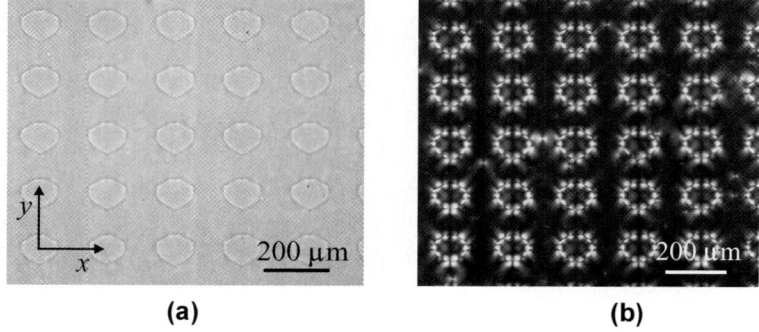

Fig. 7.19. Optical microscope image of the sample C after EFP and resist stripping (**a**) under bright field observation and (**b**) under dark field observation with crossed polarizers

Characterization of Fast Dynamic Evolution of Ferroelectric Domains in KTiOPO$_4$

Among ferroelectric crystals periodically poled KTiOPO$_4$ (PPKTP) are widely used in the field of nonlinear optics for quasi-phase-matched (QPM) frequency conversion processes [56]. Also for such material most of the investigation methods presented in literature are based on *ex-post* approaches, involving atomic force microscopy [57], polarity selective etching [58] or X-ray investigations [59]. An electro-optic (EO) based technique has been presented for *in-situ* visualization of periodic EFP in flux-grown KTP samples along the *b* crystal axis [60]. However, due to rather long optical path and associated large EO induced phase-shifts, this configuration does not allow unambiguous imaging of ferroelectric domain kinetics.

By DH based technique, also the fast dynamic evolution of ferroelectric domains during electric field poling in flux grown KTiOPO$_4$ crystals can monitored online. The dependence of the ferroelectric domain kinetics on the electric field temporal wave form and poling history was investigated. High-speed imaging by means of a complementary metal-oxide-semiconductor image sensor camera (CMOS) allowed *in-situ* visualization and measurement of the domain wall propagation speed under different poling conditions.

Flux grown KTP samples (20 × 20 × 0.5) mm³ sized have been obtained by dicing single domain 500 μm thick crystal wafers polished on both sides. The electrode area is about 5 mm in diameter. The samples have been mounted into the plexiglas holder to perform *in-situ* investigation of 180° ferroelectric domain reversal during EFP performed by applying high voltage pulses exceeding the coercive field of the material for KTP is about 2 kV/mm.

As usual the plexiglas holder is inserted into one arm of a MZ type interferometer as shown in Fig. 7.10 with a frequency doubled Nd:YAG laser emitting at 532 nm. The fringe pattern is digitised by a CMOS camera with (512 × 512) square pixels 12 μm sized. The field of view captured by the CMOS, slightly larger than (5 × 5) mm², allows to image the whole electrode area.

In case of KTP samples, $r_{13} = 9.5$ pm/V [6]; $n_0 = 1.78$; $d_{33} \sim 22 \times 10^{-12}$ m/V. Moreover d is the sample thickness, $n_w = 1.33$ is the refractive index of water and $U = 1.25$ kV is the external applied voltage. Although the pixel of the CMOS camera is (12×12) μm² sized, the lateral resolution of the reconstructed phase shift images obtained here (∼16 μm) is about three times better than in case of the RGI set-up in [55] (∼55 μm). This is basically due to the shorter reconstruction distances available in case of the MZ set-up and to the shorter source wavelength, thus providing smaller reconstruction pixel according to (7.2).

Domain growth and propagation in KTP under a steady-state field was studied. Virgin samples were subjected to 100 ms long square electrical pulses. If necessary, more than one pulse was applied until complete polarization reversal was obtained. Figure 7.20 shows sequences of selected frames of the reconstructed 2D phase shift distribution for the first forward poling under a steady-state field. The spontaneous polarization axis is normal to the image plane. The crystallographic axis and the time and length scales have been added to the images.

No significant domain growth is observed during the first 20 ms after the field is turned on. Then, wedge-shape domains start appearing at one side of the edge under the electrode. This edge of the sample corresponds to the low conductive edge. These wedge-shape domains propagate towards the center at an average speed of 4 μm/ms in the a direction and they broaden with approximately 12 μm/ms in the b direction during the next 20–22 ms. Their speed then drops to zero and no significant growth is observed until the voltage is turned off and then on again. Once the field is applied a second time, the domains continue to grow for 30 ms with approximately the same velocity, then the growth stops again until the field is turned off 10 ms after the beginning of the third pulse (frame 213 ms) a cluster of domains start to form in the middle of the sample. They primarily merge in the a direction, forming a serrate wall, and then advance in the b direction at more or less constant velocity of 75 μm/ms. This movement continues for 25 ms; then it drops to nearly zero. With the next pulse the domains continue to grow again. At the same time, some other

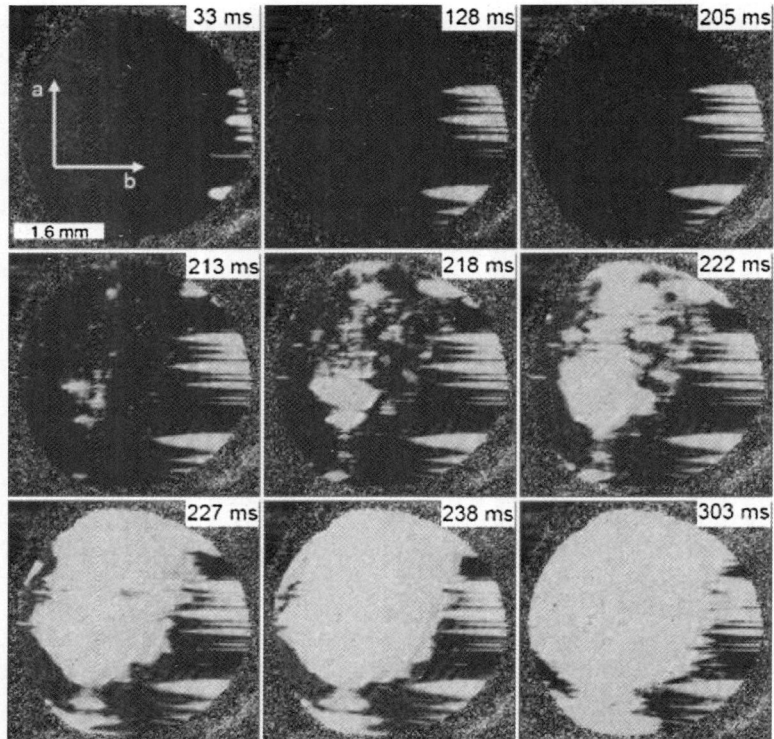

Fig. 7.20. Sequences of selected frames of the reconstructed two-dimensional map of the phase shift distribution for first forward poling for steady-state electric field. The light color represents the area of the growing inverted domains, while the darker color is the original domain. The time scale refers to the accumulated time that the sample has been subjected to the electric field, hence four pulses were applied in this case

domains emerge from the opposite edge of the electrode. Finally, the main serrated wall advances towards the wedge-shape domains until they merge together. The total switching time is 320 ms. If the sample is switched back and forth i.e., first reverse poling, second forward poling, etc. the switching times and domain wall speeds are more than one order of magnitude larger than for the first forward poling.

These results contribute to the understanding of domain switching under an applied electric field in KTP and can be used to find an optimized wave form and poling procedure for the fabrication of periodically poled samples of flux grown KTP.

7.3.2 Investigation of the Electro-Optic Effect and Internal Fields

In this section will be described experimental procedures to obtain a 2D mapping of the electro-optic coefficient and how to investigate the internal electric field (IF) by means of Digital Holographic methods. Both static and elastic components of IF are measured in a congruent sample and in an off-congruent one in order to investigate its origin. Finally, it is shown that IF derives from the presence of non-stoichiometric defects in the crystal lattice.

Study of a congruent lithium niobate crystal in two different conditions as been performe: in a virgin state, during a rising voltage ramp application, to measure the electro-optic (EO) coefficient, and after the ferroelectric domain inversion, with a rising voltage ramp below the coercive field value, to investigate the internal field behaviour.

In both cases, the sample is inserted in one arm of a RGI interferometer [55]. When an external electric field is applied across the sample, the fringe pattern changes due to electro-optic and piezoelectric effect. A CCD camera continuously records the evolving fringe pattern. Then the digital holograms are used to numerically reconstruct the wavefield phase and amplitude maps [53, 54].

Therefore DH analysis provides sequence of quantitative two dimensional spatially resolved phase maps. In this way two different analysis on LN crystals are performed.

Mapping Electro-Optic Coefficient

First of all a 2D mapping of EO coefficient has been performed. The sample is fixed into a transparent holder to permit minimal wave-front deformation and the application of an electric field via liquid electrodes. A reference digital hologram h_r with the sample in the interferometer arm, at zero voltage is recorded. Then, a sequence of digital holograms h_i (i from 1 to 30) are recorded while a linear voltage ramp ranging from 0 to 5 KV is applied across the sample, with a recording frame rate of 10 holograms per second. The phase retardation experienced by the object beam with respect to the reference beam, for a uniform virgin sample, is given by $\varphi_r = (2\pi/\lambda)(n_0 - 1)d + \Psi_0$ where λ is the laser wavelength, d is the sample thickness, n_0 is the sample ordinary refractive index and Ψ_0 is a constant phase due to the sample holder, liquid electrolyte and interferometer. With the applied voltage, the phase retardation variation $\Delta\phi_i$ occurred in the sample for each hologram h_i obtained by subtracting from the phase map ϕ_i of the current hologram the phase retardation ϕ_r of the reference hologram h_r. Therefore results

$$\Delta\varphi_i(x,y) = \varphi_i(x,y) - \varphi_r = \frac{2\pi}{\lambda}\left[\Delta n_i(x,y)d + (n_0 - n_W)\Delta d_i(x,y)\right]$$

$$= \frac{2\pi}{\lambda}\left[-\frac{1}{2}n_0^3 r_{13}(x,y) + (n_0 - n_W)d_{33}(x,y)\right] \times \Delta V_i \quad (7.9)$$

where $\Delta n_i(x,y)$ is the refractive index-change and $\Delta d_i(x,y)$ the piezoelectric thickness change due to the applied voltage ΔV_i while the hologram h_i is recorded; $r_{13}(x,y)$ is an electro-optic coefficient, $d_{33}(x,y)$ is a component of the piezoelectric strain tensor and n_W is the refractive index of water, used as liquid electrode. For each point (x,y) of the phase map a linear fit has been performed using (7.1).

Assuming that the errors on $d_{33}(x, y)$, V and indexes of refraction are small, the distribution of the linear eletro-optic coefficient $r_{13}(x, y)$ is determined in the reconstruction plane with a spatial resolution given by the width of the reconstruction pixel. A 2D map of $r_{13}(x, y)$ is shown in Fig. 7.1, concerning a congruent, uniformly 0.5 mm thick wafer crystal.

The sensitivity of this technique in measuring the EO coefficient an be appreciated by calculating the noise of phase map. To this aim is calculated the root mean square of a hologram recorded with no sample in the arm of our interferometer. The result is 0.07 rad which gives a sensitivity of 0.2 pm/V in the electro-optic coefficient measurement. In order to calculate the average r_{13} value, an uniform area of the sample has been chosen. The data inside the area that corresponded to the reconstructed phase map for all the digitized holograms were averaged, and linear best fits at incremental values of voltage were made. From the linear fitted data it is obtained $r_{13} = 10.0 \pm 0.1$ pm/V, in agreement with the value specified by the supplier of the sample.

From the phase map it is clear that EO phase retardation is not uniform over the sample shown in Fig. 7.1. In fact, the calculated value of r_{13} on a 4.5 mm × 31.8 mm area shows a rms value of 2 pm/V, corresponding to a spatial variation of as much as 20% with respect to the average value. This spatial distribution of the EO coefficient is due to the well known lack of uniformity in the crystal structure.

Investigation of the Internal Field

Further application is given to analyse an engineered crystal sample. Two opposite ferroelectric domain structure has been created by poling process explained in the previous paragraph.

This structure is simple to investigate and is widely used in electro-optical beam switching devices, where the boundary between the domain regions should be free from any residual index difference, which should affect the switching contrast ratio. In fact, after poling process when no electric field is applied, a residual index of refraction difference exists, due to the internal electric field (IF) of the crystal. An interferometric quantitative analysis of IF in absence of any external voltage, by measuring the optical path difference (OPD) in opposite ferroelectric domains in LN samples is performed. The phase step localized across the domain boundary is given by the formula

$$\Delta\phi = (2\pi/\lambda)\big(d\Delta n_0 + \Delta d(n_0 - 1)\big) = (2\pi/\lambda)\big(-r_{13}n_0^3 + 2d_{33}\big(n_0^{-1}\big)\big)d \cdot E_{\text{int}}.$$

The mean value of the phase map along the row of the data matrix at constant y of the in-focus image (framed area in Fig. 7.21 upper side) is calculated. The result is an averaged profile section [see Fig. 7.21(bottom)] with a step shape giving a phase difference 1.34 rad between opposite ferroelectric domains. After, the sample has been annealed at 240°C for 5 h and the measurement repeated. A lower difference in the opposite domain phase difference has been measured and shown in Fig. 7.22(b), but residual optical path difference is still visible. A second heating at 240°C for 4 h

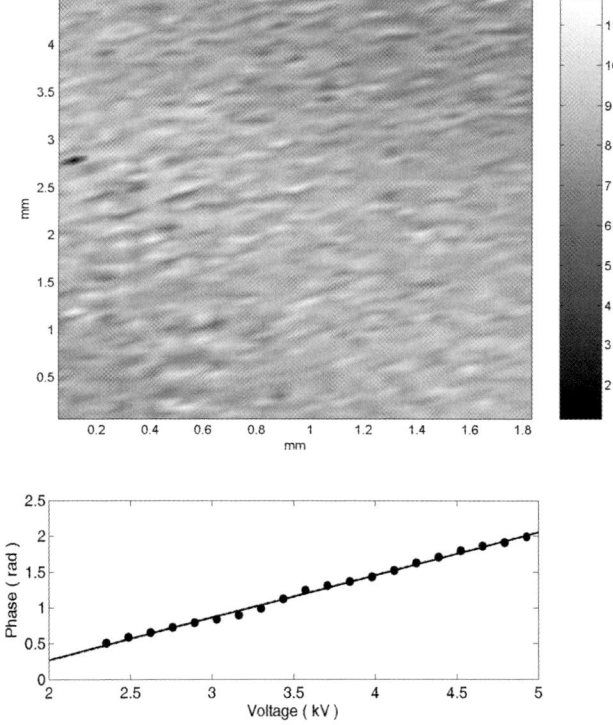

Fig. 7.21. 2D map of $r_{13}(x, y)$ of a 0.5 mm wafer of congruent lithium niobate. In the insert the plot shows the experimental points and the fitting line from which the r_{13} coefficient has been obtained by averaging the data inside the area corresponding to the reconstructed phase-map for all of the digitized holograms

has been performed and a residual phase difference is still visible between opposite domains. IF disappears completely after 16 months as shown in Fig. 7.22(d) where a zero phase difference is measured between opposite domains.

The magnitude of IF is defined by half the difference between the field required for forward poling E_f and that required for reverse poling E_r ($E_{\text{int}} = (E_f - E_r)/2$) and depends on the delay time between the two poling processes and on its temperature. In fact, as observed, IF value decreases with time and temperature by means of annealing process [61, 62].

It will be shown that the presence of IF affects also the EO response in antiparallel domains. In a partially poled sample, as shown in Fig. 7.23(a), (7.1) has to be changed to describe the different behaviour in the two opposite domain areas, as follows:

$$\Delta \varphi_{iA,B}(x, y) = \varphi_{iA,B}(x, y) - \varphi_{rA,B}$$

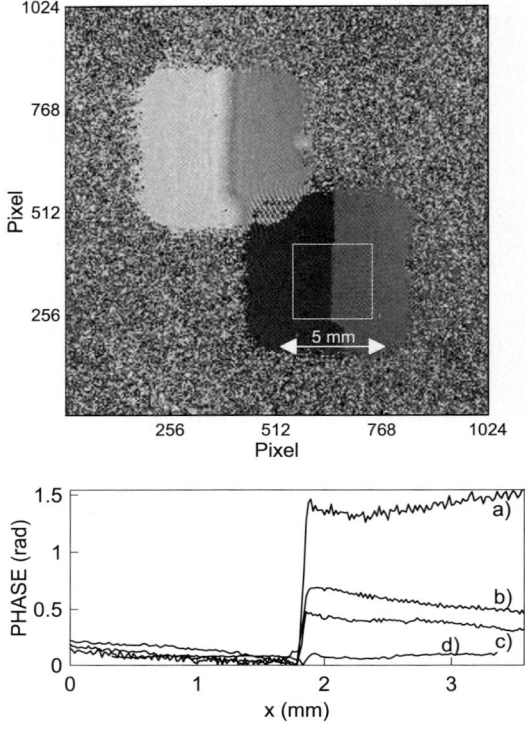

Fig. 7.22. (*Up*) Phase map showing the optical path difference between domains; (*down*) averaged profile section of the phase map before annealing (**a**), after 4 h annealing at 240°C (**b**), after second annealing at 240°C (**c**), after 16 months (**d**)

$$= \pm \frac{2\pi}{\lambda}\left(-\frac{1}{2}n_0^3 r_{13}(x,y) + (n_0 - n_W)d_{33}(x,y)\right)_{A,B} \times \Delta V_i \quad (7.10)$$

where A is the non-inverted area and B is the reversed domain area. The mean value of the phase maps in the two regions A and B, versus the applied voltage, is calculated. If phase retardation is measured just after electric poling, it ultimately has opposite sign and also different magnitude in the areas with opposite ferroelectric polarization, as shown in Fig. 7.23(b). The asymmetric behaviour exhibited by the two regions is clearly visible. The presence of this phase difference means that the virgin and the poled areas reply in a different way to the applied voltage.

By measured values results a relative phase difference of 16%. In contrast, if the sample is thermally annealed after poling (4 h at 500°C), the magnitude of phase retardation is the same in the two antiparallel domain areas (see Fig. 7.23(c)).

Regarding the physical interpretation of the different EO behaviour in a sample with two antiparallel domains before and after the thermal annealing, that probably is related to the presence of IF. On this subject, it important to remark that the asymmetry in the EO response between two antiparallel domain areas comes out

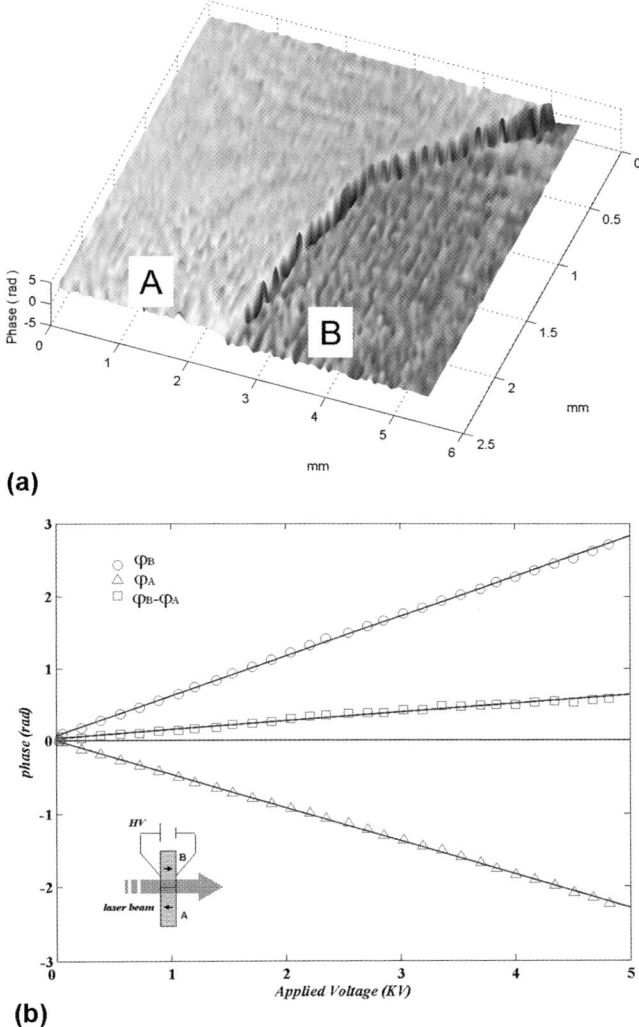

Fig. 7.23. Phase map reconstruction of the engineered crystal at a fixed applied voltage, A is the virgin area and B the reversed domain area (**a**); measured phase on $A(\triangle)$ and $B(\circ)$ areas as function of the applied voltage in a partially poled sample just after poling (**b**) and after thermal annealing (**c**). The algebraic sum ($\square$) of these phases is shown, too

only in samples that are not annealed, where IF is present, whereas the asymmetry disappears if the sample is thermally annealed, i.e., when IF has zero value. Several studies suggest that it originates from the non-stoichiometric defects associated with crystal growth from a congruent melt.

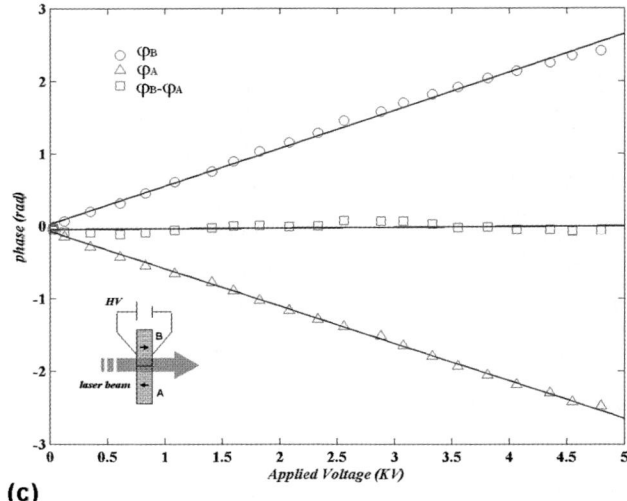

(c)

Fig. 7.23. *Continued*

According to a defect model supported by several experimental studies, a defect complex comprises a niobium antisite $(Nb_{Li})^{4+}$ and four lithium vacancies $(V_{Li})^{-}$. (See [62–64] and references therein.) The defect complex is associated with an electrical dipole moment which has two contributions: one is due to the electrical dipole arising only from the $(Nb_{Li})^{4+}$ antisite defect, the other one comes from the electrical dipole arising from the relative arrangement of lithium vacancies $(V_{Li})^{-}$ around niobium antisite defect $(Nb_{Li})^{4+}$. Before any external field is applied, defect dipoles are aligned, thus stabilizing the polarization state of the virgin crystal. After an electrical domain reversal, the two components of the defect polarization present a different behaviour, in fact the inversion of one component is faster than the other one, that requires a thermal annealing to occur [62, 65]. The stabilization time of ferroelectric inverted domains is related to the fast-switching defect dipole while the slow-switching defect dipole causes the asymmetry in the hysteresis loop, i.e. the IF.

According to defect model, the presence of defect-complexes causes local distortions of the structure deforming the lattice cell [66]. An analysis, made by means of X-rays, of structural distortions caused by the niobium antisite defect, reveals the presence of a "contraction" of the nearest three oxygen ions and a displacement from the Z axis of the nearest ^{93}Nb nuclei [67].

Therefore the electrical dipoles associated with defect complexes that give rise to conventionally known IF in LN (i.e. $E_{int} = (E_f - E_r)/2$ that was named name *static* IF) have elastic dipole components, too. Consequently, it has been supposed that the observed asymmetry of the EO behaviour of two anti-parallel domains, for a sample not annealed, depends on the elastic dipole components that give rise to a previously unknown component of the IF, that was named *elastic*.

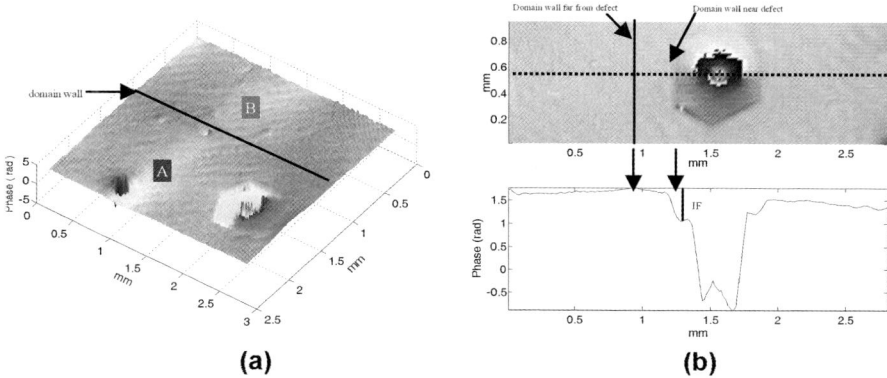

Fig. 7.24. (a) In the optical phase retardation map, a null phase step is noticeable across the domain boundary (*black line*) between the reversed (*B*) and un-reversed (*A*) regions whereas a phase step is visible in proximity of defects. (b) (*Upper portion*) Close-up of the phase map in which a hexagonally shaped reversed area is evident around a defect. (*Lower portion*) Plot showing the phase profile along the dashed line

With the aim to validate this hypothesis on the origin of the IF, it has been measured directly on some clusters of defects embedded in a stoichiometric matrix. The IF value is found to grow near defects and vanish far from them, confirming the hypothesis.

Just after electric field poling, it was possible to measure the "static" IF, i.e., with no external voltage applied to the sample, by the reconstructed phase maps. To this aim, optical phase retardation was obtained by double exposure holographic interferometry [61]. Figure 7.24(a) shows the numerically reconstructed phase-map of the off-congruent sample. It is clearly shown that a residual phase-step was localized at the wall boundaries, close to the defects. By contrast, the phase-difference appears to be almost zero across the domain wall far from the defect, at the image centre (Fig. 7.24(a)). The phase-profile along a row of the data matrix (Fig. 7.24(b), top) allows us to compare quantitatively the phase difference across the two kinds of boundaries (Fig. 7.24(b), bottom). Null phase results across the domain wall far from defect, while a phase difference of ∼0.9 rad was measured between opposite ferroelectric domains across the inverted hexagon.

By evaluating this phase step, (indicated by the arrow in Fig. 7.24(b) down) across the hexagonal boundary, it was possible to measure the IF value. An IF equal to ∼735 V/mm was measured close to the defect, and, therefore, these measurements unequivocally demonstrate that the IF was caused by defects.

Then, it was measured the *elastic* IF across the two different kinds of domain walls from the same phase map, i.e. one near the defect and the other one far from it. In this way it is possible definitely demonstrate that elastic IF depends on defect presence as does static IF.

Figure 7.25(a) shows a frame of the movie obtained collecting the two-dimensional map of the optical phase retardation, calculated for each hologram during

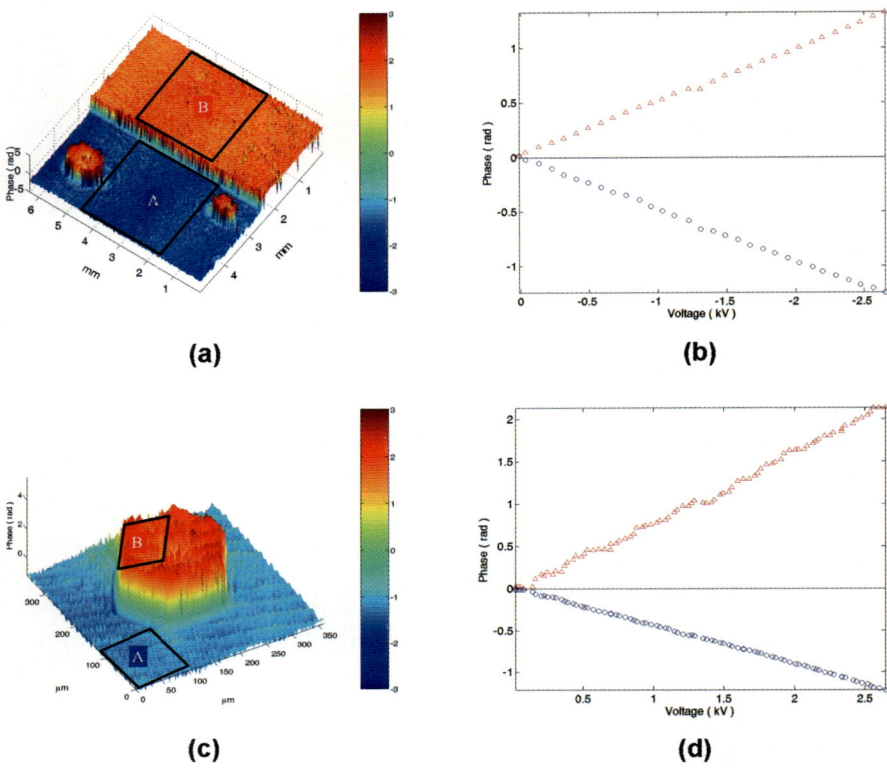

Fig. 7.25. (**a**) Phase map of the sample at fixed applied voltage (2.7 kV). A phase step is visible across the domain boundary between the reversed (*B*) and un-reversed (*A*) areas and also across reversed areas in proximity of defect. (**b**) Plot of the averaged phase on virgin $A(\circ)$ and poled $B(\triangle)$ areas as function of the applied voltage. No differences are noticed between the electro-optic behaviour of the two areas. (**c**) Phase map with higher magnification showing the poled zone around the defect. (**d**) Plot of the averaged phase on $A(\circ)$ and $B(\triangle)$ areas versus the applied voltage. The asymmetric behaviour exhibited by the two regions is clearly visible

application of an external voltage. In this case, all domain walls are clearly visible due to the opposite signs of the electro-optic and piezoelectric coefficients in the two domains. The inverted hexagon was not well defined because of the low spatial resolution of this map. The mean value of the phase maps in the two framed regions A and B, versus the applied voltage, is shown in Fig. 7.25(b). No differences are noticed between the virgin (A) and the poled (B) area since they are far from defects, as happen for *static* IF, i.e., when no external voltage was applied. Indeed, optical phase retardation values have opposite signs but equal magnitude in the two anti-parallel domains areas.

To measure the *elastic* IF close to the defect the spatial resolution to micrometric scale to resolve the hexagon (Fig. 7.25(c)) has been improved. The mean value of the phase maps in virgin (A) and poled area (B) around the defect, versus the applied voltage, is shown in Fig. 7.25(d), where the asymmetric behaviour exhibited by the two regions is clearly visible.

7.3.3 Evaluation of Optical Birefringence at Ferroelectric Domain Wall in LiNbO$_3$

The reversal domain process produce in Lithium niobate high stresses across each domain wall, that is the transition region between neighbouring anti-parallel domains. Such stresses induce, through the photo-elastic constants, a local birefringence that it is shown here can be investigated through a full-field polarimetric method [68]. A measure of domain wall width and determined the direction of the principal axes of stress-induced birefringence can be obtained.

A polarimetric method has been adopted [69–71] for investigating the domain wall in LN. To be precise, a measurement of the whole-field map of isoclinic angle α, that is the angle between a principal stress direction and a reference axis, by a phase-shifting based technique is reported. In this way it is possible to obtain a measure of domain wall width, that is the width of the birefringent region near a domain wall, and determined the direction of the principal axes of stress-induced birefringence.

The tested sample is a z-cut, 0.5 mm thick, LN crystal of congruent composition with the top and bottom surfaces optically polished. Ferroelectric polarization of a small is reversed applying an external electric field above the coercive field of the material. The poling process is monitored in real time as explained in previous section. In this way it is possible to stop the application of voltage when only few domains are grown. The reversed domains have the typical hexagonal shape characteristic of the three-folded symmetry group 3 m to which LN belongs. Then the sample is placed in a plane polariscope in dark field configuration as shown in Fig. 7.26. A He–Ne laser emitting at 632.8 nm provides the light source and an in-focus real image of it is acquired by means of a CCD.

The transmitted intensity is expressed via the formula:

$$I = I_0 - I_0 \sin^2 2\alpha \sin^2 \frac{\delta}{2} = I_0 \left[1 - \sin^2 \frac{\delta}{2}(1 - \cos^2 2\alpha) \right]$$
$$= I_0 \left[1 - \frac{1}{2} \sin^2 \frac{\delta}{2}(1 - \cos 4\alpha) \right] = I_B + V \cos 4\alpha \quad (7.11)$$

where α is the isoclinic angle while δ is a phase difference $\delta = (2\pi/\lambda)|n_x - n_y|d$ with the x and y axes considered in optical set-up coordinate system and d the sample thickness.

Four images from the plane polariscope in dark field configuration are recorded while rotating the polarizer and analyzer in steps of $\pi/8$. When the whole polar-

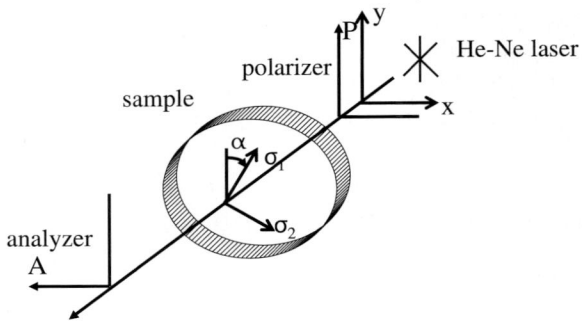

Fig. 7.26. Schematic view of the set-up: σ_1 and σ_2 indicate the directions of the principal axes of stress-induced birefringence, while α is the angle between a principal stress direction and the vertical reference axis

iscope is rotated by β_i, the intensity transmitted can be expressed as

$$I_i = I_B + V \cos 4(\alpha - \beta_i). \tag{7.12}$$

The isoclinic parameter α is obtained by means of a four-step algorithm with the intensity data obtained at $\beta_i = (i-1)\pi/8$ for $i = 1, 2, 3, 4$ through the relation

$$\tan 4\alpha = \frac{I_4 - I_2}{I_3 - I_1}. \tag{7.13}$$

In Fig. 7.27 are shown the intensity maps I_i for $i = 1, 2, 3, 4$. Then the corresponding isoclinic angle map of the sample, represented in the range $[-\pi/4, \pi/4]$, is shown in Fig. 7.28.

By this kind of analysis it is possible to measure only the direction of the principal axis that belongs to the range $[-\pi/4, \pi/4]$. The other one will be in the normal direction.

From the map shown in Fig. 7.28 it is possible to assess that principal axes of stress result to be parallel and ortogonal to domain walls of reversed exagon. It is interesting to note that on the left side there are irregular domain walls probably due to the edge of the electrolyte used to apply external voltage during the poling process. They exhibited significant optical birefringence contrast as expressed by the intense mixed red and blue colour in the map. To appreciate the sensitivity of this technique in measuring the isoclinic angle α, calculating the rms of the angle map in an uniform area, that results 0.03 rad.

A measure of the domain wall width can be obtained by the isoclinic angle maps considering the width of the birefringence regions adjacent to a wall. By detailed measures, the wall width value ranges from 10 microns to about 50 microns. The measured variations occur both between walls of different hexagons and between domain frontiers of the same hexagon. By polarimetric method it is possible to appreciate the spatial variations of domain wall width thanks to two-dimensional resolved measurements, in contrast to other tecniques that have a submicron spatial

7 Visual and Quantitative Characterization   203

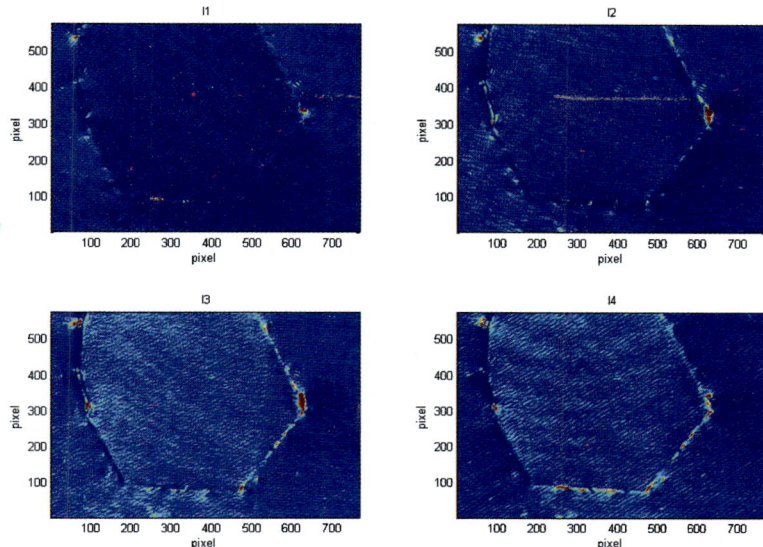

Fig. 7.27. Four images of congruent z-cut LiNbO$_3$ crystal acquired with a β angle equal to $\beta = 0$, $\beta = \pi/8$, $\beta = 2\pi/8$, $\beta = 3\pi/8$, rispectively

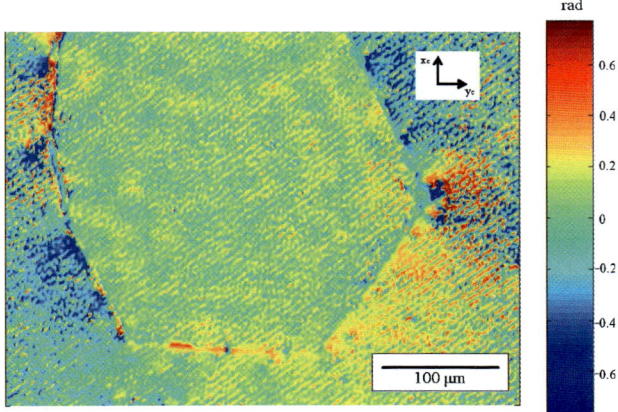

Fig. 7.28. Full-field isoclinic angle phase map

resolution, but are point-wise methods, i.e. AFM or NSOM analysis. Phase-shift polarimetry, as shown here, allows investigation of optical birefringence near ferroelectric domain walls in LiNbO$_3$.

References

1. U. Simon, F.K. Tittel, in *Atomic, Molecular and Optical Physics: Electromagnetic Radiation*, vol. 29C, ed. by F.B. Dunning, R.G. Hulet (Academic Press, New York, 1997), p. 231
2. D. Mazzotti, P. De Natale, G. Giusfredi, C. Fort, J. Mitchell, L. Hollberg, Opt. Lett. **25**, 350 (2000)
3. N. Picqué, P. Cancio, G. Giusfredi, P. De Natale, Opt. Soc. Am. B **5**, 692 (2001)
4. K. Fradkin, A. Arie, P. Urenski, G. Rosenman, Opt. Lett. **25**, 743 (2000)
5. R.G. Hunsperger, *Integrated Optics*, 4th edn. (Springer, Berlin, 1995)
6. S. Breer, K. Buse, Appl. Phys. B: Lasers Opt. **66**, 339 (1998)
7. S. Breer, H. Vogt, I. Nee, K. Buse, Electron. Lett. **34**, 2419 (1998)
8. M.M. Fejer, G.A. Magel, D.H. Jundt, R.L. Byer, IEEE J. Quantum Electron. **28**, 263 (1992)
9. C. Becker, A. Greiner, T. Oesselke, A. Pape, W. Sohler, H. Suche, Opt. Lett. **23**, 1194 (1998)
10. B. Andreas, K. Peithmann, K. Buse, Modification of the refractive index of lithium niobate crystals by transmission of high energy $4He^{2+}$ and D^+ particles. Appl. Phys **84**, 3813–3815 (2004)
11. D.-C. Su, C.-C. Hsu, Method for determining the optical axis and (n_e, n_o) of a birefringent crystal. Appl. Opt. **41**, 3936–3940 (2002)
12. Y.-C. Huang, C. Chou, M. Chang, Direct measurement o refractive indices (n_e, n_o) of a linear birefringent retardation plate. Opt. Commun. **133**, 11–16 (1997)
13. R.P. Shukla, G.M. Perera, M.C. George, P. Venkateswarlu, Measurement of birefringence of optical materials using a wedged plate interferometer. Opt. Commun. **78**, 7–12 (1990)
14. M.-H. Chiu, C.-D. Chen, D.-C. Su, Method for determining the fast axis and phase retardation of a wave plate. J. Opt. Soc. Am. A **13**, 1924–1929 (1996)
15. G.E. Jellison Jr., F.A. Modine, L.A. Boatner, Measurementof the optical functions of uniaxial materials by twomodulator generalized ellipsometry: rutile (TiO_2). Opt. Lett. **22**, 1808–1810 (1997)
16. J.D. Hecht, A. Eifler, V. Riede, M. Schubert, G. Krauss, V. Krämer, Birefringence and reflectivity of single-crystal $CdAl_2Se_2$ by generalized ellipsometry. Phys. Rev. B **57**, 7037–7042 (1998)
17. D.F. Heller, O. Kafri, J. Krasinnski, Appl. Opt. **33**, 3037 (1985)
18. J.C. Bhattacharya, Appl. Opt. **40**, 1658 (2001)
19. P.S.K. Lee, J.B. Pors, M.P. van Exter, J.P. Woerdeman, Simple method for accurate haracterization of birefringent crystals. Appl. Opt. **44**, 866–870 (2005)
20. S. De Nicola, P. Ferraro, A. Finizio, P. De Natale, S. Grilli, G. Pierattini, A Mach–Zehnder interferometer system for measuring the refractive indices of uniaxial crystals. Opt. Commun. **202**, 9–15 (2002)
21. L.H. Takeda, S. Kobayashy, J. Opt. Soc. Am. **72**, 156 (1982)
22. T. Fukano, L. Yamaguchi, Simultaneous measurements of thickness and refractive indices of multiple layers bu low coherence confocal coherence interference microscope. Opt. Lett. **21**, 1942–1944 (1986)
23. G.J. Tearney, M.E. Brezinski, J.F. Southern, B.E. Bouma, M.R. Hee, G. Fujimoto, Determination of the refractive index of light scattering human tissue by optical coherence tomography. Opt. Lett. **20**, 2258–2260 (1995)
24. M. Haruna, M. Ohmi, Y. Mitsuyama, H. Tajiri, H. Maruyama, M. Hashimoto, Simultaneous measurement of the phase and group indices and the thickness of transparent plates by low coherence interferometry. Opt. Lett. **23**, 966–968 (1998)

25. J.C. Martinez-Anton, E. Bernabeu, Simultaneous determination of film thickness and refractive index by interferential spectrogoniometry. Opt. Commun. **132**, 312–328 (1996)
26. G. Coppola, P. Ferraro, M. Iodice, S. De Nicola, Method for measuring the refractive index and the thickness of transparent plates with a lateral-shear, wavelength-scanning interferometer. Appl. Opt. **42**, 3882–3887 (2003)
27. L. Byer, Nonlinear optics and solid-state lasers: 2000. IEEE J. Select. Top. Quantum Electron. **6**, 911–930 (2000)
28. N.G.R. Broderick, G.W. Ross, H.L. Offerhaus, D.J. Richardson, D.C. Hanna, Hexagonally poled lithium niobate: a two-dimensional nonlinear photonic crystal. Phys. Rev. Lett. **84**, 4345–4348 (2000)
29. S.J. Holmgren, V. Pasiskevicius, S. Wang, F. Laurell, Three-dimensional characterization of the effective second-order nonlinearity in periodically poled crystals. Opt. Lett. **28**, 1555–1557 (2003)
30. M. Yamada, N. Nada, M. Saitoh, K. Watanabe, First-order quasi-phase matched $LiNbO_3$ waveguide periodically poled by applying an external field for efficient blue second-harmonic generation. Appl. Phys. Lett. **62**, 435–436 (1993)
31. S. Grilli, P. Ferraro, S. De Nicola, A. Finizio, G. Pierattini, P. De Natale, M. Chiarini, Investigation on reversed domain structures in lithium niobate crystals patterned by interference lithography. Opt. Express **11**, 392–405 (2003). http://www.opticsexpress.org/abstract.cfm?URI=OPEX-11-4-392
32. M.J. Missey, S. Russell, V. Dominic, R.G. Batchko, K.L. Schepler, Real-time visualization of domain formation in periodically poled lithium niobate. Opt. Express **6**, 186–195 (2000)
33. V. Gopalan, T.E. Mitchell, Wall velocities, switching times, and the stabilization mechanism of 180° domains in congruent $LiTaO_3$ crystals. J. Appl. Phys. **83**, 941–954 (1998)
34. K. Terabe, M. Nakamura, S. Takekawa, K. Kitamura, S. Higuchi, Y. Gotoh, Y. Cho, Microscale to nanoscale ferroelectric domain and surface engineering of a near-stoichiometric $LiNbO_3$ crystal. Appl. Phys. Lett. **82**(3), 433–435 (2003)
35. G. Rosenman, P. Urenski, A. Agronin, Y. Rosenwaks, M. Molotskii, Submicron ferroelectric domain structures tailored by high voltage scanning probe microscopy. Appl. Phys. Lett. **82**(1), 103–105 (2003)
36. M. Mohageg, D. Strekalov, A. Savchenkov, A. Matsko, V. Ilchenko, L. Maleki, Calligraphic poling of lithium niobate. Opt. Express **13**, 3408–3419 (2005). http://www.opticsinfobase.org/abstract.cfm?URI=oe-13-9-3408
37. K. Nassau, H.J. Levinstein, G.M. Loiacono, Ferroelectric lithium niobate. 1. Growth, domain structure, dislocations and etching. J. Phys. Chem. Solids **27**, 983–988 (1966)
38. V. Gopalan, M.C. Gupta, Origin of internal field and visualization of 180° domains in congruent $LiTaO_3$ crystals. J. Appl. Phys. **80**, 6099–6106 (1996)
39. M. Flörsheimer, R. Paschotta, U. Kubitscheck, Ch. Brillert, D. Hofmann, L. Heuer, G. Schreiber, C. Verbeek, W. Sohler, H. Fuchs, Second-harmonic imaging of ferroelectric domains in $LiNbO_3$ with micron resolution in lateral and axial directions. Appl. Phys. B **67**, 593–599 (1998)
40. T.J. Yang, V. Gopalan, P.J. Swart, U. Mohideen, Direct observation of pinning and bowing of a single ferroelectric domain wall. Phys. Rev. Lett. **82**, 4106–4109 (1999)
41. J. Wittborn, C. Canalias, K.V. Rao, R. Clemens, H. Karlsson, F. Laurell, Nanoscale imaging of domains and domain walls in periodically poled ferroelectrics using atomic force microscopy. Appl. Phys. Lett. **80**, 1622–1624 (2002)
42. M.C. Wengler, M. Müller, E. Soergel, K. Buse, Poling dynamics of lithium niobate crystals. Appl. Phys. B **76**, 393–396 (2003)

43. V. Gopalan, S.S.A. Gerstl, A. Itagi, T.E. Mitchell, Q.X. Jia, T.E. Schlesinger, D.D. Stancil, Mobility of 180° domain walls in congruent LiTaO₃ measured using real-time electro-optic imaging microscopy. J. Appl. Phys. **86**, 1638–1646 (1999)
44. V. Gopalan, T.E. Mitchell, *In situ* video observation of 180° domain switching in LiTaO₃ by electro-optic imaging microscopy. J. Appl. Phys. **85**, 2304–2311 (1999)
45. V. Goapalan, Q.X. Jia, T.E. Mitchell, *In situ* observation of 180° domain kinetics in congruent LiNbO₃ crystals. Appl. Phys. Lett. **75**, 2482–2484 (1999)
46. S. Kim, V. Gopalan, K. Kitamura, Y. Furukawa, Domain reversal and nonstoichiometry in lithium tantalate. J. Appl. Phys. **90**, 2949–2963 (2001)
47. M.J. Missey, S. Russell, V. Dominic, R.G. Batchko, K.L. Schepler, Real-time visualization of domain formation in periodically poled lithium niobate. Opt. Express **6**, 186–195 (2000)
48. M. Müller, E. Soergel, K. Buse, Visualization of ferroelectric domains with coherent light. Opt. Lett. **28**, 2515–2517 (2003)
49. M. Müller, E. Soergel, K. Buse, Light deflection from ferroelectric domain structures in congruent lithium tantalate crystals. Appl. Opt. **43**, 6344–6347 (2004)
50. M. Müller, E. Soergel, M.C. Wengler, K. Buse, Light deflection from ferroelectric domain boundaries. Appl. Phys. B **78**, 367–370 (2004)
51. S. Grilli, P. Ferraro, S. De Nicola, A. Finizio, G. Pierattini, R. Meucci, Whole optical wavefields reconstruction by digital holography. Opt. Express **9**, 294–302 (2001). http://www.opticsexpress.org/abstract.cfm?URI=OPEX-9-6-294
52. S. De Nicola, P. Ferraro, A. Finizio, G. Pierattini, Correct-image reconstruction in the presence of severe anamorphism by means of digital holography. Opt. Lett. **26**, 974–976 (2001)
53. P. Ferraro, S. De Nicola, G. Coppola, Digital hoplography: recent advancements and prospective improvements for applications in microscopy, in *Optical Imaging Sensors and Systems for Homeland Security Applications*, ed. by B. Javidi (Springer, New York, 2006), pp. 47–84, Chap. 3
54. P. Ferraro, S. De Nicola, G. Coppola, Controlling image recostruction process, in digital holography, in *Digital Holography and Three-Dimensional Display, Principles and Applications*, ed. by T.-C. Poon (Springer, Berlin, 2006), pp. 173–212
55. S. Grilli, P. Ferraro, M. Paturzo, D. Alfieri, P. De Natale, *In-situ* visualization, monitoring and analysis of electric field domain reversal process in ferroelectric crystals by digital holography. Opt. Express **12**, 1832–1842 (2004). http://www.opticsexpress.org/abstract.cfm?URI=OPEX-12-9-1832
56. C. Canalias, S. Wang, V. Pasiskevicius, F. Laurell, Nucleation and growth of periodic domains during electric field poling in flux-grown KTiOPO₄ observed by atomic force microscopy. Appl. Phys. Lett. **88**, 032905 (2006)
57. C. Canalias, J. Hirohashi, V. Pasiskevicius, F. Laurell, Polarization-switching characteristics of flux-grown KTiOPO₄ and RbTiOPO₄ at room temperature. J. Appl. Phys. **97**, 124105 (2005)
58. Z.W. Hu, P.A. Thomas, W.P. Risk, Studies of periodic ferroelectric domains in KTiOPO₄ using high-resolution X-ray scattering and diffraction imaging. Phys. Rev. B **59**, 14259–14264 (1999)
59. J. Hellström, R. Clemens, V. Pasiskevicius, H. Karlsson, F. Laurell, Real-time and *in-situ* monitoring of ferroelectric domains during periodic electric field poling of KTiOPO₄. J. Appl. Phys. **90**, 1489–1495 (2001)
60. C. Canalias, V. Pasiskevicius, F. Laurell, S. Grilli, P. Ferraro, P. De Natale, *In-situ* visualization of domain kinetics in flux grown KTiOPO₄ by digital holography. J. Appl. Phys. **102**, 064105 (2007)

61. M. de Angelis, P. Ferraro, S. Grilli, S. De Nicola, A. Finizio, M. Paturzo, G. Pierattini, Evaluation of the internal field in lithium niobate ferroelectric domains by an interferometric method. Appl. Phys. Lett. **85**, 2785 (2004)
62. S. Kim, V. Gopalan, K. Kitamura, Y. Furukawa, J. Appl. Phys. **90**, 2949 (2001)
63. H. Donneberg, S.M. Tomlinson, C.R.A. Catlow, O.F. Schirmer, Phys. Rev. B **40**, 11909 (1989)
64. A.V. Yatsenko, E.N. Ivanova, N.A. Sergeev, Physica B **240**, 254 (1997)
65. J.-H. Ro, M. Cha, Appl. Phys. Lett. **77**, 2391 (2000)
66. G. Arlt, H. Neumann, Ferroelectrics **87**, 109 (1988)
67. A.V. Yatsenko, Phys. Solid State **40**, 109 (1998)
68. M. Paturzo, L. Aiello, F. Pignatiello, P. Ferraro, P. De Natale, M. de Angelis, S. De Nicola, Investigation of optical birefringence at ferroelectric domain wall in LiNbO$_3$ by phase-shift polarimetry. Appl. Phys. Lett. **88**, 151918–151920 (2006)
69. T.J. Yang, U. Mohideen, Phys. Lett. A **250**, 205 (1998)
70. T.Y. Chen, C.H. Lin, Opt. Lasers Eng. **30**, 527 (1998)
71. A. Asundi, L. Tong, C.G. Boay, Appl. Opt. **38**, 5931 (1999)

8 New Insights into Ferroelectric Domain Imaging with Piezoresponse Force Microscopy

T. Jungk, Á. Hoffmann, and E. Soergel

Ferroelectric domain patterns are intensively investigated due to their increasing practical importance, e.g., for frequency conversion [1, 2] or high-density data storage [3]. For their characterization, a visualization technique with high lateral resolution is required. Among the wealth of techniques [4], piezoresponse force microscopy (PFM) has become a standard tool for visualizing micron-sized domain structures [5]. This is mainly due to its easy use without any specific sample preparation and the high lateral resolution of a few 10 nm. The vertical resolution of PFM reaches even the sub-picometer regime. Despite these impressive numbers, there are, however, only a few publications reporting quantitative data obtained with PFM. This seems to be amongst others due to a lack of knowledge on scanning force microscopy issues.

In this chapter we present an overview on PFM imaging restricted to the detection of ferroelectric domains on the polar faces of single crystals. We intend to provide a deeper insight into PFM imaging and thereby a more reliable interpretation of PFM images.

8.1 Introduction

We will start with a short survey on ferroelectricity to the extent it is necessary for further understanding of PFM imaging (Sect. 8.1.1). One of the most prominent examples for a ferroelectric crystal is lithium niobate ($LiNbO_3$), enabling very promising applications and thus being intensively investigated. We therefore briefly summarize the major properties of $LiNbO_3$ with regard to PFM (Sect. 8.1.2).

8.1.1 Ferroelectrics

From the 32 crystal classes ferroelectrics belong to the group of pyroelectrics, which represent a subgroup of the piezoelectric family [6]. Thus, ferroelectric crystals are piezo- and pyroelectric as well.

If stress is applied to a piezoelectric charges of different sign are generated on opposite crystal faces, which is called the direct piezoelectric effect. Due to thermodynamics, the inverse piezoelectric effect exists as well, which implies that the crystal deforms when an electrical field is applied to it. A subgroup of the piezoelectrics is the pyroelectric crystal class that includes 10 members (1, 2, 3, 4, 6, m, mm2, 3m, 4mm, 6mm). Even without an external electrical field or mechanical stress, these crystals exhibit a spontaneous polarization P_S due to a dislocation of the positive and negative ions in the crystal lattice. A change of the temperature causes a deformation of the crystal along its polar axis and vice versa, which is called the pyroelectric effect.

Several members of the pyroelectric crystal class show a reversible spontaneous polarization with two or more stable states that can be reached via an external electrical field or via mechanical stress. A spatial region with a homogeneous polarization is called ferroelectric domain. Ferroelectric domains only exist in well-defined crystallographic orientations [7], such as 180° domains along a polar axis. Moreover, the relation between polarization and applied electrical field is given by a hysteresis loop. In contrary to piezo- and pyroelectricity, the ferroelectric crystals cannot be predicted by symmetry considerations but have to be discovered experimentally. A detailed description of ferroelectricity can be found in the literature [8–10].

8.1.2 Lithium Niobate (LiNbO₃)

In its ferroelectric phase, LiNbO$_3$ belongs to the point group 3m, i.e., it is a uniaxial crystal with a 3-fold rotation symmetry. The orientation of the crystallographic c-axis, which is defined as z-axis in Cartesian coordinates, is given by the position of the cations Li$^+$ and Nb^{5+} relative to the position of the oxygen anions [11]. The domain formation is based on displacement polarization along the z-axis, thus antiparallel 180° domains develop. We primarily utilized congruently melting LiNbO$_3$ produced by the Czochralski method [12], which was prepared in the desired domain configuration by electrical field poling. Quite often we investigated periodically poled lithium niobate (PPLN).

If the crystal is poled from one domain orientation into the other, all crystal properties described by odd rank tensors have to switch sign, such as the piezoelectric tensor which is the basis for PFM. In the case of LiNbO$_3$, the application of the von Neumann principle to the point group 3m gives 4 independent components (d_{113}, d_{222}, d_{311}, and d_{333}) for the piezoelectrical tensor. Note that the longitudinal piezoelectrical effect described by the diagonal tensor elements d_{iii} is independent of the crystal thickness. An exact mathematical description of the piezoelectrical effect can be found in [13].

For stoichiometric LiNbO$_3$, several material parameters change [14], as it is the case for doped LiNbO$_3$ too [15]. A detailed summary of the properties of LiNbO$_3$ can be found in [16].

8.2 Principles of Scanning Force Microscopy (SFM)

The main attention in this section is focussed on a sound understanding of PFM investigation of single crystals. Therefore a deeper insight into scanning force microscopy (SFM) in general is required. In most PFM publications a detailed description of SFM is missed out, that is why we will pay special attention to this topic. After reviewing the tip-surface interactions relevant for the investigation of ferroelectric crystals (Sect. 8.2.1), we will describe the resulting responses of the cantilever (Sect. 8.2.2). A separate section is dedicated to the problem of cross-talk between different cantilever movements (Sect. 8.2.3). This issue is of major importance as cross-talk can give rise to pretended signals bereft of any physical origin. Finally, special attention is payed to the calibration of the SFM with regard to PFM applications, since an accurate calibration is mandatory to allow quantitative measurements (Sect. 8.2.4).

In the following, we assume a standard commercial scanning force microscope equipped with a laser-beam deflection readout [17]. The microscope is operated in contact mode. Furthermore an electrical connection allowing the application of moderate voltages to the tip is provided.

8.2.1 Tip-Cantilever-Surface Interactions

Mechanical Parameters of the Tip and Cantilever

The dimensions of a typical commercially available SFM probe used for PFM are the following: cantilever width $w = 25\text{--}35\,\mu\text{m}$, cantilever length $l = 100\text{--}300\,\mu\text{m}$, tip height $h \approx 15\,\mu\text{m}$, and tip radius $r \approx 50\,\text{nm}$ (Fig. 8.1). As it is obvious from these dimensions, the lateral resolution is dominated by the apex of the tip. Any additional contribution of the cantilever can only resolve features of the size of the

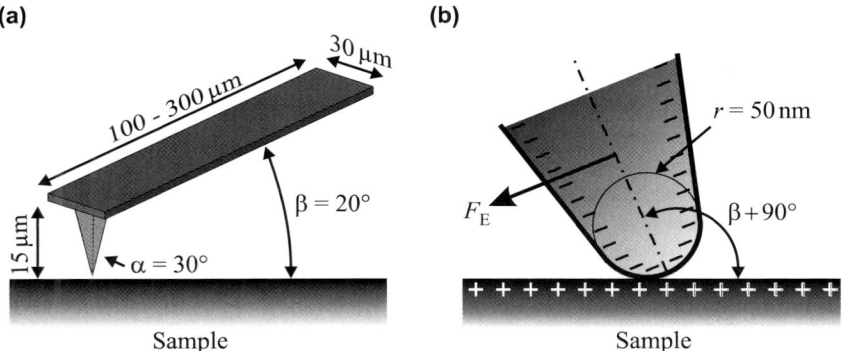

Fig. 8.1. (a) Typical dimensions of a commercially available cantilever and tip. (b) Closer view to the conical apex of the tip from which the tip radius r is defined. Possible electrostatic forces F_E due to an inclination angle β are depicted

cantilever itself. Thus, in the case of the visualization of μm-sized structures, the cantilever has no impact on the imaging process.

Electrostatic Interaction between Tip & Cantilever and Surface

Since the spontaneous polarization P_s leads to a surface charging, the electrostatic interaction between tip & cantilever and the surface are of major interest. Again an interaction between the cantilever and the surface cannot account for any SFM imaging with a lateral resolution better than the cantilever size. Because the tip is in contact with the surface, electrostatic forces, although present, cannot result in a vertical displacement of the tip. Note that for electrostatic force microscopy [18], the SFM is operated in noncontact mode, so the situation is completely different. One might think that due to the inclined position of the tip with respect to the surface, the asymmetric field distribution could lead to a torsion (Fig. 8.1(b)). However, a rough estimate of the torsional moment excludes such a contribution. Furthermore, we performed a series of experiments investigating the PFM contrast on PPLN samples for different inclination angles β ranging from 0° to 60° by mounting the sample on different wedges. Indeed, the contrast was observed to be independent of the angle β.[1]

8.2.2 Cantilever Movements

Depending on the relative orientation of the driving force with respect to the axis of the cantilever, the latter can perform three mostly independent movements: (i) deflection if the force acts along the tip axis, (ii) torsion if the force acts perpendicular to the tip and to the cantilever, and (iii) buckling if the force acts perpendicular to the tip but along the cantilever (Fig. 8.2). Whereas deflection and torsion can be found in every standard textbook [17], buckling is not that much known, that is why a more detailed description seems appropriate.

Firstly, it is important to note that a buckling of the cantilever gives rise to a signal in the vertical channel and can thus not be easily distinguished from a deflection of the cantilever. From the side view in Fig. 8.2(c) it can be seen that the readout of buckling is very sensitive to the position of the laser beam on the back side of the cantilever: displacing the laser beam from the very end on top of the tip towards the middle of the cantilever results in an inversion of the readout signal. Thus, it is evident that buckling cannot be observed if the laser spot is too large, but also an inappropriate adjustment of the laser spot on top of the anti-node can discard the buckling signal.

---

[1] These statements are not in contradiction to the lateral signal explained later via electrostatic forces. Here we discuss an electrostatic contribution to the PFM signal on domain faces, i.e., the surface underneath the tip is homogeneously charged. In the case of lateral forces, however, the signals occur only at the domain boundaries and thus at the places where an electric field in plane with the surface is present.

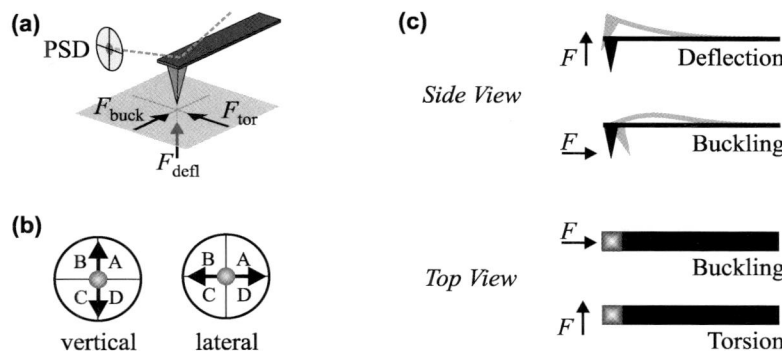

Fig. 8.2. Forces acting on the tip with the resulting movements of the cantilever (**a**, **c**). In (**b**) the definition of the readout channels of the position sensitive detector (PSD) is given

It is also worth to note that the deflection signal measures a height displacement of the tip (in m), whereas both torsion and buckling measure a force acting on the tip (in N). A detailed calibration procedure for SFM will be given below (Sect. 8.2.4).

8.2.3 Cross-Talk

Although often ignored, cross-talk between the vertical and the lateral readout channel is a serious source for erroneous signals. Whereas electronic cross-talk (due to an imperfect electric shielding) and mechanical cross-talk (primarily arising only on samples with pronounced topography) can in general not be influenced by the user, the cross-talk originating from a misalignment of the optical detection system can be eliminated by a simple electronic circuit [19]. This cross-talk arises if the segmented photodetector is rotated with respect to the plane of the readout laser beam. Depending on the specific SFM, the mounting of the cantilever and thus the subsequent laser-beam adjustment, the cross-talk between the vertical and the lateral channel can amount up to 30%. A pure deflection of the cantilever thus leads to a pretended torsional signal, bereft of any physical origin.

To obtain reliable data it is mandatory to compensate for the cross-talk. The amount of cross-talk, and thus the necessity to compensate for it, can be estimated in a very simple way: put the tip far away from the surface and excite the cantilever to vibrate at the first deflection mode, the "standard" resonance frequency. Any signal in the lateral readout channel must originate from cross-talk, since the torsional vibration modes have a much higher resonance frequency [20].

8.2.4 Calibration

Usually, the calibration of the SFM for PFM measurements is performed using an α-quartz sample as calibration standard as, e.g., described in [21]. However, due to the system-inherent background (Sect. 8.4), that calibration procedure fails. This is described in detail elsewhere [22]. Instead, the calibration should be performed with a piezoelectric sample exhibiting a large piezoelectric coefficient; even if the latter is not known with high accuracy, it can easily be determined within the calibration procedure. For example, a lead zirconate titanate (PZT) disc is well suited. In brief, for a reliable calibration of the SFM and especially for PFM applications, three steps have to be accomplished:

- Calibration of the z-scanner of the SFM.
- Measurement of thickness change Δt_{PZT} of the calibration sample for a specific voltage U_{tip}. This measurement is performed with the so-called height-mode of the SFM. That is why the piezoelectric constant of the calibration sample has to be large in order to yield measurable thickness changes Δt at moderate voltages.
- Read-out of the output of the lock-in amplifier P_{PZT} while disabling the feedback-loop of the SFM (with the same sample and otherwise unchanged settings as used in the step before).

This procedure allows to determine the calibration constant $k = \Delta t_{PZT}/P_{PZT}$, which can then be used to measure any other material.

8.3 Principles of Piezoresponse Force Microscopy (PFM)

The aim of PFM measurements is to detect a deformation of the sample due to the converse piezoelectric effect. Therefore an alternating voltage U_{tip} is applied to the tip and the response, i.e., the thickness change Δt of the sample due to the converse piezoelectric effect is measured via lock-in detection. Although straightforward at first sight, PFM measurements gave rise to a wealth of discussions mainly due to their strong frequency dependence [23, 24] which lead to alternative explanations of the origin of the domain contrast [25–27]. Above all, another name was introduced for one and the same method – dynamic contact electrostatic force microscopy [25].

Within this section, our aim is to recall the generally used setup and standard settings for PFM measurements in order to specify them in a closed form (Sect. 8.3.1). We will then present an analysis of PFM measurements taking into account the system-inherent background (Sect. 8.3.2) followed by its vectorial description (Sect. 8.3.3).

8.3.1 PFM Setup & Standard Settings

To utilize an SFM for piezoresponse force microscopy requires mainly two instrumental features: (i) an electrical connection to the tip and (ii) direct access to the signals of the position sensitive detector recording the movement of the cantilever. Furthermore, a lock-in amplifier is necessary for sensitive readout of the cantilever movement. Figure 8.3 shows the standard setup for PFM. In the following, the crucial parts of the experimental setup are described in order to define the parameters and denotations used further on. In addition, our experimental settings for PFM operation are given:

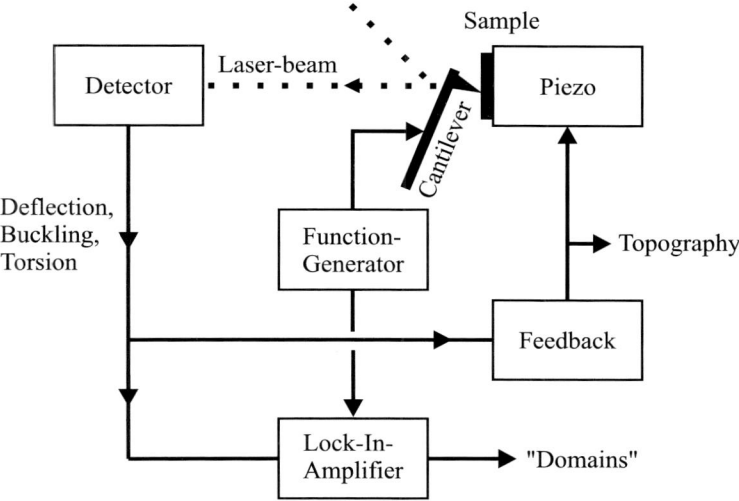

Fig. 8.3. Setup for PFM measurements

- Tip of the SFM: for PFM operation, the tip must be conductive and electrically connected to allow the application of voltages. The resonance frequency of the cantilever is not crucial; it should always be far away from the frequency of the alternating voltage applied to the tip. Typically, cantilevers with resonance frequencies $f_0 > 100\,\text{kHz}$ are utilized. The alternating voltage applied to the tip is usually chosen to have a frequency between 10 kHz and 100 kHz with an amplitude $U \leq 20\,\text{V}_{pp}$. The time constant of the feedback-loop of the SFM must be large compared to the period of modulation of the applied voltage to avoid a compensation of the signal.
  We mostly utilize Ti–Pt coated tips (MicroMasch) with resonance frequencies $f_0 = 150–400\,\text{kHz}$, spring constants $k = 3–70\,\text{N/m}$ and apply an alternating voltage of 30–60 kHz with an amplitude of $\sim 10\,\text{V}_{pp}$.
- Sample: in large part, PFM measurements are performed with crystals exhibiting antiparallel domains only. For investigation, the samples are cut in such a way that the domain boundaries are perpendicular to the surface to be studied. We will restrict ourselves to such a configuration exclusively.
- SFM: generally all scanning force microscopes are suited for PFM operation as long as they allow application of voltages to the tip and separate readout of the cantilever movement. The scanning velocity has to be adapted to the rise time of the lock-in amplifier.
  We use a SMENA SFM (NT-MDT) modified to apply voltages to the tip and upgraded with an additional interface board for readout of the cantilever movement. Typical scanning velocity is about 1 µm/s.
- Lock-in amplifier: most PFM setups use dual-phase lock-in amplifiers which allow to chose between two output schemes: (i) in-phase output (also denoted as

X-output) and orthogonal output (Y-output) or (ii) magnitude $R = \sqrt{X^2 + Y^2}$ and phase $\theta = \arctan(Y/X)$. These output signals of the lock-in amplifier will be named PFM signals: **P** on a positive $+z$ domain face and **N** on a negative $-z$ domain face. To specify the output (and thus the component of the particular vector), the adequate symbol (X, Y, R, or θ) will be added as a subscript. For example, P_X denotes the in-phase output signal of the lock-in amplifier on a positive $+z$ domain face.

The experiments presented here are performed with a SR830 lock-in amplifier (Stanford Research Systems). Typical settings are 1 mV for the sensitivity and 1 ms for the time constant.

8.3.2 System-Inherent Background in PFM Measurements

According to the physical principles of the converse piezoelectric effect, PFM signals have to meet the following requirements:

- Depending on the orientation of the polar axis, the piezoresponse must be either in phase or out of phase by 180° with respect to the alternating voltage applied to the tip.
- The amplitude of the PFM signal must be same on $+z$ and $-z$ domain faces.
- No frequency dependence of the PFM signals is expected. This holds true for frequencies < 100 kHz as piezomechanical resonances usually occur at much higher frequencies [28].

Interestingly, these physically mandatory properties for PFM signals are not generally fulfilled [23, 24, 29–33]. That is why a series of alternative explanations [25–27] or at least additional contributions to the PFM signal [34, 35] have been discussed. However, an accurate analysis of the situation reveals the presence of a system-inherent background, strongly affecting the measurements [36, 37]. As a consequence, the PFM signals turn out to be a superposition of the background signal **B** and the piezoresponse signal **d** of the sample.

To prove the existence of the system-inherent background, we performed comparative measurements on PPLN samples and on standard microscope glass slides, the latter being not piezoelectric. We firstly determined the frequency dependence on the $+z$ and $-z$ faces thus recording P_X and N_X. Due to the requirements listed above, the sum of those two signals must be free of any piezoelectric contribution. The result, however, can be seen in Fig. 8.4(a), where we plotted $\frac{1}{2}(P_X + N_X)$. As pronounced frequency dependence is present, however, the origin of this signal must be independent of the piezoelectric properties of the sample. This was demonstrated by comparative measurements on a microscope glass slide, showing the same frequency dependence (Fig. 8.4(b)). The small difference of both scans is shown in Fig. 8.4(c), the vertical scale being expanded by a factor of ten.

8.3.3 Vectorial Description

In order to clarify the statements about the system-inherent background presented above, we show a vector diagram illustrating the case for two different frequencies

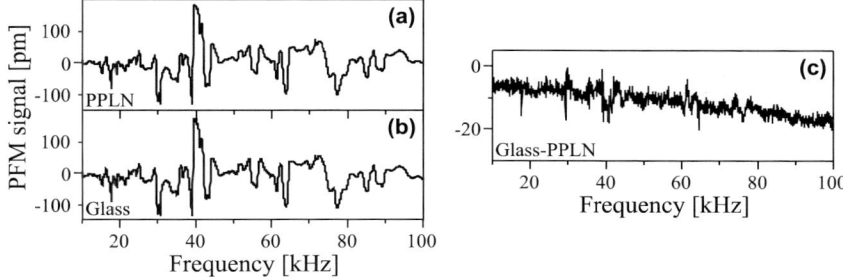

Fig. 8.4. Frequency dependence of the in-phase PFM background signal on a PPLN surface (**a**), on a glass surface (**b**), and their difference (**c**)

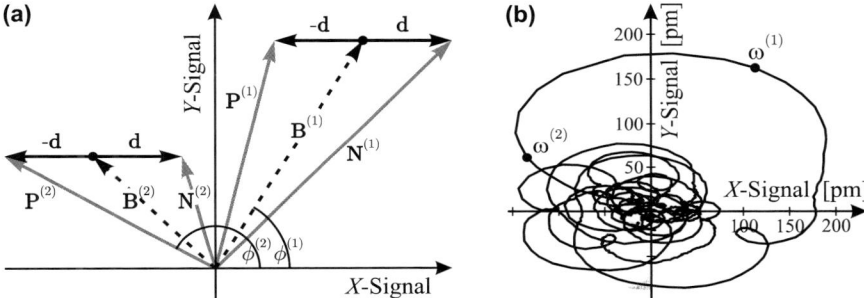

Fig. 8.5. (**a**) Vector diagram showing the PFM signals on the $\pm z$ domain faces of a ferroelectric sample (X and Y: in-phase and orthogonal output of the lock-in amplifier) for two different frequencies $\omega^{(1)}$ and $\omega^{(2)}$. The superscripts indicate the corresponding signals: the PFM signal $\mathbf{P}$ ($\mathbf{N}$) measured on a $+z$ ($-z$) domain, the background $\mathbf{B}$, and its corresponding phase ϕ. The piezoresponse signal from a $\pm z$ domain face is denoted by $\mp \mathbf{d}$. (**b**) Frequency dependence (10–100 kHz) of the background $\mathbf{B}$ determined on a PPLN surface with 10 V applied to the tip

$\omega^{(1)}$ and $\omega^{(2)}$ of the alternating voltage applied to the tip (Fig. 8.5(a)). At a specific frequency $\omega^{(1)}$, a background PFM signal $\mathbf{B}^{(1)}$ is present. The piezoresponse of the sample contributes $\mathbf{d}$ for the $-z$ face and $-\mathbf{d}$ for the $+z$ face to the PFM signal, both of same amplitude with a 180° phase shift in between. This results in the measurement of $\mathbf{N}^{(1)} = \mathbf{B}^{(1)} + \mathbf{d}$ for the $-z$ face and $\mathbf{P}^{(1)} = \mathbf{B}^{(1)} - \mathbf{d}$ for the $+z$ face. It is important to note that the phasing between $\mathbf{P}^{(1)}$ and $\mathbf{N}^{(1)}$ is not 180° and that their amplitudes are unequal and larger than expected. The same considerations apply for any other frequency $\omega^{(2)}$. It is obvious from Fig. 8.5(a) that although $\mathbf{d}$ is unchanged for both frequencies, the PFM signals measured at different frequencies differ with respect to amplitude and phase. This can be verified experimentally with an oscilloscope in X–Y-display mode when applying different frequencies of the alternating voltage to the tip while scanning across a domain wall.

To underline the arbitrariness of the background Fig. 8.5(b) shows an X–Y-diagram of its frequency dependence from 10 kHz to 100 kHz. Note that the big slope covers a frequency span from 39 kHz to 42 kHz only.

8.4 Consequences of the System-Inherent Background

The origin of the system-inherent background is not clear until now, and therefore getting rid of it seems quite difficult. In general one can assume, however, that the whole SFM head acts as a mechanical resonance box. Therefore minimum changes like the readjustment of the optical readout changes the frequency spectrum. Also the elongation of the tube scanner was found to have an impact on the background. The background shows the very same linear dependence on the applied voltage as the piezoresponse signal. Thus, a separation of the two signals via a tricky choice of the applied voltage fails. A main reason for renouncing to invest into a efficient suppression of the background is its amplitude: typical values are several 10 pm/V (similar to piezoelectric coefficients of many ferroelectric crystals). The high sensitivity of the lock-in amplifier makes these vibrations easily accessible, however, suppressing a complex mechanical setup from oscillating with such small amplitudes is not quite realistic. Of course the investigation of ceramics such as PZT with much larger piezoelectric coefficients [38] is only marginally influenced by the system-inherent background.

In this section we want to point out the strong influences of the background on the PFM measurements (Sect. 8.4.1). We also give a – fortunately very easy – solution for background-free PFM imaging (Sect. 8.4.2).

8.4.1 Background-Induced Misinterpretations

Several surprising features concerning the domain contrast as well as the shape and location of domain boundaries turn out to possibly originate from the system-inherent background [37]. Of course, also physical effects can influence the domain contrast and the domain boundaries, however, a careful analysis of the measured data is mandatory to avoid misinterpretation. In the following, we exemplify possible consequences of the background:

- Enhancement of the domain contrast;
- Nulling of the domain contrast;
- Inversion of the domain contrast;
- Arbitrary phase difference between $\pm z$ domains;
- Shift of the domain boundary;
- Change of the shape of the domain boundary.

Detailed considerations based on the vector diagram of Fig. 8.5(a) allow one to understand the influence of the background on the domain contrast and the phase

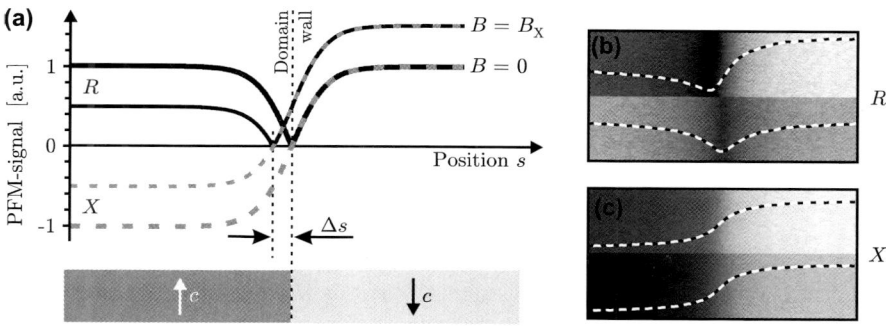

Fig. 8.6. Influence of the read-out settings of the lock-in amplifier on the detected domain wall. (**a**) Model predictions of the expected PFM signals, where *black lines* correspond to the R-signal and *grey dashed lines* to the X-signal without ($B = 0$) and with the presence of a background signal ($B = B_X$). The model is confirmed by PFM measurements of a single domain boundary in LiNbO$_3$. During image acquisition, the frequency of the applied voltage was changed, thereby adding a background signal B_X. Using the R-output leads to a pretended shift of the domain boundary (**b**), whereas the PFM image recorded with the X-output just becomes brighter (**c**). The line scans are averages over 20 image lines. The image size is $1 \times 0.5\,\mu m^2$

signal of the lock-in amplifier. The consequences of the background signal on the shape and location of the domain boundaries in PFM measurements when using the magnitude output of the lock-in amplifier, however, need a more careful analysis. In order to clearly expose the contributions of the background to the PFM signal, it is reasonable to discuss two special cases separately: the phasing between background and piezoresponse signal being (i) 0° or 180° and (ii) 90° or 270°. In (i) the background lies on the X-axis in Fig. 8.5(a) and in (ii) parallel to the Y-axis. In general the phasing will be arbitrary leading to a superposition of the phenomena described in the following.

Background B along X

The consequences of a background along the X-axis can be seen in Fig. 8.6(a), where scan lines across a domain boundary for both the X- and R-outputs of the lock-in amplifier are simulated. In the case of no background signal ($B = 0$, thick lines), both readout signals show the domain boundary at its real position $s = 0$, in the R-signal as a minimum and in the X-signal as the inflection point of the slope. When adding the background B_X, the minimum of the R-signal is shifted by Δs pretending the domain boundary to be at a different location. Moreover, a distinct change of the domain contrast can be observed. Figures 8.6(b, c) show images of a single domain boundary recorded simultaneously with the X- and R-outputs of the lock-in amplifier. After the first half of the image, the frequency of the alternating voltage is changed in order to alter the background. Whereas in the X-signal the location of the domain boundary is not affected (Fig. 8.6(c)), the image taken with

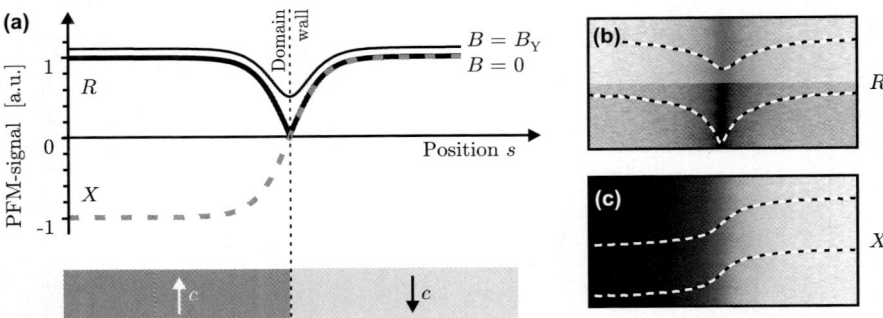

Fig. 8.7. Influence of the read-out settings of the lock-in amplifier on the detected domain boundary. (**a**) Model predictions of the expected PFM signals, where *black lines* correspond to the R-output and *grey dashed lines* to the X-output. The model is confirmed by PFM measurements of a single domain boundary in LiNbO$_3$. During image acquisition, the frequency of the voltage applied to the tip was changed, thereby adding a background B_Y. Using the R-output leads to a pretended broadening of the domain boundary (**b**), whereas the image of the X-output stays unchanged (**c**). The line scans are averages over 20 image lines. The image size is $1 \times 0.5\,\mu m^2$

the R-output shows a distinct shift of the domain boundary (Fig. 8.6(b)). As a further consequence of the in-phase background, a broadening and also an asymmetry of the detected domain wall is pretended.

Background B along Y

The situation for a background along Y can be explained in a similar way. As can be seen from Fig. 8.7(a), PFM images recorded with the R-output show the domain boundaries only. Their full width at half maximum W broadens with increasing background. At the same time, the contrast of the domain boundary decreases. Figures 8.7(b,c) show PFM images of a single domain boundary recorded simultaneously with the X- and R-outputs of the lock-in amplifier. After recording half of the image, the frequency of the alternating voltage is changed in order to alter the background. Whereas in the X-signal no changes can be observed (Fig. 8.7(c)), the image taken with the R-output shows a distinct broadening of the domain wall as well as a faded contrast (Fig. 8.7(b)).

8.4.2 Background-Free PFM Imaging

As can be seen from the Figs. 8.6 and 8.7, reliable experimental data in PFM imaging can be obtained when using the X-output of the lock-in amplifier for data acquisition.

From the experimental side, however, there is an additional problem arising: although the piezoelectric response must be in-phase with the alternating voltage applied to the tip (at least for frequencies $<100\,kHz$), there is always a small phase

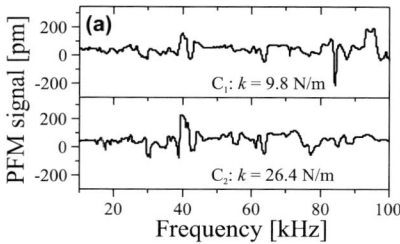 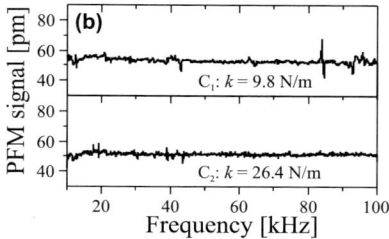

Fig. 8.8. Frequency spectrum on a $-z$ domain face of LiNbO$_3$ before (**a**) and after (**b**) background subtraction for two different cantilevers (C$_1$ and C$_2$) with a voltage of 10 V$_{pp}$ applied to the tip

shift ($<10°$) originating from the read-out electronics. In the vector diagram of Fig. 8.5(a), the piezoresponse signal **d** would show up slightly tilted. To extract nevertheless correct data from PFM measurements, the easiest solution is to set the phase of the lock-in amplifier such that no domain contrast is visible in the Y-output of the lock-in amplifier. This corresponds to a rotation of the vector diagram in the coordinate system in Fig. 8.5(a). An equivalent (and even more precise) solution is a rotation of the coordinate system after image acquisition such that the standard deviation of the of the image recorded with the Y-output is minimized.

8.5 Quantitative Piezoresponse Force Microscopy

With the basic knowledge acquired above and having performed a reliable calibration of the microscope, it is now possible to record quantitative data for the PFM amplitude (Sect. 8.5.1) and the domain wall width seen by PFM (Sect. 8.5.2).

8.5.1 Amplitude of the PFM Signal

To obtain quantitative data of the piezoelectric deformation of the sample underneath the tip, the PFM background must be corrected. The background can be determined with the help of a PPLN crystal, measuring the piezoresponse on both domain faces (P_X and N_X) and then calculating $B = \frac{1}{2}(P_X + N_X)$. Another way to determine B consists in using a nonpiezoelectric sample (glass, metallized surface) as a reference. Figure 8.8(a) shows the uncorrected PFM signal and (b) the result after background correction for a frequency scan from 10 kHz to 100 kHz. The amplitude of the PFM signal becomes frequency independent, as it is required from the physics of the converse piezoelectric effect in this frequency regime. Furthermore, the obtained value for the piezoelectric coefficient of $\sim$6 pm/V is in first order consistent with other measurements. Interestingly, for this value, the agreement is not very good in literature, the published data varying between 6 pm/V and 23 pm/V (e.g., see references within [39]).

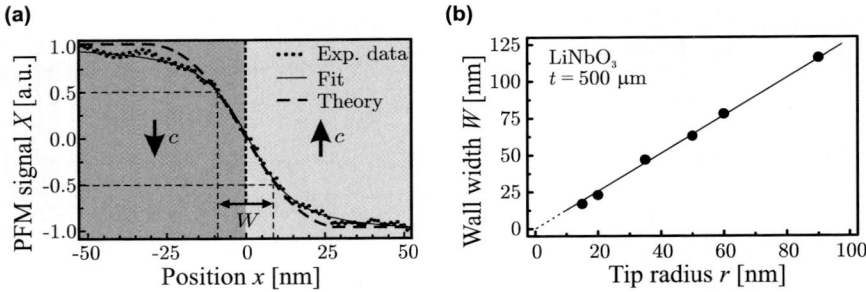

Fig. 8.9. (a) Measured PFM signal line scan (•) across a 180° domain wall in LiNbO$_3$ recorded with a tip of $r = 15$ nm radius. The domain wall width W is determined fitting the data using (8.1). For comparison, a line scan is shown that was calculated with the theoretical model for a tip of the same radius. (b) Measured domain wall width W as a function of the nominal tip radius r. The straight line was calculated using an analytical model

One explanation for obtaining such small values with PFM lies in the strongly inhomogeneous electric field of the tip inside the crystal. We therefore performed comparative measurements with single crystals exploring two different configurations [39]: (i) standard PFM with the tip acting as top-electrode and (ii) large-area electrode via metallization of the top face of the crystal. We have found that for the case (ii) with a homogeneous electric field inside the crystal, the measured value was larger by roughly a factor of three, i.e., 20 pm/V. In order to sustain this result, we performed comparative measurements with α-quartz and KTiOPO$_4$ [40]. All samples exhibited the same reduction of the signal by a factor of three using standard PFM. Obviously the strong restriction of the electric field within a few µm^3 leads to clamping inside the crystal thus limiting the deformation to roughly 1/3 of the expected value. Hence, PFM is not suited to yield reliable piezoelectrical coefficients [22].

8.5.2 Domain Wall Width

With the correction of the system-inherent background, the lateral resolution of PFM can reliably be answered. As a measure of the wall width seen by PFM (not to be confused with the real wall width which is expected to be few lattice units [41]), we determined the full width at half maximum W of the slope of the PFM signal across a domain boundary. Figure 8.9(a) shows an example for a measurement performed on LiNbO$_3$ using a tip with a radius of $r = 15$ nm only. We then fitted the curve with

$$X(x) = A \tanh\left(\frac{x}{w}\right) + B \arctan\left(\frac{x}{w}\right) \tag{8.1}$$

to determine W. In addition the graph also shows the theoretical curve calculated with a simple analytical model. A detailed description, however, would go beyond the scope of this contribution and can be found elsewhere [42]. Note that

Table 8.1. Domain wall width W measured for different samples with tips of radius r. c-LiNbO$_3$: congruently melting and s-LiNbO$_3$: stoichiometric lithium niobate respectively

Sample	Domain wall width W [nm]		Dielectric anisotropy
	$r = 15$ nm	$r = 35$ nm	$\gamma = \sqrt{\varepsilon_z/\varepsilon_r}$
BaTiO$_3$	19	46	0.17
KNbO$_3$	18	45	0.34
KTiOPO$_4$	17	46	1.16
c-LiNbO$_3$	17	46	0.58
s-LiNbO$_3$	17	45	0.58
Mg:LiNbO$_3$	18	47	0.58
LiTaO$_3$	18	45	1.10
Sr$_{0.61}$Ba$_{0.39}$Nb$_2$O$_6$	18	48	1.52
Pb$_5$Ge$_3$O$_{11}$	18	45	1.40

$W = 17$ nm is the so far highest lateral resolution achieved with PFM in bulk single crystals.

In order to establish a relation between tip radius r and lateral resolution, we determined W for a series of tips with different tip radii. The result depicted in Fig. 8.9(b) shows a clear linear dependence, which is consistent with an infinite sharp domain boundary.

We also compared W for different crystals using two tips of different radius. All samples show the same width W for a specific tip of radius $r = 15$ nm or $r = 35$ nm within an error of ± 1 nm, although their dielectric anisotropy $\gamma = \sqrt{\varepsilon_z/\varepsilon_r}$ differ as listed in Table 8.1. A detailed calculation of the electric field distribution inside the crystal shows, however, that the latter is independent of γ, thus although the amplitude of the PFM signal differs, the slope of W has to be same for all crystals [42].

8.6 Ferroelectric Domain Imaging by Lateral Force Microscopy

The contrast mechanism for the detection of ferroelectric domain boundaries with lateral force microscopy was generally assumed to be caused by the deformation of the sample at the domain boundaries due to the converse piezoelectric effect [43, 44]. The tip was expected to be deflected sidewise due to the slope of the surface. In this case, however, the amplitude of the lateral signal should scale with the load of the tip. This was not observed [45]. A quantitative analysis of the measured forces shows that the electrostatic interaction between the charged tip and the electric fields arising from the surface polarization charges causes the contrast (Sect. 8.6.1). We therefore call the detection technique lateral electrostatic force microscopy (LEFM). Ferroelectric domain structures in a single crystal turned out to be an ideal sample to show the different movements of the cantilever (deflection, torsion, buckling) without additional topographic features (Sect. 8.6.2).

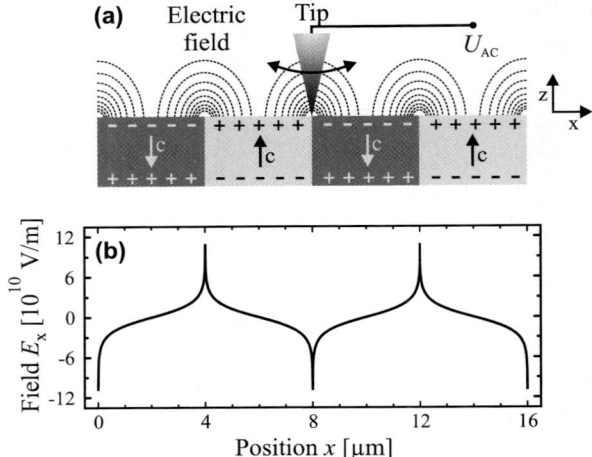

Fig. 8.10. (a) Schematic drawing of the static electric fields above the z face of a periodically poled ferroelectric crystal. Here c denotes the polar axis, U_{AC} the alternating voltage applied to the tip. (b) Electrical field component E_x 35 nm above the surface of a PPLN crystal with a period length of 8 µm calculated using (8.2)

8.6.1 Origin of the Lateral Signal

The (uncompensated) surface polarization charge density for LiNbO$_3$ is $\sigma = 0.71$ C/m^2 [46]. Figure 8.10 shows a sidewise sketch of a PPLN crystal. Because of the surface polarization charges, electric fields build up whose strength parallel to the surface is most at the domain boundaries. The electric field $E_x(x, z)$ with x being the axis parallel to the surface and perpendicular to the domain boundaries and z denoting the distance from the sample surface for an infinite PPLN structure is given by [45]

$$E_x(x, z) = \frac{\sigma}{4\pi\varepsilon_0} \ln\left[\prod_{n=-\infty}^{\infty} \frac{[(x+2na)^2 + z^2]^2}{[(x+2na+a)^2 + z^2]^2}\right] \quad (8.2)$$

with a denoting the domain size (PPLN period: $\Lambda = 2a$) and n the number of domains being included. For the PPLN sample investigated ($\Lambda = 8$ µm), electric field strengths of 10^{11} V/m are theoretically expected if no compensation of the surface charges is assumed.

8.6.2 Application to PPLN

In Fig. 8.11 the experimental results for deflection (a) and torsion (b) images of the end of a poled stripe of PPLN are shown with the corresponding scan lines in (d). The orientation of the cantilever was chosen to be parallel to the stripe (see also inset of Fig. 8.11(a, b)). At first sight it is obvious that the deflection image

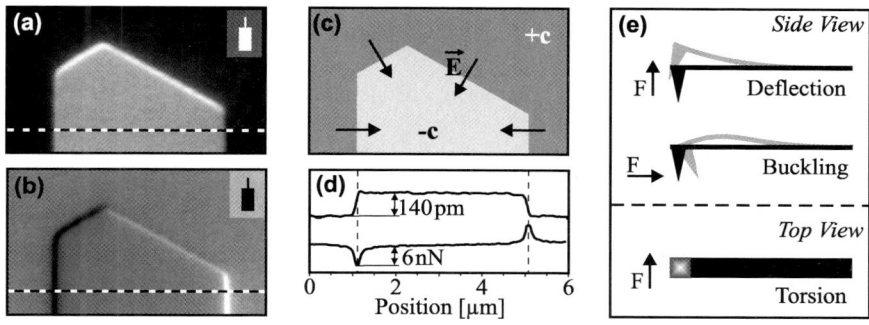

Fig. 8.11. Deflection (**a**) and torsion (**b**) images simultaneously recorded on a LiNbO$_3$ crystal (image size 6 × 3.5 µm^2) with the corresponding scanlines (**d**) with 10 V$_{pp}$ applied to the tip. The orientation of the chip with the cantilever is shown as *insets* in (**a**) and (**b**). Schematic drawing (**c**) of the electric field distribution **E**. In (**e**) the possible movements of the cantilever are depicted. The deflection image (**a**) shows deflection (PFM) and buckling (LEFM), the torsion image (**b**) the twisting of the cantilever (LEFM)

(a) shows the domain faces (due to the converse piezoelectric effect), whereas the torsion image (b) only shows the domain boundaries, at the left edge as a dark stripe and at the right edge as a bright stripe. The contrast inversion is due to the change of the direction of the electric field (see also Fig. 8.11(c)). The contrast is reduced when the electric field vector perpendicular to the cantilever becomes smaller as it can be seen on the tilted edges of the domain. As the cross-talk between vertical and lateral signals was suppressed [19], the level of the torsion signal within and outside the domain is same. Looking more closely at Fig. 8.11(a), at the top edges of the domain, a bright stripe is visible. When comparing with the schematic drawing of the electric field configuration in Fig. 8.11(c), at these edges, the electric field has a component along the axis of the cantilever. This also leads to lateral forces acting on the tip which result in a buckling of the cantilever. Quantitatively comparing the measured lateral forces with the ones expected from a simple model suggests a compensation of the surface polarisation charges by a factor of 100–1000. However, this is in full agreement with previously published data [47, 48].

To proof that electrostatic forces are responsible for the lateral signals at the domain boundaries, we performed comparative experiments with periodically poled KTiOPO$_4$ crystals. Because the piezoelectric coefficient d_{333} of KTiOPO$_4$ [49] is comparable to the one of LiNbO$_3$, the expected tilting of the surface at the domain boundaries should be identical. The measured lateral forces, however, are smaller by a factor of 2.6 with respect to those on LiNbO$_3$. This agrees with an electrostatic origin of the lateral forces as the surface polarization charge density is approximately three times smaller for KTiOPO$_4$ than for LiNbO$_3$ [40, 46].

8.7 Conclusions

Our aim in this chapter was to clear out inconsistencies and surprising features that were reported previously in connection with PFM measurements. In the end, it turns out, that the discovery of the system-inherent background leads to a simplification of the interpretation of the observed signals and enables to record quantitative data with PFM. All signals detected on the polar faces of multi-domain single crystals can be explained by the converse piezoelectric effect for the deflection (on top of the domain faces) and electrostatic interaction for torsion and buckling (at the domain boundaries).

References

1. M.M. Fejer, G.A. Magel, D.H. Jundt, R.L. Byer, Quasi-phase-matched 2nd harmonic-generation – tuning and tolerances. IEEE J. Quantum Electron. **28**, 2631–2654 (1992)
2. L.E. Myers, R.C. Eckardt, M.M. Fejer, R.L. Byer, Quasi-phase-matched optical parametric oscillators in bulk periodically poled $LiNbO_3$. J. Opt. Soc. Am. B **12**, 2102–2116 (1995)
3. H. Ishiwara, M. Okuyama, Y. Arimoto, *Ferroelectric Random Access Memories: Fundamentals and Applications*, vol. 93 (Springer, Berlin, 2004)
4. E. Soergel, Visualization of ferroelectric domains in bulk single crystals. Appl. Phys. B **81**, 729–752 (2005)
5. M. Alexe, A. Gruverman, *Nanoscale Characterisation of Ferroelectric Materials* (Springer, Berlin, 2004)
6. R.E. Newnham, *Properties of Materials: Anisotropy, Symmetrie, Structure* (Oxford University Press, Oxford, 2005)
7. J. Erhart, Domain wall orientations in ferroelastics and ferroelectrics. Phase Transitions **77**, 989–1074 (2004)
8. M.E. Lines, A.M. Glass, *Principles and Applications of Ferroelectrics and Related Materials* (Oxford University Press, New York, 2001)
9. B.A. Strukov, A.P. Levanyuk, *Ferroelectric Phenomena in Crystals* (Springer, Berlin, 1998)
10. J.A. Gonzalo, B. Jiménez (eds.), *Ferroelectricity: The Fundamentals Collection* (Wiley-VCH Verlag, Weinheim, 2005)
11. R.S. Weis, T.K. Gaylord, Lithium niobate: summary of physical properties and crystal structure. Appl. Phys. A **37**, 191–203 (1985)
12. J.C. Brice, *Crystal Growth Processes* (Halsted, New York, 1986)
13. J.F. Nye, *Physical Properties of Crystals* (Oxford University Press, Oxford, 1985)
14. V. Gopalan, T.E. Mitchell, Y. Furukawa, K. Kitamura, The role of nonstoichiometry in 180° domain switching of $LiNbO_3$ crystals. Appl. Phys. Lett. **72**, 1981–1983 (1998)
15. M.C. Wengler, B. Fassbender, E. Soergel, K. Buse, Impact of ultraviolet light on coercive field, poling dynamics and poling quality of various lithium niobate crystals from different sources. J. Appl. Phys. **96**, 2816–2820 (2004)
16. K.K. Wong, *Properties of Lithium Niobate* (INSPEC, London, 2002)
17. D. Sarid, *Scanning Force Microscopy* (Oxford University Press, London, 1994)
18. E. Soergel, W. Krieger, V.I. Vlad, Charge distribution on photorefractive crystals observed with an atomic force microscope. Appl. Phys. A **66**, S337–S340 (1998)

19. Á. Hoffmann, T. Jungk, E. Soergel, Cross-talk correction in atomic force microscopy. Rev. Sci. Instrum. **78**, 016101 (2007)
20. M. Reinstaedtler, U. Rabe, V. Scherer, J.A. Turner, W. Arnold, Imaging of flexural and torsional resonance modes of atomic force microscopy cantilevers using optical interferometry. Surf. Sci. **532–535**, 1152–1158 (2003)
21. S.V. Kalinin, A. Gruverman, *Scanning Probe Microscopy: Electrical and Electromechanical Phenomena at the Nanoscale* (Springer, Berlin, 2006)
22. T. Jungk, Á. Hoffmann, E. Soergel, Challenges for the determination of piezoelectric constants with piezoresponse force microscopy. Appl. Phys. Lett. **91**, 253511 (2007)
23. M. Labardi, V. Likodimos, M. Allegrini, Force-microscopy contrast mechanisms in ferroelectric domain imaging. Phys. Rev. B **61**, 14390–14398 (2000)
24. A. Agronin, M. Molotskii, Y. Rosenwaks, E. Strassburg, A. Boag, S. Mutchnik, G. Rosenman, Nanoscale piezoelectric coefficient measurements in ionic conducting ferroelectrics. J. Appl. Phys. **97**, 084312 (2005)
25. J.W. Hong, K.H. Noh, S. Park, S.I. Kwun, Z.G. Khim, Surface charge density and evolution of domain structure in triglycine sulfate determined by electrostatic-force microscopy. Phys. Rev. B **58**, 5078–5084 (1998)
26. M. Shvebelman, P. Urenski, R. Shikler, G. Rosenman, Y. Rosenwaks, M. Molotskii, Scanning probe microscopy of well-defined periodically poled ferroelectric domain structure. Appl. Phys. Lett. **80**, 1806–1808 (2002)
27. K. Takata, Comment on "Domain structure and polarization reversal in ferroelectrics studied by atomic force microscopy". J. Vac. Sci. Technol. B **14**, 3393–3394 (1996). [J. Vac. Sci. Technol. B **13**, 1095 (1995)]
28. H. Ogi, Y. Kawasaki, M. Hirao, H. Ledbetter, Acoustic spectroscopy of lithium niobate: elastic and piezoelectric coefficients. J. Appl. Phys. **92**, 2451–2456 (2002)
29. O. Kolosov, A. Gruverman, J. Hatano, K. Takahashi, H. Tokumoto, Nanoscale visualization and control of ferroelectric domains by atomic force microscopy. Phys. Rev. Lett. **74**, 4309–4312 (1995)
30. M. Labardi, V. Likodimos, M. Allegrini, Resonance modes of voltage-modulated scanning force microscopy. Appl. Phys. A **72**, S79–S85 (2001)
31. S. Hong, H. Shin, J. Woo, K. No, Effect of cantilever-sample interaction on piezoelectric force microscopy. Appl. Phys. Lett. **80**, 1453–1455 (2002)
32. C. Harnagea, M. Alexe, D. Hesse, A. Pignolet, Contact resonances in voltage-modulated force microscopy. Appl. Phys. Lett. **83**, 338–340 (2003)
33. C.H. Xu, C.H. Woo, S.Q. Shi, Y. Wang, Effects of frequencies of AC modulation voltage on piezoelectric-induced images using atomic force microscopy. Mater. Charact. **52**, 319–322 (2004)
34. S.V. Kalinin, D.A. Bonnell, Imaging mechanism of piezoresponse force microscopy of ferroelectric surfaces. Phys. Rev. B **65**, 125408 (2002)
35. L.M. Eng, H.-J. Güntherodt, G. Rosenman, A. Skliar, M. Oron, M. Katz, D. Eger, Nondestructive imaging and characterization of ferroelectric domains in periodically poled crystals. J. Appl. Phys. **83**, 5973–5977 (1998)
36. T. Jungk, Á. Hoffmann, E. Soergel, Quantitative analysis of ferroelectric domain imaging with piezoresponse force microscopy. Appl. Phys. Lett. **89**, 163507 (2006)
37. T. Jungk, A. Hoffmann, E. Soergel, Consequences of the background in piezoresponse force microscopy on the imaging of ferroelectric domain structures. J. Microsc. Oxford **227**, 76–82 (2007)
38. W. Heywang, H. Thomann, Tailoring of piezoelectric ceramics. Ann. Rev. Mater. Sci. **14**, 27–47 (1984)

39. T. Jungk, A. Hoffmann, E. Soergel, Influence of the inhomogeneous field at the tip on quantitative piezoresponse force microscopy. Appl. Phys. A **86**, 353–355 (2007)
40. G. Rosenman, A. Skliar, M. Oron, M. Katz, Polarization reversal in $KTiOPO_4$ crystals. J. Phys. D **30**, 277–282 (1997)
41. J. Padilla, W. Zhong, D. Vanderbilt, First-principles investigation of 180° domain walls in $BaTiO_3$. Phys. Rev. B **53**, R5969–R5973 (1996)
42. T. Jungk, A. Hoffmann, E. Soergel, Impact of the tip radius on the lateral resolution in piezoresponse force microscopy. New J. Phys. **10**, 013019 (2008)
43. D.A. Scrymgeour, V. Gopalan, Nanoscale piezoelectric response across a single antiparallel ferroelectric domain wall. Phys. Rev. B **72**, 024103 (2005)
44. J. Wittborn, C. Canalias, K.V. Rao, R. Clemens, H. Karlsson, F. Laurell, Nanoscale imaging of domains and domain walls in periodically poled ferroelectrics using atomic force microscopy. Appl. Phys. Lett. **80**, 1622–1624 (2002)
45. T. Jungk, A. Hoffmann, E. Soergel, Detection mechanism for ferroelectric domain boundaries with lateral force microscopy. Appl. Phys. Lett. **89**, 042901 (2006)
46. K.-H. Hellwege (ed.), *Landolt-Börnstein: Numerical Data and Functional Relationships in Science and Technology. New Series*, vol. III/16 (Springer, Berlin, 1981)
47. V. Likodimos, M. Labardi, M. Allegrini, N. Garcia, V.V. Osipov, Surface charge compensation and ferroelectric domain structure of triglycine sulfate revealed by voltage-modulated scanning force microscopy. Surf. Sci. **490**, 76–84 (2001)
48. S.V. Kalinin, D.A. Bonnell, Local potential and polarization screening on ferroelectric surfaces. Phys. Rev. B **63**, 125411 (2001)
49. H. Graafsma, G.W.J.C. Heunen, S. Dahaoui, A.El. Haouzi, N.K. Hansen, G. Marnier, The piezoelectric tensor element d_{33} of $KTiOPO_4$ determined by single crystal X-ray diffraction. Acta Crystallograph. B **53**, 565–567 (1997)

9 Structural Characterization of Periodically Poled Lithium Niobate Crystals by High Resolution X-Ray Diffraction

M. Bazzan, N. Argiolas, C. Sada, and P. Mazzoldi

9.1 Introduction

The exploding demand for Internet access, telecommunications, broadband service has led to a push for a greater lightwave transmission capacity and compactness. In order to meet that demand, a number of sophisticated optical components are required: as far as device opportunities are concerned, several technologies are therefore contenders for implementation such as bulk-type, fiber-type and planar type devices respectively. When complex functions are integrated in a planar geometry, reduced cost is achieved with batch processing of wafers and fewer manual interconnections, which also brings enhanced reliability. Since many of these applications would benefit from the availability of lasers operating at the blue or shorter wavelength, there is a great effort to achieve this directly by production of semiconductor lasers from II–VI compounds, or alternatively, by mixing frequencies from existing lasers through highly efficient second order non-linear crystals. Despite of the large band of frequencies that can be accessed by this solution, in the last case the phase mismatch effect due to bulk material dispersion must be overcome. In order to gain reasonable intensities, the periodic modulation of the non-linear polarizability demonstrated to be the most efficient solution. This task can be accomplished in ferroelectric crystals such as $LiNbO_3$ by periodically inverting the sign of the spontaneous polarization P of the material (procedure that gives Periodically Poled Lithium Niobate or PPLN crystals). A wide number of treatments involving chemistry, heat and electric fields are commonly used to realize this inversion that usually requires the deposition of metal electrodes and the application of high electric field (of the order of kV at room temperature) to obtain the domain inversion. In this case periods of less than 4 μm over thicknesses grater then 500 μm are difficult to be achieved without imperfections due to physical limitations in the electrodes deposition and in the spontaneous polarization inversion. Moreover, the inversion of the polarization through the application of an electric field is limited to thicknesses compatible with the breakdown field of the material. When sub-micrometric PPLN structures are needed, as for photonic band gap applications, a different approach must be used as proposed by V. Shur and coworkers [1]. In this case the poling process is carried on in the overpoling regime, so that the final domain period is determined taking into account a partial reduction of the domain thickness induced by the back switching of the freshly poled regions due to a residual internal field which is always present in non-stoichiometric LN samples at room temperature [2]. Moreover, it has been reported that the combination of interferential lithography techniques with electric field poling performed in the overpoling regime [3] can be exploited to obtain good quality sub micrometer periodic domain structures. On the other hand, PPLN crystals can also be directly grown by not conventional Czochralski technique, i.e. by off-centering the temperature field axis with respect to the growth one, starting from a melt doped with rare earth ions and pulling along the crystallographic [2−1.0] direction. The periodicity of the PPLN structures depends on the growth parameters and can be tailored to obtain ferroelectric domains in the micrometer range – periods as small as than 2 μm are feasible but not straightforward. These directly-grown PPLNs present a periodic patterning through the whole volume of the crystal but the domain shape is bent, following the growth interface curvature [4, 5]. Independently of the preparation technique, anyway, a PPLN crystal is generally described as a sequence of single domains of reversed polarization: the spontaneous polarization profile is typically modelled as a square or rectangular wave, depending on the size of the opposite domains respectively. Moreover in the vast majority of theoretical studies, the structure of the crystalline matrix is taken to be perfect, i.e. equal to that of a perfect crystal, ignoring the presence of structural defects induced by the poling process. This assumption is an oversimplification of the real case. Due to the intrinsic tendency of lithium niobate to non-stoichiometry, in fact, a certain number of randomly distributed polar defects is generally present inside the crystal matrix, which are known to affect substantially the ferroelectric properties of the samples [2, 6]. Moreover, the modelling of a PPLN crystals with a nearly zero-thickness of the domain walls [8] represents a strong approximation. As a matter of fact at the domain wall a polarization gradient, a local modification of the optical properties and elastic deformations can be observed [2, 9], as sketched in Fig. 9.1.

It was pointed out that the situation in real crystals is much more complicated: domain walls appear to have a finite thickness [10, 11] and the presence of long-range strains were detected [12]. As they can hardly influence the optical properties of the PPLN crystals and the long term stability of the ferroelectric domain pattern, a measure of the structural modifications occurring within the PPLN structures is mandatory. When the crystal lattice parameters vary along the crystal volume, in fact, the physical properties of the material can change even drastically: this is true especially for the case of piezoelectric and electro-optical materials, because various thermodynamic fields can couple together giving rise to complicated physical situations. This for example is the case for domain walls: as they are at the interface between two region with opposite polarization, they are characterized by a structural change in the crystal lattice. Across a domain wall there exists a polarization

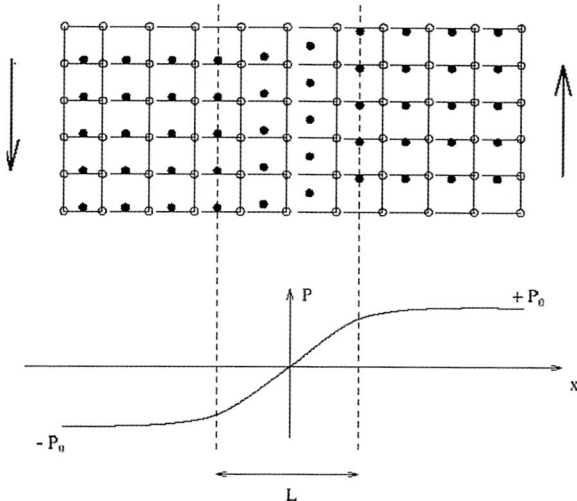

Fig. 9.1. A schematic picture of the domain wall. The polarization P changes across the border domain, taking the P_0 value in a positive domain and P_0 value in a negative one. L indicates the domain width

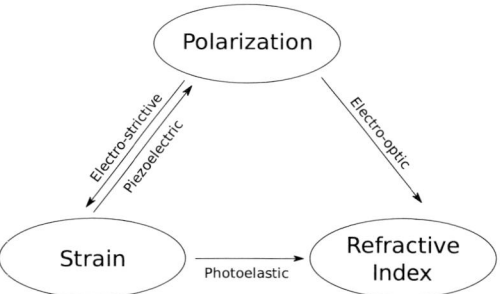

Fig. 9.2. Scheme of the coupling between different fields and physical quantities in a domain wall

gradient, a local modification in the optical properties and elastic deformations [6] coupled together as sketched in Fig. 9.2.

As the spatial distribution of lattice parameters in a crystal determines the development of strain fields, the structural investigation has a key importance to characterize the material.

Apart from giving a complete picture of the PPLN real structure, a detailed comprehension of the phenomena occurring during the poling process would benefit from the investigation of the structural properties of the material. Literature reports about investigations on PPLN crystals by means of imaging techniques such as X-ray topography which have a spatial resolution limited to some microns [12, 13]. This technique can therefore give interesting information on long period structures as those obtained by the electric field poling. When sub-micron periodic structures are realized, instead, its limited lateral resolution prevents from obtaining spatially resolved information. Recently, High Resolution X-ray Diffraction (HRXRD) technique was presented as an alternative method to investigate the structural properties of PPLN crystals and in particular, demonstrated to be a valid non-destructive method to get information on the sub-micrometric PPLN structures. Although its potentialities were already investigated on PPLN with larger periods [14], in the case of sub-micrometric PPLN crystals it is easier to extract a whole of structural information. As a matter of fact, apart from the quantitative determination of the domain period, domain shape and the eventual domain wall inclination, HRXRD can be exploited to investigate on the spontaneous polarization profile and on the presence of lattice distortions. In this case a theoretical description of the scattering process from a PPLN is however necessary. In the following HRXRD technique will be presented focusing the attention on the characterization of PPLNs structures. They, in fact, can be viewed as crystals with a periodic modulation of the inner cell structure and a shift of the cells due to the presence of both static deformations induced by the poling process and random strains due to defects. Some special cases of scientific interest will be reviewed in order to evidence both the technique potentialities and the limits. In particular the structural characterization of sub-micrometric PPLN crystals realized by interference technique and of PPLN grown by the off-center technique will be presented as applications.

9.2 The Principle of the XRD Technique

When an electromagnetic monochromatic plane wave with wavelength in the X-ray spectral range impinges on a crystal, diffraction occurs. If only elastic scattering phenomena are considered, constructive interference of waves reflected by the crystallographic planes with Miller indices $[h\,k\,l]$ and interplanar distance $d_{[h\,k\,l]}$ takes place. A plane wave with wavelength λ is scattered away from the sample provided that the Bragg conditions is fulfilled [15]:

$$m\lambda = 2d_{[h\,k\,l]}\sin(\theta) \tag{9.1}$$

where m is an integer and θ represents the angle between the beam direction and the $[h\,k\,l]$ plane (see Fig. 9.3).

The Bragg condition expressed in (9.1) means that diffraction occurs provided that the wavevector $\boldsymbol{K}_0$ of the incident X-ray beam and the wavevector $\boldsymbol{K}$ of the diffracted beam satisfy the following relation:

$$\boldsymbol{K} - \boldsymbol{K}_0 = \boldsymbol{g}_{[h\,k\,l]} \tag{9.2}$$

where $\boldsymbol{g}_{[h\,k\,l]}$ is a vector indicating a given reciprocal lattice point that is associated to that set of crystallographic planes with Miller indices $[h\,k\,l]$. In particular $g_{[h\,k\,l]} = \frac{2\pi}{d_{[h\,k\,l]}}$ and $K_0 = \frac{2\pi}{\lambda} = K$. In principle, by varying the incident and diffracted wavevectors $\boldsymbol{K}$ and $\boldsymbol{K}_0$ and recording the diffracted intensity, it is therefore

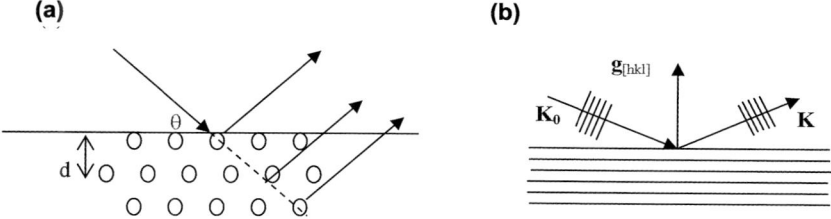

Fig. 9.3. (a) Sketch of the Bragg condition in the X-ray diffraction from a general crystal: the atoms are represented by *open circles*, *d* being the interplanar distance. (b) The Bragg condition expressed in the reciprocal space

possible to gain information on the sample structure. If structural modifications occur in the crystals, they can be detected and deeply investigated, as they modify the fine structure of the reciprocal lattice points $g_{[hkl]}$. In practise, this structural investigation is more complicated: in the following the key points of the structural characterization will be briefly revised, focusing the attention on the high resolution configuration.

9.2.1 The Theory of High Resolution X-Ray Diffraction

Perfect Crystals

The theoretical description of the scattering phenomena of the X-ray radiation occurring within the crystalline sample can be performed by using two different approaches:

- The dynamical scattering theory that describes the scattering of X-rays by solving the Maxwell equations in a medium characterized by a dielectric polarizability which is periodic along the three spatial directions, as it is the case for a crystalline matrix. This is the most rigorous approach for the description of the diffraction phenomenon and correctly predicts the functional form of the diffraction peaks, in particular their finite width and their intensity and other experimentally observed features such as the Borrmann effects [15].
- The kinematical scattering theory, in which the diffracted wavefield is viewed simply as the "sum" of the waves diffracted by each atom inside the crystal volume. This theory is generally a good approximation of the dynamical theory when multiple reflections inside the sample can be neglected, for example with very thin or highly defective epitactic layers [16] or laterally patterned structures.
  The kinematical theory can apply also to high quality crystalline samples by adding some "artificial" corrections in order to take into account typical dynamical effects, such as the limited extinction depth of X-rays inside matters.

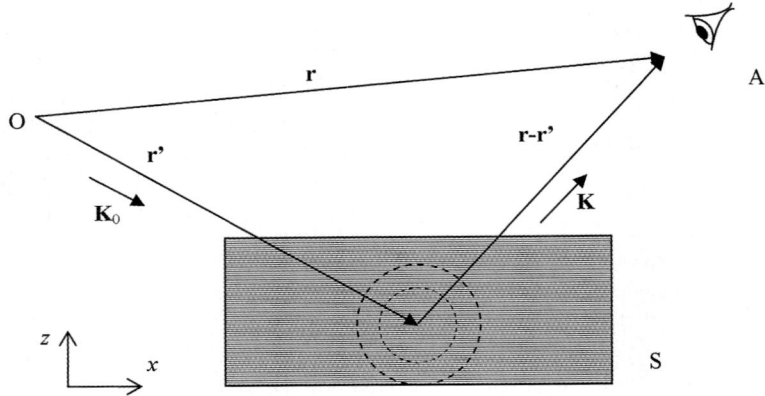

Fig. 9.4. Reference system of (9.3). S indicates the sample, A the observer and O origin of the reference system

In the following the kinematical theory will be exploited as a first attempt to the description of scattering from a ferroelectric sample with periodic domain structures, as it is easier to be understood and implemented and is able to catch the essential features of the experimental results.

Omitting the time dependence, a plane X-ray beam can be described as $E_0 e^{i(K_0 \cdot r)}$ where E_0 indicates its amplitude. If this wave impinges on a perfect crystal, the field $E(r)$ scattered by the crystal is given by [15]:

$$E(r) = E_0(r) \sin(\phi) r_{el} \int d^3 r' \, \rho(r') \Omega(r') e^{i K_0 \cdot r'} \frac{e^{iK(r-r')}}{|r-r'|} \quad (9.3)$$

where (see Fig. 9.4):

- r is the observation point; r' is the position of the scattering center (electrons in the case of X-ray interaction with matter);
- $\rho(r')$ is the electron density in the whole diffracting volume;
- $r_{el} = \frac{e^2}{4\pi \varepsilon_0 m_e c^2}$ (classical electron radius), e = electron charge, m_e = electron mass, c = vacuum light speed, ε_o is the vacuum permittivity;
- ϕ is the angle between E_0 and $r - r'$ respectively;
- $\Omega(r')$ is the shape function of the diffracting volume, which is equal to 1 inside it and 0 elsewhere.

In the following, for sake of simplicity we will focus on the σ polarization configuration, i.e. the primary (E_0) and the scattered (E) waves amplitudes are perpendicular to the plane defined by the vectors K_0 and $r - r'$ (the scattering plane) respectively. In this case $\sigma = \pi/2$, and we can take (9.3) as a scalar one.

In an infinite perfect crystal, $\rho(r')$ is a periodic function of the position and therefore can be written as a Fourier series over the reciprocal lattice with lattice

vectors g:
$$\rho(r) = \Sigma \rho_g e^{ig \cdot r} \tag{9.4}$$

where
$$\rho_g = \frac{1}{V_{cell}} \int_{V_{cell}} d^3r \, \rho(r) e^{-ig \cdot r} \tag{9.5}$$

V_{cell} being the volume of the unit lattice cell. In the theoretical treatment of X-ray scattering, however, it is customary to assume the classical electronic polarization mechanism and instead of the electron density $\rho(r)$, use the crystal polarizability $\chi(r)$:

$$\chi(r) = -r_{el} \frac{\lambda}{\pi} \rho(r). \tag{9.6}$$

In an infinite perfect crystal, $\chi(r)$ is therefore a periodic function of the position and therefore can be written as a Fourier series over the reciprocal lattice with lattice vectors g

$$\chi(r) = \Sigma \chi_g e^{ig \cdot r}, \tag{9.7}$$

$$\chi_g = \frac{1}{V_{cell}} \int_{V_{cell}} d^3r \, \chi(r) e^{-ig \cdot r}. \tag{9.8}$$

In (9.3) by expanding the term $\frac{e^{iK(r-r')}}{|r-r'|}$ in a superposition of plane waves in which $K_{\parallel}$ represents the parallel component of K and K_z the z-component respectively, one obtains:

$$\frac{e^{iK(r-r')}}{|r-r'|} = \frac{i}{2\pi} \int d^2K_{\parallel} \frac{e^{iK \cdot (r-r')}}{K_z}. \tag{9.9}$$

As a consequence, it emerges that it is always possible to express the scattered amplitude E as follows:

$$E(r) = -\frac{iK^2}{8\pi^2} E_0 \sum_g \int d^2K_{\parallel} \frac{e^{iK \cdot r}}{K_z} F_g(K - K_g) \tag{9.10}$$

where $K_g = K_0 + g$ and

$$F_g(K - K_g) = \int d^3r' \, \Omega(r') \chi_g e^{i(K - K_g) \cdot r'} \tag{9.11}$$

represents the so called structure factor of the crystal. Equation (9.10) shows that the diffracted wavefield is given by the sum of a discrete number of integrals, one for each reciprocal lattice vector g. Around each lattice point, the diffracted intensity is given by a superposition of plane waves $e^{iK \cdot r}$ weighted by the structure factor $F_g(K - K_g)$ which is a function of the distance in reciprocal space between the generic wavevector K and K_g, which is the wavevector of the wave satisfying exactly the Bragg condition (9.2). For a perfect, infinite crystal, we can see from (9.11) that the structure factor becomes simply a Dirac delta function, so that in

(9.10) for every reciprocal lattice vector g, only one plane wave survives in the integral and has an amplitude proportional to χ_g. In the reciprocal space therefore the scattered intensity presents peaks corresponding to those wave vectors that satisfy the Bragg condition (see Fig. 9.3). In experimental analysis generally one focuses on the analysis of a chosen reciprocal lattice point $g_{[hkl]}$ per time. For the sake of clarity we can define the vector:

$$q = K - K_g \qquad (9.12)$$

q representing the diffraction vector measured with respect to a reference system having the origin at the maximum of the reciprocal lattice point (see Fig. 9.5). As a consequence, near the g Bragg reflection of a perfect crystal, the reciprocal space intensity distribution can be expressed as follows:

$$J(q) = C_1 |F_g(q)|^2 \qquad (9.13)$$

where all the constants have been grouped in C_1. For a perfect crystal the intensity peaks at $q = 0$ (that is as expressed by (9.2)). A crystal is generally imperfect: as a consequence the structure factor given by (9.11) is no more a delta function but develops a non-trivial dependence on q. Consequently, the $J(q)$ peaks are broadened, containing the fingerprint of this structural deformation (Fig. 9.6). A map of the intensity $J(q)$ in an area Σ_q of the q-space is called reciprocal map. It may be obtained by recording the diffracted intensity varying q, which can be done by selecting the directions of primary K_0 and diffracted K beams in a suitable way.

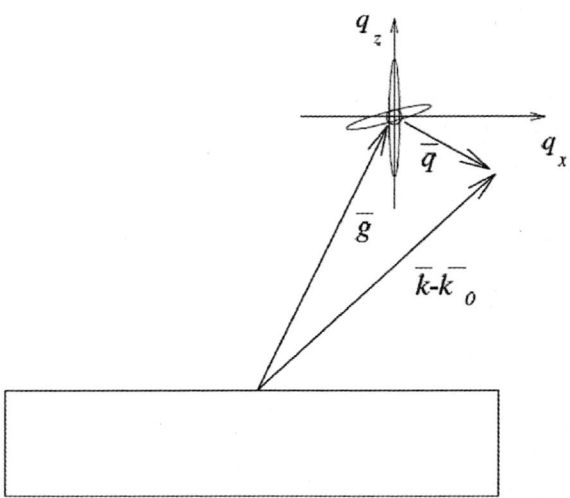

Fig. 9.5. Definition of the reciprocal lattice vectors: g is the position of the reciprocal lattice point, q is the momentum transfer vector expressed in the reference system where in g is set as the origin

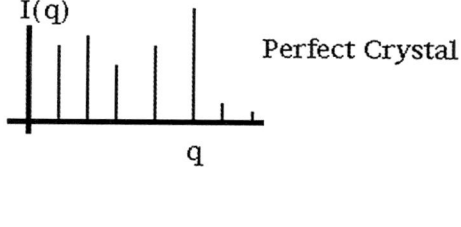
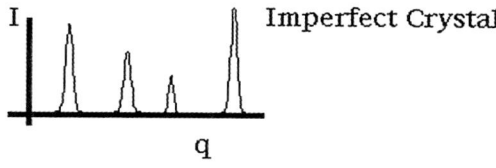

Fig. 9.6. Sketch of the shape of the intensity $J(q)$ for a perfect and an imperfect crystal respectively

Imperfect Crystals

In practice, a crystal is commonly imperfect: the presence of the surfaces by itself breaks the translational perfection of the sample. Moreover, structural defects and the consequent deformations of the crystal matrix induce some modifications of the shape of the reciprocal lattice points. In the following the general theory describing these effects will be outlined.

If the structural modifications perturbing the crystalline perfection of a sample are not too significant, they can be described in the framework of the kinematical theory in terms of two effects [17], both depending on the position r:

1. A modification of the positions of the atoms inside the unit cells with respect to their position in a perfect, reference crystal.
2. A shift of the cell as a whole with respect to the position it should occupy in an undistorted lattice.

From the viewpoint of the X-ray diffraction theory, the sample structure is described by the polarizability $\chi(r)$. If the disturbance of the structure is not too strong, we can express the polarizability of an imperfect sample by using a modified Fourier series [16]:

$$\chi(r) = \sum_g \chi_g(r) e^{ig \cdot (r - u(r))}. \tag{9.14}$$

In (9.14), the shift of the cells is described by the vector field $u(r)$, which is called displacement field (see Fig. 9.7). It represents the displacement of a given cell at a position r in an undistorted lattice which, after being subjected to some external influence such as the poling process, moves to a new position $r' = r - u(r)$. This vector field is a function of the position, since it may change according to different positions inside the sample.

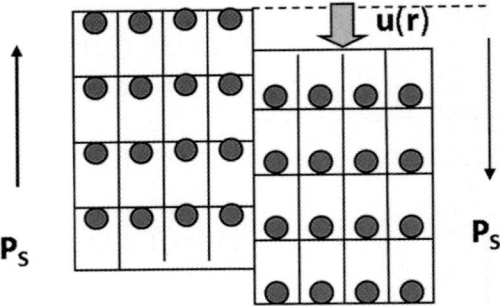

Fig. 9.7. Sketch representing the effect of the modification of the atoms position within the unit cell and the shift of the cell with respect to its position in a perfect crystal

From the displacement field, the deformation tensor $e^{(2)}(r)$ can be calculated as $e_{ij} = \frac{\partial u_i}{\partial x_j}$. The strain tensor $\varepsilon^{(2)}(r)$ and the rotation tensor $\varpi^{(2)}(r)$ are the symmetric and the antisymmetric part of the deformation tensor respectively: $\varepsilon_{ij} = \frac{1}{2}(\frac{\partial u_i}{\partial x_j} + \frac{\partial u_j}{\partial x_i})$ and $\varpi_{ij} = \frac{1}{2}(\frac{\partial u_i}{\partial x_j} - \frac{\partial u_j}{\partial x_i})$ so that when $u(r)$ is known, the deformation state of the sample is completely characterized.

The smooth dependence upon r of the Fourier coefficients $\chi_g(r)$ expresses the change in the positions of the atoms inside the unit cell. If the deformation field $u(r)$ and the modifications of the inner cell structure χ_g are known, the structure factor $F_g(K - K_g)$ in (9.11), can be calculated as well as the reciprocal space intensity distribution (see (9.13)) in order to compare it to experimental results of reciprocal space mapping measurements. It is worth mentioning that, in order to improve the comparison between the experimental data and the theory when the kinematical approach is used, it is important to take into account that the incident X-ray beam intensity decreases while propagating inside the crystal. This can be done by defining an effective absorption coefficient [18] that considers the dynamical extinction of the X-rays within the material (ξ being the correspondent extinction coefficient) and the absorption due to the photovoltaic effect (μ being the absorption coefficient). The effective penetration length L_p, therefore contains the contribute of the extinction length $L_e = \sin(\alpha_i)/\xi(\alpha_i)$ and the absorption length $L_a = \sin(\alpha_i)/\mu$ respectively where α_i is the angle between the direction of the primary beam and the sample surface (see Fig. 9.8). As a consequence, L_p depends on the q_x and q_z values:

$$\frac{1}{L_p(q_x, q_z)} = \frac{1}{L_a(q_x, q_z)} + \frac{1}{L_e(q_x, q_z)}. \qquad (9.15)$$

In summary, the finite vertical size of the diffracting volume is determined by the reduction of the primary beam intensity that undergoes an exponential decay with an effective penetration length L_p.

The analysis of the diffracted intensity therefore can give many information on the material structure: in the special case of PPLN crystals, by the high resolution X-ray diffraction it is therefore possible to determine the domain period and shape,

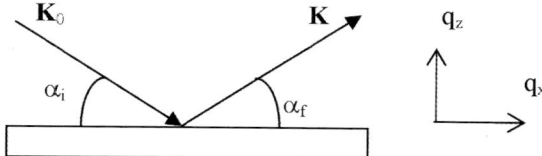

Fig. 9.8. Sketch of the analyzed configuration: α_i indicates the angle between the direction of the primary beam and the sample surface. α_f indicates the angle between the direction of the scattered beam and the sample surface

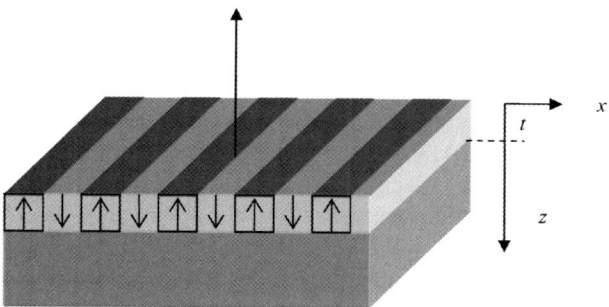

Fig. 9.9. Sketch of a typical PPLN structure: the domain pattern develops in the x direction through a thickness t

the eventual domain wall inclination, the spontaneous polarization profile and the presence of lattice distortions. In the following section this subject will be treated in more detail.

9.2.2 The HRXRD Applied to PPLN Crystals

The scattering from a PPLN crystals differs from that observed in a bulk crystal as in this case periodic modulations in the structural properties develop within the material. Let us consider a reference system where the z-axis is parallel to the c-axis of $LiNbO_3$, the domains are parallel to the y direction and therefore the grating runs in the x direction (see Fig. 9.9). In particular, depending on the preparation technique the domains can run over all the sample thickness (typically less than 1 mm in the case of the standard electric poling technique) or being confined only near the surface (in a thickness t close to few tenth of micrometers, as those prepared by the technique presented in [1]). In principle, domain walls can have tailored shapes: in the case of the conventional electric field poling they are straight. Domain walls, however, can be curved, as it happens in the PPLN crystals grown by the off-center Czochralski (CZ) technique.

Due to the periodic reversal of the spontaneous polarization, the unit cell of the crystal changes passing from one polarization state to the other by a shift of the Li sublattice, according to the polarity of the domain where the cell is contained (see Fig. 9.1).

According to the above discussion, this effect can be modelled by a positional dependence of the Fourier coefficients of the polarizability $\chi_g(r)$. This means that the structure factor of the crystal is not only a function of the reciprocal lattice vectors g as in the case of a homogeneous, single domain crystal, but it depends also on the position along the sample. By assuming that the structure factor of PPLN sample depends only on x (i.e. spanning the grating direction), the function describing the structure factor for a given Bragg reflection g is periodic along x with a period Λ. It is therefore possible to associate to it the fundamental grating vector $K_x = \frac{2\pi}{\Lambda}$. On the other hand, it is known from X-ray topographic analyses [12] that, depending on the preparation conditions, the periodic reversal of the polarization is often accompanied by the presence of structural deformations. These distortions have been attributed to a piezoelectric interaction of the crystal lattice with the internal field generated by polar defects. Moreover, the presence of lattice defects such as dislocations and charged point defects can induce a disordering of the lattice which can be modelled as a random deformation with zero average of the lattice. If these deformations are small, they should not further modify the inner cell structure and their effect can be described only by a shift of the unit cells as a whole (Takagi approximation). In order to describe the deformations which can be linked directly to the periodic domain structure, it is convenient to define [20] the displacement field $u(r)$ as the sum of two independent contributions:

$$u(r) = \delta t(r) + t(r) \qquad (9.16)$$

where $\delta t(r)$ is a random displacement of the unit cells through the whole crystal due to the presence inside the matrix of randomly distributed structural defects. As it is demonstrated in [16], this random strain field gives rise to a large diffuse scattering peak which sums up to the intensity distribution produced by periodic or non-random features of the sample. It can be considered as a background to be subtracted from the maps. Moreover, $t(r)$ is a displacement produced by the poling process and therefore has a non-random character. Being determined by the domain structure, $t(r)$ can be considered as periodic along x with a period Λ. Since the system is invariant along y, we can take $t(r)$ to be independent of this direction. Inserting $\chi_g(r)$ and $t(r)$ inside (9.11) and subsequently in (9.13), it is possible to calculate the intensity distribution in reciprocal space [20]. As mentioned above, the random part of the deformation can be treated separately from the coherent part of the scattered field, being responsible of the diffused scattered intensity which adds up to the coherent part of the scattering. For sake of simplicity, in the following this diffused intensity will be treated as a background to be removed. Focusing on the deterministic part of the structural deformation, i.e. the one described by $\chi_g(r)$ and $t(r)$, it is possible to calculate the shape of the reciprocal space maps. The key point in the derivation of the diffracted intensity distribution is that the shape of reciprocal lattice points is connected to the features in real space by a Fourier transformation. Since both $\chi_g(r)$ and $t(r)$ are periodical along x with the same period Λ, it may

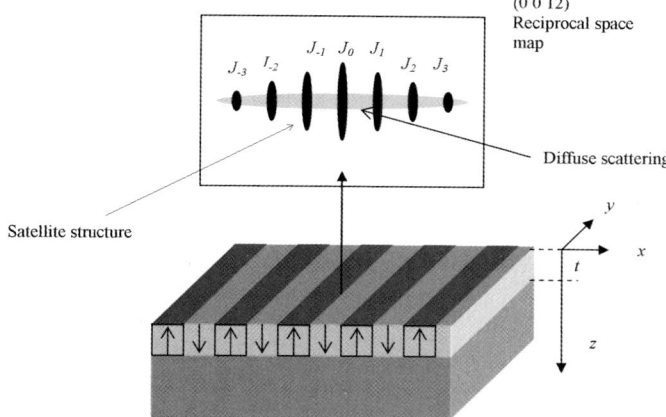

Fig. 9.10. Sketch of the sample reference system and the associated reciprocal lattice point features. The J_n's label the satellites. The thickness t of the domain grating is also indicated

be demonstrated that the reciprocal lattice points develops a fine structure made up of satellites. Their intensity and shape are determined by the Fourier transforms of $t(\mathbf{r})$ and $\chi_g(\mathbf{r})$ respectively, while their spacing is given by multiples of the domain grating wavevector $K_x = \frac{2\pi}{\Lambda}$ (see Fig. 9.10).

As we will see in Sect. 9.4, if the domain walls are not perpendicular to the surface, but they are tilted, the domain grating wavevector $\mathbf{K}_\Lambda$ is no more parallel to the sample surface. The satellite pattern will therefore be inclined with respect to the q_x direction: this is the case, for instance, of the PPLN prepared by Cz-off-center technique. If $t(\mathbf{r})$ is periodical along x and has no constrains in z, the superstructure is made up of a series of vertical "streaks", parallel to the q_z direction (see Fig. 9.9). When the separation between the streaks along q_x (that is equal to $K_x = \frac{2\pi}{\Lambda}$) is large enough to neglect interference effects between adjacent satellites, the reciprocal space intensity distribution $J(\mathbf{q})$ is given simply by the sum of the intensity distributions relative to every single satellite as individuated by multiples of K_x:

$$J(\mathbf{q}) = J_0(\mathbf{q}) + C \sum_{K_x} J_{K_x}(\mathbf{q}) \qquad (9.17)$$

where $J_0(\mathbf{q})$ is the intensity distribution of the zero-order peak, corresponding to the average structure of the sample, C is a constant and:

$$J_{K_x}(\mathbf{q}) = \left|\delta(q_x - K_x)\right| \left|[A_g + B_g(q_z)]\right|^2 \qquad (9.18)$$

is the intensity distribution along the streak located at a given K_x. It is worth mentioning that (9.18) contains the sum of two terms A_g and $B_g(q_z)$, both depending on the reflection $\mathbf{g}$ probed, multiplied by a delta function which allows for a non-zero value of the intensity only for $q_x = K_x$, that is along the streak as required from the assumed periodicity of the structure along x. The A_g term depends on $F^g_{K_x}$, which is the Fourier component of the structure factor for the **g** Bragg reflection, considered as a function of the position along the crystal as discussed above:

$$A_g = \delta(q_x) F_{K_x}^g \qquad (9.19)$$

A_g is therefore non-zero only for $q_z = 0$, as required by the fact that $F_g(r)$ does not depend on y and z. The second term $B_g(q_z)$ is due to the presence of the non-random displacement field $t(r)$ connected to the domain grating:

$$B_g(q_z) = F_0^g [\mathbf{g} \cdot \mathbf{t}]^{\mathrm{FT}}(K_x, q_z) * \Omega^{\mathrm{FT}}(q_z) \qquad (9.20)$$

where F_0^g is the spatial average over all the crystal of the structure factor for the **g** Bragg reflection. We remind that t is the depth of the region where the domain structure is present so that in PPLN crystals Ω, i.e. the shape function of the crystal, assumes the following expression: $\Omega(z) = 1$ for $t < z < 0$ and $\Omega(z) = 0$ elsewhere. The superscript FT stems for the Fourier transform operation and the symbol $*$ indicates the convolution operation. The function $B_g(q_z)$ has non-zero values also for $q_z \neq 0$ and therefore describes the intensity distribution along the tails of the vertical streaks. Equation (9.20) shows that this intensity is determined by the Fourier spectrum of $\mathbf{g} \cdot \mathbf{t}$, the projection of the displacement field along the reciprocal lattice vector of the reflection chosen. Equations (9.17)–(9.20) express the fact that, apart from the contribution of A_g at $q_z = 0$ and the convolution with $\Omega^{\mathrm{FT}}(q_z)$, the reciprocal space map carries information on the Fourier representation of that part of the static displacements which are parallel to the diffraction vector chosen. If the depth of domain inverted region is large enough, the Fourier transform of the shape function Ω is a very narrow peak function and the convolution operation in (9.20) is negligible. On the other hand, if the domain structure is present only on a thin layer at the surface of the sample, the Fourier transform of would give rise to a series of fringes equally spaced along q_z, the so-called thickness fringes.

9.3 Experimental Set-Up for Structural Characterization by HRXRD

As discussed in the previous sections, the High resolution X-rays diffraction technique consists basically in recording the diffracted intensity as a function of the direction of the primary and of the diffracted wavevectors. As a consequence the main experimental task is to realize an equipment able to irradiate the sample with a well-defined monochromatic and collimated beam and to scan the diffracted beam with a narrow angular acceptance. Obviously, the system has to be mechanically stable in order to allow for the tight angular accuracy needed to perform the measurements. The departure from one of these conditions can be described as the convolution of the true reciprocal space maps with a resolution function which broadens the experimental details if this function is not enough narrow. X-ray investigations are performed using commercial or home-made diffractometers, or also synchrotron radiation facilities. The general scheme of an X-rays apparatus is depicted in Fig. 9.11.

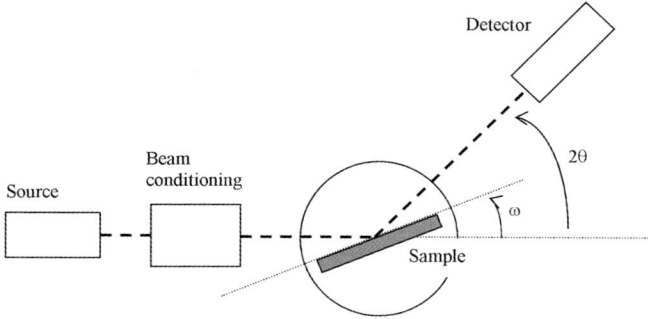

Fig. 9.11. General scheme of an X-rays apparatus

Basically it consists of a primary beam generation stage, where the beam coming from a source (typically a sealed X-ray tube or a synchrotron storage ring) is conditioned by a series of X-rays optical elements, generally nearly perfect semiconductor crystals, which act as diffraction grating. By the combination of dispersive and non-dispersive geometries, the beam coming from the source can be collimated and monochromatized up to the experimental requirements imposed by the sample. Typical values are a divergence of few thousandths of degrees and a spectral purity of one part over 10^5. Analogously, the detection part is conceived to accept only a very narrow angular window, which can be done again by using a series of Bragg reflections on nearly perfect analyzer crystals. The sample and the detector are mounted on two precise rotational stages which guarantees an accurate definition of the incident and diffracted wavevectors. Usually those goniometers are controlled by a computer, so that a set of points in reciprocal space can be probed by a suitable choice of angular positions. By defining α_i and α_f as the angles between the incident beam and the diffracted beam respectively with the sample surface, ω as the angle of the goniometer in the sample stage with respect to the primary beam and 2θ as the angle of the detector goniometer with respect to the primary beam again, the following relations holds:

$$\omega = \alpha_i, \tag{9.21}$$

$$2\theta = \alpha_i + \alpha_f, \tag{9.22}$$

$$K_x = K(\cos\alpha_i - \cos\alpha_f), \tag{9.23}$$

$$K_z = K(\sin\alpha_i + \sin\alpha_f). \tag{9.24}$$

So that by programming the movements of the goniometers, a given region of the reciprocal space can be probed.

In order to perform a diffraction experiment able to investigate a given reciprocal space volume, it is necessary to estimate the reciprocal space resolution of the instrument used. A non-zero divergence of the primary beam, its non-zero spectral width and a non-zero angular width of the acceptance of the analyzer smear out the reciprocal space distribution of the scattered intensity. Then, the measured intensity corresponds to a convolution of the true intensity distribution with a function having a certain width. If we focus on the typical case of two dimensional maps, corresponding to experiments performed with no resolution in the q_y direction, the width of the instrumental function corresponds to a definite area in reciprocal space, called resolution area. It is easy to give a rough estimate of this area. By differentiating (9.24), we get the displacements δq_x and δq_z in the reciprocal space caused by a variation of the parameters α_i, α_f and K, referred to the reciprocal lattice point under investigation:

$$\delta q_x = K[\Delta \alpha_i \sin \alpha_i - \Delta \alpha_f \sin \alpha_f] + \Delta K[\cos \alpha_i - \cos \alpha_f], \quad (9.25)$$

$$\delta q_z = K[\Delta \alpha_f \cos \alpha_f + \Delta \alpha_i \cos \alpha_i] + \Delta K[\sin \alpha_i + \sin \alpha_f]. \quad (9.26)$$

Assuming statistical independence of the various sources of broadening we can calculate approximately the broadening in the q_x and q_z directions by squared sum:

$$\Delta q_x = K \left[\Delta \alpha_i^2 \sin^2 \alpha_i + \Delta \alpha_f^2 \sin^2 \alpha_f + \left(\frac{\Delta K}{K}\right)^2 (\cos \alpha_f - \cos \alpha_i)^2 \right]^{1/2}, \quad (9.27)$$

$$\Delta q_z = K \left[\Delta \alpha_i^2 \cos^2 \alpha_i + \Delta \alpha_f^2 \cos^2 \alpha_f + \left(\frac{\Delta K}{K}\right)^2 (\sin \alpha_f + \sin \alpha_i)^2 \right]^{1/2}. \quad (9.28)$$

When probing laterally a patterned sample, an important parameter is the coherence length of the primary radiation on the sample. In fact, in order to probe the effect of the superstructure it is necessary that the coherence length is longer than the period of the domain grating. It may be demonstrated [16] that the total coherence length of the X-rays radiation is given by the following relation:

$$L_{tot} = \lambda \left[\frac{1}{\Delta \alpha_i \sin \alpha_i} + \frac{1}{\Delta \alpha_f \sin \alpha_f} + \frac{\lambda}{2\Delta \lambda} \left(\frac{1}{\cos \alpha_i} + \frac{1}{\cos \alpha_f} \right) \right]. \quad (9.29)$$

The coherence length therefore depends on the reflection under investigation and on the characteristic of the instrument used. In Fig. 9.12 we report the HRXRD configurations used to investigate the structural properties of PPLN samples that we present in Sect. 9.4.

The first setup is a commercial Philips MRD diffractometer with a sealed Cu anode source (40 kV acceleration voltage at 40 mA), equipped with a parabolic multilayer mirror for enhanced beam intensity (at open detector the statistics reaches 1.3×10^7 cps). The primary beam operates at $\lambda = 0.154056$ nm with a spectral purity $\frac{\Delta \lambda}{\lambda} = 2 \times 10^{-5}$ and a divergence of 0.0039 deg. It is collimated and monochromatized using a four-bounce (220) channel – cut Ge monochromator. The detector was a Xe proportional counter with a linear response up to 6×10^5 cps that can be equipped with a three bounce Ge (220) analyzer whose acceptance is 0.0039 degrees. In this case the high resolution (HR) configuration is installed. When the analyser is excluded, instead, the detectors is therefore open allowing the recording

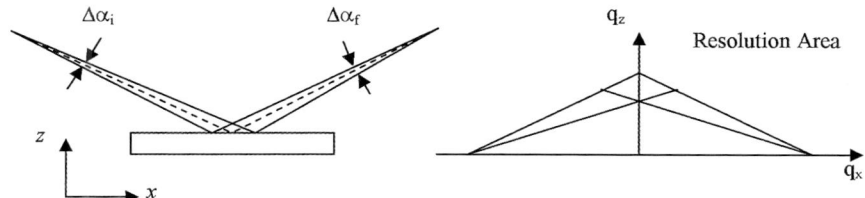

Fig. 9.12. HRXRD configurations used to investigate the structural properties of PPLN samples that we present in Sect. 9.4

Table 9.1. Resolutions in reciprocal space and correlation length L calculated at the sample surface using (9.28) and (9.29)

$(h\,k\,l)$	$\Delta q_x^{\text{optical}}$ $(10^{-3}\,\text{nm}^{-1})$	$\Delta q_z^{\text{optical}}$ $(10^{-3}\,\text{nm}^{-1})$	$\Delta q_x^{\text{mechanical}}$ $(10^{-3}\,\text{nm}^{-1})$	$\Delta q_x^{\text{mechanical}}$ $(10^{-3}\,\text{nm}^{-1})$	L (μm)
(0 0 6)	1.3	3.7	0.3	0.7	23
(0 0 12)	2.6	2.9	0.5	0.5	18
(2 2 12)	2.5	2.7	0.7	0.3	9

of angle integrated reflectivity curve (the so called Rocking curve, RC). In order to avoid artifacts during the measurements due to thermal drifts, the temperature of the measure chamber is stabilized at $(25.0 \pm 0.1)\,°C$ and the final accuracy of the instrument when absolute measurements of the angle are needed, is about 0.001 deg. In Table 9.1, the resolutions in reciprocal space of the diffractometer are calculated together with the correlation length at the sample surface using (9.28) and (9.29).

The second setup used was implemented at the ID 10 B Troika II" beamline at the European Synchrotron Radiation Facility (E.S.R.F.) in Grenoble. In this case the reciprocal space resolution was increased in order to study samples with closely spaced features in reciprocal space. The energy of the primary beam was settled at 7.76 keV by a C (1 1 1) diamond monochromator and the beam spectral purity was improved up to about 10^{-5} by inserting a channel-cut Si (3 3 3) monochromator. The beam collimation is guaranteed by the sample-to-source distance, which for a synchrotron source is generally very high (40 m), leading to a primary beam divergence of 2.52×10^{-4} degrees. The angular acceptance of the detector is defined by another Si (3 3 3) channel-cut detector, and is estimated in 5.56×10^{-4} degrees. The detector was a proportional scintillation detector. In this case the reciprocal space resolution can be estimated, for the (0 0 12) reflection in $\Delta q_x = 3.14 \times 10^{-4}\,\text{nm}^{-1}$, $\Delta q_z = 6.43 \times 10^{-4}\,\text{nm}^{-1}$, about one order of magnitude better than a laboratory HRXRD equipment.

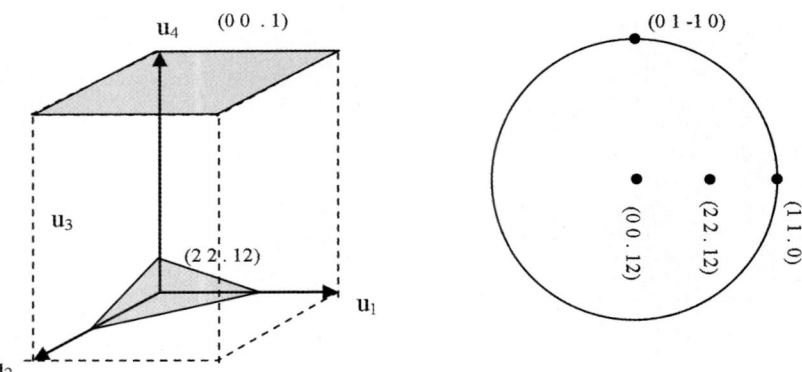

Fig. 9.13. *Left*: Crystallographic planes – sketched with respect the LiNbO$_3$ lattice cell – probed by the symmetrical (0 0 12) and the asymmetrical (2 2 12) reflections. *Right*: Stereographic projection of the employed reflections on a z-cut sample

9.4 Applications

In the following we will present how high resolution X-ray diffraction can be exploited to investigate two different types of PPLN samples: a PPLN obtained by the conventional electric field poling technique and a PPLN prepared by direct growth with the Czochralski off-center technique.

9.4.1 Investigation of Sub-Micrometric PPLN Crystals

A sub-micrometric PPLN crystal can be realised using the approach described in [3] that consists in resist patterning a congruent z-cut LN sample using an interferential lithographic process. The patterned sample is subjected to a room-temperature electric-field overpoling process and after poling, the resist is removed. In the following, the structural characterization of a PPLN sample with nominal period close to 700 nm will be presented. Owing to the small periodicity of the PPLN sample, the satellite structure near the reciprocal lattice points is expected to have a period of about 10^{-2} nm^{-1}, so that they can be well resolved with a standard apparatus. The X-ray investigations were performed using a Philips MRD diffractometer as described in Sect. 9.3. Two reflections were investigated in reciprocal space mapping mode [19]: the symmetrical (0 0 12) and the asymmetrical (2 2 12) respectively (see Fig. 9.13).

Due to the very large acceptance of the detector on the vertical plane, the three dimensional reciprocal space maps are integrated onto the x–z plane: the measured maps are shown in Fig. 9.14.

The reciprocal maps show a periodic structure made of a series of satellite peaks (up to the fourth order), as previously mentioned in the theoretical treatment outlined in Sect. 9.2.2. A large diffuse scattering peak is clearly visible in both reflections. The latter clearly indicates the presence of some kind of incoherent scattering

9 Structural Characterization of PPLN cryslals by HXRD   247

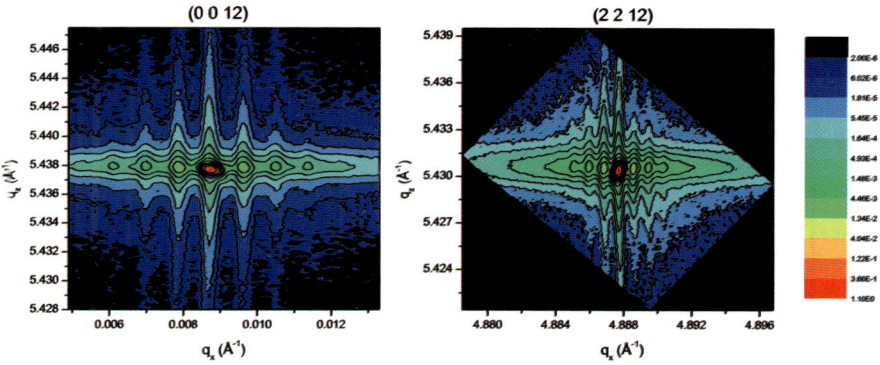

Fig. 9.14. Reciprocal map relative to the symmetrical (0 0 12) and asymmetrical (2 2 12) reflections

Table 9.2. Fitting parameters for the diffuse scattering function

Refl	A (counts × Å^2)	b_x	b_z	w (Å^{-1})
(0 0 12)	4000	1.2	30.3	1.1×10^{-3}
(2 2 12)	5000	0.9	25.0	1.0×10^{-3}

processes. The diffuse scattering background can be fitted with analytical function and therefore easily subtracted [20]. In particular, for the reciprocal maps shown in Fig. 9.14 a good fitting function was:

$$f(q_x, q_z) = A \frac{w^2}{b_x q_x^2 + b_z q_z^2 + w^2} \quad (9.30)$$

for the (0 0 12) and for the (2 2 12) reflections respectively. The parameters A, b_x, b_z, w were determined by a least – squares fit (see Table 9.2).

The constant A depends upon the peak intensity of the reflection, while b_x, b_z, w were found to be significantly close for the two reflections [20]. After the removal of the background (Fig. 9.15), information on the periodicity of the PPLN structure can be easily extracted.

The diffraction vector $\mathbf{h}$ relative to the (0 0 12) reflection (Fig. 9.15(a)) is perpendicular to the sample surface, therefore the satellite streaks give information only on the distribution of Fourier components of the static displacements perpendicular to the surface, according to (9.16)–(9.19). As it can be seen, the experimental maps are in good qualitative agreement with the expectations of the theory. The period of the PPLN structure can be extracted and is given by $\lambda = (714 \pm 6)$ nm. Moreover, the centers of the satellites are aligned along the q_x direction, indicating that the domain walls are perpendicular to the sample surface. According to (9.20), the vertical streak presents at each satellite is given by the convolution of the Fourier transform of the shape function Ω^{FT} of the domain region with the Fourier transform of the vertical displacement field $[\mathbf{g} \cdot \mathbf{t}]^{FT}$. Since no modulations are detected,

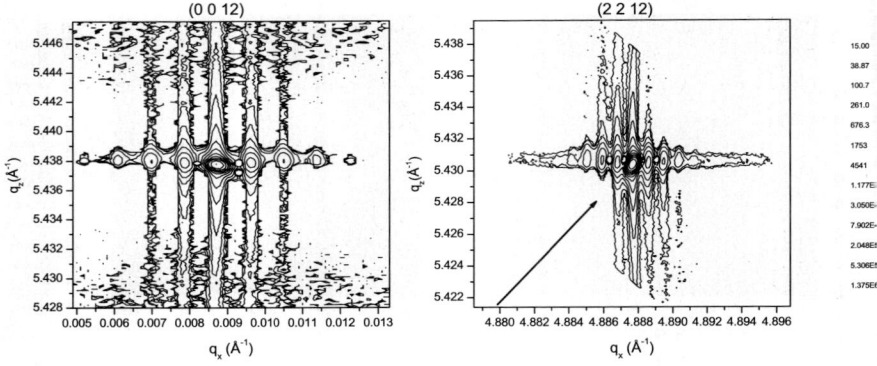

Fig. 9.15. Reciprocal maps as in Fig. 9.14 after the subtraction of the background contribute

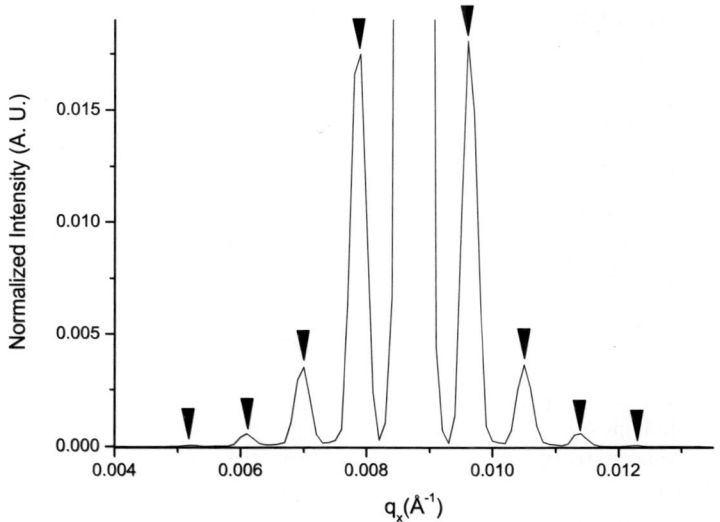

Fig. 9.16. Projection of the reciprocal map on the q_x axis by integrating along q_z

we may conclude that the thickness fringes, if present, are below the experimental resolution and therefore the structure has to be thicker than several microns. The intensity distribution about the (0 0 12) reciprocal lattice point shows a symmetrical shape, the spectrum of the vertical displacement being evenly distributed along q_z. For this reason, the reciprocal map can be projected on the q_x axis by integrating along q_z (see Fig. 9.16).

As stated by (9.16)–(9.19), the intensity at the peaks positions is due both to the Fourier components of the polarizability and to the static displacements. It is worth mentioning that when the peak intensity is not proportional to $|\frac{\sin(q_x)}{q_x}|^2$, the PPLN structure cannot be described simply as a square wave with duty cycle equal to one as predicted by (9.18). In particular, in that case, the odd orders of the superstructure should be absent. However, since both the A_h and B_h contributes play a role (see (9.16)–(9.20)), it is not possible to decide whether this effect is due to an unexpected behaviour of the polarization profile or of the deformations or both. Because the first diffraction order is several times higher than the others, the modifications induced by the periodic poling should have a nearly sinusoidal dependence on the position. In the case of the reflection (2 2 12), the Fourier components of the static displacements inclined of about 42 degrees with respect to the sample surface are investigated. The reciprocal map is therefore sensitive also to the lateral component of fluctuations of static displacements. The analysis of the (2 2 12) reciprocal map shows that the satellite peaks are aligned along q_x and the peaks intensities present the same behavior of those observed in the (0 0 12) reflection. This fact confirms that the profile of the modifications depends only on the x coordinate and has an almost sinusoidal character. However, in this map the intensity distribution along the satellite streaks follow an asymmetric distribution. As discussed in Sect. 9.2.2, the intensity distribution along the streaks is connected to the Fourier spectrum of static displacements. From the symmetrical (0 0 12) reflection we observed that the vertical part of displacements has a Fourier spectrum which is symmetrical along q_z. If the displacements had no lateral components, from (9.20) we shall expect a symmetrical shape also of the (2 2 12) maps. In this particular case, instead, a lateral component of displacements must be present. By comparing the (0 0 12) and (2 2 12) reciprocal maps, it is therefore possible to conclude that:

1. the $t(r)$ term is periodic along x as it can be deduced from the presence of vertical streaks equally spaced and parallel to z. Owing to the elongated shape of the streaks, we can infer that the vertical displacements devolp;
2. the satellite peaks are aligned along q_x, indicating that domain walls are perpendicular to the sample surface. In other cases, in fact, the reciprocal space map would result in a satellite structure not parallel to the q_x direction;
3. as thickness oscillations are not detected along the satellite streaks, the domain structure extends well below the surface for several microns;
4. from the measurement of the satellite spacing $K_x = \frac{2\pi}{\Lambda}$, it is possible to obtain the period of the structure as (714 ± 6) nm;
5. in both the experimental maps a pronounced diffuse scattering peak is present with iso-intensity curves in the shape of ellipses. It is worth mentioning that the diffuse scattering has to be removed in order to extract the coherent part of the intensity and get information on the distribution of fluctuations of structure factor and static displacements. A detailed description of the this procedure is described in [20].

9.4.2 Investigation of Micrometric PPLN Crystals with Bent Domain Walls

In a PPLN crystal prepared by the Czochralski off-center technique the domains pattern is present inside all the crystal volume. The domains shape, however, is determined by the solid–liquid interface of the growing crystal boule and generally

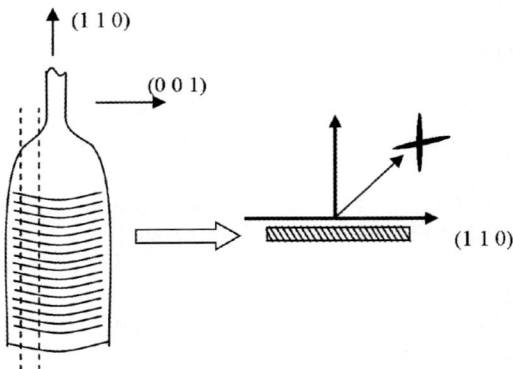

Fig. 9.17. Domain pattern in a PPLN crystal grown by the Cz-off-center

is curved. Details on the preparation technique are reported elsewhere: we recall, however, that the key point relays on the shift of the growth axis with respect to the temperature field one [5]. Due to fluctuations inside the growth chamber, the domain structure is affected by random errors in the position of the domain boundaries [21], which reduce the overall quality of the grating. When an as-grown crystal is processed, several slabs are obtained, with domains running as depicted in Fig. 9.17, with typical periodicities ranging from 2 to 8 μm. The samples were characterized by recording the reciprocal maps of two symmetrical reflections – namely (0 0 6), (0 0 12) – and one asymmetrical (2 2 12) respectively. In Table 9.1 the reciprocal space resolution and the coherence length (as determined by (9.28)) are reported.

If an X-rays diffraction experiment similar to those illustrated in the previous sections is performed on such a sample, the domain curvature is not detectable, because the curvature radius of the domains is much larger than the lateral correlation length of the X-rays beam. Also in this case, therefore, the domain grating can be approximated with a plane wave; however in this case, the wavevector of the grating is no more parallel to the sample surface. If the measure is performed striking the sample with the X-rays beam about at its center, we may well take the domain grating wavevector to lay in the scattering plane.

In Fig. 9.18 a symmetrical (0 0 12) and an asymmetrical (2 2 12) reciprocal space maps for a PPLN sample with nominal period equal to two microns are shown. In this case the reciprocal space resolution of the diffractometer is not able to resolve the satellite structure, that therefore is seen as a nearly horizontal "streak", inclined parallel to the domain grating wavevector and with the same angle for both reflections. These maps therefore allows for a direct measure of the tilt angle of the domain structure, which for the sample reported in Fig. 9.18 is equal to (5.2 ± 0.3) degrees. The measurement of this quantity is of great importance for domain physics, since it is directly connected to the presence of polarization charges located on the domain walls surface.

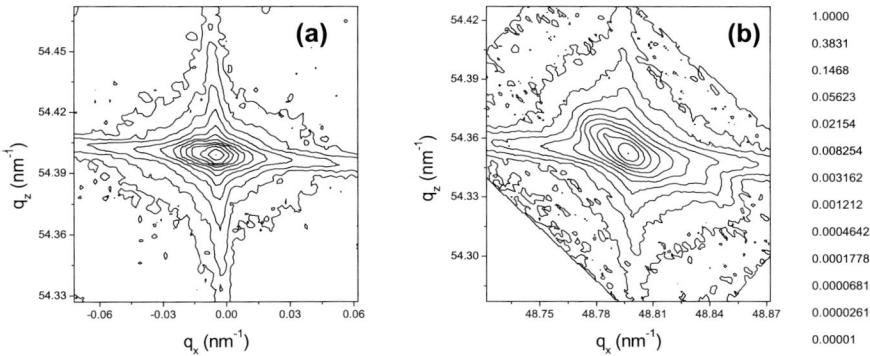

Fig. 9.18. (a) Symmetrical (0 0 12) reciprocal space map for a PPLN sample with nominal period equal to two microns. (b) Asymmetrical (2 2 12) reciprocal space map for a PPLN sample with nominal period equal to 2 μm. In both reflections a small inclination of the horizontal streak can be seen

The overall effect on the diffracted intensity that in the reciprocal space map the nearly horizontal streak is inclined at an angle β with respect to the q_x direction (see Fig. 9.18).

It is worth mentioning that in both the maps, a significant diffuse scattering peak with an ellipsoid shape is present, with the major axis is perpendicular to the vector joining the reciprocal lattice point to the origin of the reciprocal space. Moreover, no vertical broadening of the "domain streak" is detectable along the q_z direction suggesting that the eventual presence of a strain gradient is below the detection limit of the technique. Figure 9.18 should be compared with Fig. 9.14, showing the same experiment performed with the same instrument but with a sample prepared by the electric field poling process. Apart from the presence of the satellites, it is evident that in the case of the PPLN prepared by Cz-off-center, the vertical elongation of the satellites, which is attributed to vertical deformations (see (9.20)) is absent in this case. This difference is due to the preparation techniques: in the grown samples, the domain pattern is formed at very high temperature, so that all the system is able to relax by building up its internal field after the domain formation and parallel to the spontaneous polarization, so that no piezoelectric interactions is present. Moreover at high temperature lithium niobate is conductive, allowing for the presence of free charges that screen the polar defects. On the other hand, the electric field poling process is performed at room temperature and its effects on the sample structure and quality are much more strong. It is worth mentioning that when the coherence length has the same order of magnitude of the domain width, the presence of coherent scattering coming from the domain patterns can be excluded: if present, it will, instead, introduce a superstructure on the reciprocal map as already pointed out in Sect. 9.4.1. If the coherence length has the same order of magnitude of the domain width, therefore, the scattering process can be described as an incoherent superposition of scattering from a given number of single domains so that the information

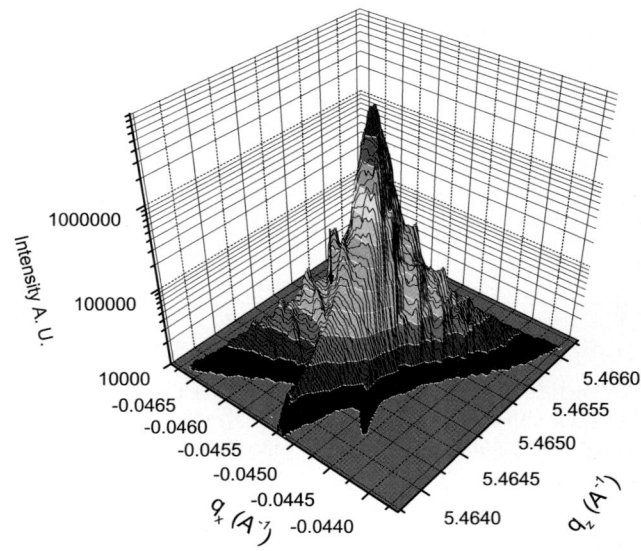

Fig. 9.19. Reciprocal space map of the (0 0 12) reflection for a PPLN sample with nominal period 2 μm obtained at the ID 10 line, E.S.R.F.

that can be gained from HRXRD investigation refer to a single domain. In order to confirm the interpretation given to the measurements obtained on PPLN sample prepared by the Cz-off-center technique, the measurements were repeated using the very high resolution setup at the ID 10 line at the European Synchrotron Radiation Facilities (E.S.R.F.) described in the previous sections, which also offers a greater coherence length. The results are reported in Fig. 9.19 where a reciprocal space map of the (0 0 12) reflection for a sample with nominal period 2 μm, prepared by the Cz-off-center technique, is shown.

The presence of the satellite structure is clearly visible. It can be seen that in this case they are much less pronounced that in the case of PPLN crystals prepared by electric field poling and that no streaks parallel to the q_z direction are present. Since the intensity of the satellites at their center is determined both by the Fourier components of the polarizability and by those of the deformation fields (see (9.18)–(9.20)), we have a confirmation that for this kind of sample the structural modifications are much weaker than in the case of an electric field poled one. Again, practically no strain is detectable in the domain structure. A large diffuse scattering peak is present also in this sample, indicating that some incoherent scattering process is taking place. However from these high resolution measurements we can see that the diffuse scattering peak is not simply given by the convolution of the satellite structure with the instrumental contribution, because in this case the latter is negligible. This means that the diffuse scattering peak has an elongated shape which is parallel to the domain grating direction (as in the case of the diffuse scattering peak observed in electric field poled samples). This may be an indication that this feature is originated from the domain walls. Further studies are needed to confirm this claim.

9.5 Conclusions

The real structure of PPLN crystals is different in several aspects from that of an ideal, perfect lithium niobate substrate. Those differences are originated both from the presence of ferroelectric domains and intrinsic defects of the material, but they also can be induced by the preparation process used to create the domain pattern. An accurate characterization of the real structure of a PPLN is therefore of great importance, both for a better comprehension of the fundamental physical phenomena underlying the ferroelectric domain formation, and also from a technological point of view, in order to tailor the preparation conditions to the best performance of the final device. The High Resolution X-ray diffraction technique, employed in the reciprocal space mapping mode, was proven to be highly informative when applied to the study of periodically poled materials, as it allows for an accurate and non-destructive measure of several aspects such as domain grating period, domain wall inclination, presence of structural deformations and so on. Its applicability is regardless of the preparation condition of the PPLN structure, provided that the experimental resolution in reciprocal space is adequate to the domain period. The shorter the period, in fact, the lower is the resolution needed. In particular, for sub-micrometric structures, as those of interest for several photonic applications such as photonic band gap crystals or frequency mixing at short wavelengths, a standard laboratory HRXRD equipment is able to resolve the reciprocal space features of the investigated samples. The presence of the periodic domain structure was shown to change remarkably the shape of the reciprocal lattice points. This effect can be described in the framework of the kinematical theory of X-ray diffraction, by assuming that the domain structure affects the crystal lattice in two independent ways. On one hand, the polarization reversal affects the arrangement of the atoms inside the elementary unit cell. On the other hand, instead, the presence of static deformations can be described by a rigid shift of the whole cell with respect to the position they should occupy into a reference, undistorted crystal. By combining these two effects, the theoretical description of the reciprocal space intensity distribution can be developed. It was shown that the spontaneous polarization profile gives rise to a periodic satellite structure around each reciprocal lattice point, whose position and inclination permits a direct measurement of the domain grating period and domain walls inclination. The presence of static distortions adds some new features to the satellite structure: if the deformations are induced by the periodic poling operation (as in the case of PPLN prepared by electric field poling), they share the same periodicity of the polarization profile and, therefore, they give rise to a modification of the periodic satellite structure. Otherwise, if deformations are induced by the presence of random modifications of the lattice structure, they produce a broad diffuse scattering peak which superimposes to the satellite structure.

The applicability of the HRXRD technique was tested experimentally on two PPLN samples prepared under very different conditions and different domain periodicities: one by electric field poling on a congruent lithium niobate substrate with a domain period of about 700 nm, the other by Czochralski off-center technique on Er-doped $LiNbO_3$, with a domain period of about 2 µm. Due to the very different periodicities, in the second case a very high resolution setup was needed and was realized at the ID 10 B beamline at the E.S.R.F. In both cases the periodic structure was successfully revealed and the domain period and shape measured with a relative accuracy of about 1%. Moreover, the technique was able to distinguish clearly the presence of stronger structural deformations in the sample prepared by electric field poling, as expected. Finally, a strong diffuse scattering peak was detected in both samples which, according to the theoretical treatment outlined above, is due to the presence of a random displacement field present is both samples. Some experimental findings indicate the domain wall regions as a possible origin of this random field. Further investigations are in progress on this interesting aspects. The approach here presented is very promising for the study of periodically poled crystals because of its accuracy and its non-destructive nature. Presently the main task is to improve the theoretical treatment of experimental data in order to extract the maximum amount of information by comparing them with simulations, as it is customary in other fields of research of the material science involving X-rays diffraction experiments.

References

1. R.G. Batchko, V.Ya. Shur, M.M. Fejer, R.L. Byer, Appl. Phys. Lett. **75**(12), 1673 (1999)
2. V. Gopalan, T.E. Mitchell, Y. Furukawa, K. Kitamura, Appl. Phys. Lett. **72**, 1981 (1998)
3. S. Grilli, P. Ferraro, P. De Natale, B. Tiribilli, M. Vassalli, Appl. Phys. Lett. **87**, 233106 (2005)
4. N.B. Ming, J.-F. Hong, D. Feng, J. Mater. Sci. **17**, 1663 (1982)
5. C. Sada, N. Argiolas, M. Bazzan, Appl. Phys. Lett. **79**(14), 2163 (2001)
6. S. Kim, B. Steiner, A. Gruverman, V. Gopalan, On domain wall broadening in ferroelectric lithium niobate and tantalate. In *Fundamental Physics of Ferroelectrics 2002*, ed. by R.E. Cohen (American Institute of Physics, Washington, 2002)
7. S. Kim, V. Gopalan, A. Gruverman, Appl. Phys. Lett. **80**, 2740 (2002)
8. J. Padilla, W. Zhong, D. Vanderbilt, Phys. Rev. B **53**, R5969 (1996)
9. M. Bazzan, Domain structures in $LiNbO_3$ crystals grown by Czochralski off-center technique for non-linear optical application. PhD thesis, University of Padova, Padova, Italy (2004)
10. D.A. Scrymgeour, V. Gopalan, Phys. Rev. B **72**, 024103 (2005)
11. S. Kim, V. Gopalan, Mater. Sci. Eng. B **120**, 91 (2005)
12. T. Jach, S. Kim, V. Gopalan, S. Durbin, D. Bright, Phys. Rev. B **69**, 064113 (2004)
13. Z.W. Hu, P.A. Thomas, J. Webjorn, J. Appl. Crystallogr. **29**, 279 (1996)
14. Z.W. Hu, P.A. Thomas, J. Webjorn, J. Phys. D Appl. Crystallogr. **28**, 189 (1995)
15. A. Authier, *Dynamical Theory of X-Ray Diffraction*, IUCr Monograph on Crystallography (Oxford University Press, Oxford, 2001)
16. V. Holy, U. Pietsch, T. Baumbach, *High-Resolution X-Ray Scattering from Thin Films and Multilayers*, 1st edn. (Springer, Berlin, 1999)

7. M.A. Krivoglaz, *Theory of X-Ray and Thermal-Neutron Scattering by Real Crystals*, 1st edn. (Plenum Press, New York, 1969)
8. E. Zolotoyabko, J. Appl. Phys. **81**, 2143 (1997)
9. P.F. Fewster, X-ray diffraction from low-dimensional structures. Semicond. Sci. Technol. **8**, 1915 (1993)
20. M. Bazzan, N. Argiolas, C. Sada, P. Mazzoldi, S. Grilli, P. Ferraro, P. De Natale, L. Sansone, Ferroelectrics **352**, 25 (2007)
21. M. Bazzan, N. Argiolas, A. Bernardi, P. Mazzoldi, C. Sada, Mater. Charact. **51**, 177 (2003)

Part III

Applications

10 Nonlinear Interactions in Periodic and Quasi-Periodic Nonlinear Photonic Crystals

A. Arie, A. Bahabad, and N. Habshoosh

10.1 Introduction

Nonlinear photonic crystals (NLPC) are materials in which the second order susceptibility $\chi^{(2)}$ is modulated in an ordered fashion, usually in one or two dimensions, while the linear susceptibility remains constant. This modulation can be done, for example, by periodic electric field poling of ferroelectric crystals [1] such as $LiNbO_3$ or $KTiOPO_4$, or by orientation pattering of semiconductors such as GaAs [2]. NLPC are significantly different than the more common photonic crystals, in which the linear susceptibility is modulated, bringing forth optical processes involving several frequencies. For example, one dimensional (1D) modulation of the nonlinear susceptibility is widely used nowadays for quasi-phase-matched nonlinear frequency conversion [3, 4]. The extension to two-dimensional (2D) modulation [5] allows further design flexibility, and is useful for non-collinear second harmonic generation (SHG) [4], for simultaneous wavelength interchange [6], for third and fourth harmonic generation [5, 7], and proposed for realization of all optical effects, e.g. all optical deflection and splitting [8]. Note that other types of nonlinear interactions are possible in photonic crystals, including third-order ($\chi^{(3)}$) nonlinearities in "standard" photonic crystals [9], as well as photonic crystals in which both the linear and the second-order nonlinear susceptibilities are modulated [10]. These interactions will not be considered in this chapter.

An NLPC can be modeled as a convolution between a periodic lattice and a nonlinear motif. In the case of a 1D structure, the lattice is a set of equally spaced parallel lines (the line distance is the lattice period Λ), and the motif is a strip having a nonlinear coefficient with a different sign than the background (Fig. 10.1).

On the other hand, 2D periodic structures [11] can be classified by five Bravais lattices: Hexagonal, square, rectangular, centered rectangular and oblique, as can be seen in Fig. 10.2(a)–(e). In order to convert a lattice into a 2D NLPC, each one of the lattice points is convolved with a nonlinear motif – i.e. some two dimensional geometrical shape – circle, hexagon, rectangle etc., having a nonlinear coefficient that is different that the background. Note that in the photonic crystals community, the hexagonal lattice is identified as a "triangular" lattice, while "hexagonal" or "honey-comb" lattice (in this community) is actually a hexagonal lattice with a

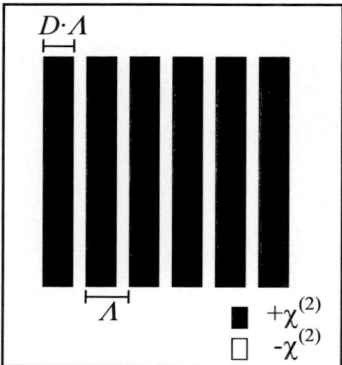

Fig. 10.1. A 1D periodic lattice with duty cycle of $D = 75\%$

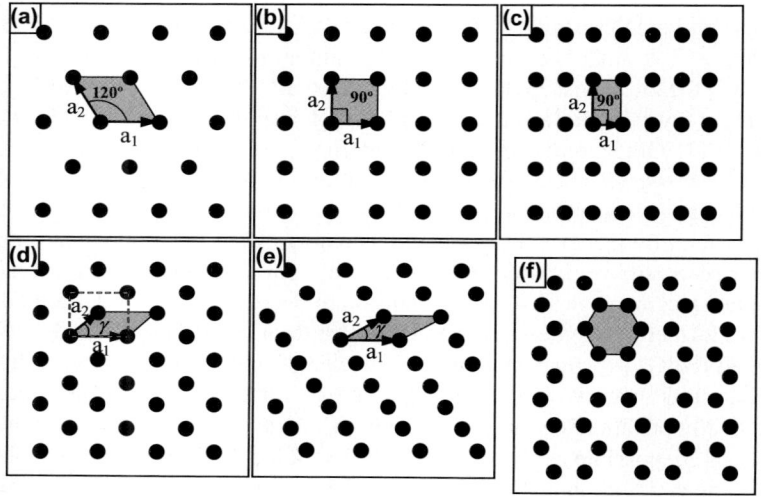

Fig. 10.2. The six lattice types, five of them (**a**)–(**e**) are Bravais lattices: (**a**) hexagonal, (**b**) square, (**c**) rectangular, (**d**) centered-rectangular, where the dashed lines form a rectangle, (**e**) oblique. Panel (**f**) is a honeycomb lattice. The gray area in each one of the lattices refers to its unit cell

missing point in the middle of each hexagon (see Fig. 10.2(f)). For completeness both the five Bravais lattices and the "honey-comb" lattice would be analyzed.

One limitation of the nonlinear periodic structures is that they usually enable to phase match only processes whose mismatch vectors correspond to integer multiples of a single vector (in the 1D case) or to a vectorial sum of only two base vectors (in the 2D case). However, the modulation of the nonlinear susceptibility does not have to be periodic, in which case it can still be efficient and provide greater design flexibility for phase matching several different processes. To overcome these limitations ad-hoc generalizations of the quasi-phase-matching procedure were developed, based either on non-collinear processes in one-dimensional periodic structures [12] or on specific quasiperiodic structures in one [13–17] and two [3] dimensions. A general method to design frequency converters that will phase match any set of interacting waves, either in a 1D or in a 2D configuration, is provided by the so-called generalized dual grid method (DGM) [6, 18]. This method is well known in quasi-crystalline research, and is used for constructing tiling models of quasi-crystals.

In order to optimally design and use an NLPC, one needs to know the conversion efficiency for any given lattice, its dependence on the shape and size of the motif, and its dependence on the quasi phase matching (QPM) orders. In this chapter an analysis of a nonlinear interaction between optical waves in nonlinear photonic crystals is presented. The approach is general, and is applied to both periodic and quasi-periodic modulations of the nonlinear susceptibility, in either one or two dimensions.

A general analysis of quasi-phase-matched interactions in a nonlinear photonic crystal is given in Sect. 10.2. Section 10.3 includes a mathematical description of the real and reciprocal periodic lattice in 1D and 2D, and analysis of the effect of the lattice, motif and interaction area on the generated electric field and intensity. Quasi periodic structures are discussed in Sect. 10.4. The results are discussed and summarized in Sect. 10.5.

10.2 Wave Equations in NLPC

Consider now the case of second harmonic generation in a NLPC. The results shown here can be easily generalized to other three wave mixing processes, e.g. sum frequency generation and difference frequency generation. Assuming that a plane wave of frequency ω propagates in the transverse plane of the NLPC, this wave generates a second harmonic wave owing to the second order susceptibility of the material. Assuming that the fundamental frequency is linearly polarized along one of the NLPC axes, and considering a specific linear polarization of the generated second harmonic wave, the coupling between the two beams is given by the appropriate element of the nonlinear susceptibility tensor $d_{ij} = \chi_{ij}^{(2)}/2$, where i and j are contracted Cartesian indices [19].

The relevant component of the second harmonic wave can be written as

$$\tilde{E}_{2\omega}(r, t) = \frac{1}{2} E_{2\omega}(r) \exp[i(2\omega t - k_{2\omega} \cdot r)] + \text{c.c.} \tag{10.1}$$

Assuming that the nonlinear conversion efficiency is low, the pump amplitude can be assumed constant throughout the entire interaction length (non-depletion approximation). It is further assumed that the slowly varying envelope approximation [19] applies for the second harmonic wave. Under these assumptions it can be shown that the variation of the second harmonic wave amplitude is determined by the nonlinear polarization (in M.K.S):

$$\mathbf{k}_{2\omega} \cdot \nabla E_{2\omega}(\mathbf{r}) = -2\mathrm{i}\frac{\omega^2}{c^2} E_\omega^2 d_{ij}(\mathbf{r}) \exp[\mathrm{i}(\mathbf{k}_{2\omega} - 2\mathbf{k}_\omega) \cdot \mathbf{r}]. \quad (10.2)$$

When the structure's nonlinearity is ordered (i.e. it is a periodic or quasi-periodic function of space), the nonlinearity can be written as a Fourier series. Explicitly, 1D periodic structures with building blocks of length Λ that are subdivided into two sub-blocks with opposite sign of the nonlinear coefficient, are represented as

$$d_{ij}(x) = d_{ij}\mathrm{sign}\left[\mathrm{saw}\left(\frac{x}{\Lambda} - D\right) - \mathrm{saw}\left(\frac{x}{\Lambda}\right)\right] = d_{ij} \sum_{m=-\infty}^{m=\infty} G_m \mathrm{e}^{\mathrm{i}k_m x}, \quad (10.3)$$

where saw is the saw-tooth function $\mathrm{saw}(x) = 2[x - \mathrm{floor}(x)] - 1$, D is the duty cycle (defined as the ratio of the positive domain length to the period) $G_m = 2\sin(m\pi D)/m\pi$ are the Fourier coefficients and $k_m = 2\pi m/\Lambda$ are the reciprocal wave-vectors. The highest Fourier coefficient, obtained for first order ($m = 1$) process, and for a duty cycle of $D = 50\%$, is $2/\pi = 0.6366$.

This can also be expanded for a 2D periodic structure (or for more general quasi-periodic structures). For the case of infinitely large interaction area, the nonlinearity is given by:

$$d_{ij}(\mathbf{r}) = d_{ij} \sum G_{mn} \exp(-\mathrm{i}\mathbf{K}_{mn} \cdot \mathbf{r}) = d_{ij} g(\mathbf{r}), \quad (10.4)$$

where $g(\mathbf{r})$ is a normalized and dimensionless function, representing the space dependence of the nonlinear coefficient function. Significant build-up of the second harmonic wave requires phase matching, i.e. $\mathbf{k}_{2\omega} - 2\mathbf{k}_\omega - \mathbf{K}_{mn} \approx 0$. This vectorial phase-matching condition is just a crystal-momentum conservation law: the required momentum balance for the interaction is accomplished through a reciprocal lattice vector (RLV). Usually it can be assumed that if the phase matching condition is achieved by some order (m, n), it would be the only order which contributes to the build-up of the second harmonic while all the other orders contributes negligible oscillating terms. The analysis for two-dimensional periodic structures can now be continued by rewriting the wave (10.2) as:

$$\mathbf{k}_{2\omega} \cdot \nabla E_{2\omega}(\mathbf{r}) = -2\mathrm{i}\frac{\omega^2}{c^2} E_\omega^2 d_{ij} G_{mn} \exp[\mathrm{i}(\mathbf{k}_{2\omega} - 2\mathbf{k}_\omega - \mathbf{K}_{mn}) \cdot \mathbf{r}]. \quad (10.5)$$

Note that these results can be easily adapted for the one-dimensional case, by replacing the two-dimensional Fourier coefficients and reciprocal vectors with their one dimensional counterparts.

This process can also be analyzed in Fourier space by integrating (10.5) over a rectangular area $a(\mathbf{r})$ of length L and width W (see an example [20]). The result is the second harmonic amplitude after an interaction length of L:

$$E_{2\omega}(\Delta k) = \frac{-2\mathrm{i}\omega^2 E_\omega^2 d_{ij}}{k_{2\omega}c^2 W} \iint_{a(\mathbf{r})} g(\mathbf{r}) \exp(-\mathrm{i}\Delta \mathbf{k} \cdot \mathbf{r}) \, da, \quad (10.6)$$

where $\Delta \mathbf{k} = \mathbf{k}_{2\omega} - 2\mathbf{k}_\omega$ is the phase-mismatch vector and $a(\mathbf{r}) = \mathrm{rect}(x/L) \cdot \mathrm{rect}(y/W)$ is the integration area. Setting $g(\mathbf{r})$ to zero outside the NLPC, the integration limits can be extended to infinity and so:

$$E_{2\omega}(\Delta k) = \frac{\kappa}{W} G(\Delta k), \qquad (10.7)$$

where $G(\Delta k)$ is just the two-dimensional Fourier transform of $g(r)$ and κ is a constant defined as:

$$\kappa = \frac{-2i\omega^2 E_\omega^2 d_{ij}}{k_{2\omega} c^2} = \frac{-i\omega E_\omega^2 d_{ij}}{n_{2\omega} c}. \qquad (10.8)$$

From (10.7) it can be seen that the field amplitude evaluation for some specific phase mismatch value $\Delta k = \Delta k_0$ is proportional to $|G(\Delta k_0)|$.

If phase matching condition is achieved by some order (m, n) then the integral above is dominated by this order and so:

$$E_{2\omega}(\Delta k = K_{mn}) \cong \kappa L G_{mn} \exp\left[-i\left(\frac{\Delta k_{mn,x} L}{2} + \frac{\Delta k_{mn,y} W}{2}\right)\right]$$
$$\times \operatorname{sinc}\left(\frac{\Delta k_{mn,x} L}{2\pi}\right) \operatorname{sinc}\left(\frac{\Delta k_{mn,y} W}{2\pi}\right), \qquad (10.9)$$

where $\Delta k_{mn} = \Delta k - K_{mn} = \Delta k_{mn,x} \hat{x} + \Delta k_{mn,y} \hat{y}$.

For perfect quasi-phase-matching $\Delta k_{mn,x} = \Delta k_{mn,y} = 0$ and so:

$$E_{2\omega}(\Delta k = K_{mn}) \cong \kappa L G_{mn}. \qquad (10.10)$$

The fundamental and second harmonic amplitudes are related to the corresponding intensities by:

$$I_\omega = \frac{1}{2} n_\omega \sqrt{\frac{\varepsilon_0}{\mu_0}} |E_\omega|^2, \qquad I_{2\omega} = \frac{1}{2} n_{2\omega} \sqrt{\frac{\varepsilon_0}{\mu_0}} |E_{2\omega}|^2. \qquad (10.11)$$

Hence, the intensity of the second harmonic for the case of perfect quasi-phase-matching after an interaction length L is:

$$I_{2\omega} = \frac{2\omega^2 d_{ij}^2 |G_{mn}|^2}{n_{2\omega} n_\omega^2 c^3 \varepsilon_0} I_\omega^2 L^2, \qquad (10.12)$$

and so the interaction efficiency is proportional to the absolute square of the relevant Fourier coefficient $|G_{mn}|^2$. As such, $|G_{mn}|^2$ would be related to as the normalized efficiency.

10.3 Analysis of a Periodic Nonlinear Photonic Crystal

10.3.1 The Real Lattice

A one-dimensional nonlinear lattice is defined by one primitive vector representing the lattice period, usually parallel to one of the main axes of the crystal $a = \Lambda \hat{x}$. On the other hand, a two-dimensional nonlinear lattice is defined by two primitive, non-parallel vectors a_1 and a_2. Each lattice point is given by

$$\begin{aligned} \text{1D:} & \quad x_m = ma, \\ \text{2D:} & \quad r_{mn} = ma_1 + na_2. \end{aligned} \quad (10.13)$$

The lattice is represented by a set of distributed Dirac delta functions:

$$\begin{aligned} \text{1D:} & \quad u(r) = u(x) = \sum_m \delta(x - x_m) = \sum_m \delta(x - m\Lambda\hat{x}), \\ \text{2D:} & \quad u(r) = \sum_{m,n} \delta(r - r_{mn}) = \sum_{m,n} \delta(r - ma_1 - na_2). \end{aligned} \quad (10.14)$$

The lattice can be converted into a nonlinear photonic crystal by convolving the lattice points with a suitable nonlinear optical motif as shown in the two examples of Fig. 10.3. In the case of 1D lattice, the structure is a set of equally spaced parallel lines, and the motif is a strip having a nonlinear coefficient with a different sign than the background, whereas a 2D lattice can have for example a hexagonal motif, defining a positive value for the nonlinear coefficient, centered at each one of the lattice points. These motifs are surrounded by a background with a different nonlinear coefficient. The pattern outside the motifs, may be linear (zero nonlinearity, as for example is the case for patterns made of poled and un-poled glass), or may have an opposite sign of the nonlinear coefficient (as is the case in domain inverted ferroelectric crystals).

If the background is nonlinear (with opposite sign to the motif nonlinear coefficient) the motif function denoted $s(r)$, has values of 1 and (-1) instead of 1 and 0. This amount in a DC shift in the Fourier transform of the overall structure function. The result, as implied through (10.10) in Sect. 10.2, is doubling of the electric

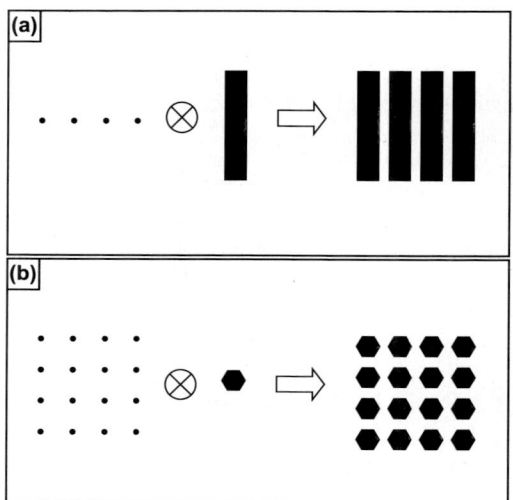

Fig. 10.3. Examples of convolution for NLPC: (**a**) 1D lattice, (**b**) a square 2D lattice with hexagonal motif

field conversion efficiency for a QPM process. To simplify the following analysis it would be assumed that the background has a zero nonlinear coefficient, while later (in Sect. 10.3.3) the final results would be adjusted to account for non-zero background. Note that the lattice area is still restricted by the rectangular interaction area $a(\mathbf{r})$.

The relevant Cartesian component of the normalized nonlinear dielectric tensor as a function of position in a nonlinear photonic crystal can be therefore expressed mathematically as

$$g(\mathbf{r}) = a(\mathbf{r}) \times (u(\mathbf{r}) \otimes s(\mathbf{r})), \tag{10.15}$$

where $\otimes$ is the convolution operator.

In the case of 1D nonlinear photonic crystal, the relevant component is described in (10.3), Sect. 10.2.

10.3.2 The Reciprocal Lattice

Similar to the analysis of crystals in solid-state physics [11], it is useful to define a reciprocal lattice for $u(\mathbf{r})$ using one or two primitive vectors (depending on the lattice dimensionality) that together with the direct lattice primitive vectors obey the following orthogonality relation:

$$\mathbf{a}_i \cdot \mathbf{b}_j = 2\pi \delta_{ij}, \tag{10.16}$$

for a 1D lattice the orthogonality relation becomes $\mathbf{a} \cdot \mathbf{b} = 2\pi$, therefore $\mathbf{b} = 2\pi/\Lambda \hat{x}$, and the reciprocal lattice points are $\mathbf{K}_m = m\mathbf{b}$.

The 2D reciprocal lattice points are given by

$$\mathbf{K}_{mn} = m\mathbf{b}_1 + n\mathbf{b}_2. \tag{10.17}$$

Although the following expressions apply to the case of a 2D lattice, they can easily be adapted for the case of a 1D lattice. The reciprocal lattice function is the two-dimensional Fourier transform of the direct (or "real") lattice function:

$$U(\mathbf{f}) = \frac{1}{A_{\text{UC}}} \sum_{m,n} \delta\left(\mathbf{f} - \frac{m\mathbf{b}_1}{2\pi} - \frac{n\mathbf{b}_2}{2\pi}\right) = \frac{(2\pi)^2}{A_{\text{UC}}} \sum_{m,n} \delta(\mathbf{K} - m\mathbf{b}_1 - n\mathbf{b}_2)$$

$$= \frac{(2\pi)^2}{A_{\text{UC}}} \sum_{m,n} \delta(\mathbf{K} - \mathbf{K}_{mn}), \tag{10.18}$$

were $A_{\text{UC}} = |a_{1x}a_{2y} - a_{1y}a_{2x}|$ is the area of the unit cell [21], $\mathbf{f}$ is the spatial frequency in the two-dimensional Fourier space, $\mathbf{K} = 2\pi \mathbf{f}$ and $\mathbf{a}_1 = (a_{1x}, a_{1y})$, $\mathbf{a}_2 = (a_{2x}, a_{2y})$. For the following discussion, it is also useful to calculate the Fourier transform of $g(\mathbf{r})$ (given by (10.15)):

$$G(\mathbf{f}) = \text{FT}\{g(\mathbf{r})\} = U(\mathbf{f}) \otimes A(\mathbf{f}) \times S(\mathbf{f}), \tag{10.19}$$

where $A(f)$ and $S(f)$ are the Fourier transform functions of the area function and motif function, respectively. For some specific motif functions, $S(f)$ is known analytically.

In the case of infinite area, $G(f)$ consists of a distributed set of Dirac delta functions (Bragg peaks), located at integral combinations of the reciprocal lattice base vectors. Furthermore, using (10.19), the Fourier coefficient becomes

$$G_{mn} = \frac{1}{A_{UC}} S\left(\frac{K_{mn}}{2\pi}\right). \tag{10.20}$$

As shown in (10.12), the conversion efficiency is proportional to $|G_{mn}|^2$. Equation (10.20) shows the combined effect of the lattice (through the unit cell area), the motif (through its Fourier transform function S) and the QPM orders m, n on the nonlinear process.

10.3.3 Conversion Efficiency for Specific Types of 2D Periodic Structures

General Expressions for Fourier Coefficients

The primitive vectors of the real lattice are (without loss of generality):

$$\boldsymbol{a}_1 = (a_{1x}, a_{1y}), \qquad \boldsymbol{a}_2 = (a_{2x}, a_{2y}). \tag{10.21}$$

Using the orthogonality condition defined in (10.16), the reciprocal lattice vectors (RLVs) are:

$$\begin{aligned}\boldsymbol{b}_1 = (b_{1x}, b_{1y}) &= \frac{2\pi}{A_{UC}}(a_{2y}, -a_{2x}), \\ \boldsymbol{b}_2 = (b_{2x}, b_{2y}) &= \frac{2\pi}{A_{UC}}(-a_{1y}, a_{1x}).\end{aligned} \tag{10.22}$$

The conversion efficiency of the six possible 2D lattices (see Fig. 10.2) would now be analyzed. For each one of the five Bravais lattice, Table 10.1 displays the primitive vectors, reciprocal lattice vectors (RLVs) and the unit cell areas (A_{UC}). Without loss of generality, the coordinate system can be defined so that one of its axes is in the same direction as that of the $\boldsymbol{a}_1$ primitive vector, and γ is the angle between the two primitive vectors.

As was shown in the previous section, the nonlinear conversion efficiency depends on the Fourier coefficient G_{mn} of the normalized nonlinear susceptibility. These Fourier coefficients depend on the motif function $s(\boldsymbol{r})$, as can be seen from (10.15). Several motifs (shown in Fig. 10.4), their $s(\boldsymbol{r})$ function and their Fourier transform function $S(f)$, are displayed in Table 10.2.

Efficiency for Specific Motif and QPM Orders

For each one of the motifs shown in Table 10.2, one can find the relevant Fourier coefficient G_{mn}. This enables to determine the highest possible efficiency for a given

Table 10.1. Primitive vectors, RLVs and unit cell area for the five lattice types

Lattice types	Primitive vectors	RLVs	Unit cell area
Hexagonal $\gamma = 120$ deg	$\boldsymbol{a}_1 = a(1, 0)$ $\boldsymbol{a}_2 = \frac{a}{2}(-1, \sqrt{3})$	$\boldsymbol{b}_1 = \frac{2\pi}{a\sqrt{3}}(\sqrt{3}, 1)$ $\boldsymbol{b}_2 = \frac{4\pi}{a\sqrt{3}}(0, 1)$	$A_{UC} = \frac{a^2\sqrt{3}}{2}$
Square $\gamma = 90$ deg	$\boldsymbol{a}_1 = a(1, 0)$ $\boldsymbol{a}_2 = a(0, 1)$	$\boldsymbol{b}_1 = \frac{2\pi}{a}(1, 0)$ $\boldsymbol{b}_2 = \frac{2\pi}{a}(0, 1)$	$A_{UC} = a^2$
Rectangular $\gamma = 90$ deg	$\boldsymbol{a}_1 = a_1(1, 0)$ $\boldsymbol{a}_2 = a_2(0, 1)$	$\boldsymbol{b}_1 = \frac{2\pi}{a_1}(1, 0)$ $\boldsymbol{b}_2 = \frac{2\pi}{a_2}(0, 1)$	$A_{UC} = a_1 a_2$
Centered-Rectangular $0 \leq \gamma \leq 90$ deg	$\boldsymbol{a}_1 = a(1, 0)$ $\boldsymbol{a}_2 = \frac{a}{2}(1, \tan \gamma)$	$\boldsymbol{b}_1 = \frac{2\pi}{a \tan \gamma}(\tan \gamma, -1)$ $\boldsymbol{b}_2 = \frac{4\pi}{a \tan \gamma}(\tan \gamma, 1)$	$A_{UC} = \frac{a^2}{2} \tan \gamma$
Oblique $0 \leq \gamma \leq 180$ deg	$\boldsymbol{a}_1 = a_1(1, 0)$ $\boldsymbol{a}_2 = a_2(\cos \gamma, \sin \gamma)$	$\boldsymbol{b}_1 = \frac{2\pi}{a_1 \tan \gamma}(\tan \gamma, -1)$ $\boldsymbol{b}_2 = \frac{2\pi}{a_2 \sin \gamma}(0, 1)$	$A_{UC} = a_1 a_2 \sin \gamma$

structure and the required dimensions and shape of the motif. Furthermore, it allows determining motif shapes that will completely null the nonlinear conversion efficiency (which, for example, can be useful to nullify unwanted processes).

As an example, Table 10.3 displays the Fourier coefficient for a circular motif for each one of the six lattice types. Note that the Fourier coefficients in Table 10.3 are suitable for the case in which the background has an opposite nonlinear coefficient with respect to that of the motif (as is the case for domain-inverted ferroelectrics). If the background has zero-nonlinearity, the Fourier coefficients shown in Table 10.3 should be multiplied by 1/2. The effects of other motifs can be calculated in a similar way using (10.20) and Table 10.2. It is worth mentioning that the rectangular motif was analyzed in detail in [22].

The normalized efficiency is examined as a function of the normalized radius R/a (the ratio between the circle radius (R) and a primitive vector length (a)) for three specific QPM orders: $(m, n) = (1, 0), (1, 1)$ and $(1, 2)$, in three out of the six lattice types, namely the hexagonal, square and honeycomb. For the other three lattices (rectangular, centered-rectangular and oblique), the motif dimension and the efficiency will depend on the two primitive vectors and the angle γ between them.

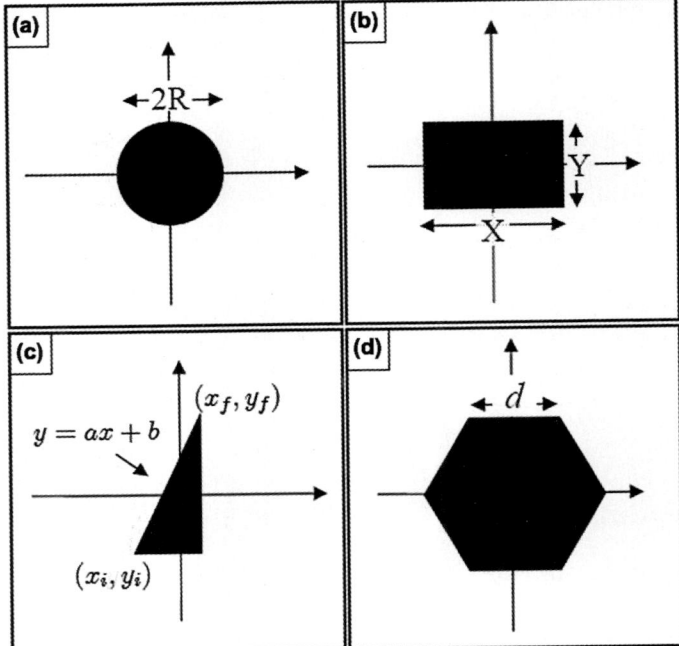

Fig. 10.4. Several motifs, all of them except the triangular are centered around point (0,0): (**a**) circular motif where R is the radius, (**b**) rectangular motif where X and Y are the length and width respectively, (**c**) triangular motif (90 deg angle), (**d**) regular-hexagonal motif where d is the edge length

The first QPM order, $(1, 0)$ is usually the most efficient process in a 2D nonlinear structure, however it relies on only one of the two primitive vectors. The second QPM order, $(1, 1)$ is usually the most efficient process that relies on both primitive vectors, although in some cases, the $(1, 2)$ order (as shown in Fig. 10.7) and the $(2, -1)$ order [22], can be more efficient.

In the following analysis for each one of the different lattices, the motif size is limited in order to avoid overlap between motifs of adjacent lattice points, therefore the normalized radii do not exceed 0.5 (Figs. 10.5, 10.6, 10.7). The insets in those figures simply show the 2D NLPC for specific R/a ratios and lattice, the black circles represent a certain sign of the nonlinear coefficient, whereas the white areas represent the opposite sign.

Generally, the square lattice can provide the highest efficiency between the three lattices, when a specific motif size is chosen, due to smallest unit cell area (see (10.18)). Figure 10.5 shows that for order $(m, n) = (1, 0)$, the square lattice with a motif size of $R/a = 0.383$, provides the highest efficiency $|G_{10}|^2 = 0.158$. A higher efficiency can be achieved in a rectangular lattice with two specific primitive vectors [22], whereas lower efficiency is obtained for the hexagonal and honeycomb

10 Nonlinear Interactions in Periodic and Quasi-Periodic 269

Table 10.2. Motif types, motif functions and their Fourier transform functions

Motif types	$s(\mathbf{r})$	$S(\mathbf{f})$
Circular	$\mathrm{circ}\left(\frac{|\mathbf{r}|}{R}\right) \equiv \begin{cases} 1, & |\mathbf{r}| \leq R \\ 0, & \text{elsewhere} \end{cases}$	$\frac{R}{|\mathbf{f}|} J_1(2\pi R|\mathbf{f}|)$
Rectangular	$\mathrm{rect}\left(\frac{x}{X}\right) \cdot \mathrm{rect}\left(\frac{y}{Y}\right)$ $\mathrm{rect}(x) = \begin{cases} 1, & |x| \leq \frac{1}{2} \\ 0, & \text{elsewhere} \end{cases}$	$XY\,\mathrm{sinc}(f_x X)\,\mathrm{sinc}(f_y Y)$
Triangular 90 deg	$\begin{cases} 1, & y_i \leq y \leq ax+b \\ & x_i \leq x \leq x_f \\ 0, & \text{elsewhere} \end{cases}$	$\frac{\mathrm{i}(x_{f-i})}{2\pi f_y}\exp[-\pi\mathrm{i}f_x(x_{f+i})]\{\exp[-\pi\mathrm{i}f_y(y_{f+i})]\mathrm{sinc}[(f_x+f_y a)(x_{f-i})]$ $-\exp[-2\pi\mathrm{i}f_y y_i]\mathrm{sinc}[f_x(x_{f-i})]\}$ $x_{f-i} = x_f - x_i$ $x_{f+i} = x_f + x_i$ $y_{f+i} = y_f + y_i$
Hexagonal	⬡	$d^2\sqrt{3}\,\mathrm{sinc}(f_x d)\mathrm{sinc}(f_y d\sqrt{3}) + \frac{3d^2 f_x}{2 f_y}\mathrm{sinc}(\frac{d}{2}f_x)\mathrm{sinc}(\frac{3d}{2}f_x)$ $-\frac{d}{2\pi f_y}\left\{\mathrm{sinc}\left[\frac{d}{2}(f_x+f_y\sqrt{3})\right]\sin\left[\frac{\pi d\sqrt{3}}{2}(f_x\sqrt{3}-f_y)\right]\right.$ $\left.+\mathrm{sinc}\left[\frac{d}{2}(f_x-f_y\sqrt{3})\right]\sin\left[\frac{\pi d\sqrt{3}}{2}(f_x\sqrt{3}+f_y)\right]\right\}$

Table 10.3. Fourier coefficient of a circular motif

Lattice types	Fourier coefficient G_{mn} of a circular motif
Hexagonal	$G_{mn}^{\text{hex}} = \dfrac{2R}{a\sqrt{m^2+n^2+mn}} J_1\left(\dfrac{4\pi R}{a\sqrt{3}}\sqrt{m^2+n^2+mn}\right)$
Square	$G_{mn}^{\text{sq}} = \dfrac{2R}{a\sqrt{m^2+n^2}} J_1\left(\dfrac{2\pi R}{a}\sqrt{m^2+n^2}\right)$
Rectangular	$G_{mn}^{\text{rect}} = \dfrac{2R}{\sqrt{(ma_2)^2+(na_1)^2}} J_1\left(2\pi R\sqrt{\dfrac{m^2}{a_1^2}+\dfrac{n^2}{a_2^2}}\right)$
Centered-Rectangular	$G_{mn}^{\text{cen-rect}} = \dfrac{2R\cdot 2\cos\gamma}{a\sqrt{m^2+4n^2\cos^2\gamma-4mn\cos^2\gamma}}$ $\times J_1\left(\dfrac{2\pi R}{a\sin\gamma}\sqrt{m^2+4n^2\cos^2\gamma-4mn\cos^2\gamma}\right)$
Oblique	$G_{mn}^{\text{ob}} = \dfrac{2R}{\sqrt{a_1 a_2}\sqrt{\dfrac{m^2 a_2}{a_1}+\dfrac{n^2 a_1}{a_2}-2mn\cos\gamma}}$ $\times J_1\left(\dfrac{2\pi R}{\sin\gamma\sqrt{a_1 a_2}}\sqrt{\dfrac{m^2 a_2}{a_1}+\dfrac{n^2 a_1}{a_2}-2mn\cos\gamma}\right)$
Honeycomb	$G_{mn}^{\text{hon}} = \begin{cases} G_{mn}^{\text{hex}} - G_{\tilde{m}\tilde{n}}^{\text{hex}}(\tilde{a}), & \tilde{m} \text{ and } \tilde{n} \in \mathbb{Z} \\ G_{mn}^{\text{hex}}, & \text{elsewhere} \end{cases}$ $\begin{bmatrix} \tilde{m} = \dfrac{2m-n}{3} \\ \tilde{n} = \dfrac{m+n}{3} \\ \tilde{a} = a\sqrt{3} \end{bmatrix}$

lattices. The two latter lattices exhibit identical efficiency for the $(1, 0)$ order, as seen both in Fig. 10.5 and in Table 10.3. For the order $(m, n) = (1, 1)$, the maximum efficiency $|G_{11}|^2 = 0.04$ is achieved for a square lattice with $R/a = 0.271$ (Fig. 10.6). For comparison, the highest efficiency of a periodic 1D structure is $|G_1(D = 0.5)|^2 = (2/\pi)^2 \simeq 0.4$.

For order $(m, n) = (1, 2)$, the honeycomb lattice produces a higher effciency than the square and the hexagonal lattices (Fig. 10.7) because for this QPM order, the honeycomb lattice is a difference between two hexagonal lattices, one with a period a, and the other with a larger period $a\sqrt{3}$ (Table 10.3). The contribution of the second hexagonal lattice to the efficiency is more significant, since the relevant Fourier coefficient is the relatively large $G_{01}^{\text{hex}}(a\sqrt{3})$, corresponding to the $(\tilde{0}, \tilde{1})$ QPM order. Indeed, as shown in the inset of Fig. 10.7, for $R/a = 0.5$ (largest non-overlapping radius), where the maximum efficiency $|G_{12}|^2 = 0.076$ is obtained, the holes of the honeycomb lattice form a shape that appears very similar to an hexagonal lattice with a period $a\sqrt{3}$. Also note that the maximum efficiency achieved

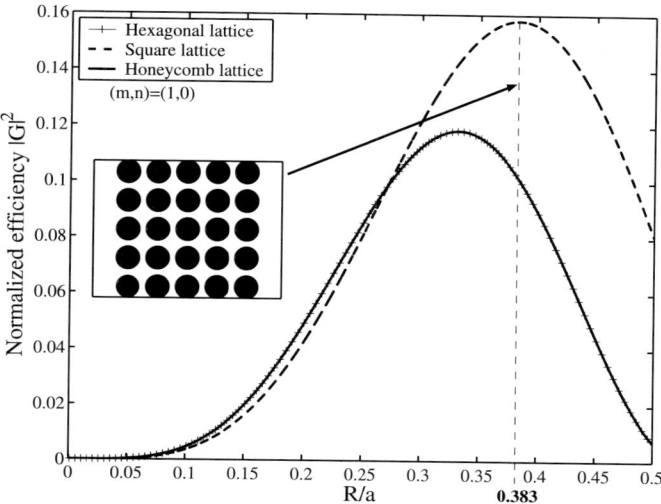

Fig. 10.5. Normalized efficiency for $(m, n) = (1, 0)$ order as a function of the motif radius to primitive vector magnitude ratio. The plus-sign, *dashed* and *solid lines* represent the efficiency curves for the hexagonal, square and honeycomb lattice, respectively. The *inset* shows the 2D NLPC for a square lattice with $R/a = 0.383$

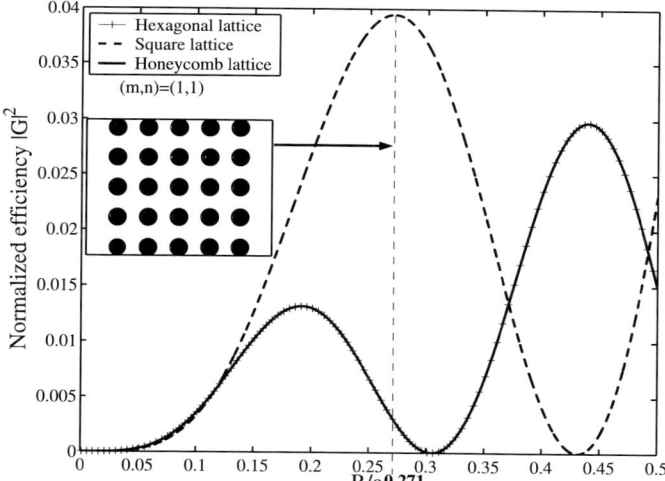

Fig. 10.6. Normalized efficiency for $(m, n) = (1, 1)$ order as a function of the motif radius to primitive vector magnitude ratio. The plus-sign, *dashed* and *solid lines* represent the efficiency curves for the hexagonal, square and honeycomb lattice, respectively. The *inset* shows the 2D NLPC for a square lattice with $R/a = 0.271$

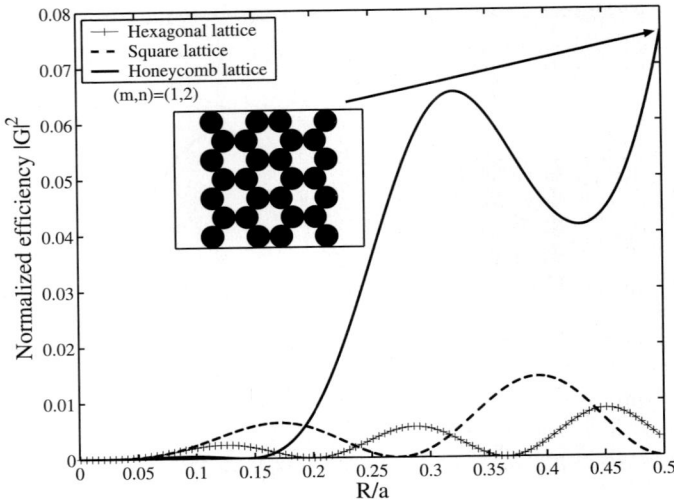

Fig. 10.7. Normalized efficiency for $(m, n) = (1, 2)$ order as a function of the motif radius to primitive vector magnitude ratio. The plus-sign, *dashed* and *solid lines* represent the efficiency curves for the hexagonal, square and honeycomb lattice, respectively. The *inset* shows the 2D NLPC for a honeycomb lattice with $R/a = 0.5$

for order $(1, 2)$ is greater than for order $(1, 1)$. This effect can be understood by examining the normalized efficiency expression in Table 10.3.

The examples discussed in this section, are for normalized radii that provide high efficiencies. However, Table 10.3 and Figs. 10.5, 10.6, 10.7 can also be used to extract normalized radii that nullify the conversion efficiency for specific orders and lattices.

10.4 Analysis of a Quasi-Periodic Nonlinear Photonic Crystal

10.4.1 Statement of the Problem

Considering an arbitrary set of D nonlinear $\chi^{(2)}$ processes, each of the processes is defined through a phase-mismatch vector $\Delta \boldsymbol{k}^{(j)}$ $[j = 1, \ldots, D]$. The objective is to create an NLPC, represented geometrically by the normalized function $g(\boldsymbol{r})$, that will simultaneously phase match all processes. The following discussion would focus on a two dimensional problem – that is, the mismatch vectors reside within some plane. However, it is important to note that the treatment is general and can be easily applied to one (see example Sect. 10.4.7) or to three dimensions. For the trivial case in which the D mismatch vectors are integrally spanned by two of them, the solution as shown in Sect. 10.3, is simply given by a periodic lattice, where the two vectors serve as a basis for a two-dimensional reciprocal periodic lattice. The direct, or real-space lattice is spanned by the basis vectors: $\boldsymbol{a}^{(1)}$ and $\boldsymbol{a}^{(2)}$ which, as mentioned previously, obey an orthogonality relation with the reciprocal lattice base vectors:

$$\boldsymbol{a}^{(i)} \cdot \Delta \boldsymbol{k}^{(j)} = 2\pi \delta_{ij}. \tag{10.23}$$

A non-trivial case exists when the set of mismatch vectors do not belong to a periodic reciprocal lattice (which is the Fourier transform of a periodic direct lattice). Some words on the desired structure: its spectrum should consist mainly from Dirac delta functions which means that the direct structure should have high degree of order or long-range correlations. For practical physical fabrication options there must be some minimal distance between the structure features. Taking these into account, some lattice can be envisioned, but it is known that in this case a periodic lattice can not solve the problem. However, an ordered lattice structure which is not necessarily periodic do exist in the form of quasi-periodic lattices, giving form to the famous physical quasicrystal [23]. Such structures do not posses translation symmetry, which means that they are not constrained to have only the specific 2, 3, 4 and 6-fold rotational symmetry, allowed for periodic structures [11].

10.4.2 Solution by Quasiperiodic Lattices

To understand how quasiperiodic lattices can solve the general problem, two different but equivalent methods for generating quasicrystal models needs to be considered. The first uses a periodic lattice in a high-dimensional space and project a subset of this lattice onto a subspace [24]. When the subspace is oriented with irrational angles compared to the high-dimensional periodic lattice primitive vectors the projection operation creates a quasiperiodic lattice upon this subspace. Another method, called the Dual Grid Method [25–27] can create quasicrystals whose tiles are parallelograms. In this method, a dual structure, called the dual grid, which contains all the topological information required to built the quasicrystal is first constructed. Then, using a simple transformation, this dual grid is transformed to a quasicrystal. These two seemingly different methods were proven to be equivalent [28], that is, any quasicrystal built with one of them can be built with the other.

To solve the problem both aspects are used [29]. To define the quasicrystal spanning vectors through the desired spectrum (containing information in a dual or transformed space) an orthogonality relation is created, similar to the one used for periodic lattices in the trivial case but this time in a higher dimensional space. This relation is then projected into the subspace that contain the problem. Then, to define the topology of the quasicrystal, that is, the information of how to connect together points spanned by the base vectors of the quasicrystal and which points to take (certainly not all points should be taken because in most cases the outcome would be dense filling of the space), a dual grid is constructed. The dual grid is also defined by the desired spectral content.

10.4.3 Establishing an Orthogonality Condition

The explicit construction of the orthogonality relation works as follows:

- The D two-dimensional mismatch vectors $\Delta \boldsymbol{k}^{(j)}$ [$j = 1, \ldots, D$] are viewed as two D-dimensional vectors $(\Delta k_\mu^{(1)}, \ldots, \Delta k_\mu^{(D)})$ [$\mu = 1, 2$]. These vectors span a two-dimensional subspace of a D dimensional vector space. Thus a D dimensional problem is formulated.
- Additional $D - 2$ D-dimensional vectors are chosen: $(q_\mu^{(1)}, \ldots, q_\mu^{(D)})$ [$\mu = 3, \ldots, D$] which must be orthogonal to the first two. These vectors span the remaining $(D - 2)$ dimensional subspace.
- Now there are enough components to go back to the initial phase-mismatch vectors and expand them into D dimensions: $\boldsymbol{K}^{(j)} = (\Delta \boldsymbol{k}^{(j)}, \boldsymbol{q}^{(j)})$ (there are D such vectors [$j = 1, \ldots, D$] which together span the entire D-dimensional space).
- And so a D-dimensional orthogonality condition can be formulated:

$$\boldsymbol{A}^{(i)} \cdot \boldsymbol{K}^{(j)} = 2\pi \delta_{ij} \tag{10.24}$$

- Each D-dimensional real-space vector is partitioned into a two-dimensional part and into a complimentary $D - 2$ part: $\boldsymbol{A}^{(j)} = (\boldsymbol{a}^{(j)}, \boldsymbol{b}^{(j)})$. As soon would be explained, the two-dimensional $\boldsymbol{a}^{(j)}$ vectors are used to span the quasi periodic lattice (they are the tiling vectors), while The $\boldsymbol{b}^{(j)}$ can be used to calculate the spectrum of the generated lattice.

All the equations implied by (10.24) can be written in matrix form. The rows of the matrix $\boldsymbol{A}$ are constructed from the $\boldsymbol{A}^{(j)}$ vectors:

$$A = \begin{pmatrix} A^{(1)} \\ A^{(2)} \\ \vdots \\ A^{(D)} \end{pmatrix}. \tag{10.25}$$

A similar $\boldsymbol{K}$ matrix is constructed out of the $\boldsymbol{K}^{(j)}$ vectors:

$$K = \begin{pmatrix} K^{(1)} \\ K^{(2)} \\ \vdots \\ K^{(D)} \end{pmatrix}. \tag{10.26}$$

And now all the equations implied by (10.24) can be written as:

$$A K^\mathrm{T} = 2\pi I. \tag{10.27}$$

Because the $\boldsymbol{K}^\mathrm{T}$ matrix is built from linear independent rows not only it has a left-hand inverse matrix (in the form of $\boldsymbol{A}$) but the same inverse is also a right-hand inverse matrix:

$$K^\mathrm{T} A = 2\pi I. \tag{10.28}$$

But this matrix equation can be in turn separated into multiple equations written as:

$$\sum_{j=1}^{D} A_{\mu}^{(j)} K_{\nu}^{(j)} = 2\pi \delta_{\mu\nu}, \qquad (10.29)$$

where μ and ν are the components of the A and K vectors. If only the first two components are taken into account (that is $\mu, \nu \in \{x, y\}$) the last equation reduces to:

$$\sum_{j=1}^{D} a_{\mu}^{(j)} \Delta k_{\nu}^{(j)} = 2\pi \delta_{\mu\nu}, \qquad (10.30)$$

which can be regarded as the orthogonality condition in the dimensionality of the problem. Put in other words it is the projection of the D-dimensional orthogonality condition onto a two-dimensional sub-space. This condition is also called the Ho condition [30]. Note that the choice of the $D - 2$ vectors $(q_{\mu}^{(1)}, \ldots, q_{\mu}^{(D)})$ is not unique, but once these are chosen the tiling vectors $a^{(j)}$ and their $(D - 2)$-dimensional extensions $b^{(j)}$ are uniquely determined by the orthogonality condition (10.24).

10.4.4 Tiling the Quasi-Periodic Lattice by the Dual Grid Construction

Establishing the orthogonality condition which gives the spanning vectors $(a^{(j)})$ for the quasi-periodic lattice, information is still needed on how to tile the lattice using these vectors. One cannot simply take all the linear integral (that is – with integers) combinations of these vectors, as is the case for periodic lattices, because the outcome would be a dense filling of space. Here, the dual grid construction [25–27] determines this information – the topology of the structure. The construction goes as follows:

- Each mismatch vector $\Delta k^{(j)}$ is associated with an infinite family of parallel lines normal to the direction of $\Delta k^{(j)}$ with separation between the lines of $2\pi/|\Delta k^{(j)}|$. Each such family is shifted from the origin (in the opposite direction of $\Delta k^{(j)}$) by an amount of $f_j L_j$ for some $0 \le f_j < 1$. These shift ensure that no more than two lines from two different families would cross each other at a given point. The arbitrariness in the value of f_j eventually (as would be shown later) manifest in a phase change in the Fourier components of the quasi-periodic lattice spectrum, but as only the absolute value of these components have any physical significance it does not affect the overall solution. The set of all families together consist the dual grid.
- Each lines family j is enumerated with a set of integers n_j.
- Each tile (space between lines) of the dual grid is associated with a set of the corresponding n_j [$j = 1, \ldots, 3$] integers: the n_j integer denotes the number of the line from family j which is the closest to the tile from the $(-\Delta \hat{k}^{(j)})$ direction (or from the positive direction, but this choice must be consistent for each lines family).

- Each tile of the dual grid is transformed into a vertex in the quasi-periodic lattice at the position $\sum_j n_j a^{(j)}$. This last step sorts the linear integral combinations of the tiling vectors that would be included in the tiling from all possible combinations.

Thus the construction of the quasi-periodic lattice is established. The dual grid is topological dual to the quasiperiodic lattice in the sense that each vertex of the dual grid corresponds to a tile of the quasiperiodic lattice and each cell in the dual grid corresponds to a vertex in the quasiperiodic lattice. To transform the lattice into an NLPC some motif (representing for example a positive value of $\chi^{(2)}$ over a negative or zero background) needs to be attached to the lattice points, or a set of motifs to the different tiles. This would be discussed shortly.

10.4.5 The Fourier Transform of the Quasi-Periodic Lattice

A very important issue that must be considered is the following: it can be shown [26] that if $\rho(r)$ is the geometrical representation of the quasi-periodic lattice (that is – a set of distributed delta function centered at the positions of the lattice points) then its Fourier transform $\rho(k)$ is non-vanishing at most on the lattice of integral linear combinations of $\Delta k^{(j)}$. If the $\Delta k^{(j)}$ are integrally independent Bragg peaks occur at these positions, thus satisfying the condition for phase matching a process whose mismatch value is one of those $\Delta k^{(j)}$. If, on the other hand, some of the mismatch vectors are integrally dependent some of the Bragg peaks may become extinguished. To avoid this situation an integrally independent set of phase mismatch vectors must always be chosen (for an arbitrary set this means excluding vectors until this requirement is satisfied).

Assuming that indeed all of the $\Delta k^{(j)}$ are integrally independent the Fourier coefficient at some $k = \sum m_j \Delta k^{(j)}$ is given by [26]:

$$\rho(k) = \frac{1}{v} e^{iq \cdot f} \int_W ds\, e^{iq \cdot s}, \qquad (10.31)$$

where $q = \sum m_j q^{(j)}$, $f = \sum f_j b^{(j)}$, W is the so-called $(D-2)$-dimensional window, given by the set of all points $\sum \lambda_j b^{(j)}$ with $0 < \lambda_j < 1$, s is a $(D-2)$-dimensional integration vector and v is the volume of the primitive cell of the D-dimensional real space lattice generated by the vectors $A^{(j)}$ (which is the determinant of the square matrix whose rows are $A^{(j)}$).

To evaluate this integral it would be beneficial to consider the contribution of each spatial component. The μ component of the window can be written as:

$$\min_{j,\lambda_j}\left\{-\sum_{j=1}^{D}\lambda_j b_\mu^{(j)} \mid 0 < \lambda_j < 1\right\} \div \max_{j,\lambda_j}\left\{-\sum_{j=1}^{D}\lambda_j b_\mu^{(j)} \mid 0 < \lambda_j < 1\right\}. \qquad (10.32)$$

For each such component the minimum value is found by: setting 1 for every λ_j multiplying a positive $b_\mu^{(j)}$ and setting 0 for every λ_j multiplying a negative $b_\mu^{(j)}$.

The maximum value is found by: setting 1 for every λ_j multiplying a negative $b_\mu^{(j)}$ and setting 0 for every λ_j multiplying a positive $b_\mu^{(j)}$.

Using this information (10.31) can be written as:

$$\rho(k) = \frac{1}{v} e^{i\sum_{\mu=1}^{D-2} q_\mu f_\mu} \int_W ds\, e^{i\sum_{\mu=1}^{D-2} q_\mu s_\mu} = \frac{1}{v} \prod_{\mu=1}^{D-2} e^{iq_\mu f_\mu} \prod_{\mu=1}^{D-2} \int_{W_\mu} ds_\mu\, e^{iq_\mu s_\mu}, \tag{10.33}$$

which can be evaluated to give:

$$\rho(k) = \frac{1}{v} \prod_{\mu=1}^{D-2} e^{iq_\mu f_\mu} \prod_{\mu=1}^{D-2} \frac{1}{iq_\mu} \left[e^{iq_\mu W_{\max\mu}} - e^{iq_\mu W_{\min\mu}} \right]. \tag{10.34}$$

10.4.6 From Lattice to a Nonlinear Photonic Crystal

After the fundamental lattice has been established, the NLPC can be constructed. Two different approaches towards this goal might be considered. The first one is accomplished by simply convolving all the lattice points with a single motif representing, say, a positive value of $\chi^{(2)}$ over a reversed or zero background. The merits of this approach are two-fold: first the fabrication of planar NLPCs with such simple motifs is already a known and accomplished practice [4, 7, 20]. Secondly, in this case writing down an analytical expression for the NLPC's Fourier coefficients is straight forward: let $g(r)$ denote the geometrical representation of the actual NLPC. If it is extended over a finite area $a(r)$ and consists of positive domain motifs of a given shape $s(r)$, positioned at the vertices of the tiling, its Fourier transform, using the convolution theorem, is given by:

$$g(k) = \Delta\chi \left(\rho(k) \otimes \int_{a(r)} e^{ik \cdot r} d^2 r \right) \int_{s(r)} e^{ik \cdot r} d^2 r, \tag{10.35}$$

where $\Delta\chi$ is the absolute difference between the normalized positive and negative (or null) values used for the relevant tensor component of $\chi^{(2)}$ (that is for a null background $\Delta\chi = 1$ and for a reversed background $\Delta\chi = 2$).

The second approach is based on choosing a specific modulation for different tiles of the fundamental lattice. In this case the specific choice of motifs is made through a numerical analysis and we are not aware of an analytical solution to give the best results. In this case simple expression for the Fourier coefficients of the structure can not be derived as in the previous approach. However in this case far better results can be established as well as options for spectral shaping as would be demonstrated in the accompanying one-dimensional NLPC example. One of the options to tackle this numeric endeavor is through the rigorous definition of a mathematical maximizing problem which must be optimized under some constraints [31]. Such an approach requires very high poling resolution, possibly exceeding the fabrication capabilities. Simpler solutions are to treat each tile to be composed of some simple filling form. For example – for one dimensional problems this reduces to finding the best positively poled length over a negative (or null) background of the length between adjacent lattice points. For two-dimensional problems it must be considered that the lattice tiles are all parallelograms, and so some simple filling rule needs to be applied to such forms – for example a small parallelogram of positively poled material over the background defined by a parallelogram tile.

10.4.7 A One-Dimensional Example – The Three Wave Doubler

Device Requirements

To illustrate this general method consider an explicit example of a one-dimensional device set up to phase match three different second harmonic generation processes of three wavelengths in the fiber telecom C-band: 1530 nm, 1550 nm and 1570 nm. The nonlinear crystal chosen is KTiOPO$_4$ (KTP) operating at (an arbitrary chosen) temperature of 100° centigrade. At these conditions the phase-mismatch values for the processes are [32, 33]: $\Delta k^{(1)} = 0.263$ μm^{-1}, $\Delta k^{(2)} = 0.256$ μm^{-1} and $\Delta k^{(3)} = 0.249$ μm^{-1} respectively. For phase matching all processes, the desired NLPC must accommodate in its spectrum significant values located at these phase mismatch locations. Alternatively the infinite lattice which is the infrastructure of the photonic crystal must contain these mismatch vectors as part of its reciprocal lattice. Remembering that the device is one dimensional – all mismatch values lies on the same line, then if all of the required mismatch values were an integral multiple of one some value say $\Delta k^{(0)}$, the real lattice with base vector (value) of $2\pi/\Delta k^{(0)}$ would solve the problem (note the fulfillment of the orthogonality condition). The mismatch values above shows that for practical purposes this is not the case. Practical, because one can argue that some comparable tiny value that will fulfill this condition with large integers can be found. However when the nonlinear photonic crystal is built upon the lattice a motif representing a specific polarity of the nonlinear polarization is attached to each lattice point. The effect of this motif in Fourier space can be approximated as a low pass filter – greatly attenuating high spatial frequency components represented by the base reciprocal vector multiplied by a large integer [22]. Thus the possible efficiency of phase matching a process using such components is very small. Excluding the option for a trivial solution made of a periodic lattice the steps to create a quasi-periodic lattice are now followed. For the following the set of required mismatch vectors are assumed to be integrally independent (as is the case), otherwise vectors must be excluded until this requirement is satisfied (the final structure would accommodate the excluded values in its reciprocal lattice).

Establishing the Orthogonality Condition

An orthogonality condition is established in a space with dimensionality equal to the number of required mismatch vectors, three dimensional space in the present case. The first step is to view the three one-dimensional mismatch vectors $\Delta k^{(j)}$ [$j = 1, \ldots, 3$] as a single vector containing three components: $\boldsymbol{k}_1 = (\Delta k^{(1)},$

$\Delta k^{(2)}$, $\Delta k^{(3)}$). This vector spans a one-dimensional subspace of a three dimensional vector space. With this vector an orthogonality relation needs to be defined in a three-dimensional space. For this purpose the newly formed vector are expanded into a 3 × 3 non-singular matrix by adding two new vectors, $\boldsymbol{q}_2$ and $\boldsymbol{q}_3$ orthogonal to the first, $\boldsymbol{q}_i \cdot \boldsymbol{k}_1 = 0$ [$i = 2, 3$], so that the newly formed matrix is:

$$\boldsymbol{K} = \begin{pmatrix} \Delta k^{(1)} & q_2^{(1)} & q_3^{(1)} \\ \Delta k^{(2)} & q_2^{(2)} & q_3^{(2)} \\ \Delta k^{(3)} & q_2^{(3)} & q_3^{(3)} \end{pmatrix} = \begin{pmatrix} \boldsymbol{K}^{(1)} \\ \boldsymbol{K}^{(2)} \\ \boldsymbol{K}^{(3)} \end{pmatrix}. \tag{10.36}$$

The matrix $\boldsymbol{A}$ is found from (10.27):

$$\boldsymbol{A} = \begin{pmatrix} \boldsymbol{A}^{(1)} \\ \boldsymbol{A}^{(2)} \\ \boldsymbol{A}^{(3)} \end{pmatrix} = 2\pi \left(\boldsymbol{K}^{\mathrm{T}} \right)^{-1}. \tag{10.37}$$

The calculated q-vectors are: $q_2 = (0.648, -0.342, -0.333)$ and $q_3 = (-0.342, 0.667, -0.324)$ (note that this selection is not unique). The different components of the $\boldsymbol{A}$ matrix are denoted in the following way:

$$\boldsymbol{A} = \begin{pmatrix} a^{(1)} & b_2^{(1)} & b_3^{(1)} \\ a^{(2)} & b_2^{(2)} & b_3^{(2)} \\ a^{(3)} & b_2^{(3)} & b_3^{(3)} \end{pmatrix}. \tag{10.38}$$

Every row is a vector of the form $\boldsymbol{A}^{(j)} = (a^{(j)}, \boldsymbol{b}^{(j)})$. The vectors $\boldsymbol{b}^{(1)} = (6.283, 0)$, $\boldsymbol{b}^{(2)} = (0, 6.283)$, and $\boldsymbol{b}^{(3)} = (-6.635, -6.453)$ can be used to calculate the Fourier coefficients of every Bragg peak in the quasicrystal spectrum [29]. Thus for an NLPC with a single motif repeated at the quasiperiodic lattice unit cells the expected efficiency for a quasi-phase-matching processes that relies on a specific peak can be calculated. The $a^{(j)}$ vectors (scalars in the present one-dimensional structure with values $a^{(1)} = 8.394$ μm, $a^{(2)} = 8.165$ μm and $a^{(3)} = 7.950$ μm) are used to span the quasiperiodic lattice using the dual grid construction as follows.

Dual Grid Construction

Essentially the algorithm outlined in Sect. 10.4.4 is used here for the present one-dimensional case. It is illustrated in Fig. 10.8a–c for unrelated but graphically convenient three phase mismatch values. The size of the quasicrystal is governed by the number of lines in the dual grid. For the present example, for a 1 cm long quasicrystal, approximately 800 lines in each family are needed.

From Lattice to Nonlinear Photonic Crystal

Here the approach of choosing a specific modulation (or building block) for the different lattice tiles was used. Each such building block can be partitioned into two parts – one with $+\chi^{(2)}$ and the other with $-\chi^{(2)}$. Each such building block thus have a specific ratio or duty cycle. This is illustrated in Fig. 10.8(d).

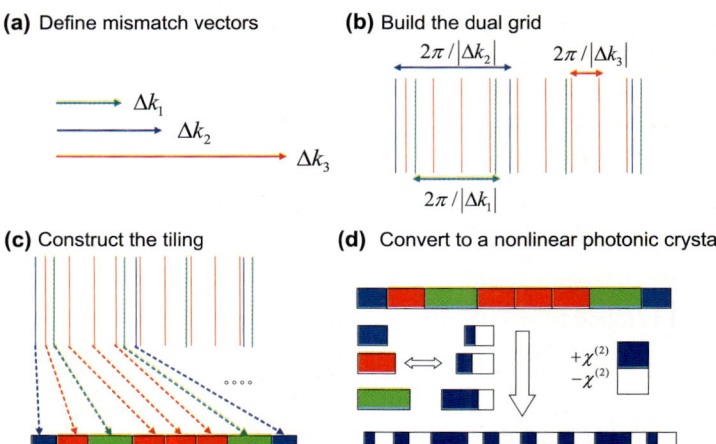

Fig. 10.8. Illustration of the solution for designing a one-dimensional NLPC for multiple collinear optical processes, using the dual grid method. (**a**) The required mismatch vectors. (**b**) The dual grid, in which each family of lines is shown with a different color. (**c**) Tiling of the real-space line according to the order in which lines of different families appear in the dual grid. (**d**) Associating a given duty cycle with each tiling vector. Positively-poled segments are shown in blue, and negatively-poled segments are shown in white

We have found out that in some cases the best results are achieved when some of the building blocks are made with 100% duty cycle while the others are with 0% duty cycle. These are also the easiest to fabricate structures in terms of required resolution. Such a selection of duty cycles might look at first glance as counter intuitive, and in fact previous one-dimensional modulations of quasi-periodic nonlinear photonic crystals did not use such a selection [13, 14, 34]. One needs to consider that even with a 0% duty cycle where instead of a building block there is a margin or space (between 100% building blocks), these margins still carry the long-range ordering of the lattice infrastructure. One can also consider that the most basic NLPC – a one-dimensional periodic structure with 50% duty cycle [18] can be treated as made out of two interchanging families of building blocks (at twice the spatial frequency) where one of them is made with 0% duty cycle and the other with 100% duty cycle.

Spectral Shaping

The effect of different duty cycles on the spectral response of the NLPC is illustrated in Fig. 10.9 where three different sets of duty cycles for the building blocks are used. The building blocks duty cycles for the fabricated NLPC are 100%, 0% and 0%. The Fourier coefficients are comparable to the $2/\pi \simeq 0.6366$ figure of merit (see Sect. 10.2). In fact as the efficiency of such processes depends on the square of the Fourier coefficient and on the square of the interaction length, the efficiency per process when three periodic NLPCs with total length of L are used goes like

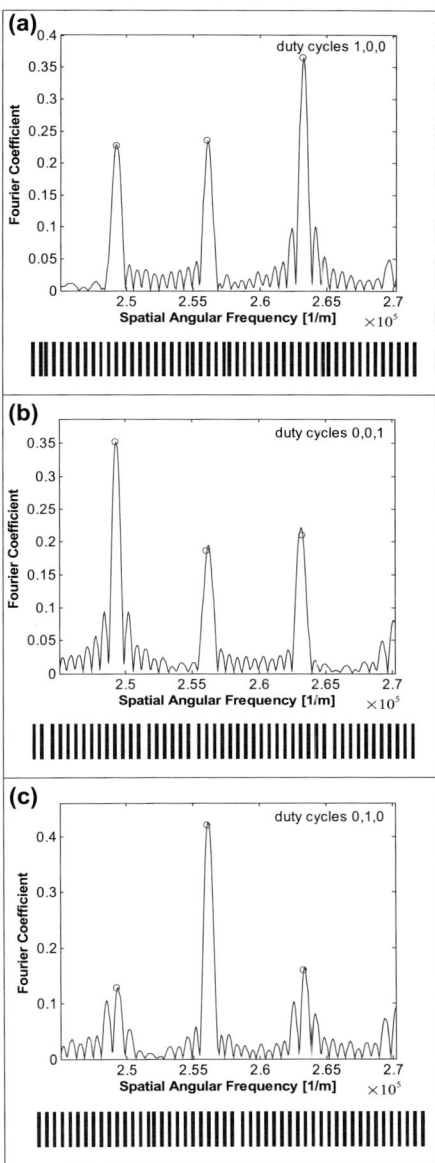

Fig. 10.9. Spectral shaping. Each panel shows the magnitude of the Fourier transform for a 1 cm long NLPC made to phase match the three collinear processes, described in the text. In each panel one of the tiling vectors is given a duty cycle of 100%, denoted as 1, and the remaining two a duty cycle of 0%, or 0. Each panel also shows a piece of the corresponding real-space representation of the NLPC, where the smallest element size is 8 μm

$(\frac{1}{3} \cdot \frac{2}{\pi})^2 = 0.045$ (assuming a length of $L/3$ is used for each period). In the present case the lowest process efficiency scales as $(1 \cdot 0.23)^2 = 0.053$ and the highest process efficiency scales as $(1 \cdot 0.365)^2 = 0.133$, both of them being more efficient than the composite periodic structure.

To verify the above analysis an actual NLPC was fabricated and tested with experimental results that were highly correlated with the theoretical predictions [35].

The ability to shape the spectrum using different motifs can be further extended for cascaded processes (processes, which not like for the three-wave doubler example, interact with each other). For example, considering a multi-harmonic generation in which the output of one processes serves as an input to a second processes it might proof beneficial to give precedence to the efficiency of the first processes at the beginning of the NLPC and gradually, along the interaction length, change the balance towards the second process.

10.5 Discussion and Summary

In this chapter, three wave mixing processes in ordered – periodic and quasi-periodic – nonlinear structures were analyzed. The main emphasis was on one-dimensional and two-dimensional modulation of the second-order susceptibility, which is currently feasible in various nonlinear materials, e.g. ferroelectrics and semiconductors.

In the case of periodic structures, the efficiency of the nonlinear interaction is governed by the choice of the lattice and the nonlinear motif. Whereas a one-dimensional NLPC has essentially a single type of lattice type (a set of equally-spaced parallel lines) and motif (a strip having nonlinear coefficient with opposite sign than that of the background), the two-dimensional NLPCs offer much larger selection of lattice types and motif shapes. A general expression that determines the effect of the NLPC's lattice and motif is given in (10.12) and (10.20). The five possible two-dimensional Bravais lattices, as well as the commonly used honeycomb lattice were studied in detail in Sect. 3. Furthermore, several specific shapes of two-dimensional motifs – circular, rectangular, triangular and hexagonal were analyzed. This analysis enables one to optimize the NLPC shape and determine its performance.

Quasi-periodic nonlinear photonic crystals extend the phase matching possibilities beyond the basic limitations of the modulation dimensions. For example, whereas a one-dimensional periodic NLPC is efficient only for integer multiples of a fundamental phase mismatch value, any set of wave-vectors can be designed to be efficient in a quasi-periodic structure. Similarly, whereas a 2D periodic structure is efficient only for phase mismatch values that are integer combination of two fundamental vectors in a 2D periodic NLPC, a 2D quasi-periodic structure can be designed to be efficient in any arbitrary set of 2D vector combinations. The key issue in using quasi-periodic structures in nonlinear optics is how to practically design them, since the structure requirements are usually defined in the reciprocal

(Fourier) space. Unlike for periodic NLPCs, there is no simple transformation between the reciprocal space and real space. In Sect. 4 a systematic procedure for designing a real-space NLPC is presented, based on phase matching requirements in the reciprocal space [29]. This method was recently confirmed experimentally, in one-dimensionally-poled KTP [35].

In this chapter, the nonlinear interactions were analyzed under the assumptions of having plane-waves and negligible pump depletion. However, this analysis can be easily extended for the case of finite beams that may also be strongly-coupled. As long as the beam's size and the coupling length are large compared to the characteristic length of the NLPC (period, motif size, etc.), it can be assumed that the effective nonlinearity is "averaged" over the interaction area, and therefore an effective nonlinear coefficient can be defined as:

$$d_{\text{eff}} = d_{ij} G_{mn}. \tag{10.39}$$

By replacing the material nonlinearity with the effective nonlinearity, the derivations of this chapter can be applied for the cases of strongly-coupled nonlinear waves [19] or for interacting Gaussian beams [36].

Finally, it should be mentioned that there are additional ordered nonlinear photonic structures that were not discussed in this chapter. One recent example is the annular symmetry frequency converter [37]. This structure consists of concentric rings alternating between $+d_{ij}$ and $-d_{ij}$. This device has continuous rotational symmetry, and in contrary to the commonly used periodic structures, it has no translation symmetry. It possesses interesting phase matching attributes that are significantly different than those of periodic structures. In particular, it enables simultaneous phase-matched frequency conversion of the same input wave into several different directions. Moreover, it has extremely wide phase-mismatch tolerance, since a change in the phase matching conditions does not change the frequency-converted power, but only changes its propagation direction.

References

1. V. Berger, Phys. Rev. Lett. **81**(19), 4136–4139 (1998)
2. L.A. Eyres, P.J. Tourreau, T.J. Pinguet, C.B. Ebert, J.S. Harris, M.M. Fejer, L. Becouarn, B. Gerard, E. Lallier, Appl. Phys. Lett. **79**, 904–906 (2001)
3. R.T. Bratfalean, A.C. Peacock, N.G.R. Broderick, K. Gallo, R. Lewen, Opt. Lett. **30**, 424–426 (2005)
4. N.G.R. Broderick, G.W. Ross, H.L. Offerhaus, D.J. Richardson, D.C. Hanna, Phys. Rev. Lett. **84**(19), 4345–4348 (2000)
5. N.G.R. Broderick, R.T. Bratfalean, T.M. Monro, D.J. Richardson, C.M. de Sterke, J. Opt. Soc. Am. B **19**, 2263 (2002)
6. A. Chowdhury, C. Staus, B.F. Boland, T.F. Kuech, L. McCaughan, Opt. Lett. **26**, 1353 (2001)
7. S. Saltiel, Y.S. Kivshar, Opt. Lett. **25**, 1204–1206 (2000)
8. S.M. Saltiel, Y.S. Kivshar, Opt. Lett. **27**, 921 (2002)

9. S.F. Mingaleev, Y.S. Kivshar, Phys. Rev. Lett. **86**, 5474 (1996)
10. M. Bertolotti, J. Opt. A Pure Appl. Opt. **8**, S9–S32 (2006)
11. C. Kittel, *Introduction to Solid State Physics*, 7th edn. (Wiley, New York, 1995)
12. T. Ellenbogen, A. Arie, S.M. Saltiel, Opt. Lett. **32**, 262–264 (2007)
13. K. Fradkin-Kashi, A. Arie, P. Urenski, G. Rosenman, Phys. Rev. Lett. **88**(2), 023903 (2001)
14. S.N. Zhu, Y.Y. Zhu, N.B. Ming, Science **278**, 843 (1997)
15. H. Liu, S.N. Zhu, Y.Y. Zhu, N.B. Ming, X.C. Lin, W.J. Ling, A.Y. Yao, Z.Y. Xu, Appl. Phys. Lett. **81**, 3326–3328 (2002)
16. J. Liao, J.L. He, H. Liu, J. Du, F. Xu, H.T. Wang, S.N. Zhu, Y.Y. Zhu, N.B. Ming, Appl. Phys. B Lasers Opt. **78**, 265–267 (2004)
17. M.H. Chou, K.R. Parameswaran, M.M. Fejer, I. Brener, Opt. Lett. **24**, 1157–1159 (1999)
18. M.M. Fejer, G.A. Magel, D.H. Jundt, R.L. Byer, IEEE J. Quantum Electron. **28**, 2631–2654 (1992)
19. R.W. Boyd, *Nonlinear Optics*, 2th edn. (Academic Press, San Diego, 2003)
20. S.M. Russell, P.E. Powers, M.J. Missey, K.L. Schepler, IEEE J. Quantum Electron. **37**, 877 (2001)
21. C. Giacovazzo, H.L. Monaco, G. Artioli, D. Viterbo, G. Ferraris, G. Gilli, G. Zanotti, M. Catti, *Fundamentals of Crytallography*, 2nd edn. (University Press, Oxford, 2002)
22. A. Arie, N. Habshoosh, A. Bahabad, Opt. Quantum Electron. **39**, 361–375 (2007)
23. D. Shechtman, I. Blech, D. Gratias, J.W. Cahn, Phys. Rev. Lett. **53**(20), 1951–1953 (1984)
24. V. Elser, Phys. Rev. B **32**(8), 4892–4898 (1985)
25. N. de Bruijn, Proc. K. Ned. Akad. Wet. Ser. A **84**, 39–66 (1981)
26. D.A. Rabson, T.L. Ho, N.D. Mermin, Acta Crystallogr. Sect. A **44**(5), 678–688 (1988)
27. D.A. Rabson, T.L. Ho, N.D. Mermin, Acta Crystallogr. Sect. A **45**(8), 538–547 (1989)
28. F. Gahler, J. Rhyner, J. Phys. A Math. Gen. **19**, 267–277 (1986)
29. R. Lifshitz, A. Arie, A. Bahabad, Phys. Rev. Lett. **95**(13), 133–901 (2005)
30. T.L. Ho, Phys. Rev. Lett. **56**(5), 468–471 (1986)
31. A.H. Norton, C.M. de Sterke, Opt. Lett. **28**, 188–190 (2003)
32. K. Fradkin, A. Arie, A. Skliar, G. Rosenman, Appl. Phys. Lett. **74**, 914–916 (1999)
33. S. Emanueli, A. Arie, Appl. Opt. **42**, 6661–6665 (2003)
34. K. Fradkin-Kashi, A. Arie, IEEE J. Quantum Electron. **35**, 1649–1656 (1999)
35. A. Bahabad, N. Voloch, A. Arie, R. Lifshitz, J. Opt. Soc. Am. B **24**, 1916–1921 (2007)
36. G.D. Boyd, D.A. Kleinman, J. Appl. Phys. **39**, 3596–3639 (1968)
37. D. Kasimov, A. Arie, E. Winebrand, G. Rosenman, A. Bruner, P. Shaier, D. Eger, Opt. Express **14**, 9371–9376 (2006)

11 Domain-Engineered Ferroelectric Crystals for Nonlinear and Quantum Optics

M. Bellini, P. Cancio, G. Gagliardi, G. Giusfredi, P. Maddaloni, D. Mazzotti, and P. De Natale

11.1 Introduction

Nonlinear optics studies the class of phenomena occurring when an intense light field, typically from a laser source, modifies the optical properties of a transparent material in a nonlinear way [1–3]. The polarization $\mathbf{P}(\mathbf{x}, t)$ of the material can be written as a power series in the field strength $\mathbf{E}(\mathbf{x}, t)$:

$$\mathbf{P} = \chi^{(1)}\mathbf{E} + \chi^{(2)}\mathbf{E}^2 + \chi^{(3)}\mathbf{E}^3 + \cdots \qquad (11.1)$$

where $\chi^{(i)}$ is the i-order optical susceptibility of the material. Nonlinear phenomena arise from the nonzero value of the $\chi^{(2)}$ susceptibility in noncentrosymmetric crystals. A large class of nonlinear materials (among them LN, KTP, BBO, and LBO) has been studied and used since 1960's for up/down-conversion of the existing laser sources to wavelength regions which are not directly accessible otherwise [4]. Some of these materials also belong to ferroelectrics, and this feature can be exploited to engineer the orientation of their nonlinear susceptibility. One of the earliest [5] and most commonly used material is LiNbO$_3$ (LN), because of its high nonlinear coefficient ($d_{33} \approx 27$ pm/V) and its wide transparency range from the UV to the mid IR (0.3÷5 µm). A technique giving access to d_{33} in LN for optimizing nonlinear conversion processes, named *quasi-phase-matching* (QPM) [6], was thought even before the first fabrication of this material. About 20 years later, the first experimental demonstration of this idea was obtained [7], and nowadays periodic poling of ferroelectrics crystals is a widely spread technology making these devices worldwide used and commercially available.

11.1.1 Classification of Nonlinear Processes

Depending on the optical configurations in which the input/output waves interact through the $\chi^{(2)}$ susceptibility, different class of processes can be identified. Each kind of process always involves 3 optical waves named *pump*, *signal*, and *idler* (with decreasing frequencies $\nu_p > \nu_s > \nu_i$). The energy conservation law for the discrete conversion of a *pump* photon into a *signal/idler* photons pair (or viceversa) requires:

$$\nu_p - \nu_s - \nu_i = 0. \tag{11.2}$$

It is obvious that the frequency tunability of the output wave(s) as well as their linewidths are strictly bound by this law. The spatial modes and the power of the output wave(s) depend on the focusing and overlapping of the input wave(s).

When the input waves are *signal* and *idler* and the output wave is the *pump*, the process is named *sum-frequency generation* (SFG). This is an up-conversion process, in which the generated frequency is higher than the 2 generating ones. In the degenerate case where $\nu_s = \nu_i$ and $\nu_p = 2\nu_s$, the process is named *second-harmonic generation* (SHG) and consists in an optical frequency doubling.

When the input waves are *pump* and *signal* and the output wave is the *idler*, the process is named *difference-frequency generation* (DFG). This is a down-conversion process, in which the generated frequency is lower than the 2 generating ones.

When the input wave is the *pump* and the output waves are *signal* and *idler*, the process is named *optical parametric generation* (OPG). This is a down-conversion process, in which the 2 generated frequencies are lower than the generating one. When one of the generated wave, either *signal* or *idler* (or even both), resonates in an optical cavity, the process is named *optical parametric oscillation* (OPO) and, similarly to what happens in a laser cavity, a threshold *pump* power exists for the oscillation to occur.

11.1.2 Phase Matching

Equation (11.2) is not the only conservation law binding the nonlinear process. The momentum conservation law, better called phase matching (PM) condition, is another physical condition determining the "direction" of the process (direct/inverse), i.e., if photons will emerge from the crystal at the frequency of the *pump/signal/idler*:

$$\mathbf{k}_p - \mathbf{k}_s - \mathbf{k}_i = \mathbf{0} \tag{11.3}$$

that for collinear waves is equivalent to

$$n_p \nu_p - n_s \nu_s - n_i \nu_i = 0. \tag{11.4}$$

Because of the dispersion in transparent media ($n_p < n_s < n_i$), this condition is not automatically satisfied, and suitable optical properties overcoming this naturally occurring phase mismatch must be exploited.

Birefringent Phase Matching

Most of the nonlinear optical materials with a nonzero $\chi^{(2)}$ are also birefringent (apart from crystals with cubic symmetry). For example, the birefringence of uniaxial crystals can be exploited to achieve PM due to the existence of ordinary/extraordinary refraction indexes n_o and n_e along different directions with respect to the optical axis. A careful tuning of the orientation angle and temperature of the crystal can give the desired PM condition. In special cases named *noncritical PM*, where the crystal orientation angle is either 0° or 90°, an unwanted walk-off between the interacting waves is also avoided, and the efficiency is higher. With this kind of PM the polarization of the 3 waves involved in the nonlinear process cannot be all the same, since these waves are both ordinary and extraordinary. Hence birefringent PM (BPM) gives access to small off-diagonal nonlinear coefficients (i.e., the d_{ij} elements with $i \neq j$) only.

Quasi-Phase Matching

For not all nonlinear mixing processes, a proper crystal with noncritical PM at a certain temperature can be found. This can be overcome with a special class of crystals, whose ferroelectric domains are periodically poled. The phase mismatch of the interacting waves is quasi-compensated every half period by the inversion of the $\chi^{(2)}$ susceptibility. The QPM gives access also to the large diagonal nonlinear coefficients (i.e., the d_{ii} elements), increasing the efficiency of optical mixing processes by about a factor 20 with respect to BPM.

11.2 Nonlinear Optics for Spectroscopic Applications

Since its early appearance, laser spectroscopy has shown to be a powerful tool to investigate atomic and molecular physics with great precision and sensibility. While present laser sources cover most of the visible/near-IR spectrum of light, there is a lack of available coherent sources in the 2÷5 μm region. However this spectral window is of the utmost importance for trace gas detection, because here lie the fundamental ro-vibrational bands of many molecular species of atmospheric interest.

11.2.1 Coherent Sources for mid-IR Spectroscopy and Metrology

Several coherent sources exist with emission in the mid-IR. Now, let us shortly describe their features in terms of frequency tunability, power, and linewidth.

Gas lasers (e.g., He–Ne, CO, CO_2) are generally powerful but poorly tunable coherent sources: this seriously limits the probability to find coincidences between their discrete emission frequencies and molecular spectra.

Quantum cascade lasers (QCLs) [8] are particularly engineered semiconductor lasers directly emitting in the mid-IR even more than 1 W average power. The main drawback is that room-temperature and single-mode CW operation is available only in the spectral range 4.8÷9.6 μm [9, 10]. Also in this case, their tunability is rather poor ($<10\,\text{cm}^{-1}$), when compared with conventional semiconductor lasers in external-cavity configuration with a feedback grating ($>500\,\text{cm}^{-1}$), and it is only achievable with big changes in the operation temperature.

Optical parametric oscillators (OPOs) are powerful and widely tunable ($2 \div 4.5\,\mu m$) sources based on $\chi^{(2)}$ optical processes occurring in nonlinear crystals within singly or doubly resonant cavities. They often suffer from uncontrolled mode-hops and require cavity locking with servo loops.

Alternatively, coherent sources based on difference-frequency generation (DFG) rely on single-pass nonlinear optical processes and can generate narrow-linewidth and tunable radiation in the same spectral region, with IR powers ranging from the μW- up to the mW-level [11–13]. DFG source does not require resonant cavities and is intrinsically mode-hop free. Periodically-poled $LiNbO_3$ (PPLN) is one of the most efficient nonlinear crystals for the mid-IR [4]. Its transparency range allows down-conversion processes from visible/near-IR lasers to the $2 \div 4.5\,\mu m$ range.

For many years, mid-IR metrology has been limited to a few frequency references, mainly gas lasers stabilized onto molecular transitions, such as He-Ne/CH_4 at $3.39\,\mu m$ or CO_2/OsO_4 at $10.6\,\mu m$ [14]. The advent of optical frequency-comb synthesizers (OFCSs), based on mode-locked fs lasers, has suddenly led to new advances in the field of precision spectroscopy [15, 16]. By acting as a bridge between the radio-frequency (RF) and the optical domain, an OFCS allows one to count the optical cycles of a continuous-wave (CW) laser directly with respect to an absolute frequency standard, such as an atomic clock. This has represented an immediate breakthrough for accurate frequency metrology in the visible/near-IR spectrum, where the first OFCSs worked, enabling for measurements of atom energies with a relative precision approaching 10^{-15} [17, 18]. In this frame, new perspectives have been opened recently by the demonstration of frequency combs operating in the UV, based on high-order harmonic up-conversion [19–21]. On the other hand, further extension of OFCSs to the IR region is crucial for absolute frequency measurements on molecular ro-vibrational spectra. So far, direct broadening of the spectrum of fs mode-locked lasers through highly-nonlinear optical fibres has succeeded in extending combs up to a 2.3-μm wavelength [22]. For longer wavelengths, a few alternative schemes have been devised, essentially based on parametric generation processes in nonlinear crystals. A 270-nm-span frequency comb at $3.4\,\mu m$ has been realized by DFG between two spectral peaks emitted by a single uniquely-designed Ti:Sa fs laser [23]. In a different approach, the metrological performance of a Ti:Sa frequency comb generator has been transferred to the 9-μm region by using two diode lasers at 852 and 782 nm as intermediate oscillators, with their frequency difference phase-locked to a CO_2 laser. Then, the CO_2 laser has been used for saturated absorption spectroscopy to provide absolute frequency measurements of several CO_2 lines [24].

11.2.2 OFCS Extension to the mid-IR

We report two different schemes which exploit a DFG process to transfer the metrological performance of a visible/near-IR OFCS to the mid-IR. In one scheme, the DFG near-IR pumping lasers are phase-locked to their associated closest tooth in the comb. Then, the generated mid-IR radiation is used for saturated absorption spectroscopy providing absolute frequency measurements of CO_2 lines at $4.2\,\mu m$

with a relative uncertainty of about 1.4×10^{-11}. In the second scheme, a frequency comb is directly created at 3 μm by nonlinear mixing of a near-IR fiber-based OFCS with a CW laser. In the latter case, the generated comb can be employed both as a frequency ruler and as a direct source for molecular spectroscopy.

4-μm Experiment

The set-up implemented to lock the DFG radiation to the visible/near-IR OFCS, described in details in previous works [25, 26], is shown in Fig. 11.1. The pump source is a diode laser operating between 830 and 870 nm with a maximum power of 130 mW. The signal laser source is a monolithic-cavity Nd:YAG laser at 1064 nm seeding an Yb fiber amplifier with a maximum power of 5 W. The diffracted (1)-order beam coming from an acousto-optic modulator (AOM) is then used for the DFG process. The latter takes place in a PPLN crystal (with a period around 23 μm) and produces about 200 μW of idler radiation at 4.2 μm. The OFCS is based on a Kerr-lens mode-locked Ti:Sa laser and covers an octave in the visible/NIR region (500÷1 100 nm). Its repetition rate ($\nu_r = 1$ GHz) is locked to a high-stability reference oscillator. The latter consists of a 10-MHz quartz which is locked to a

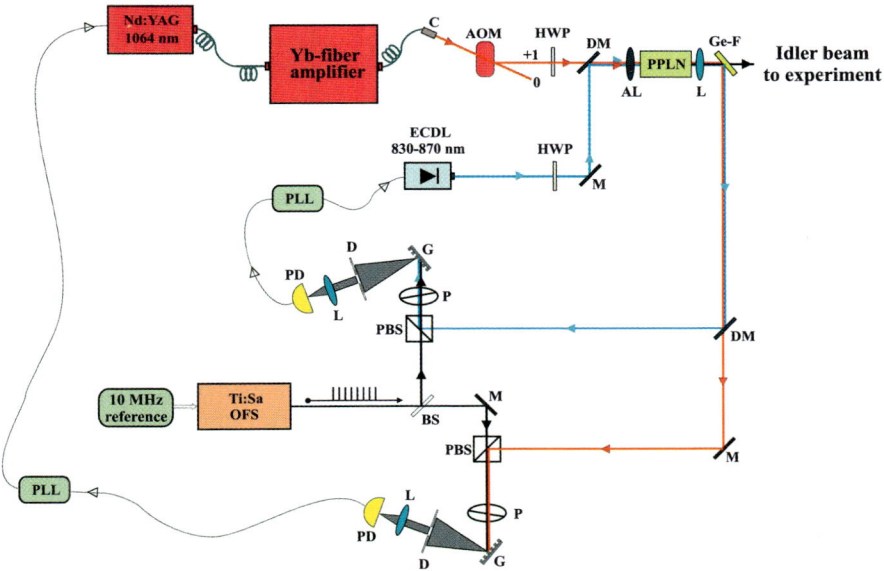

Fig. 11.1. Schematic of the experimental apparatus: C = fiber collimator, AOM = acousto-optic modulator, HWP = half-waveplate, DM = dichroic mirror, AL = achromatic lens, L = lens, Ge-F = germanium filter, M = mirror, PLL = phase-locked loop, PBS = polarizing beam splitter, G = diffraction grating, PD = photodiode. The 4-μm radiation is phase-locked to the Ti:Sa comb through the DFG pumping lasers and used either for sub-Doppler or cavity-ring-down spectroscopy providing absolute frequency measurements of ro-vibrational molecular transitions

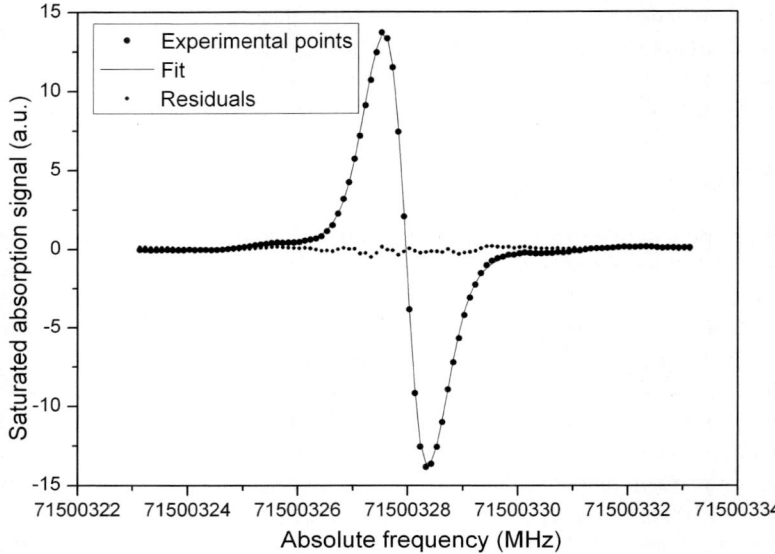

Fig. 11.2. Experimental recording of the CO_2 $(00^01\text{--}00^00)$ R(60) saturated-absorption line at $2\,384.994\,\text{cm}^{-1}$. The enhancement optical cavity is filled with pure gas at a pressure of 27 µbar, and the first-derivative signal is obtained by modulation of the Nd:YAG laser frequency. The experimental data points, the fitting curve, and the residuals are shown. The line center is measured by fitting the experimental line shape with a theoretical model taking into account the various broadening effects. This yields the value $71\,500\,327.968(1)\,\text{MHz}$ corresponding to a relative uncertainty of 1.4×10^{-11}

GPS-disciplined Rb frequency reference. The measured stability of such a system against a Cs-fountain-disciplined H maser limits the OFCS precision to 6×10^{-13} at 1 s and its accuracy to 2×10^{-12}. After nonlinear mixing, RF beat notes ($\Delta \nu_p$ and $\Delta \nu_s$) are generated between the pumping beams and their associated closest tooth (N_p and N_s) in the comb and used to phase-lock the mid-IR radiation to the OFCS. For this purpose, two phase-locked-loop (PLL) circuits are used to feed appropriate frequency corrections back to the lasers. Then, the frequency of the generated idler radiation is given by

$$\nu_i = (N_p - N_s)\nu_r \pm \Delta\nu_p \pm \Delta\nu_s \qquad (11.5)$$

and its stability is limited by the OFCS. Frequency scans across a molecular resonance are performed by sweeping the beat-note frequency of one of the pumping lasers. Simultaneously, the beat-note frequencies $\Delta\nu_p$ and $\Delta\nu_s$ for each data point in the spectrum are recorded to yield an absolute frequency scale following (11.5). Then, the line center absolute frequency is measured by fitting a suitable theoretical function to the experimental line shape. For transitions with a sufficiently high dipole moment, precision of measurements can be further improved by performing saturated-absorption spectroscopy which reduces the observed linewidth $\Delta\nu$ thus increasing the quality factor $Q = \nu\Delta\nu$. An example of spectrum is given in

Fig. 11.2, which shows the Lamb-dip profile for the (00^01-00^00) R(60) CO_2 transition at 2384.994 cm^{-1} recorded by a liquid-N_2-cooled InSb detector. The first-derivative signal was obtained by modulation of the Nd:YAG laser frequency at a rate of a few kHz. Due to the limited DFG power, in this experiment, the 4-μm beam was coupled into a confocal Fabry–Perot enhancement cavity (FSR = 1.3 GHz, finesse ≈ 500) filled with pure gas at a pressure of 27 μbar. The reflection signal from the cavity was also detected by a second InSb detector and used to actively control its length (by means of a PZT transducer) in order to keep the cavity mode resonant with the IR frequency during the scan. The line center was measured by fitting the experimental line shapes with a theoretical model taking into account several broadening effects, the main contributions coming from collisions (∼100 kHz) and transit time (∼400 kHz). The measured value is 71 500 327.968(1) MHz, which corresponds to a relative uncertainty of 1.4×10^{-11}. Moreover, this uncertainty can be further reduced by repeating the above procedure many times over a long period (a few months) and taking the weighted-average value [25]. Absolute frequency measurements can also be extended to very weak transitions by using the comb-referenced DFG radiation in high-sensitivity detection schemes. Indeed, cavity ring-down spectroscopy (CRDS) has also been performed by coupling the 4-μm beam to a high-finesse optical cavity (FSR = 150 MHz, finesse > 24 000). In this configuration, when a given threshold for the intra-cavity photon filling is achieved, a digital oscilloscope starts acquiring the signal transmitted through the cavity, and the corresponding trigger signal makes the acousto-optic modulator on the Nd:YAG laser rapidly (∼1 μs) switch off the DFG beam. As an example, in Fig. 11.3 we show the Doppler-broadened (05^51-05^50) P(19e) ro-vibrational transition of $^{13}CO_2$ at $2 209.109 \text{ cm}^{-1}$ with linestrength $S = 4.1 \times 10^{-27}$ cm, recorded with a gas pressure of 9 mbar. The Gaussian fit curve is also shown, giving for the line center the absolute frequency value of 66 227 614(1) MHz. In this case, the higher relative uncertainty (1.5×10^{-8}) is basically due to the lower quality factor Q of measurements performed in Doppler broadening regime.

3-μm Experiment

The method described above can be readily applied to different spectral windows by a proper choice of the DFG pumping sources and the nonlinear crystal. In this regard, a more powerful and tunable DFG apparatus can produce absolute frequency measurements on several molecular species and in much simpler configurations. For this purpose, a novel DFG source operating from 2.9 and 3.5 μm with a maximum output power of 5 mW has been realized and used for sub-Doppler molecular spectroscopy with no need of enhancement optical cavities [13]. Indeed, by a simple pump-and-probe scheme, saturation Lamb-dips have been observed for a number of ro-vibrational transitions belonging to the CH_4 ν_3 fundamental band. An example is shown in Fig. 11.4 for the $A_1^{(2)}$ R(4) line at $3 067.300 \text{ cm}^{-1}$, recorded in a 50-cm-long cell filled with pure gas at a pressure of 40 μbar. Then, the above comb-referencing scheme is able to provide the absolute frequency of the observed

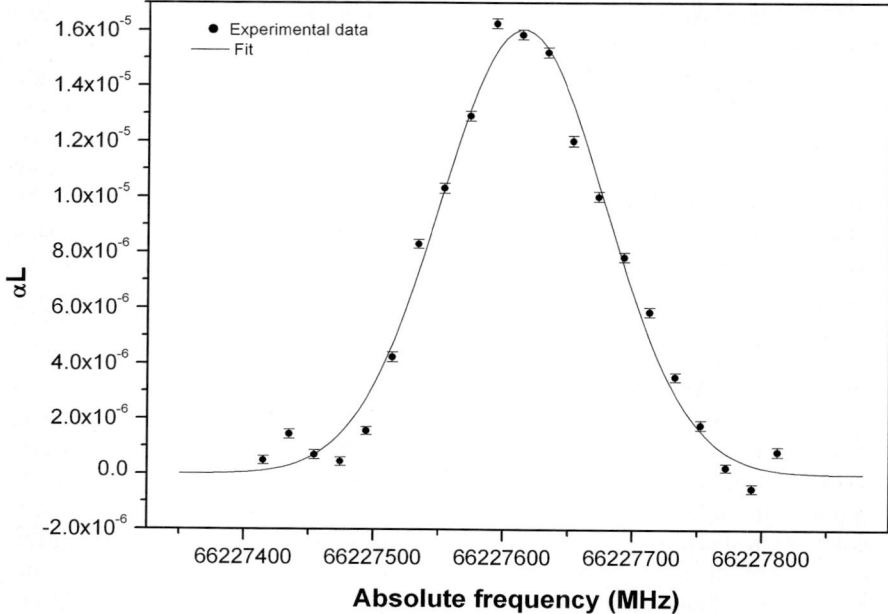

Fig. 11.3. Doppler-broadened (05^51-05^50) P(19e) ro-vibrational transition of $^{13}CO_2$ at $2\,209.109\,\text{cm}^{-1}$ (linestrength $S = 4.1 \times 10^{-27}$ cm), recorded by means of CRDS at a gas pressure of 9 mbar. Here, L is the cavity mirror spacing and α the absorption coefficient. The Gaussian fit gives the line-center absolute frequency with a relative uncertainty of 1.5×10^{-8}

transitions. However, one drawback of this approach is the impossibility of comb-referencing for direct laser sources operating in the mid-IR, such as QCLs. In this section, we demonstrate a novel scheme, based on DFG, which directly realizes a mid-IR optical frequency comb. The nonlinear down-conversion process occurs in a PPLN crystal (with a period around 30 μm) between a near-IR (OFCS) and a CW tunable laser. The generated mid-IR frequency comb covers the region from 2.9 to 3.5 μm in 180-nm-wide spans with a 100-MHz mode spacing and keeps the same metrological performance as the original comb source. Such a scheme can be easily implemented in other spectral regions by use of suitable pumping sources and nonlinear crystals. The apparatus devised to create the 3-μm frequency comb, reported in a previous work [27], is shown in Fig. 11.5. The DFG signal radiation comes from a near-IR OFCS based on an Er doped fiber laser which utilizes passive mode locking to provide ultra-short pulses (∼100 fs). The following spectral broadening through a nonlinear fiber makes the OFCS cover an octave from 1 050 to 2 100 nm. Its repetition rate (100 MHz) and carrier-envelope offset frequency are locked to a reference oscillator. The signal beam is then provided by feeding a fraction (25 mW) of the fs fiber laser system output (before the spectral broadening stage), covering the 1 500–1 625 nm interval, to an external Er-doped fiber amplifier (EDFA). The power spectral distribution, resulting from the convolution with

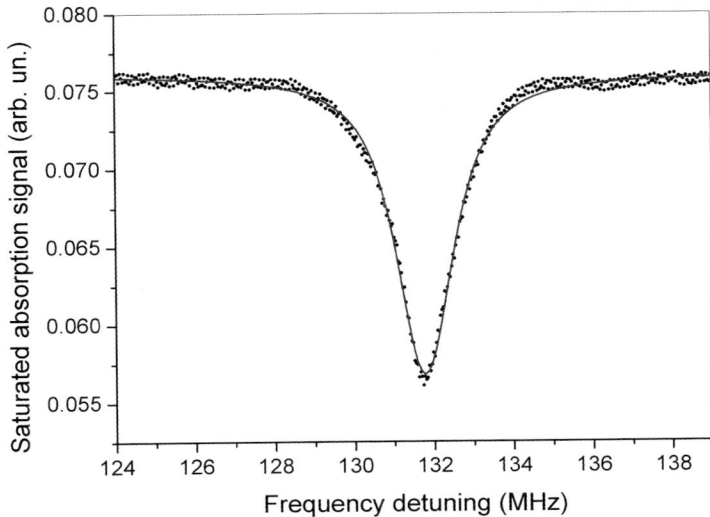

Fig. 11.4. Lamb-dip detection on the Doppler profile of the $A_1^{(2)}$ R(4) transition at $3\,067.300\,\text{cm}^{-1}$ obtained by means of a 3-mW DFG beam used in a simple pump-and-probe scheme. By phase-locking the DFG pumping lasers to the near-IR OFCS, absolute frequency measurement of the observed line is possible. The line is a Lorentzian fit to the experimental points

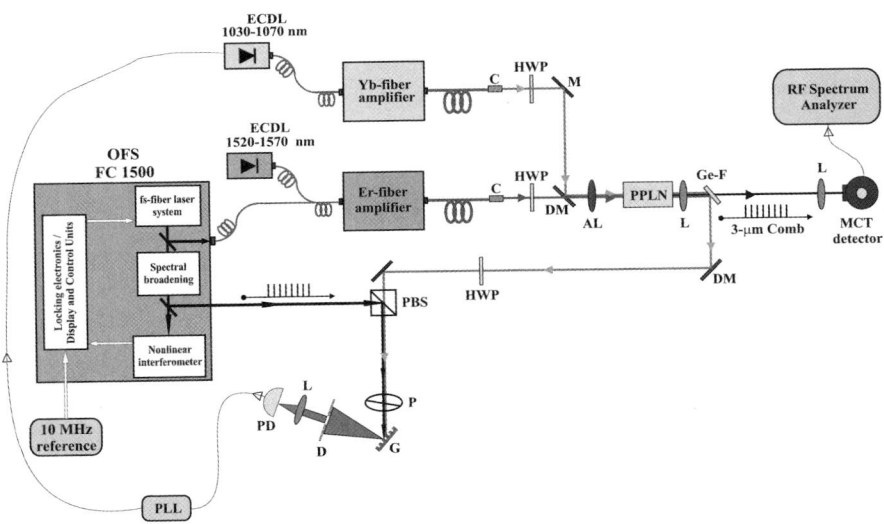

Fig. 11.5. Layout of the optical table. A 3-μm frequency comb is created by difference-frequency-generation in a PPLN crystal between a near-IR OFCS and a CW laser. A fast, 100-μm-diameter MCT detector is used to characterize the generated mid-IR comb

the EDFA gain curve, is measured by an optical spectrum analyzer. The amplified comb beam has an overall power of 0.7 W and spans from 1 540 to 1 580 nm with a 100 MHz spacing, corresponding to nearly $N_t = 50 000$ teeth (i.e., about 14 µW per tooth). The pump beam is generated by an external-cavity diode laser (ECDL) emitting in the range 1 030–1 070 nm and is amplified by an Yb-doped fiber amplifier which delivers up to 0.7 W, preserving the linewidth of the injecting source (less than 1 MHz). Afterwards, the two laser beams are combined onto a dichroic mirror and focused by a near-IR achromatic lens into a temperature-controlled, antireflection-coated PPLN crystal. The latter consists of an array of 9 channels, with different poling periods ranging from 29.6 to 30.6 µm. Once the wavelength of the pump source is fixed (1 055 nm), the channel and temperature value (around 340 K) are properly chosen to satisfy the QPM condition with the center wavelength of the near-IR comb (1 560 nm). The teeth on both sides are involved in as many DFG processes, with a conversion efficiency decreasing according to the well-known $sinc^2$ law [13]. The 3-µm comb is detected by filtering the DFG idler beam from the unconverted near-IR light and focusing it on to a liquid-N_2-cooled, 150-MHz bandwidth HgCdTe (MCT) detector. In this way, an RF beat note at $\nu_r = 100$ MHz is recorded by a spectrum analyzer, which is the sum of the beat signals between all pairs of consecutive teeth in the generated comb. The latter has a bandwidth of 180 nm (5 THz) centered near 3.3 µm and its measured overall power is about $P = 5$ µW. This value corresponds to a power of nearly $P/N_t = 100$ pW per mode of the IR comb. Since the linewidth of the ECDL is around 1 MHz, the DFG comb lines are significantly wider than those of the near-IR OFCS. This can be partially overcome by locking the ECDL to a tooth of the near-IR comb (see Fig. 11.5), which also cancels out the carrier-envelope phase offset present in the original frequency comb. As discussed in the previous section, if the optical comb were used as a frequency ruler, its metrological performance would be transferred to a CW laser by phase-locking the latter to the closest comb tooth. In order to demonstrate that such a scheme is possible even in a hardly accessible spectral region, like the 2.9–3.5 µm range, a CW DFG beam is simultaneously produced for characterization. This is accomplished by simultaneously seeding the Er-fiber amplifier with an ECDL emitting in the 1 520–1 570 nm interval (having a linewidth lower than 500 kHz). In this configuration, a CW 1.5-µm beam is also produced by the EDFA, which gives rise to a second DFG process with the pump radiation thus producing a CW idler beam around 3 µm with a power between 1.5 and 3 mW, depending on the wavelength. As a consequence, two additional RF beat notes at $\nu_1 = \nu_{CW} - \nu_n$ and $\nu_2 = \nu_{n+1} - \nu_{CW}$ are detected between the DFG CW radiation at ν_{CW} and its two closest mid-IR comb teeth at ν_n and ν_{n+1}, respectively (see Fig. 11.6). The signal-to-noise ratio (SNR) for such beat notes can be measured as the 1.5-µm ECDL wavelength is tuned from 1 540 to 1 570 nm (ν_{CW} from 3.22 to 3.35 µm). The SNR value reaches a maximum of 35 dB at the center wavelength (1 555 nm), while decreases almost symmetrically down to less than 20 dB at the upper and lower edges. This configuration limits to about 130 nm the interval which is suitable for use in optical phase-locked systems. Actually, the 180-nm

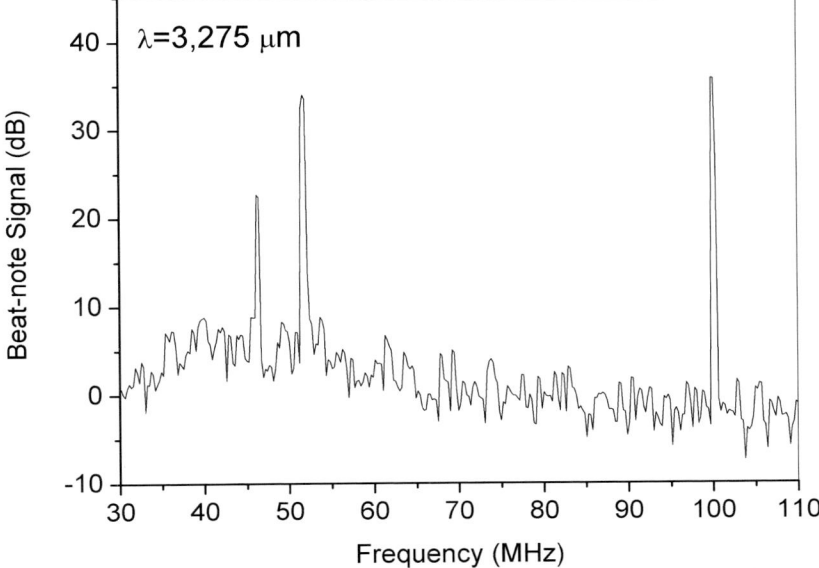

Fig. 11.6. Beat signals recorded by the RF spectrum analyzer at the center of the 3-μm comb span. The peak at $\nu_r = 100$ MHz is the sum of the beat signals between all pairs of consecutive teeth in the generated comb, while the peaks at ν_1 and ν_2 correspond to the beat notes between the DFG CW radiation and its two closest comb teeth. Resolution and video bandwidth are 3 kHz and the sweep time is about 1 s

span can be fully exploited as higher beat notes are expected when only an external 3-μm mW-power source is used during normal operation (i.e., in absence of the simultaneous CW DFG beam coming from the same EDFA which subtracts power from the DFG comb). Moreover, SNR levels can be further improved selecting a smaller number of teeth by using an IR diffraction grating. Finally, by tuning the 1-μm laser wavelength, the center frequency of the DFG comb is tuned from 3.1 to 3.4 μm, without any need to adjust the QPM conditions. This is shown in Fig. 11.7, where the peak signal of the beat note at 100 MHz is plotted as a function of the 1-μm wavelength, the upper limit being set by the laser tunability range. By also tuning the QPM conditions, higher conversion efficiencies and further extension of the span (from 2.9 to 3.5 μm) can be accomplished. Thus, such a generated comb might be strategic for future metrological applications of novel lasers under development [28].

11.2.3 Future Perspectives

We have presented two schemes for performing absolute frequency measurements in the mid-IR spectral region. OFSs are used either to directly create a mid-IR frequency comb through a DFG process with a cw laser or to reference the DFG radia-

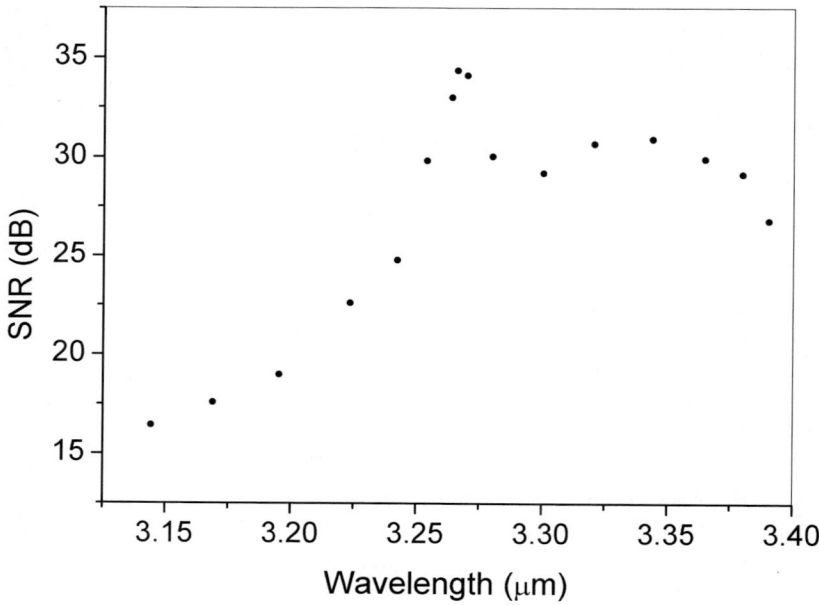

Fig. 11.7. SNR for the beat note at $v_r = 100\,\mathrm{MHz}$ as a function of the DFG wavelength, recorded by tuning the pump source from 1 040 to 1 070 nm. Each point represents a frequency comb with a span of about 150 nm. The asymmetric behavior with respect to the central wavelength is caused by the decrease in the optical power of the Yb fiber amplifier

tion to the Cs primary standard by phase-locking of the pumping sources. This opens new perspectives for absolute frequency measurements on ro-vibrational molecular transitions for determination of molecular constants and frequency grids with improved accuracy. As a direct spectroscopic source, the generated mid-IR comb may be used for coherent coupling to high-finesse cavities to provide sensitive molecular detection across 100 nm [29]. In the time domain, the realization of phase-coherent mid-IR pulses offers a novel tool for Fourier-transform molecular spectroscopy [30] and coherent control of molecular reactions in a previously-uncovered spectral region [31]. Finally, use of advanced fiber-based devices, which benefit from the continuous progress in telecom technology, may give, in the future, additional advantages for the realization of more and more effective set-ups [32].

11.3 Structured Nonlinear Crystals for Quantum Optics

Quantum mechanical phenomena, besides their importance for our understanding of the fundamental structure of Nature, have the potential of enormously improving the performances in a variety of emergent technologies. Since the theoretical beginnings, dating back to the eighties, the exploitation of quantum effects in the field of information processing has seen an explosive growth, both in the number of theoretical proposals and in the first experimental realizations [33]. The new field of quantum information science has so far identified several important objectives, ranging from quantum computation [34] and cryptography [35] to quantum-enhanced metrology [36]. Quantum computation may result in computation faster than any computation possible with classical means. Quantum cryptography, and in particular, quantum key distribution, makes intrinsically secure sharing of cryptographic keys possible against any possible attack of an eavesdropper. Quantum metrology allows one to attain an unsurpassable precision in the measurement of a physical quantity by beating the standard limits due to shot noise.

The practical realization of such applications is particularly well suited to optical systems, where the basic quantum states can be simply prepared, manipulated, and detected, and where some of the basic quantum operators are readily implemented. Photons are ideally suited for the transmission of quantum information and can be made relatively immune to decoherence, i.e., to the loss of their quantum character. In order to efficiently pursue such objectives, photonic technologies are asked to provide reliable sources of quantum light states and high-efficiency photon counting detectors.

11.3.1 Quantum Light Sources

Squeezed Light

In general, squeezing refers to the reduction of quantum fluctuations in one observable below the standard quantum limit (the minimal noise level of the vacuum state, or shot-noise) at the expense of an increased uncertainty of the conjugate variable. The suppressed quantum noise of squeezed light can thus improve the sensitivity of optical measurements (e.g., by increasing the precision in the measurement of phase shifts in an interferometer) [37]. Other applications involve quantum information processing with continuous variables [38], where squeezed states are used to generate entanglement and perform quantum teleportation [39].

Most of the experimental realizations of squeezed light have involved the process of parametric amplification and deamplification of vacuum field fluctuations in a nonlinear crystal. Either narrowband CW cavity-enhanced OPOs [40] or single-pass pulsed schemes [41] have been frequently used to demonstrate squeezing. Since the possible use of such squeezed sources critically depends on the amount of noise suppression available, efforts have concentrated in improving the squeezing level by increasing the strength of the nonlinear interaction between the pump field and the crystal.

The use of periodically-poled crystals and waveguides was demonstrated already in 1995 with short pump pulses [42, 43], and it proved an efficient way to improve the nonlinear interaction (by using the d_{33} nonlinear coefficient) and the longitudinal (thanks to QPM) and transverse mode matching (thanks to the waveguide) between the pump and the generated fields. Both KTP at 830 nm [43] and LN at 1 064 nm [42] achieved squeezing levels of the order of 10–15%. More re-

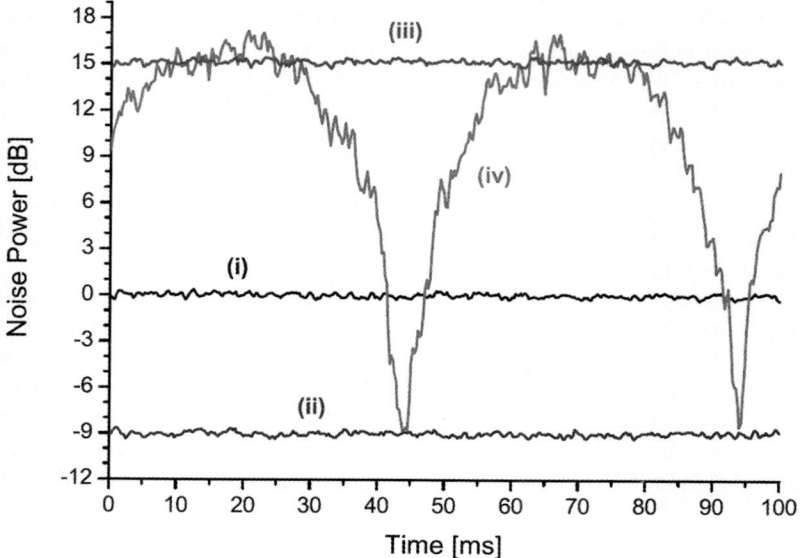

Fig. 11.8. Experimental noise levels for a quadrature-squeezed vacuum state of CW light generated with a sub-threshold optical parametric oscillator containing a periodically-poled KTP crystal. (i) Shot noise level; (ii) noise level for the squeezed quadrature; (iii) noise level at the antisqueezed quadrature; (iv) noise level as recorder while the phase is scanned. A squeezing level of -9.01 dB is observed. Figure taken from [46]

cently, single-pass parametric amplification in periodically-poled KTP has resulted in squeezing of about -3 dB [44].

Nowadays, parametric down-conversion in subthreshold optical parametric oscillators is often employed for the generation of CW squeezed light. Although squeezing at a level of -6 dB has been observed with bulk nonlinear crystals in noncritical PM conditions [45], the advent of QPM periodically-poled materials has allowed a dramatic increase in the efficiency and in the range of available wavelengths. Efforts in this direction have recently brought to impressive results, with up to 9 dB of noise suppression below the shot noise achieved in PPKTP at 860 nm (see Fig. 11.8) [46]. The recently demonstrated possibility of generating narrowband CW highly-squeezed light resonant with atomic transitions [47] will open new perspectives in the use of atomic media as a possible way to delay and store quantum information.

Single and Entangled Photon Sources

Many of the proposed schemes for quantum communication and cryptography involve light sources capable of emitting fully characterized individual photons on demand. Unfortunately, such sources do not currently exist, and one has to sacrifice either the purity of the single-photon states or their deterministic production. Single

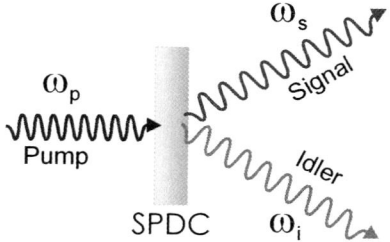

Fig. 11.9. Scheme of the process of spontaneous parametric down-conversion, where a pump photon of frequency ω_p is split into two lower-energy photons at frequencies ω_s and ω_i such that $\omega_s + \omega_i = \omega_p$. The detection of the idler photon can be used to *herald* the presence of the *twin* signal photon in a well-defined mode

emitters, like quantum dots [48], isolated fluorescence molecules [49], or nitrogen vacancy color centers in diamond [50], have proved capable of emitting indistinguishable single photons almost on demand after their pulsed optical excitation, but their use is not straightforward and there are problems related to their broad bandwidth and low out-coupling efficiency which do not allow a precise characterization of the output mode.

The historically most used source of single and entangled photons is however the process of spontaneous parametric down-conversion (SPDC) of light in $\chi^{(2)}$ nonlinear crystals [51] (see Fig. 11.9). In such a process, a photon of high energy (usually produced by frequency doubling a laser field and named *pump*) is split into two longer-wavelength photons (normally named *signal* and *idler*) whose energies sum up to that of the parent. Besides energy conservation, also momentum conservation must be obeyed in the process, so that the directions where the two photons are emitted are strictly related. As the emission only takes place in pairs, the detection of the idler photon can be used to *herald* the presence of the signal photon, which can then be used for applications.

This kind of source is nondeterministic, since one cannot precisely know when the single photon will be emitted but, once the idler photon is detected in a well-defined spectral/spatial mode, also the emission mode of the signal single photon is exactly determined by the energy/momentum correlations imposed by conservation rules (or PM conditions) [52]. This enables the conditional production of single photons in tightly defined modes which highly facilitates their coupling to subsequent optical processing and detection units [53–56].

Besides conditionally generating single photons in well-defined modes, the correlations existing between the twin photons emitted in SPDC are of an intrinsic quantum nature and lead to entanglement in one or more degrees of freedom between the photon pairs. Entanglement is the essence of quantum physics and dictates that, although the individual properties of the two parties may be totally (quantum-mechanically) undetermined, their relative value is perfectly fixed in a nonlocal fashion. Entangled states of light are a critical resource for the realization of many quantum information protocols, such as teleportation, and for improving the security of quantum cryptographic schemes. Polarization entanglement has been deeply analyzed and is the most used kind entanglement for demonstrating quantum properties [57], however time/energy and time-bin kinds of entanglement are receiving increasing attention and will probably prove more immune to decoherence for long-distance quantum communication [58, 59].

An optimal quantum source obviously requires high efficiency in the conversion of the pump photons into down-converted photon pairs in order to obtain higher signal-to-noise ratios and shorter measurement times. Conversion efficiency in bulk materials is limited by the choice of available crystals, so the engineering of the crystal structure may bring significant advantages. Periodic poling allows one to take advantage of crystals (like lithium niobate) with higher nonlinear susceptibilities, thus helping in significantly enhancing the conversion efficiency. Furthermore, the presence of a waveguiding structure in the material can also greatly enhance the emission in well-defined spatial modes which are much easier to collect and couple into single-mode fibers.

The use of periodically-poled crystals in the generation of entangled photon pairs is rather recent but it has already shown its exceptional potential. Already in the first works of 2001, an increase in the efficiency of about four orders of magnitude compared with bulk crystals was demonstrated. A PPLN waveguide with a period of 12.1 μm was used in that case, for type-I down-conversion of CW light at 657 nm into degenerate photon pairs at 1315 nm, which are suitable for long distance fiber communications. Both energy-time and time-bin entanglement of the emitted photon pairs were demonstrated [60, 61] (see Fig. 11.10).

A simple separation of the photons of the entangled pair was later obtained by using nondegenerate down-conversion. In this case, CW pump photons at 712 nm were converted into pairs at 1.55 and 1.31 μm, the best wavelengths for fiber transmission [62]. However efficient single-photon detectors are not available at these wavelengths (see later), and efforts have also been devoted to the generation of entangled photon pairs closer to the visible region, around 800 nm.

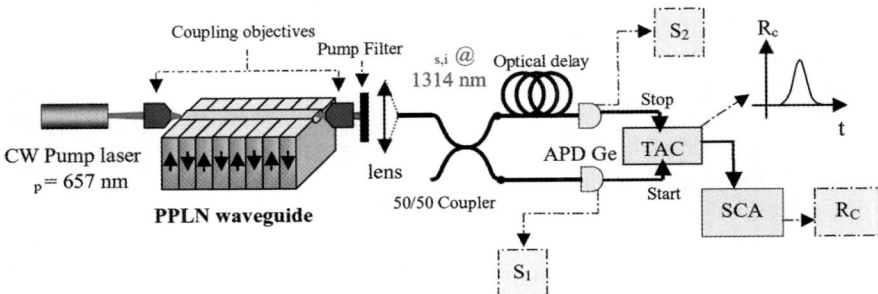

Fig. 11.10. Schematics to characterize the down-conversion and photon-pair production efficiency of a PPLN waveguide. TAC is a time to amplitude converter, and SCA is a single-channel analyzer. S_1, S_2, and R_c denote single count rates in the two detectors or coincidences. Figure taken from [60]

The use of a PPLN waveguide for the generation of a pump beam at 427 nm by SHG of the laser diode emission at 854 nm and then for its down-conversion back to 854 nm was reported in 2001. Efficient CW conversion (of the same order of that obtained with bulk crystals with a thousand time greater pump power) took place with a poling period of only 3.2 µm [63].

Ultrashort pump pulses were used in conjunction with PPKTP for type-I down-conversion [64], while the production of 800 nm orthogonal-polarization photons (hence much more easily separable by a simple polarizing beamsplitter) was reported by means of type-II SPDC in a PPKTP waveguide (with a 8.7 µm period) [65]. In the latter case, the use of a much weaker d_{24} nonlinear element (compared to the usual d_{33}) was compensated by the long interaction allowed by the limited cross section of the waveguide, and resulted in the high-fidelity conditional production of fiber-coupled single photons.

The use of ultrashort pulses at about 800 nm from a Ti:sapphire laser in combination with a periodically-poled structure (and possibly waveguiding) in Lithium Niobate, both for frequency doubling and for subsequent SPDC, would greatly benefit of its higher nonlinear coefficient but is currently hard to realize. The main problem is due to the limitations in the realization of small scale periods (about 2.6 µm for operation at room temperature) with the available technology.

The use of highly efficient structured crystals has the potential to release many of the requirements on the power of the existing pump sources for SPDC and will facilitate the transition towards compact diode-laser-based systems, which will finally bring to completely integrated systems for quantum information processing on a chip.

11.3.2 Single-Photon Detectors

The ability to detect photons with high efficiency and to distinguish the number of photons in an incident quantum state is of very high importance in quantum information science. Detection efficiencies approaching unity are required for a loophole-free test of the violation of Bell's inequality (that would definitely prove the nonlocal character of Nature) and for the realization of a scalable linear-optics quantum computer. Distinguishing the number of photons in a state would allow one to reliably produce many-photon entangled states and many other exotic states of light.

Unfortunately, photon number resolution has only been obtained with superconducting bolometric detectors so far [66]. Apart from the inconvenience of working at cryogenic temperatures, these devices still suffer from low detection efficiency and low counting rates.

If one relaxes the requirement of photon number resolution, visible photons can be conveniently detected with silicon avalanche photodiodes (APDs), which exhibit good quantum efficiency (50–70%), a low number of dark events and a high counting rate. The situation in the infrared however is much worse. Here, InGaAs avalanche photodiodes are available, but their efficiency is much lower, and their high dark counts can only be eliminated by working in a time-gated configuration.

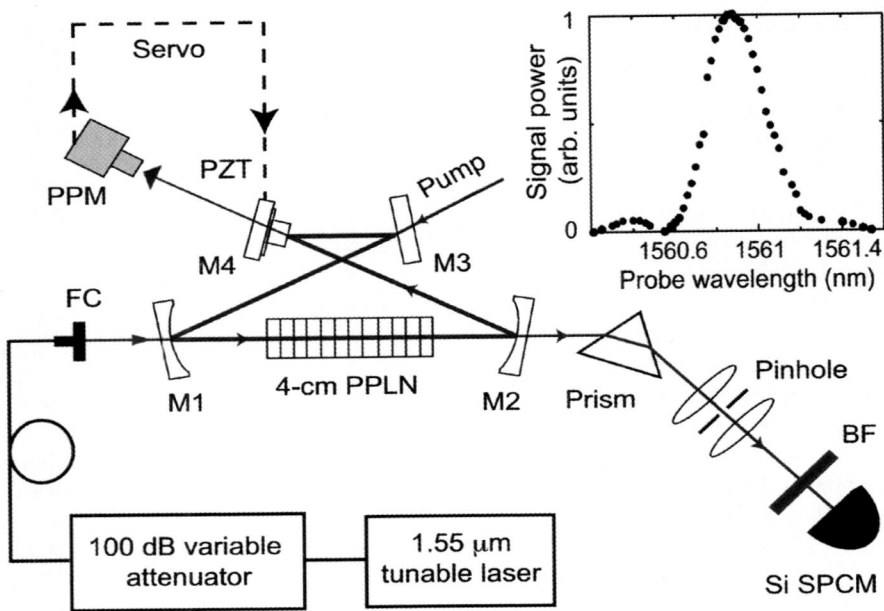

Fig. 11.11. Experimental setup for single-photon detection at 1.55 μm by means of up-conversion. FC, fiber-optic collimator; BF, 10-nm interference filter at 633 nm, and 1 064-nm HR mirror; PPM, pump power monitor; PZT, piezoelectric transducer. *Inset*, wavelength PM curve at a PPLN temperature of 229°C (0.3-nm bandwidth). Figure taken from [67]

Since IR wavelengths are the most interesting for the transport of quantum information via the existing telecom fiber network, due to the low absorption and dispersion associated to the 1.31 and 1.55 μm regions, an efficient way to detect IR photons is highly desirable. Periodically-poled crystals have been recently shown to conveniently up-convert the frequency of IR photons to the visible, where single photon detection by standard silicon APDs is more efficient. If a strong pump is available, a weak IR input signal can be up-converted with near unity efficiency. Different configurations have already been demonstrated: the single-pass up-conversion (with an efficiency of 90%) of a 1.55 μm photon to 630 nm by means of a PPLN crystal in a pump (at 1 064 nm) enhancement cavity [67] (see Fig. 11.11); the single-pass up-conversion by means of a pulsed pump has also demonstrated similar efficiency for the same wavelengths [68]. Recently, by means of the single-pass up-conversion of 1.312 μm photons to 712 nm in a PPLN waveguide pumped by a CW Er-doped fiber amplified laser at 1.56 μm, the transfer of quantum information has been clearly demonstrated [62].

Future quantum information networks made of telecom-wavelength transport channels and frequency-conversion stations will permit to efficiently manipulate, store (in atomic memories) and detect quantum light states in the spectral regions, where these processes offer the best performances.

References

1. N. Bloembergen, *Nonlinear Optics*, 4th edn. (World Scientific, Singapore, 1996)
2. R.W. Boyd, *Nonlinear Optics*, 2nd edn. (Academic Press, Amsterdam, 2003)
3. Y.R. Shen, *The Principles of Nonlinear Optics* (Wiley, New York, 2002)
4. D.N. Nikogosyan, *Nonlinear Optical Crystals* (Springer, Berlin, 2005)
5. G.D. Boyd, R.C. Miller, K. Nassau, W.L. Bond, A. Savage, $LiNbO_3$: an efficient phase matchable nonlinear optical material. Appl. Phys. Lett. **5**, 234 (1964)
6. J.A. Armstrong, N. Bloembergen, J. Ducuing, P.S. Pershan, Interactions between light waves in a nonlinear dielectric. Phys. Rev. **127**, 1918 (1962)
7. D. Feng, N.-B. Ming, J.-F. Hong, Y.-S. Yang, J.-S. Zhu, Z. Yang, Y.-N. Wang, Enhancement of second-harmonic generation in $LiNbO_3$ crystals with periodic laminar ferroelectric domains. Appl. Phys. Lett. **37**, 607 (1980)
8. J. Faist, F. Capasso, D.L. Sivco, C. Sirtori, A.L. Hutchinson, A.Y. Cho, Quantum cascade laser. Science **264**, 553 (1994)
9. T. Aellen, R. Maulini, R. Terazzi, N. Hoyler, M. Giovannini, J. Faist, S. Blaser, L. Hvozdara, Direct measurement of the linewidth enhancement factor by optical heterodyning of an amplitude-modulated quantum cascade laser. Appl. Phys. Lett. **89**, 091121 (2006)
10. J.S. Yu, S. Slivken, S.R. Darvish, A. Evans, B. Gokden, M. Razeghi, High-power, room-temperature, and continuous-wave operation of distributed-feedback quantum-cascade lasers at $\lambda \sim 4.8\,\mu m$. Appl. Phys. Lett. **87**, 041104 (2005)
11. D. Mazzotti, P. De Natale, G. Giusfredi, C. Fort, J.A. Mitchell, L. Hollberg, Difference-frequency generation in PPLN at 4.25 µm: an analysis of sensitivity limits for DFG spectrometers. Appl. Phys. B **70**, 747 (2000)
12. S. Borri, P. Cancio, P. De Natale, G. Giusfredi, D. Mazzotti, F. Tamassia, Power-boosted difference-frequency source for high-resolution infrared spectroscopy. Appl. Phys. B **76**, 473 (2003)
13. P. Maddaloni, G. Gagliardi, P. Malara, P. De Natale, A 3.5-mW continuous-wave difference-frequency source around 3 µm for sub-Doppler molecular spectroscopy. Appl. Phys. B **80**, 141 (2005)
14. A. Clairon, B. Dahmani, A. Filimon, J. Rutman, Precise frequency measurements of CO_2/OsO_4 and $He–Ne/CH_4$-stabilized lasers. IEEE Trans. Instrum. Meas. **34**, 265 (1985)
15. D.J. Jones, S.A. Diddams, J.K. Ranka, A. Stentz, R.S. Windeler, J.L. Hall, S.T. Cundiff, Carrier-envelope phase control of femtosecond mode-locked lasers and direct optical frequency synthesis. Science **288**, 635 (2000)
16. T. Udem, S.A. Diddams, K.R. Vogel, C.W. Oates, E.A. Curtis, W.D. Lee, W.M. Itano, R.E. Drullinger, J.C. Bergquist, L. Hollberg, Absolute frequency measurements of the Hg^+ and Ca optical clock transitions with a femtosecond laser. Phys. Rev. Lett. **86**, 4996 (2001)
17. R. Holzwarth, T. Udem, T.W. Hänsch, J.C. Knight, W.J. Wadsworth, P.S.J. Russell, Optical frequency synthesizer for precision spectroscopy. Phys. Rev. Lett. **85**, 2264 (2000)
18. S.T. Cundiff, J. Ye, Colloquium: Femtosecond optical frequency combs. Rev. Mod. Phys. **75**, 325 (2003)
19. S. Witte, R.T. Zinkstok, W. Ubachs, W. Hogervorst, K.S.E. Eikema, Deep-ultraviolet quantum interference metrology with ultrashort laser pulses. Science **307**, 400 (2005)
20. R. Jason Jones, K.D. Moll, M.J. Thorpe, J. Ye, Phase-coherent frequency combs in the vacuum ultraviolet via high-harmonic generation inside a femtosecond enhancement cavity. Phys. Rev. Lett. **94**, 193201 (2005)

21. C. Gohle, T. Udem, M. Herrmann, J. Rauschenberger, R. Holzwarth, H.A. Schuessler, F. Krausz, T.W. Hänsch, A frequency comb in the extreme ultraviolet. Nature **436**, 234 (2005)
22. I. Thomann, A. Bartels, K.L. Corwin, N.R. Newbury, L. Hollberg, S.A. Diddams, J.W. Nicholson, M.F. Yan, 420-MHz Cr:forsterite femtosecond ring laser and continuum generation in the 1–2-μm range. Opt. Lett. **28**, 1368 (2003)
23. S.M. Foreman, A. Marian, J. Ye, E.A. Petrukhin, M.A. Gubin, O.D. Mücke, F.N.C. Wong, E.P. Ippen, F.X. Kärtner, Demonstration of a He–Ne/CH_4-based optical molecular clock. Opt. Lett. **30**, 570 (2005)
24. A. Amy-Klein, H. Vigué, C. Chardonnet, Absolute frequency measurement of $^{12}C^{16}O_2$ laser lines with a femtosecond laser comb and new determination of the $^{12}C^{16}O_2$ molecular constants and frequency grid. J. Mol. Spectrosc. **228**, 206 (2004)
25. D. Mazzotti, P. Cancio, G. Giusfredi, P. De Natale, M. Prevedelli, Frequency-comb-based absolute frequency measurements in the mid-IR with a difference-frequency spectrometer. Opt. Lett. **30**, 997 (2005)
26. D. Mazzotti, P. Cancio, A. Castrillo, I. Galli, G. Giusfredi, P. De Natale, A comb-referenced difference-frequency spectrometer for cavity ring-down spectroscopy in the 4.5-μm region. J. Opt. A **8**, S490 (2006)
27. P. Maddaloni, P. Malara, G. Gagliardi, P. De Natale, Mid-infrared fiber-based optical comb. New J. Phys. **8**, 262 (2006)
28. F. Capasso, C. Gmachl, R. Paiella, A. Tredicucci, A.L. Hutchinson, D.L. Sivco, J.N. Baillargeon, A.Y. Cho, H.C. Liu, The Doppler-splitting method for the ground vibrational state. IEEE J. Sel. Top. Quantum Electron. **6**, 931 (2000)
29. M.J. Thorpe, K.D. Moll, R. Jason Jones, B. Safdi, J. Ye, Broadband cavity ring-down spectroscopy for sensitive and rapid molecular detection. Science **311**, 1595 (2006)
30. F. Keilmann, C. Gohle, R. Holzwarth, Time-domain mid-infrared frequency-comb spectrometer. Opt. Lett. **29**, 1542 (2004)
31. A. Marian, M.C. Stowe, J.R. Lawall, D. Felinto, J. Ye, United time-frequency spectroscopy for dynamics and global structure. Science **306**, 2063 (2004)
32. P. Maddaloni, P. Malara, G. Gagliardi, P. De Natale, Two-tone frequency modulation spectroscopy for in-situ trace gas detection using a portable difference-frequency source. Appl. Phys. B **85**, 219 (2006)
33. D. Bouwmeester, A. Ekert, A. Zeilinger (eds.), *The Physics of Quantum Information* (Springer, Cambridge, 2000)
34. E. Knill, R. Laflamme, G.J. Milburn, A scheme for efficient quantum computation with linear optics. Nature **409**, 46 (2001)
35. N. Gisin, G.G. Ribordy, W. Tittel, H. Zbinden, Quantum cryptography. Rev. Modern Phys. **74**, 145 (2002)
36. A. Migdall, Correlated-photon metrology without absolute standards. Phys. Today **52**(1), 41 (1999)
37. C.M. Caves, Quantum-mechanical noise in an interferometer. Phys. Rev. D **23**, 1693 (1981)
38. S.L. Braunstein, P. van Loock, Quantum information with continuous variables. Rev. Mod. Phys. **77**, 513 (2005)
39. A. Furusawa, J.L. Sorensen, S.L. Braunstein, C.A. Fuchs, H.J. Kimble, E.S. Polzik, Unconditional quantum teleportation. Science **282**, 706 (1998)
40. L.-A. Wu, M. Xiao, H.J. Kimble, Squeezed states of light from an optical parametric oscillator. J. Opt. Soc. Am. B **4**, 1465 (1987)
41. R.E. Slusher, P. Grangier, A. LaPorta, B. Yurke, M.J. Potasek, Pulsed squeezed light. Phys. Rev. Lett. **59**, 2566 (1987)

42. D.K. Serkland, M.M. Fejer, R.L. Byer, Y. Yamamoto, Squeezing in a quasi-phase-matched $LiNbO_3$ waveguide. Opt. Lett. **20**, 1649 (1995)
43. M.E. Anderson, M. Beck, M.G. Raymer, J.D. Bierlein, Quadrature squeezing with ultrashort pulses in nonlinear-optical waveguides. Phys. Rev. Lett. **20**, 620 (1995)
44. T. Hirano, K. Kotani, T. Ishibashi, S. Okude, T. Kuwamoto, 3 dB squeezing by single-pass parametric amplification in a periodically poled $KTiOPO_4$ crystal. Opt. Lett. **30**, 1722 (2005)
45. K. Schneider, M. Lang, J. Mlynek, S. Schiller, Generation of strongly squeezed continuous-wave light at 1064 nm. Opt. Express **2**, 59 (1998)
46. Y. Takeno, M. Yukawa, H. Yonezawa, A. Furusawa, Observation of -9 dB quadrature squeezing with improvement of phase stability in homodyne measurement. Opt. Express **15**, 4321 (2007)
47. G. Hetet, O. Gloeckl, K.A. Pilypas, C.C. Harb, B.C. Buchler, H.-A. Bachor, P.K. Lam, Squeezed light for bandwidth-limited atom optics experiments at the rubidium D_1 line. J. Phys. B **40**, 221 (2007)
48. C. Santori, D. Fattal, J. Vuckovic, G.S. Solomon, Y. Yamamoto, Indistinguishable photons from a single-photon device. Nature **419**, 594 (2002)
49. C. Brunel, B. Lounis, P. Tamarat, M. Orrit, Triggered source of single photons based on controlled single molecule fluorescence. Phys. Rev. Lett. **83**, 2722 (1999)
50. C. Kurtsiefer, S. Mayer, P. Zarda, H. Weinfurter, Stable solid-state source of single photons. Phys. Rev. Lett. **85**, 290 (2000)
51. D.C. Burnham, D.L. Weinberg, Observation of simultaneity in parametric production of optical photon pairs. Phys. Rev. Lett. **25**, 84 (1970)
52. C.K. Hong, L. Mandel, Experimental realization of a localized one-photon state. Phys. Rev. Lett. **56**, 58 (1986)
53. A.I. Lvovsky, H. Hansen, T. Aichele, O. Benson, J. Mlynek, S. Schiller, Quantum state reconstruction of the single-photon Fock state. Phys. Rev. Lett. **87**, 050402 (2001)
54. A. Zavatta, S. Viciani, M. Bellini, Tomographic reconstruction of the single-photon Fock state by high-frequency homodyne detection. Phys. Rev. A **70**, 053821 (2004)
55. A. Zavatta, S. Viciani, M. Bellini, Quantum-to-classical transition with single-photon-added coherent states of light. Science **306**, 660 (2004)
56. V. Parigi, A. Zavatta, M. Kim, M. Bellini, Probing quantum commutation rules by addition and subtraction of single photons to/from a light field. Science **317**, 1890 (2007)
57. P.G. Kwiat, K. Mattle, H. Weinfurter, A. Zeilinger, A.V. Sergienko, Y. Shih, New high-intensity source of polarization-entangled photon pairs. Phys. Rev. Lett. **75**, 4337 (1995)
58. I. Marcikic, H. De Riedmatten, W. Tittel, V. Scarani, H. Zbinden, N. Gisin, Time-bin entangled qubits for quantum communication created by femtosecond pulses. Phys. Rev. A **66**, 062308 (2002)
59. I. Marcikic, H. De Riedmatten, W. Tittel, H. Zbinden, M. Legre, N. Gisin, Distribution of time-bin entangled qubits over 50 km of optical fiber. Phys. Rev. Lett. **93**, 180502 (2004)
60. S. Tanzilli, H. De Riedmatten, W. Tittel, H. Zbinden, P. Baldi, M. De Micheli, D.B. Ostrowsky, N. Gisin, PPLN waveguide for quantum communication. Eur. Phys. J. D **18**, 155 (2002)
61. S. Tanzilli, H. De Riedmatten, W. Tittel, H. Zbinden, P. Baldi, M. De Micheli, D.B. Ostrowsky, N. Gisin, Highly efficient photon-pair source using a periodically-poled lithium niobate waveguide. Electron. Lett. **37**, 26 (2001)
62. S. Tanzilli, W. Tittel, M. Halder, O. Alibart, P. Baldi, N. Gisin, H. Zbinden, A photonic quantum information interface. Nature **437**, 116 (2005)

63. K. Sanaka, K. Kawahara, T. Kuga, New high-efficiency source of photon pairs for engineering quantum entanglement. Phys. Rev. Lett. **86**, 5620 (2001)
64. B.S. Shi, A. Tomita, Highly efficient generation of pulsed photon pairs using a bulk periodically poled potassium titanyl phosphate. J. Opt. Soc. Am. B **12**, 2081 (2004)
65. A.B. U'Ren, C. Silberhorn, K. Banaszek, I.A. Walmsley, Efficient conditional preparation of high-fidelity single photon states for fiber-optic quantum networks. Phys. Rev. Lett. **93**, 093601 (2004)
66. A.J. Miller, S.W. Nam, J.M. Martinis, A.V. Sergienko, Demonstration of low-noise near-infrared photon counter with multiphoton discrimination. Appl. Phys. Lett. **83**, 791 (2003)
67. M.A. Albota, F.N.C. Wong, Efficient single-photon counting at 1.55 μm by means of frequency up-conversion. Opt. Lett. **29**, 1449 (2004)
68. A.P. van Devender, P.G. Kwiat, High efficiency single photon detection via frequency up-conversion. J. Mod. Opt. **51**, 1433 (2004)

12 Photonic and Phononic Band Gap Properties of Lithium Niobate

M.P. Bernal, M. Roussey, F. Baida, S. Benchabane, A. Khelif, and V. Laude

12.1 Introduction

Photonic crystals (PtCs) [1, 2], also known as photonic band gap materials, are attractive optical materials for controlling and manipulating the flow of light. They are of great interest for both fundamental and applied research, and are expected to find commercial applications soon. Their structure consists basically in periodic changes of the dielectric constant on a length scale comparable to optical wavelengths. This periodical modulation of such property can be induced along one, two, or three directions in space. This has similar influence on the propagation of light as atomic crystalline potential has on electrons.

PtCs in high refractive index contrast semiconductor materials (Si, AsGa, etc.) are currently being pursued to obtain a range of forbidden frequencies (i.e. a photonic band gap) in the optical region of the electromagnetic spectrum. A simple example consists of a periodic array of voids within dielectric material. Multiple interference between scattered light waves can eventually lead to some frequencies not being allowed to propagate, giving rise to forbidden and allowed bands, analogous to the electronic bands of a semiconductor. Since the periodicity of the medium must be comparable to the wavelengths of the electromagnetic waves to inhibit their propagation, photonic band gap materials in the optical or infrared domain require sub-micron structures, which can be realized using nano-fabrication technology. The first commercial products involving 2D periodic PCs are already available in the form of photonic crystal fibers. By introducing artificial defects in a host photonic band structure, it is possible to manipulate photons by localizing the electromagnetic states and trap light [3].

Similarly, phononic crystals (PnCs) [4, 5] are novel materials that offer exceptional control over phonons, sound and other mechanical waves. Phononic crystals make use of the fundamental properties of waves, such as scattering and interference to create band gaps; ranges of wavelength within which waves cannot propagate through the structure. The existence of structures with complete phononic band gaps has obvious applications. For instance, a phononic crystal will reflect incoming sound waves with frequencies within the gap and can therefore be used as an acoustic insulator. Moreover, the introduction of defects within the structure allows sound waves with frequencies in the band gap to be trapped near a point-like defect, or guided along linear defects. Since the discovery of the phononic crystal occurred after the photonic counterpart, there are fewer research groups working in phononic structures but a continuous publication growth on the field can be observed. It should not be unexpected that more and more researchers get involved in phononic crystals also in the following years.

Phononic crystals in the hypersonic regime [6, 7] require wavelengths shorter than 10 µm that are comparable to optical wavelengths. In addition, the understanding of hypersonic phonons is crucial for many physical phenomena in materials. For example, the interaction between electrons and high frequency phonons determines the efficiency of spontaneous light emission in silicon and other semiconductor materials that have an indirect electronic band gap. Greater control over the phonons in silicon could therefore lead to highly efficient silicon-based light-emitting devices. The challenge in hypersonic crystals concerns the fabrication technology. In contrast to sonic and ultrasonic crystals, which are macroscopic and can be readily made using standard manufacturing techniques, hypersonic crystals require patterns to be created at the submicron and nanometric scales. These challenges are comparable to the ones encountered in photonic crystal fabrication technology.

Lithium niobate (LN) is our material of choice for manufacturing both photonic and phononic crystals. It is indeed well known that monocrystaline LN is a material with many interesting nonlinear properties: it is at once ferroelectric, piezoelectric, electrooptic, photorefractive, and acousto-optic. Its high purity makes it a material of choice for micro wave frequency, optical and surface acoustic wave applications. Though it is a quite chemically insensitive material, and hence quite resilient to traditional etching techniques (as opposed to semiconductor materials such as silicon or gallium arsenide), promising etching techniques are appearing for the achievement of periodic nanostructures with high filling fractions and reasonable aspect ratios. These techniques are furthermore compatible with optical and acoustic surface wave guides. Thus they open the path for the fabrication of two dimensional photonic and phononic crystals where waves are confined in all three dimensions.

This chapter is organized as follows. Section 12.2 is devoted to a presentation of lithium niobate photonic crystals. The theoretical band structures will be used to discuss the appearance of band gaps and of slow light modes. By slow light, we refer to conditions were the group velocity of optical waves is significantly reduced from the usual velocities in homogeneous materials. Then a presentation of actual LN photonic crystals will be made and a striking phenomenon of enhanced electrooptic coefficient will be shown. In Sect. 12.3, phononic crystals fabricated in lithium niobate will be discussed theoretically and experimentally. The emphasis is here on the appearance of phononic band gap properties for surface acoustic waves, i.e. for phonons that are confined close to the surface. It will be shown that despite the fact that the radiation conditions could be expected to lead to highly leaky surface modes, surface modes exist in the phononic crystal where they can be exited. We conclude in Sect. 12.4 by summarizing our results and by giving some perspectives in the development of thin layer devices and of phoXonic crystals, artificial crystals that possess simultaneous photonic and phononic band gaps.

12.2 Photonic Crystals

12.2.1 Band Structure Theory and Slow Light

As is well known, the result of the matter-light interaction greatly depends on the duration of this mutual action. Thereby, the group velocity of the light plays a key role: for a very fast energy propagation, the interaction between light and matter mainly leads to its linear response. To be efficiently excited, non linearities need a very high electromagnetic energy. This last condition can be fulfilled by a local enhancement of the electromagnetic field. Consequently, slow light propagation causes a local exaltation of the electromagnetic field and *vice versa*.

Lithium niobate, an anisotropic material, has a quite good nonlinear response, especially for the Pockels effect (or electrooptic effect). This effect can be optimized if we consider an X-cut substrate to build a PtC. Due to their dispersion properties that can almost be modified as desired, PtCs are most suitable candidates to obtain a light speed reduction. In fact, the main condition to satisfy concerns the dispersion curve of the PtC that must be flat enough for the considered spectral range and according to the desired propagation direction. Based on this choice, we can theoretically determine the PtC geometry assuming that it is made of holes in $LiNbO_3$. For this purpose, we consider and calculate the band diagrams for the most familiar 2D lattices (square, triangular and honeycomb). The dispersion diagram is presented in Fig. 12.1(a) over the irreducible Brillouin zone. A Plane Wave Expansion (PWE) calculation is done for a period over hole radius ratio of 0.27 and a square lattice of air holes engraved in $LiNbO_3$. The extraordinary effective refractive index of the lithium niobate waveguide (n_e) is considered as the background index. This value is theoretically determined by considering the waveguide in which the PtC will be fabricated in practice ($\varepsilon_b = n_e^2 = 2.143^2 = 4.5924$).

The grey rectangle on Fig. 12.1b emphasizes the zone where a flat dispersion curve is obtained. Theoretically, such eigenmodes can lead to a small group velocity for light propagating inside the PtC at the corresponding frequency or wavelength. According to the result of Fig. 12.1(b), we choose the ΓX direction as the propagation direction and we expect a very large reduction of the group velocity especially for the dashed blue line in Fig. 12.1(c). This last line corresponds to the lower edge (small values of the wavelength) of the second partial Photonic Band Gap (PBG) that will be shown further in Fig. 12.2.

In its general form, the PWE method takes into account an infinite PtC structure. Practically, the size of the PtC is finite and the band structure is slightly modified. In order to have a more realistic model for the structure, 2D FDTD calculations have been performed. The FDTD algorithm is widely used in the domain of electromagnetism [8, 9]. It is based on the direct resolution of the Maxwell curl equations by

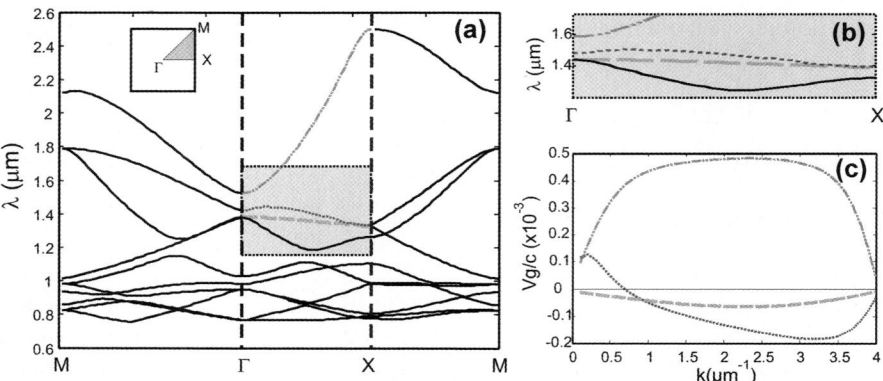

Fig. 12.1. Band diagram of the square lattice of air holes in LN for $r/a = 0.27$ and $a = 766$ nm in (**a**). (**b**) Is a zoom-in of the *dotted rectangular area* of (**a**). It shows a flat dispersion curve that corresponds to modes that provide slow light propagation. (**c**) Presents the calculated group velocity from (**b**)

discretizing and then replacing both time and space derivatives by their centered finite differences. The structure is then described by its dielectric permittivity and its magnetic permeability. For LN, which is a dielectric material, we only need to set the permittivity on each node of the spatial grid.

In our home-made FDTD code, we used an orthogonal Cartesian mesh. The resulting staircase effect typically leads to parasitic diffractive nodes that can be reduced by two ways: the first is based on the implementation of a non-uniform mesh in the code [10]. This mesh is based on a gradual variation of the spatial step in order to describe more accurately the fine details of the structure. Consequently, inside the photonic crystal a fine mesh is considered ($\delta = 18.683$ nm) while a bigger spatial step is used ($\Delta = 35$ nm) outside the PtC. The second way involves the use of a staggered grid [11]. The conventional cell, corresponding to one period of the PtC, would be defined by a grid of 41×41 nodes. Nevertheless, this is not sufficient to get a good hole definition. Thus, each point of the previous grid is considered as a cell of 20×20 points and the dielectric constant of the grid point is taken to be the average dielectric value over these 20×20 nodes of this sub-grid.

In addition we have introduced the PML technique of Bérenger [12] in order to avoid parasitical reflections on the edges of the computational window. We note that these absorbing boundary conditions (the PML) are not efficient for evanescent waves. Thus, the PML layers are placed at a distance larger than $\lambda_{max}/2$ from the outer holes of the structure, λ_{max} being the largest wavelength in the studied spectral area.

The transmission spectrum of a 15 rows long PtC is presented in Fig. 12.2. A propagation along the ΓX direction and a TE polarization are assumed. The latter corresponds to the polarization of the experimental guided mode of an annealed proton exchanged (APE) LN waveguide (see Sect. 12.2.2). This spectrum shows

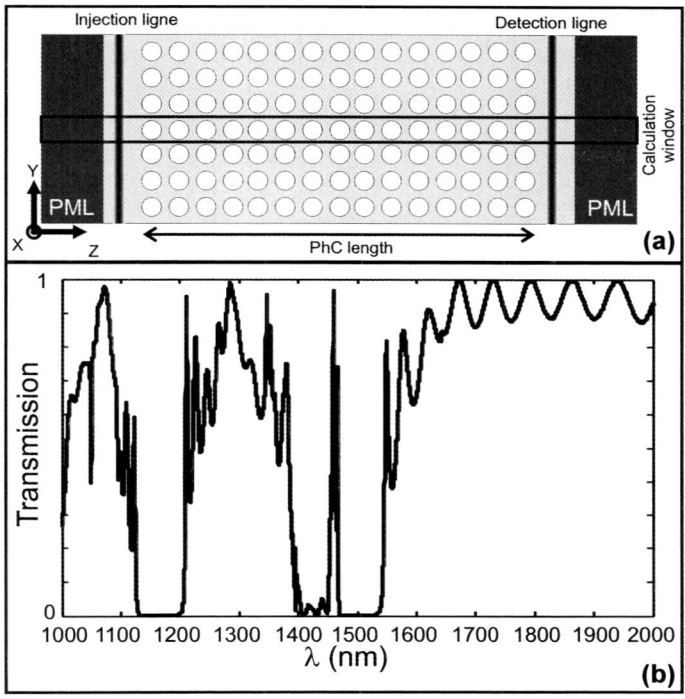

Fig. 12.2. (a) Is the scheme of the calculation window including the PML layers and (b) is the corresponding transmission spectrum of a 15 rows long LN PtC. The modeled PtC is supposed to be infinitely periodic along the transversal direction (y-axis)

three band gaps: the first is centered around $\lambda = 1\,160$ nm, the second and the third ones are located between $\lambda = 1\,400$ nm and $\lambda = 1\,550$ nm. They are separated by a very sharp transmission peak that is due to the finite size of the PtC [13].

To quantify the enhancement of the Pockels effect induced by nanostructuring, we have to point out and to calculate the local enhancement of the electromagnetic field inside the PtC. For a bulk material, the Pockels effect causes a modification of the refractive index given by:

$$\Delta n = -\frac{1}{2} \times n_e^3 \times r_{33} \times E_s, \qquad (12.1)$$

where $r_{33} \simeq 30.8$ pm/V is the highest electro-optic coefficient of LN and E_s is the external applied electric field. For a structured material, this index modification can be expressed by a similar equation:

$$\Delta n = -\frac{1}{2} \times n_e^3 \times r_{33} \times f^3 \times E_s, \qquad (12.2)$$

where f is the *local field factor* that corresponds to the enhancement of the electromagnetic field due to the nanostructuring of the substrate i.e. to the PtC. f becomes here the key parameter that must be optimized in order to enhance the Pockels effect. It can be calculated via the group velocity or via the electric field enhancement inside the PtC as follows:

$$f = \sqrt{\frac{v_g^{BULK}}{v_g^{PC}}}, \quad (12.3)$$

$$f = \frac{1}{S} \int_{PC} \frac{E_{local}^{PC}}{E_{local}^{BULK}} \, dy \, dz. \quad (12.4)$$

In these equations, v_g is the group velocity of the light inside the bulk material or inside the PtC, E_{local} is the local optical electric field and S is the surface area of the PtC structure. It is clear that the local field factor is equal to 1 outside the photonic band gap (PBG) in order to recover the bulk material properties and it must be equal to zero inside the PBG.

Using (12.3) and (12.4) and considering the square lattice studied above, we have theoretically demonstrated a maximum f value of 7 leading to a variation of the refraction index of $\Delta n = 0.33$ for an applied external electric field of $E_s = 6.15$ V/μm. This value has been obtained for the wavelength corresponding to the lower edge ($\lambda = 1\,395$ nm) of the PBG. When introducing this modified value of the refractive index in our numerical code, we observe a photonic band gap shift of about $\Delta\lambda \simeq 200$ nm.

To confirm the slowing down of light inside the PtC, we have performed an additional numerical experiment that amounts to the observation of a pulse propagation through a 75 rows long PtC that is infinite towards the y-direction. The PtC is illuminated by a plane wave pulse centred on the band gap edge ($\lambda \simeq 1\,400$ nm). The time delay of the pulse is large enough in order to cover a thin spectral interval ($\Delta\lambda = 30$ nm). A spatial average value over the z-component of the Poynting vector versus time is presented in Fig. 12.3. We mainly observe two different zones. The first one corresponds to light propagating with a group velocity approaching the bulk material value (solid white line), it is associated with wavelengths located outside the PBG. The second zone presents a slower light (dashed white line) with an average group velocity $v_g \simeq 6.1 \times 10^6$ m/s that corresponds to the wavelengths of the PBG edge. *This corresponds to a group velocity reduction by a factor of about 50.*

Experimentally, the PtC is finite both in the y and z directions. Nevertheless, the phenomenon of light enhancement still exists as it is shown in Fig. 12.4 where the modeled structure corresponds exactly to the fabricated device of Sect. 12.2.2. Both the guide and the PtC are taken into account for the 2D FDTD numerical simulation as seen in Fig. 12.4(a). The geometrical parameters were determined from the SEM image presented in Fig. 12.9(a). The light distribution is presented for three different values of the wavelength. The first one (see Fig. 12.4(b)), for $\lambda = 1\,503$ nm, corresponds to a zero transmission while the case of a large transmission (see Fig. 12.4(c)) is obtained for $\lambda = 1\,730$ nm. The last image (Fig. 12.4(d)) shows the light confinement inside the PtC obtained at the left edge of the PBG namely for $\lambda = 1\,383$ nm. This light confinement is at the origin of the enhancement of the non linear effects as it will be demonstrated further.

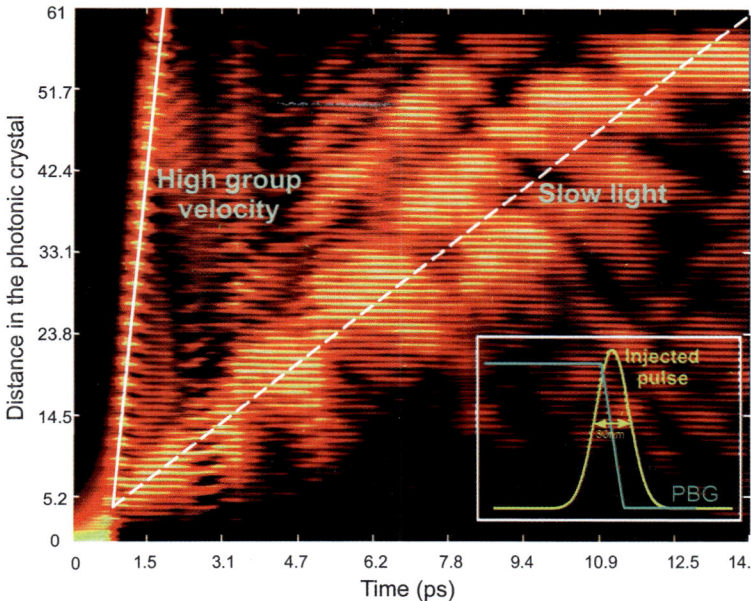

Fig. 12.3. Time evolution of the average value (over one period along the y-direction) of the z-component of the Poynting vector recorded across the PtC. The injected pulse is shown in the *inset*; it is centered around the PBG edge and is spectrally large enough to cover the two regimes of transmission (inside and outside the PBG)

12.2.2 Fabrication and Examples

Two alternative methods based on focused ion beam bombardement (FIB) to produce photonic band gap structures on LiNbO$_3$ (LN) substrates with a spatial resolution of 70 nm have been reported. The high resolution and the ability to drill holes directly from the sample surface make FIB milling one of the best candidates for designing good optical quality patterns at submicrometer scale [14]. The only constraint is that the sample surface must be metalized and grounded to avoid charge accumulation. Firstly, we describe the method for directly etching the LN substrate by FIB milling through the metal. This method has been already employed to etch sub-micrometric one-dimensional structures in LN [15]. The second related method is based on RIE etching after FIB milling of the metal layer which behaves as a mask. The advantage of this alternative solution is a lower exposure time. Another expected advantage would be a good replication of the mask shape in the whole hole depth. In both cases, the fabricated submicronic patterns are characterized by FIB imaging.

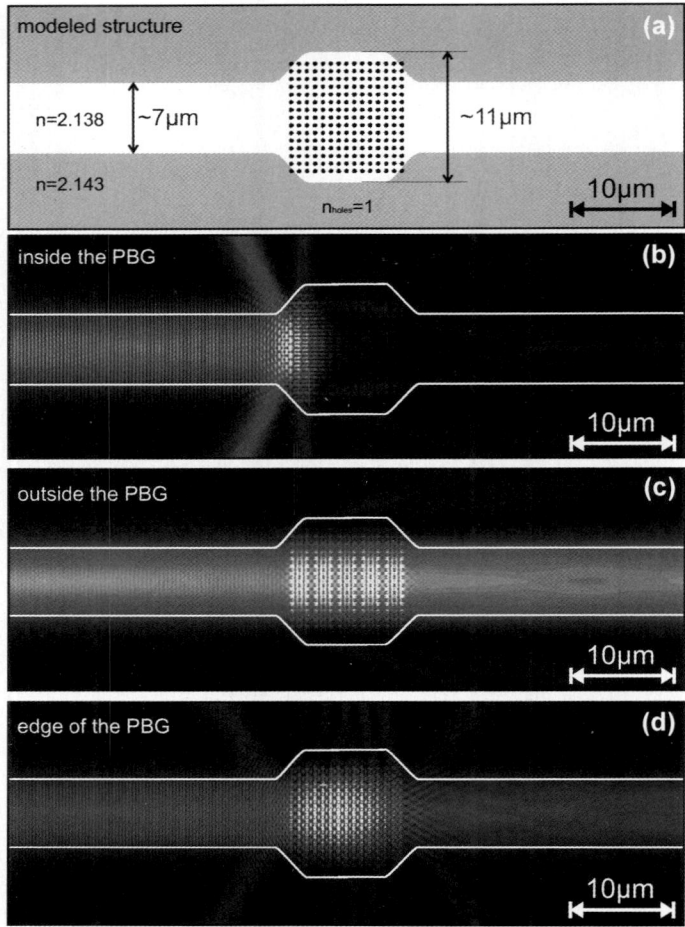

Fig. 12.4. (a) Is the scheme of the calculation window including the waveguide and the PtC; (b), (c) and (d) are the light distribution in color level calculated for three values of the wavelength corresponding to $\lambda = 1\,503$ nm, $\lambda = 1\,730$ nm and $\lambda = 1\,383$ nm respectively

12.2.3 Experimental Procedure

The two fabrication processes are schematically shown in Fig. 12.5. The first method, Fig. 12.5(a), is based on a direct etching of the LN substrate by FIB milling. The second one, Fig. 12.5(b), uses the FIB to create the metallic mask and the pattern is then transferred to the LN substrate by RIE. In both cases the sample area is 1 cm^2 and the thickness is 500 μm. A Cr layer is deposited by electron gun evaporation (Balzer, B510) and grounded with a conductive paste before introduction into the FIB vacuum chamber ($P = 2 \times 10^{-6}$ Torr). In the case of direct FIB writing the

(A)

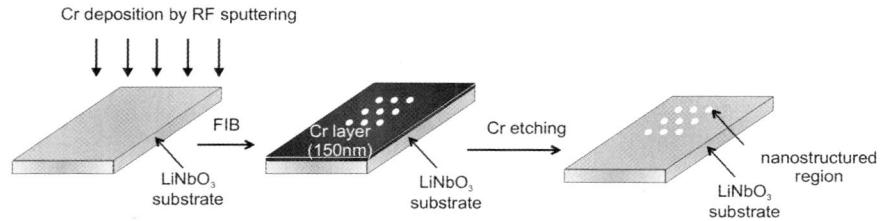

(B)

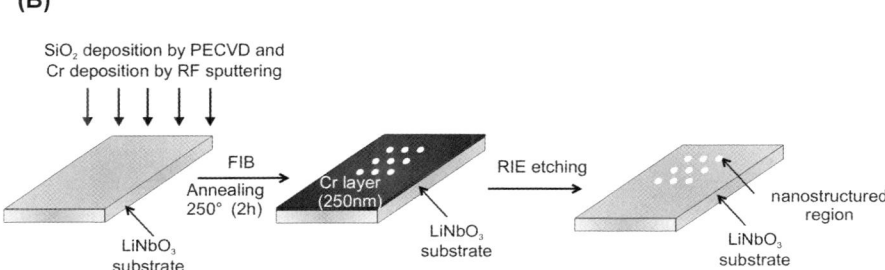

Fig. 12.5. (**a**) Fabrication steps for lithium niobate nanostructuring using FIB. (**b**) Fabrication steps for lithium niobate nanostructuring using RIE

thin Cr metal layer (150 nm) does not modify significantly the etching efficiency. In the second case a thicker Cr layer (250 nm) is deposited.

The metal-coated substrates are milled using a focused ion beam column (Orsay Physics – LEO FIB4400 for the case of FIB milling only – Fig. 12.5(a) – and a FEI Dual Beam Strata 235 for the milling of the metallic mask – Fig. 12.5(b)). This method could be directly compared with e-beam lithography. The advantage of FIB patterning of the metallic mask is its ability to selectively remove and deposit material without the use of the additional process step of developing a resist layer.

In the first case (Fig. 12.5(a)) we have fabricated an array of 4 × 4 circular holes with 540 nm diameter and 1 µm periodicity. Ga^+ ions are emitted with a current of 2 pA and accelerated by a voltage of 30 kV. The ions are focused with electrostatic lenses on the sample with a probe current of 66 pA. The pseudo-Gaussian-shaped spot size is estimated to be 70 nm on the target. The focused ion beam is scanned on the sample by a computer-controlled deflection field to produce the desired pattern (Elphy Quantum from Raith). A FIB-image cross-section of the cavities is shown in Fig. 12.6. In order to see the etching depth the sample is tilted by 30° with respect to the FIB axis. As it can be seen from the image, the 4 × 4 array exhibits well defined circular holes. The achieved etching depth is approximately 2 µm and the etching time was 12 minutes. At 1 µm deep the hole diameter is about 432 nm. This conical

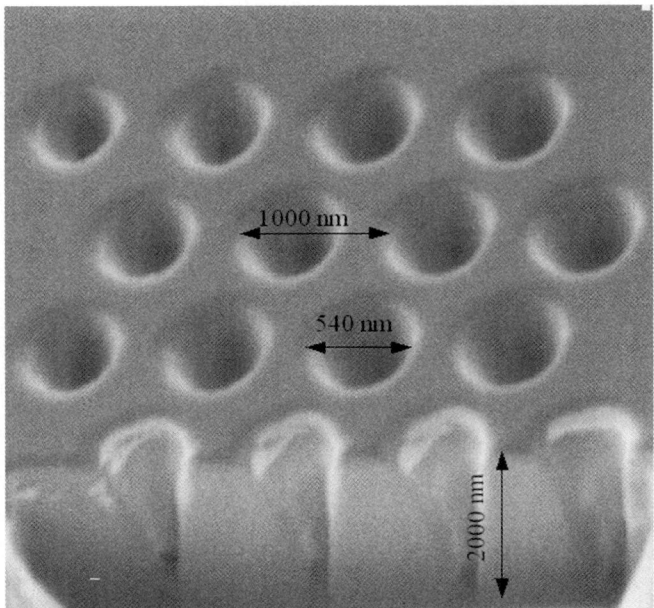

Fig. 12.6. SEM image of a 4 × 4 array of holes in lithium niobate etched by FIB

etching shape is due to material redeposition on the sidewalls while milling. In order to reduce the redeposition there are two possible solutions. If the FIB electronics is fast enough (Elphy Quantum is limited to 300 kHz) and the spot size small enough so that one can scan along the hole sidewalls longer and less on the bottom of the pit.

The second related process requires lower etching time since the desired photonic structure is fabricated at once. In this case, the FIB bombardement is used to pattern a SiO_2–Cr mask previously deposited on the LN substrate, as depicted in Fig. 12.6(b). The first step consists in depositing a 100 nm thick layer of SiO_2 by Plasma Enhanced Chemical Vapor Deposition (PECVD). A 250 nm thick chrome layer is then deposited on the substrate by sputtering. The metal is used as a mask for the RIE, while the silica layer prevents the diffusion of Chrome into the substrate during the RIE plasma processing and the increase of the optical losses. This layer is not needed in the case of direct FIB milling since the etching is done locally and the damaged area is defined by the FIB beam size. The samples are annealed at 250°C during 2 hours to release stress. The SiO_2–Cr mask is then nanostructured by FIB patterning, with a current of the sample of 100 pA. An exposure time of 3.75 s is typically required to etch a 250 nm diameter circular hole, which is 11 times less than the time required in the first process.

The pattern (an array of 24 cylindrical holes) is finally transferred to the substrate by RIE. The relevant parameters of this process are detailed in Table 12.1. It can be noticed that this process requires a very low pressure and a high RF power. In these

12 Photonic and Phononic Band Gap Properties   317

Table 12.1. RIE parameters for the Z cut of lithium niobate

Pressure (mBar)	SF$_6$ flow (sccm)	RF power (Watt)	Etching rate (nm/min)	LiNbO$_3$/Cr selectivity
3	10	150	50	0.25

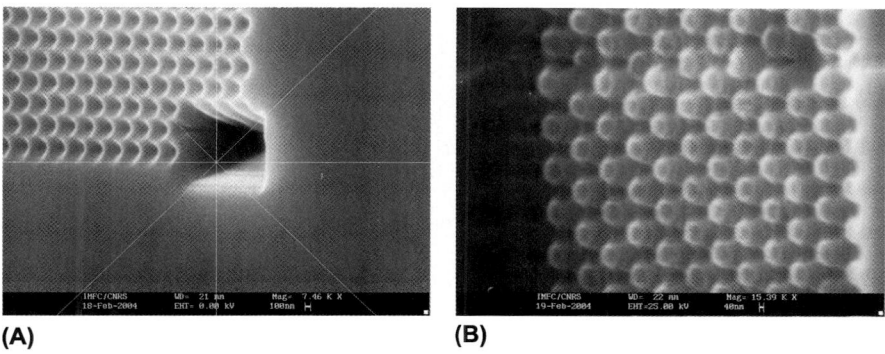

(A)   (B)

Fig. 12.7. (**a**) RIE of holes on a lithium niobate substrate. (**b**) RIE of pillars on a lithium niobate substrate

conditions, the etch rate of the mask is comparable to the etching rate of the substrate. In order to improve the selectivity of etching between the mask and the LN substrate, we start the process with an exposition of the target to a O$_2$ ionic plasma (pressure = 100 µBar, power = 60 W). The 250 nm thick layer of chrome is then more resistant to the SF$_6$-RIE. The selectivity of the mask is thus estimated to be 1 : 5 compared to the LN substrate (while the etching selectivity was measured to be of 1 : 2 without the O$_2$ ionic plasma). The etching rate of the Z-cut substrate is measured to be 50 nm/min. This process is applied to fabricate a triangular lattice of holes with $D = 250$ nm and $D = 130$ nm diameters and $p = 2D$ periodicity. Figure 12.7(a) and (b) exhibit the SEM images of the holes after FIB milling and 10 min of RIE etching. Figure 12.7(a) shows holes with good reproducibility. The etching depth is measured to be 500 nm. Figure 12.7(b) shows that the 130 nm diameter holes were transformed into 130 nm diameter rods after RIE etching, while the 250 nm diameter holes were well preserved. This is due to a higher etching rate along the sides of the triangular lattice than in the triangle center when the holes are very close to each other. We can infer from these results that the fabrication of small holes ($D < 200$ nm) requires lower RF-power to preserve the initial features.

In the next section, three examples of LN photonic crystals are shown. The first example is an hexagonal lattice of air holes in LN. Its photonic band gap (PBG) was experimentally measured. This is the first experimental evidence of a PBG in LN PtCs. The second example shows LN photonic crystals waveguides. Their optical response has been characterize with far field and scanning near field optical microscopy (SNOM). The last example shows an ultra-compact, LN photonic crystal

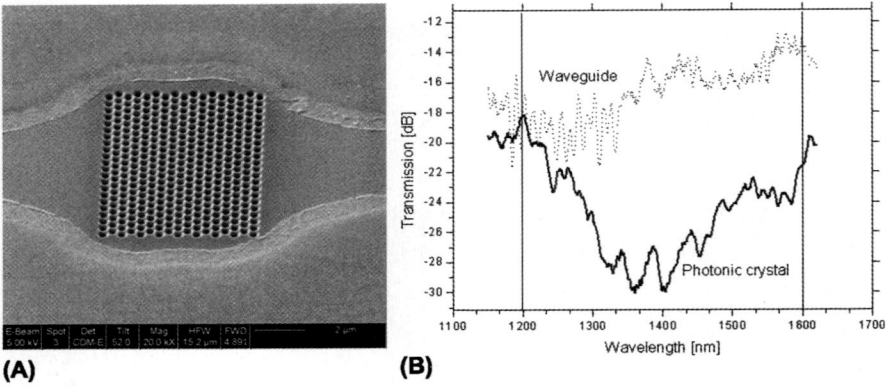

Fig. 12.8. (a) SEM image of the photonic crystal. (b) Transmission measurement of the photonic crystal

modulator based on the LN electro-optic effect. Its performance (300 times higher than for classical LN modulators) is due to slow light effects as presented previously.

12.2.4 Measurement of a PBG in a LN Photonic Crystal

To minimize the optical losses in the vertical direction, the photonic crystal array is fabricated in a lithium niobate channel waveguide. It is already well-known from the literature that in order to have the photonic band gap effect, there must be a maximum interaction between the structure and the guided mode. This requirement is particularly difficult in the case of lithium niobate optical waveguides since the optical mode is very much confined within the substrate. The photonic crystal is fabricated on a 0.3 mm thick X-cut $LiNbO_3$ wafer. In a first step, an optical gradient index waveguide was fabricated by annealed proton exchange (APE). This step was realized through a SiO_2 mask in benzoic acid at 180°C during 1.5 hours. The process was followed by an annealing of the optical waveguide at 333°C for 9 hours. These parameters were chosen to make the optical mode core as close as possible to the surface while keeping single mode propagation at 1.55 μm. Thus, the mode core is estimated to be at 1.4 μm from the surface (much better than the 5 μm depth that would be attained with a standard Ti-diffusion process).

The photonic crystal structure was fabricated in the central region on the optical channel waveguide as shown in Fig. 12.8(a). It consists of a hexagonal lattice of 21×19 circular holes. The holes are fabricated using FIB as described in the previous section. The chosen propagation direction is ΓM since theoretical simulations have shown that this direction requires only 15 rows of holes for 100% extinction ratio [16]. The etching time of the structure (21×19 hexagonal hole lattice, hole diameter = 213 nm, periodicity = 425 nm, etching depth = 1 500 nm) was 20 min. The theoretical transmission band in the ΓM direction is [1 300, 1 600] nm.

The transmission spectrum is obtained by coupling a white light fiber source into the channel optical waveguide in which the photonic crystal is located. This su-

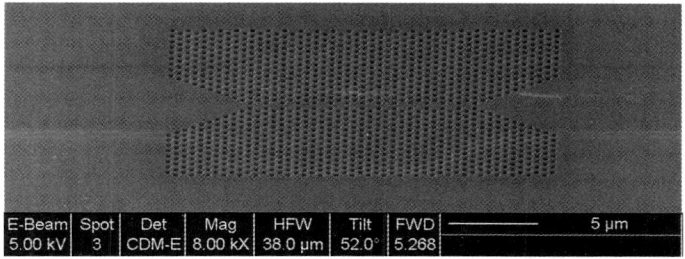

Fig. 12.9. SEM image of a PCW1 fabricated on a lithium niobate substrate with FIB

percontinuum is generated by a sub-nanosecond microchip laser emitting at 532 nm (mean output power 30 mW, FWHM pulse duration 0.4 ns). The polished optical fiber is placed in almost contact to the side entrance wall of the waveguide to decrease the optical input losses. The experimental PBG of the LN photonic crystal is shown in Fig. 12.8(b). The transmission through the photonic array (continuous line) has been compared to the transmission through a standard APE optical waveguide (discontinuous line) fabricated under the same conditions and placed in the same lithium niobate substrate than the photonic array structure. We can see that an extinction ratio of less than -12 dB is measured for the case of the photonic crystal structure and that the position of the transmission band corresponds to the theoretical predictions. The noise that is measured in both measurements is mainly due to the fact that the optical waveguide is uniquely monomode at 1.55 µm and also due to insertion losses.

12.2.5 LN PtC Waveguides: Transmission and SNOM Characterization

In this section, the possibility of guiding the light is experimentally evaluated for photonic crystal waveguides fabricated in $LiNbO_3$. Two alternative structures, based on the same array as in our previous section (ΓM direction) are fabricated. The first one has one line of defects (PCW1), and the second one three lines of defects (PCW3). The etching time of the structures PCW1 and PCW3 (48 × 26 triangular hole lattice, hole diameter = 255 nm, periodicity = 510 nm, etching depth = 1 500 nm) was 20 min each. A SEM image of the PCW1 structure is shown in Fig. 12.9.

The novel structures were first characterized by measuring their far field transmission. The light supercontinuum is generated by a sub-nanosecond microchip laser emitting at 1 064 nm with 8 µJ energy per pulse [17]. The optical transmission was measured through the two photonic waveguides, and through a standard optical waveguide, fabricated on the same wafer and in the same conditions, as described above. The experimental results are shown in Fig. 12.10. As it can be seen in the graph, optical transmissions through the photonic structures (plotted with filled triangles, empty circle, and empty square) exhibit a gap, which does not appear in the transmission through the single APE waveguide (filled circle in Fig. 12.10). In parallel, numerical simulations performed with a commercial software (Bandsolve) of the

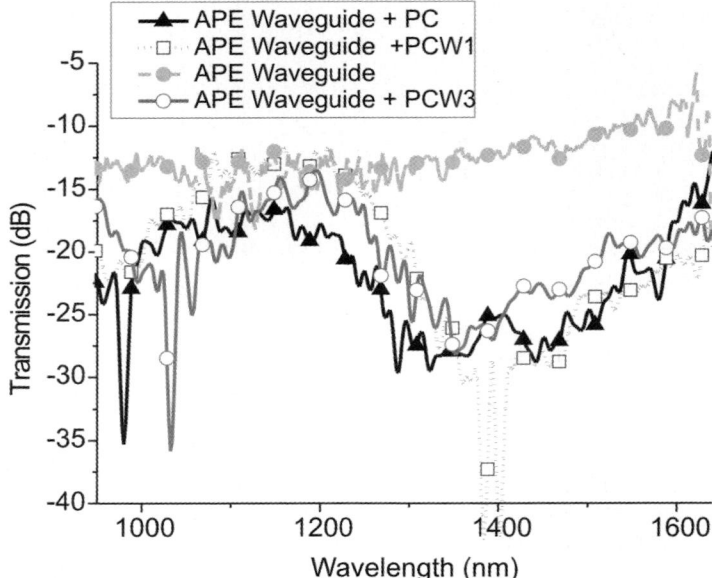

Fig. 12.10. Transmission measurements for APE waveguide only (*filled circle*), photonic crystal with no defects (*filled triangle*), PCW1 (*empty square*), and PCW3 (*empty circle*)

device without defect lines predict a band gap between 1 465 nm and 1 589 nm. The experimental gap starts in a shorter wavelength (approximately around 1 300 nm) which we believe is a consequence of fabrication imperfections. The dispersion diagram for the case of having one and three defect lines respectively predicts that for the PCW1 case, there is only one mode than is allowed to propagate. As it is intuitive, there is multimodal behavior in the case of a PCW3. Indeed, our simulations show four modes that can propagate within the band gap. Experimentally, the light propagation in the PtC waveguides is observed by an increase in the transmission inside the gap. This increase is twice more important for the PCW3 case due certainly to the multimodal behavior.

For a deeper interpretation of the propagation of the light through the structures, we have also investigated the near field behavior of the light inside the PtC waveguides. SNOM measurements are relevant in PtCs characterizations because the wave fronts of light in the photonic crystal waveguide undergo substantial modulations on length scales that are much shorter than one wavelength being impossible to resolve the spatial details of light propagation only by the far field transmission measurement described above [18–21].

The instrument used is a commercial scanning near-field optical microscope (SNOM) (NT-MDT SMENA) in collection mode [22] with a dielectric pulled fibred tip. The optical image and topography of the PCW1 at 810 nm is shown in Fig. 12.11(a) and (b). Figure 12.11(a) shows the topography of the PCW1 structure. The hole depth measured by the SNOM tip is of the order of 30 nm which is far from

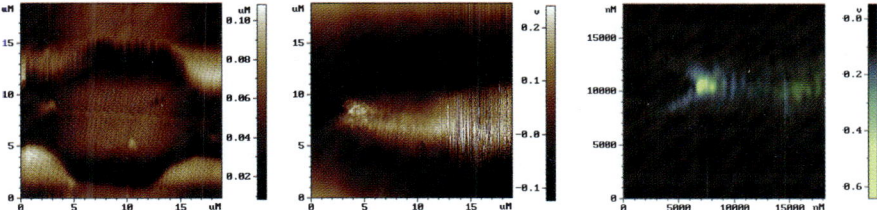

Fig. 12.11. (a) Topography of the photonic crystal structure. (b) SNOM image at 810 nm. (c) SNOM image at 1550 nm

the 1.5 μm measured in the SEM image. This is due to the fact that the hole diameter is comparable in size to the tapered fiber, being difficult for the tip to penetrate inside the holes. The signal to noise ratio (SNR) is however high (≈ 10). Figure 12.11(b) shows the optical image of the light going through the PCW1 at 810 nm. As measured by the transmission in Fig. 12.10, the optical transmission is of -20 dB since the wavelength is outside the gap of the photonic crystal.

We have also performed the near field measurements in a region inside the gap in which an optical mode propagates ($\lambda \approx 1.55$ μm, transmission ≈ -25 dB). The near field image is show in Fig. 12.11(c). The recorded signal shows clearly a confined mode that propagates through the line of defects. This propagating signal shows a distinct periodicity of λ/n_{eff} obtaining $n_{\text{eff}} \approx 2.2$, n_{eff} being the effective index of refraction corresponding to lithium niobate. With these results, we can infer that the step seen in the transmission response of the PCW1 (Fig. 12.10) around 1 500 nm is due to the existence of a guiding region.

12.2.6 A LN PtC Intensity Modulator

Tunable PtCs present special interest for integrating dense optical circuits on small surfaces. They typically consist of a periodic array of air holes on a dielectric substrate whose optical properties are modified by an external physical signal (electric or magnetic field, temperature, strain, etc.) [23–26]. One of the most suitable tuning schemes may be based on the application of an electric field, due to the technical compatibility of the PtC components with current microelectronics technology [27–31]. Indeed, one of the most promising tunable PtC configurations is a polymer-based photonic device tuned by the Pockels effect showing sub-1 V sensitivity [30]. However, up to now, electro-optical tunable photonic crystals have limited tunability because of the small attainable changes in the refractive index.

The device consists of a 15 × 15 square array of air holes etched by Focused Ion Beam (FIB) on a gradient index LN waveguide. The geometrical parameters of the PtC are fixed by theoretical calculations so that an edge of the gap corresponds to the operating wavelength of 1 550 nm. By using two-dimensional Finite Difference Time Domain (FDTD-2D) home-made calculations, we have determined that a squared arrangement of holes with a period $a = 766$ nm, and $r/a = 0.27$ ratio (r being the radius of the hole) induces a gap with an edge at 1 550 nm in the ΓX

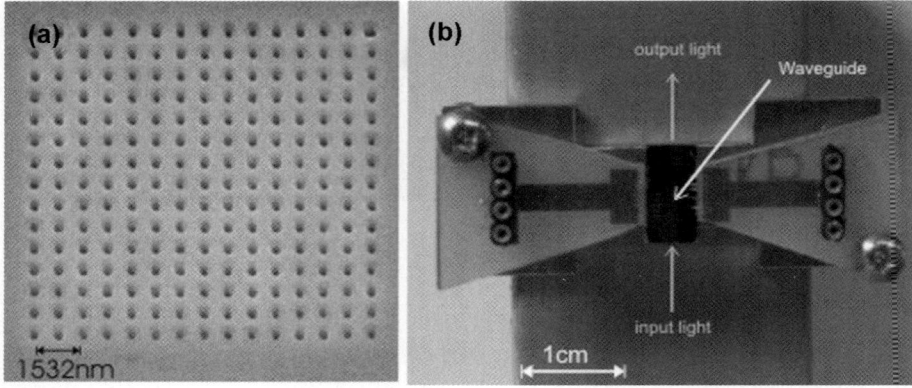

Fig. 12.12. (**a**) SEM image of the square lattice fabricated by FIB on a LN substrate. (**b**) Photograph of the final device

direction of propagation and TE polarization. By introducing the classical Pockels effect on the FDTD calculation we can predict an intensity modulation with a 24 dB extinction ratio for a 0.01 variation of the refractive index at 1 550 nm operating wavelength.

Figure 12.12(a) and (b) show the SEM image of the photonic crystal and a photograph of the fabricated tunable device, respectively. The electrodes have been fabricated by depositing 150 nm of Ti by sputtering (ALCATEL SCM 450). The distance between both electrodes is 13 µm and their length is 8 mm. The LN sample containing the electrodes and the photonic crystal is placed in a butterfly shaped electronic circuit to facilitate the electrical connection. On the circuit, two copper lines have been traced and the Ti electrodes are connected to them through wire bonding.

The transmission spectrum for different excitation voltages is shown in Fig. 12.13. At 0 V, two consecutive stop bands are observed. Their location corresponds to the theoretical prediction (first band [1 125, 1 200] nm, second band [1 400, 1 550] nm) although the measured stop bands are wider due to fabrication imperfections.

When a continuous voltage is applied the band gap shifts. The measured wavelength shift is of 2.5 nm/V. This value is 312 times bigger than the shift predicted by the Pockels based simulations (0.008 nm/V). The band-shift changes direction as we invert the applied voltage sign confirming an electric effect. To rule out photorefractive additional effects we have repeated the same experiment with the sample continuously illuminated by a blue laser beam (50 mW, $\lambda = 473$ nm) over three hours. No additional effect was appreciated. The reasons of this enhancement are explained in Sect. 12.2.1. Indeed, the effective second order susceptibility in the LN nanostructure increases, giving rise to an ultra-compact low voltage PtC modulator when it operates at its band edge (corresponding to slow photons).

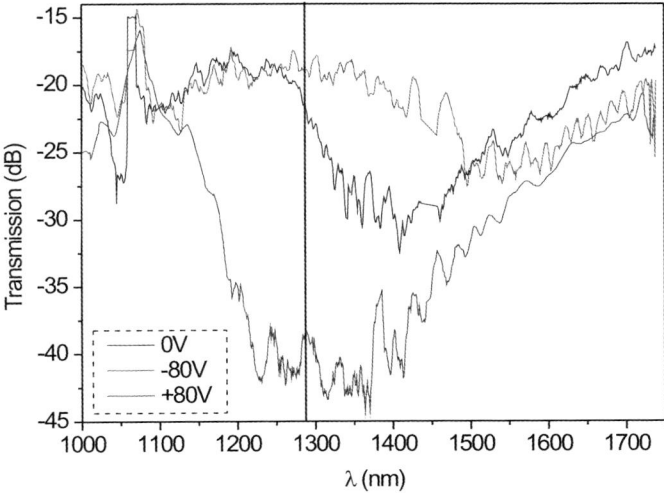

Fig. 12.13. Transmission results at 0 V, −80 V and +80 V

12.3 Phononic Crystals

12.3.1 Theory

Finite Element Method

A bulk wave finite element model can be used to calculate phononic crystal band diagrams. This bulk wave approach can be taken as it has been shown by some of the authors elsewhere [32] that band gap position and width for surface waves generally coincides with those for bulk waves, at least in the case of air holes milled in a lithium niobate substrate, as considered here. Although plane wave expansion models have previously been proved to be relevant to predict the band gap position and width by computing the band diagrams of the considered structures, even in the frame of an air/solid phononic crystal, they can appear as rather inaccurate when dealing with extreme geometrical conditions (really high filling fractions, for instance). Finite element methods not only allow to get rid of these limitations, but also present the noticeable advantage of being capable of taking into account the abrupt changes in the propagation conditions that occur at the interface between the matrix material and the inclusions.

The phononic crystal is assumed to be infinite and arranged periodically in the X and Y directions. The whole domain is split into successive unit cells, consisting of a single hole surrounded with the matrix material and indexed by (m, p). The unit cell is meshed and divided into elements connected by nodes. The structure is excited by a plane wave and the incidence is characterized by the real wave vector $\mathbf{k} = (k_x, k_y)$. According to the Bloch-Floquet theory, all fields obey a periodicity law, yielding for instance the following mechanical displacements u for the ith node:

$$u_i(x + ma_1, y + pa_2, z) = u_i(x, y, z) \exp(-j(k_x ma_1) + (k_y pa_2)), \quad (12.5)$$

where k_x and k_y are the components of the Bloch wavevectors in the X and Y directions respectively and a_1 and a_2 are the pitches of the structure. Using this relation allows us to reduce the model to a single unit cell which can be meshed using finite elements using in our case a mechanical displacement and electrical potential formulation scheme. Considering a monochromatic variation of mechanical and electrical fields with a time dependence in $\exp(j\omega t)$ where ω is the angular frequency, the general piezoelectric problem with no external applied force can then be written as:

$$\begin{bmatrix} K_{uu} - \omega^2 M_{uu} & K_{u\phi} \\ K_{\phi u} & K_{\phi\phi} \end{bmatrix} \begin{pmatrix} u \\ \phi \end{pmatrix} = \begin{pmatrix} 0 \\ 0 \end{pmatrix}, \quad (12.6)$$

where K_{uu} and M_{uu} correspond to the stiffness and mass matrices of the purely elastic part of the problem, $K_{u\phi}$ is the piezoelectric-coupling matrix, $K_{\phi\phi}$ represents the purely dielectric part and u and ϕ are respectively the nodal displacement and electrical potential. As the angular frequency ω is a periodical function of the wave vector, the problem can be reduced to the first Brillouin zone. The dispersion curves are eventually built by varying the wave vector on the first Brillouin zone for a given propagation direction. The full band diagram is then deduced using the structure symmetries.

Phononic Crystal Design

Practical interest in the fabrication of phononic crystals usually dwells in the obtaining of the largest possible band gaps. Once the structure type of symmetry is set as a first step, the next critical parameter to take into account is the lattice filling fraction. It has been shown that in an anisotropic and piezoelectric material such as lithium niobate, numerical plane wave expansion simulations predict a fractional band gap width up to 34% for a 64% filling fraction for any propagation direction and polarization along the complete anisotropic Brillouin zone [32]. The diameter over pitch (d/a) ratio of the structure is then around 0.9. Figure 12.14 displays the band diagram obtained for a structure with a different filling fraction, namely 0.94. Here again, a full band gap clearly opens, with a normalized center frequency $f \times a$ around 200 MHz and a fractional bandwidth of about 35.4%.

If band gap materials with high fractional bandwidths can be obtained in configurations exhibiting high diameter over pitch ratios, crystals with lower d/a values would certainly prove to be easier to fabricate. The point is then to determine how the fractional band gap width scales with the filling fraction of the structure. Figure 12.15 shows this last parameter variation as a function of d/a. The band gap width tends to rapidly decrease with the filling fraction of the phononic crystal and is eventually reduced to zero when this latter turns out below 50%, which corresponds to $d/a = 0.8$ and does not really allow for a large margin in terms of crystal fabrication. We have hence set our choice on the fabrication of crystals with diameter over pitch ratios higher than 0.9.

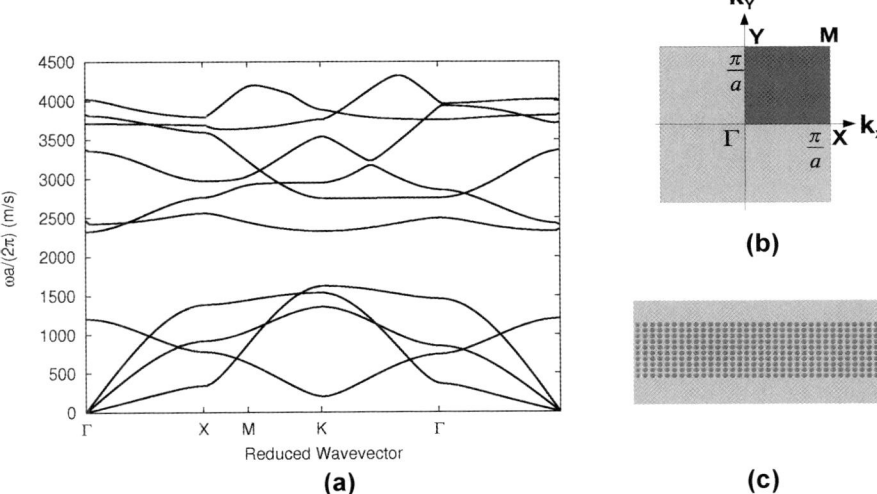

Fig. 12.14. Theoretical band structure for bulk waves propagating in the plane of a square and void/lithium niobate phononic crystal with a 0.94 diameter over pitch ratio. (**b**) Sketch of the first Brillouin zone in a square lattice. (**c**) Scanning Electron Microscope photographs of a 10 μm deep, 9.4 μm diameter hole etched in a lithium niobate substrate, before removal of the electroplated nickel mask

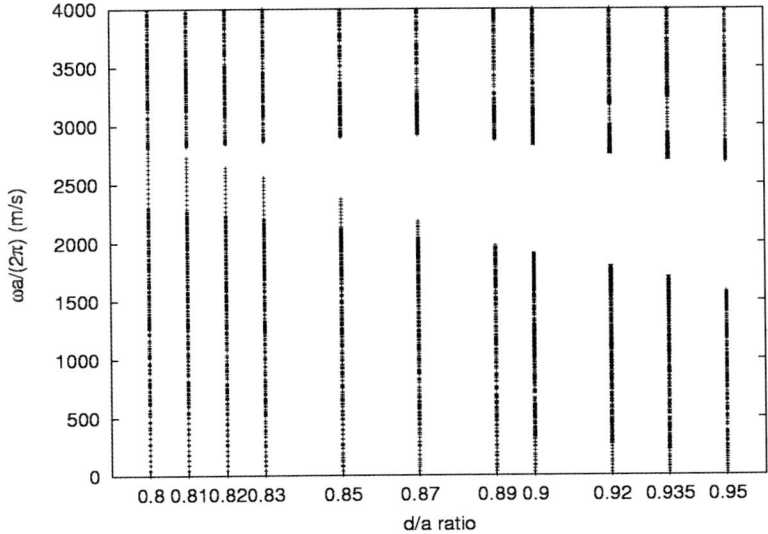

Fig. 12.15. Band gap width versus diameter over pitch ratio for a void/lithium niobate square lattice phononic crystal. The Y cut of lithium niobate is considered

12.3.2 Fabrication and Examples

The phononic crystals have been fabricated on a 500 µm thick, Y-cut lithium niobate wafer [7]. The material cut has been chosen because of its strong piezoelectric coupling coefficients, however, it is interesting to notice that our fabrication process is quite insensitive to the material cut, and that it is perfectly conceivable to fabricate the same type of phononic crystal in an optical cut of lithium niobate (e.g. the X-cut) using the very same fabrication method as developed in the following subsection. To be able to use standard optical lithography for the patterning of the crystal structure, we have chosen to work at a center operating frequency of about 200 MHz. This frequency is besides a relevant one in terms of acoustic components operating frequency. This requires hole diameters around 9 µm for a 10 µm pitch.

Lithium Niobate Etching

The point was then to mill as deep as possible holes to prevent leakage of surface waves below the crystal. The holes also had to have as vertical walls as possible to avoid surface to volume mode coupling. If this has proven quite easy to achieve in a material such as silicon, where techniques like Deep Reactive Ion Etching are readily available, the problem is more difficult to solve for lithium niobate, as micromachining techniques for this material are far from being well spread. In our case, the wafer has been processed using standard SF_6 based Reactive Ion Etching (RIE). The etching rate is considerably low, around 50 nm min^{-1}, and the etching slope is quite significant (around 17% for a 10 µm diameter hole). This greatly limits the achievable depth and aspect ratios, as this last parameter seems to be limited to values around 1.5, and creates the need for a particularly selective mask to allow for long etching times. Metallic masks generally satisfy this condition, however, considerably high thicknesses need to be deposited (around 1 µm) to obtain maximum etching depths around 10 µm. In any case, this etching depth is close to the limit set by the etching slope, but the selectivity of the mask turns out not to be sufficient to ensure a good enough surface state for surface wave transduction, as the process duration generally leads to an almost total depletion of the metallic mask. In this case, the surface of the lithium niobate wafer starts being locally etched, resulting in a deterioration of the overall surface state. The alternative we have chosen is to use a highly selective, 1 µm thick electroplated Nickel mask which enables to obtain the 10 µm depth we are aiming at with a satisfying hole profile. The mask selectivity would even allow for a deeper etch, which is an interesting result in the frame of applications requiring drilling of larger patterns. The chosen depth of 10 µm should anyway remain high enough for surface (Rayleigh) waves to feel the influence of the crystal, due to their limited penetration depth in the substrate. Figure 12.14(c) shows examples of holes obtained in a Y-cut lithium niobate wafer before removal of the electroplated Nickel mask. The initial thickness of the mask was around 1 µm, and one can notice that almost half of this initial thickness is still preserved after the etching process. These figures and further reactive ion etching tests unreported here seem to point out that the etching slope increases when the hole diameter decreases which would limit fabrication of higher frequency phononic crystals using the same RIE process. In any case, the surface state inside the holes is highly satisfying.

Crystal Characterization

Several possibilities are then offered for the characterization of our phononic crystal structure. Optical techniques like Brillouin Light Scattering (BLS), as described for instance in [6] allow for a full retrieval of the considered elastic modes band diagrams. It is also possible to perform phase sensitive measurements which give extensive information on the acoustic field profiles [33, 34]. In our case, and for now, we will limit ourselves to an electrical characterization of the crystal. This latter can be performed by direct excitation and detection of surface acoustic waves traveling through the crystal using for instance, but not necessarily, two identical IDT's, one as a transmitter, and the second one as a receiver. This has the advantage of putting us in a realistic device configuration. However, bandgap measurements produce the need for large bandwidth surface acoustic waves sources, For instance, in the present case, a large frequency range, from 150 to 30 MHz at least, needs to be covered to be able to observe the band gap phenomenon. Simple interdigital transducers unfortunately do not allow to fulfill this requirement. Indeed, increasing the bandwidth of a single IDT can only be done by reducing the number of electrodes of the device. But reducing this number is generally coupled with a loss in the electro-acoustic coupling of the device and with greater interference with bulk waves propagating in the substrate, hence leading to a small dynamic device response. To overcome this problem, a solution proposed in the literature as for instance in [35, 36] and applied in the recent experiments of Wu and coworkers [37] is to use slanted finger interdigital transducers. However we will see later on that this approach cannot actually be used here. Instead, we here propose an alternative consisting in working with a series of IDT's of varying mechanical period instead. This allows to cover the frequency range of interest by juxtaposing neighboring responses. The main drawback of such a method is that it implies the fabrication of several identical phononic crystals instead of working with a single one. This hence requires some means of "mass-production" of holey structures. This last point can be a limiting factor in some cases but remains fortunately compatible with the optical lithography combined with RIE process described in the previous section. Eight IDT's with ten digit pairs with an emission wavelength ranging from 12.2 μm to 26 μm were needed to cover the frequency range of interest. The IDT's have been realized by patterning a 150 nm thick aluminum layer, and oriented for elastic wave propagation along the ΓX, ΓM and ΓY directions of the first Brillouin zone. Figure 12.16 shows a scanning electron microscope image of some of the phononic crystals for surface acoustic waves manufactured on a single Y-cut lithium niobate wafer.

Figure 12.17 displays band diagrams for bulk waves in the $d/a = 0.94$ configuration for these three propagation directions. These latter encompass the main points of symmetry of the first Brillouin zone, and are relevant directions to consider

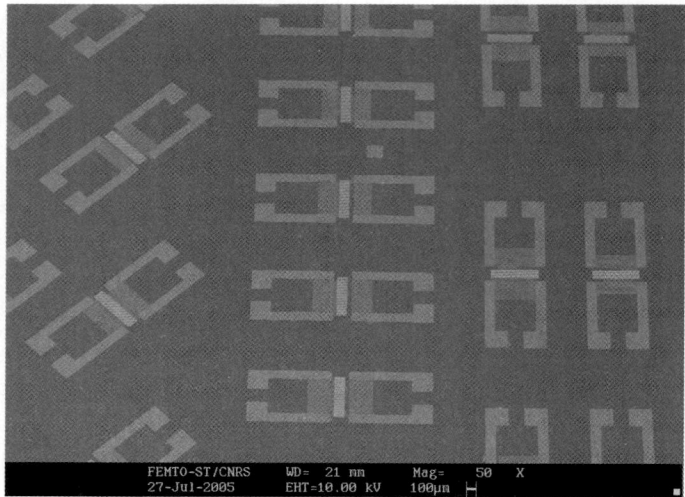

Fig. 12.16. Scanning electron microscope image of some of the phononic crystals for surface acoustic waves manufactured on a single Y-cut lithium niobate wafer

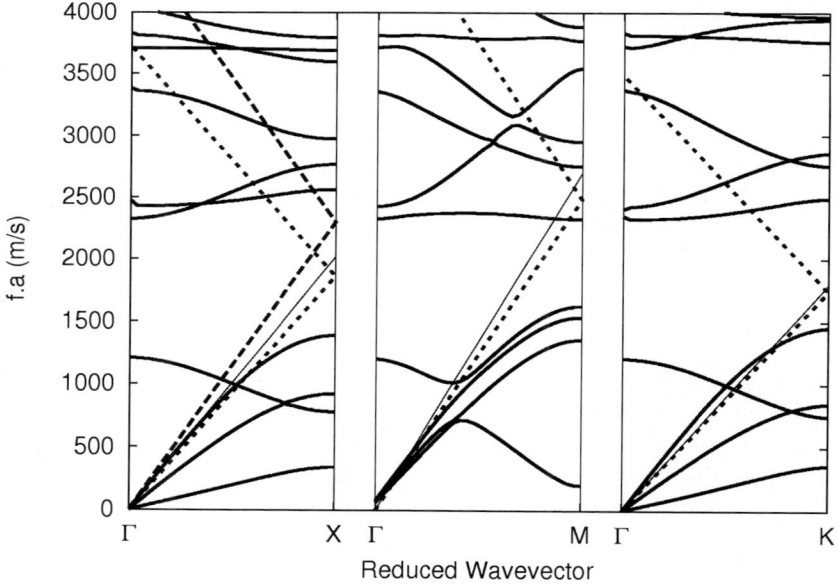

Fig. 12.17. Theoretical band diagrams for bulk waves propagating in the plane of a square lattice void/lithium-niobate phononic crystal with a 69% filling fraction ($d/a = 0.94$) along the ΓX, ΓM and ΓY directions of the first Brillouin zone. In the greyed regions, above the *soundlines*, there is coupling with the radiation modes of the substrate and surface modes become leaky. The dispersion relations on a free surface are also indicated for the Rayleigh (*short dashed line*) and the leaky (*long dashed line*) surface waves

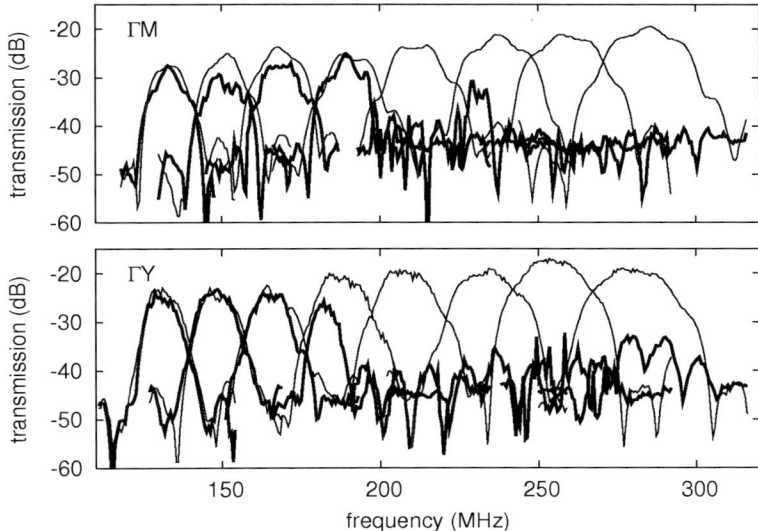

Fig. 12.18. Measurements of the 8 reference devices (*dotted line*) and the 8 phononic band gap devices (*solid line*) along the ΓM and ΓY directions

in the frame of an experimental investigation of an anisotropic phononic structure. In every case, two series of IDT's have been fabricated, one without any holey structure between the transmitter and the receiver to set a reference, the second one with a phononic crystal as described above. The electrical transmission was then measured in terms of scattering parameters using a network analyzer.

Both transmission (S_{12} or S_{21}, as our device is symmetric) measurements can provide us with relevant information on the Rayleigh waves behavior through the crystal. We expect identical electrical responses for the reference and the phononic devices at frequency below 180 MHz and above 230 MHz, which correspond respectively to the lower and upper frequency limits of the theoretically predicted band gap. Between these frequency values, we anticipate an almost complete extinction of the signal. As far as reflexion is concerned, we expect in the same way to get a similar behavior between reference and phononic crystal signal except for an enhanced reflexion phenomenon inside the band gap.

Figure 12.18 displays the transmission spectra for both the reference and the phononic crystal devices along the ΓY (a) and the ΓM (b) directions. The dotted lines stand for the reference signal, i.e. for the transmission response of the delay line constituted by two IDT's separated with free space, while the solid lines give the signal transmission through the phononic crystal. In both cases, these measurements clearly show that at frequencies below 180 MHz along ΓY, and below 200 MHz along ΓM, reference and crystal signals almost perfectly overlap: the Rayleigh wave propagation is not affected by the presence of the crystal. Above these frequencies, the transmitted signal experiences a strong attenuation, with losses estimated to be around 20 dB for a ten-period long crystal. Surprisingly, there is no increase in transmission for Rayleigh waves at frequencies higher than 230 MHz, as it would be expected from the band diagram previously introduced in Fig. 12.14. The signal does not recover at all along ΓY, and only a rather narrow transmission peak – about 5 MHz large – around 230 MHz is detected along ΓM, but there does not seem to be any higher frequency mode propagating in the structure. Any higher frequency mode propagating in the structure seems to be inhibited.

Similar results are observed along the ΓX direction. Let us here notice that, in the case of Y-cut lithium niobate, in addition to the pure Rayleigh modes aimed at, some leaky surface modes, the so-called pseudo-surface waves (PSAW), also propagate, though their range of existence is limited to angles around the X crystallographic axis. These transversally polarized waves are usually subject to stronger attenuation while propagating, as it is the case along the ΓY and ΓM directions previously discussed where their attenuation coefficients are too high to allow for their propagation and detection, but tend on the other hand to penetrate far deeper into the substrate than the pure Rayleigh wave does. This basically means that because of the very limited depth of the holes, and hence of the phononic crystal structure, the pseudo-surface waves are not expected to experience the substrate periodicity in a significant enough way to be attenuated and to give evidence of a band gap effect. Their propagation velocity is here relatively close to the Rayleigh modes velocity (4 600 versus 3 700 m s^{-1}), which leads to a partial overlap of the electrical responses of the two types of waves. This makes the use of wide bandwidth slanted fingers IDT's ambiguous for transmission measurements. The overlap between pseudo-surface waves and Rayleigh waves makes it difficult to plot the obtained transmission signals over the whole considered frequency range, as previously done for the two other directions. Thus, for a better readability of the figure, we will hence limit ourselves and present only results obtained for four of the eight IDT's, as shown in Fig. 12.19. The corresponding devices have an emission wavelength of 26, 20.6, 16.4 and 14.6 µm, respectively. In the case of Fig. 12.19(a) for instance, the first lobe centered around 145 MHz corresponds to the Rayleigh surface wave, while the second lobe, around 170 MHz corresponds to the pseudo-surface wave. This latter seems to be slightly affected by the structure, but there is no relevant drop in the transmission value. However, if closer attention is paid to Fig. 12.19(c), which represents a device with an operating frequency located inside the theoretically predicted band gap, while the pseudo-surface wave remains almost unaltered, the Rayleigh wave experiences the same sharp 20 dB attenuation. Here again, there is no increase in transmission for the Rayleigh wave for a system operating at frequencies higher than the band gap boundary. The cut-off frequency now lies around 190 MHz.

Figure 12.20 gives a summary of the obtained measurements. If a full band gap has definitely opened, with the crystal causing a 20 dB loss in the transmission value for pure surface modes, there does not seem to be any way to retrieve a high amplitude signal at high frequencies. To account for this phenomenon, let us come back to the fact that the actual crystal does not correspond to the ideal, two-dimensional

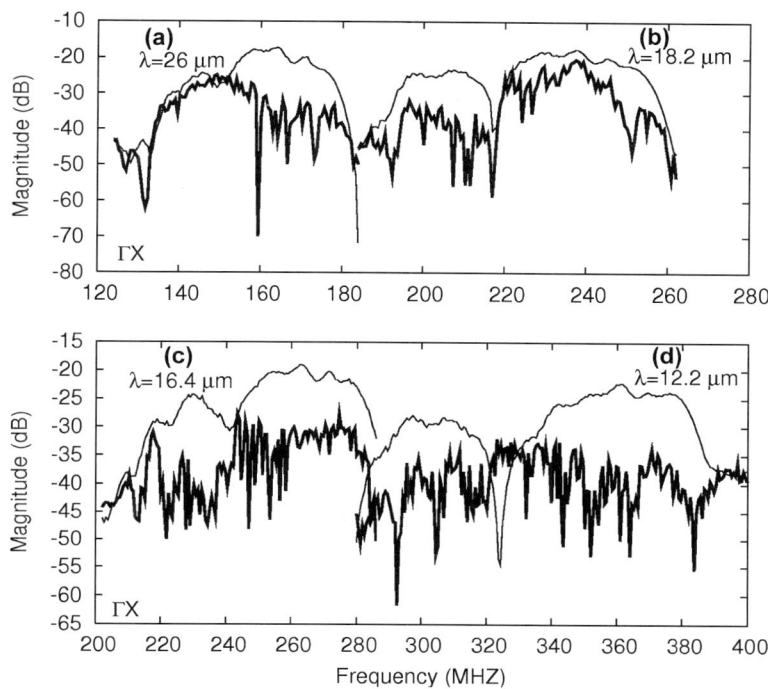

Fig. 12.19. Measurements of 4 reference devices (*dotted line*) and 4 phononic band gap devices (*solid line*) along the ΓX direction. Both the Rayleigh and the leaky surface waves exist simultaneously and give rise to adjacent frequency responses

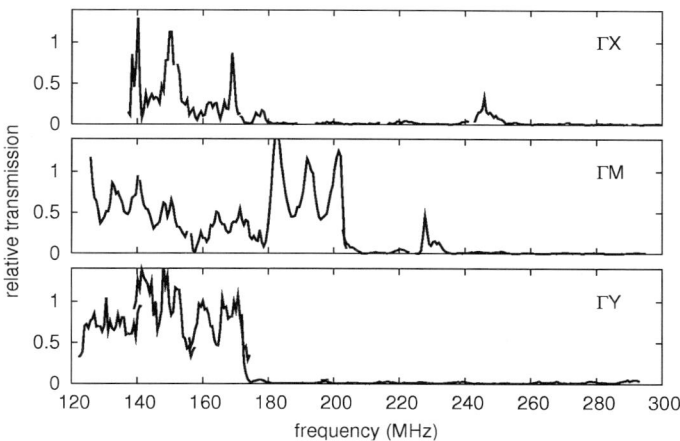

Fig. 12.20. Relative transmission, defined as the ratio of the transmittance with and without a phononic crystal. The complete band gap extends from 203 to 226 MHz

phononic crystal needed to fully retrieve the theoretically predicted results. The surface waves penetration depth in the substrate remains of the same order, or even larger than the hole depth. The experimental structure can then be seen as a stratified medium, with a void/lithium niobate phononic crystal slab on bulk lithium niobate. This, combined with the conicity of the holes tends to favor out-of-plane scattering, leading to coupling with bulk modes of the substrate, which results in propagation losses for the initial surface waves. To tackle this problem theoretically, an approach quite similar to the one applied to the so-called 2.5D photonic crystals can be taken [3]. These generally consist of a 2D photonic crystal etched through a high-index guiding layer on a lower index substrate. Guided modes as well as radiation modes into the substrate are expected to exist, and the limit between these modes is set by the so-called light line, defined by $\omega = ck$ on the usual $\omega(k)$ dispersion diagram. The straight lines plotted in Fig. 12.17 correspond to the dispersion relation of the bulk modes of the substrate exhibiting the lowest velocity, and which are then the more likely to be coupled into by the scattered surface waves. At frequencies higher than the upper edge of the band gap, the propagating modes lie above this *sound line*, meaning that they are highly lossy, radiation modes. Only modes below the *sound line*, i.e. lying at low frequencies can be transmitted, which agrees well with the experimentally observed results. In the case of the Rayleigh surface waves, this *sound line* accounts for both the total extinction of the signal at high frequencies along the ΓX and ΓY directions, but also for the peak observed along ΓM which corresponds to modes belonging to the upper part of the band diagram but still lying below this line. The pseudo surface acoustic wave being faster than the slowest shear bulk wave, it is always above the sound line, and hence always experiences strong radiation to the interior of the substrate, in agreement with the experimental observations.

12.4 Conclusion

The photonic and phononic band gap properties of lithium niobate were investigated both theoretically and experimentally. The computed photonic band structures were used to discuss the appearance of band gaps and of slow light modes. Actual lithium niobate photonic crystals were presented and a striking phenomenon of enhanced electro-optic coefficient for slow light was emphasized. Phononic crystals fabricated in lithium niobate were then discussed theoretically and experimentally, with an emphasis on phononic band gap properties for surface acoustic waves, i.e. for phonons that are confined close to the surface.

So far, all the research works (from our team and elsewhere) reported on the fabrication of photonic or phononic structures in lithium niobate have been done on bulk material which implies the use of only a small fraction of the surface volume of the material. In the case of LN photonic crystals holes have to be deeply etched in order to reach the optical mode. Optical losses due to inefficient hole-mode interaction and fabrication imperfections could be decreased if the structures were fabricated on thin layers of lithium niobate. Because oxides like lithium niobate tend to be chemically very inert, there are only a very limited number of surface modification tools that can be used for fabrication purposes. With the deposition of thin layers of lithium niobate nanostructures will be easily etched or ablated, permitting the fabrication of photolithographically defined two-dimensional and three-dimensional structures on a planar substrate. Some works on epitaxial lithium niobate have been already published in literature [38–40].

Although high-quality films have been fabricated by these techniques, many of the electrical and electro-optical properties reported are generally not comparable to those for bulk single-crystal material. One technique based of the selective He-ions implantation called smart cut has been developed mainly by Prof. Osgood at Columbia University and the ETH in Zurich [41–43]. It produces thin layers (of several microns in thickness) of high quality LN. The problem of this technique comes from the technology needed: Implantation doses of 5×10^{16} ions/cm^2 are needed. There are only few implantation machines in the world that can provide such a high density.

Once it is realized that despite vastly different frequencies, phonons and photons can share comparable wavelengths, it becomes clear that the simultaneous appearance of photonic (optical waves) and phononic (acoustic waves) band gaps in a single periodic nanostructure can be used to confine simultaneously photons and phonons. This novel phenomenon could have a strong impact on the nature and the strength of photon-phonon interactions. Long term applications include the development of a novel generation of active optical signal processing devices. We name such artificial structures phoXonic crystals. The perspectives of these novel materials are enormous and could lead to breakthroughs in the field of acousto-optics. Appropriately designed, defect structures in deaf and blind structures can lead to simultaneous confinement of light and sound, a result that can have strong influence on photon-phonon interaction and the design of a new class of acousto-optical devices that can integrate the management of elastic and electromagnetic waves. A judicious choice of material is however needed in order to have a simultaneous photonic and phononic band gap. Maldovan et al. [44] have put forward theoretically the use of silicon, a material which has obvious advantages when it comes to etching sub micrometer holes. Provided further technological improvements are made in the direction of nano structuration, we believe that lithium niobate could be a winning option, as the electro-optic and the piezoelectric effects can be exploited with profit in active devices.

Acknowledgments

The authors are grateful to N. Courjal Bodin for fruitfull discussions and to R. Salut, L. Robert, W. Daniau, J.-Y. Rauch, and G. Ulliac for assistance in the various technological operations. This work was supported by the Action Concerté Incitative under project NANO #37 COBIAN.

References

1. S. John, Strong localization of photons in certain disordered dielectric superlattices. Phys. Rev. Lett. **58**(23), 2486–2489 (1987)
2. E. Yablonovitch, Inhibited spontaneous emission in solid-state physics and electronics. Phys. Rev. Lett. **58**(20), 2059–2063 (1987)
3. J.D. Joannopoulos, R.D. Meade, J.N. Winn, *Photonic Crystals* (Princeton University Press, New Jersey, 1995)
4. M.S. Kushwaha, P. Halevi, L. Dobrzynski, B. Djafari-Rouhani, Acoustic band structure of periodic elastic composites. Phys. Rev. Lett. **71**(13), 2022–2025 (1993)
5. M.M. Sigalas, E.N. Economou, Band structure of elastic waves in two dimensional systems. Solid State Commun. **86**(3), 141–143 (1993)
6. T. Gorishnyy, C.K. Ullal, M. Maldovan, G. Fytas, E.L. Thomas, Hypersonic phononic crystals. Phys. Rev. Lett. **94**, 115501 (2005)
7. S. Benchabane, L. Robert, J.-Y. Rauch, A. Khelif, V. Laude, Evidence for complete surface wave band gaps in a piezoelectric phononic crystal. Phys. Rev. E **73**, 065601(R) (2006)
8. A. Taflove, S.C. Hagness, *Computational Electrodynamics, the Finite-Difference Time–Domain Method*, 2nd edn. (Artech House, Norwood, 2005)
9. K. Sakoda, *Optical Properties of Photonic Crystals* (Springer, Berlin, 2001)
10. J. Seidel, F.I. Baida, L. Bischoff, B. Guizal, S. Grafstrom, D. Van Labeke, L.M. Eng, Coupling between surface plasmon modes on metal films. Phys. Rev. B **69**, 121405 (2004)
11. C.T. Chan, Q.L. Yu, K.M. Ho, Order-n spectral method for electromagnetic waves. Phys. Rev. B **51**(23), 16635–16642 (1995)
12. J.-P. Berenger, A perfectly matched layer for the absorption of electromagnetic waves. J. Comput. Phys. **114**, 185–200 (1994)
13. M. Roussey, F. Baida, M.-P. Bernal, Experimental and theoretical observation of the slow light effect on a tunable photonic crystal. J. Opt. Soc. Am. B **24**, 1416 (2007)
14. C.G. Bostan, R.M. de Ridder, V.J. Gadgil, L. Kuipers, A. Driessen, Interactions with a photonic crystal micro-cavity using an AFM in contact or tapping mode operation. In *Proceedings Symposium IEE/LEOS Benelux Chapter*, IEE/LEOS, Enschede, The Netherlands, p. 25 (2003)
15. S. Yin, Lithium niobate fibers and waveguides: fabrications and applications. Proc. IEEE **87**, 1962 (1999)
16. M. Roussey, M.-P. Bernal, N. Courjal, F.I. Baida, Experimental and theoretical characterization of a lithium niobate photonic crystal. Appl. Phys. Lett. **87**, 24110 (2005)
17. K.P. Hansen, R.E. Kristiansen, *Supercontinuum Generation in Photonic Crystal Fibers*. Crystal Fibre (www.crystal-fibre.com)
18. E. Flück, M. Hammer, A.M. Otter, J.P. Korterik, L. Kuipers, N.F. van Hulst, Amplitude and phase evolution of optical fields inside periodic photonic structures. J. Lightwave Technol. **21**, 1384 (2003)
19. S.I. Bozhevolnyi, V.S. Volkov, T. Sondergaard, A. Boltasseva, P.I. Borel, M. Kristensen, Near-field imaging of light propagation in photonic crystal waveguides: explicit role of Bloch harmonics. Phys. Rev. B **66**, 235204 (2002)
20. H. Gersen, T.J. Karle, R.J.P. Engelen, W. Bogaerts, J.K. Korterik, N.F. van Hulst, T.F. Krauss, L. Kuipers, Real-space observation of ultraslow light in photonic crystal waveguides. Phys. Rev. Lett. **94**, 073903 (2005)

21. D. Gérard, L. Berguiga, F. de Fornel, L. Salomon, C. Seassal, X. Letartre, P. Rojo-Romeo, P. Viktorovitch, Near-field probing of active photonic-crystal structures. Opt. Lett. **27**, 173 (2002)
22. D.W. Pohl, Optical near-field scanning microscope. US patent number 4604520 (1986)
23. H. Takeda, Yoshino, Tunable refraction effects in two-dimensional photonic crystals utilizing liquid crystals. Phys. Rev. E **69**, 016605 (2004)
24. Y. Jiang, W. Jiang, L. Gu, X. Chen, R.T. Chen, 80-micron interaction length silicon photonic crystal waveguide modulator. Appl. Phys. Lett. **87**, 221105 (2005)
25. S.L. Kuai, G. Bader, P.V. Ashrit, Tunable electrochromic photonic crystals. Appl. Phys. Lett. **86**, 221105 (2005)
26. D. McPhail, M. Straub, M. Gu, Electrical tuning of three-dimensional photonic crystals using polymer dispersed liquid crystals. Appl. Phys. Lett. **86**, 051103 (2005)
27. M. Schmidt, M. Eich, U. Huebner, R. Boucher, Electro-optically tunable photonic crystals. Appl. Phys. Lett. **87**, 121110 (2005)
28. H.M.H. Chong, R.M. De La Rue, Tuning of photonic crystal waveguide microcavity by thermooptic effect. IEEE Photonics Technol. Lett. **16**, 1528 (2004)
29. M.T. Tinker, J.B. Lee, Thermo-optic photonic crystal light modulator. Appl. Phys. Lett. **86**, 221111 (2005)
30. W. Park, J.B. Lee, Mechanically tunable photonic crystal structure. Appl. Phys. Lett. **85**, 4845 (2004)
31. C.W. Wong, P.T. Rakich, S.G. Johnson, M. Qi, H.I. Smith, E.P. Ippen, L.C. Kimerling, Y. Jeon, G. Barbastathis, S.-G. Kim, Strain-tunable silicon photonic band gap microcavities in optical waveguides. Appl. Phys. Lett. **84**, 1242 (2004)
32. V. Laude, M. Wilm, S. Benchabane, A. Khelif, Full band gap for surface acoustic waves in a piezoelectric phononic crystal. Phys. Rev. E **71**, 036607 (2005)
33. R.L. Jungerman, J.E. Bowers, J.B. Green, G.S. Kino, Fiber optic laser probe for acoustic wave measurements. Appl. Phys. Lett. **40**, 313–315 (1982)
34. J.V. Knuuttila, P.T. Tikka, M.M. Salomaa, Scanning Michelson interferometer for imaging surface acoustic wave fields. Opt. Lett. **25**(9), 613 (2000)
35. C.K. Campbell, Y. Ye, J.J. Sferazza, Wide-band linear phase SAW filter design using slanted transducer fingers. IEEE Trans. Sonics Ultrason. **29**, 224 (1982)
36. H. Yatsuda, Y. Takeuchi, S. Yoshikawa, New design techniques for SAW filters using slanted-finger IDTs. In *IEEE Ultr. Symp.* (1990), p. 61
37. T. Wu, L. Wu, Z. Huang, Frequency band-gap measurement of two-dimensional air/silicon phononic crystals using layered slanted finger interdigital transducers. J. Appl. Phys. **97**, 094916 (2005)
38. A.A. Werberg, H.J. Gysling, A.J. Filo, T.N. Blanton, Epitaxial growth of lithium niobate thin films from a single-source organometallic precursor using metalorganic chemical vapour deposition. Appl. Phys. Lett. **62**, 946 (1993)
39. R.S. Feigelson, Epitaxial growth of lithium niobate thin films by the solid source MOCVD method. J. Crystal Growth **166**, 1 (1996)
40. Y. Shibata, K. Kaya, K. Akashi, M. Kanai, T. Kawai, S. Kawai, Epitaxial growth and surface acoustic wave properties of lithium niobate films grown by pulsed laser deposition. J. Appl. Phys. **77**, 1498 (1995)
41. M. Levy, R.M. Osgood, R. Liu, L.E. Cross, G.S. Cargill III, A. Kumar, H. Bakhru, Fabrication of single-crystal lithium niobate films by crystal ion slicing. Appl. Phys. Lett. **73**, 2298 (1998)
42. A.M. Radojevic, M. Levy, R.M. Osgood, A. Kumar, H. Bakhru, C. Tian, C. Evans, Large etch-selectivity enhancement in the epitaxial liftoff of single-crystal $LiNbO_3$ films. Appl. Phys. Lett. **74**, 3197 (1999)

43. P. Rabiei, P. Gunter, Optical and electro-optical properties of submicrometer lithium niobate slab waveguides prepared by crystal ion slicing and wafer bonding. Appl. Phys. Lett. **85**, 4603 (2004)
44. M. Maldovan, E.L. Thomas, Simultaneous localization of photons and phonons in two-dimensional periodic structures. Appl. Phys. Lett. **88**, 251907 (2006)

13 Lithium Niobate Whispering Gallery Resonators: Applications and Fundamental Studies

L. Maleki and A.B. Matsko

13.1 Introduction

Optical whispering gallery modes (WGMs) are closed circulating electromagnetic waves undergoing total internal reflection inside an axio-symmetric body of a transparent dielectric that forms a resonator. Radiative losses are negligible in these modes if the radius of the resonator exceeds several tens of wavelengths, and surface scattering losses can be made small with surface conditioning techniques. Thus, the quality factor (Q) in crystalline WGM resonators is limited by material losses that are, nevertheless, extremely small in optical materials. WGM resonators made of $LiNbO_3$ have been successfully used in optics and microwave photonics. The resonators are characterized by narrow bandwidth, in the hundred kilohertz to gigahertz range. A proper choice of highly transparent and/or nonlinear resonator material, like lithium niobate, allows for realization of a number of high performance devices: tunable and multi-pole filters, resonant electro-optic modulators, photonic microwave receivers, opto-electronic microwave oscillators, and parametric frequency converters, among others. For example, an approach to create coupling between light and microwave fields in a whispering gallery mode resonator (WGR) was recently proposed [1, 2]. A resonant interaction of several optical WGMs and a microwave mode was achieved by engineering the shape of a micro-strip microwave resonator coupled to a WGR. Based on this interaction electro-optic modulators as well as photonic microwave receivers have been realized [3–10]. Fabrication of optical WGRs with lithium niobate and tantalate [11] has also led to the demonstration of high-Q tunable filters with linewidth of less than a megahertz. A crystalline WGR was used as a Lorentzian microwave filter with a tuning range in excess of tens of gigahertz [12]. A miniature resonant electro-optically tunable third-order Butterworth filter based on cascaded lithium niobate WGRs has been demonstrated as well [13]. Optical parametric frequency conversion has been realized in a periodically poled lithium niobate (PPLN) WGR [14]. A daisy-shaped poling of $LiNbO_3$ domains, optimal for WGRs, was also proposed [7] and realized [15]. Solid state WGRs are promising not only in applied sciences and engineering but also in fundamental sciences, e.g. in material science and nonlinear optics. The appeal of WGM resonators for nonlinear optics is in their high Q-factors as well as small mode volumes, that result in a substantial enhancement of the optical nonlinearities. An advantage of crystalline WGRs in materials studies is that the resonators are mechanically shaped samples cut out of a nonlinear material. This provides the opportunity to investigate the properties of the resonator material while avoiding the undesirable contributions of complimentary optical components, such as mirrors in conventional resonators. For instance, the resonators have been used to study photorefractivity of lithium niobate in the infrared [16–18]. Photorefractivity is one of the basic properties of optical crystals possessing quadratic nonlinearity, and manifests itself as a long-term change of refractive index under light illumination. This effect, observed shortly after the discovery of laser [19], results from the imperfection of real optical crystals, unlike other fundamental second-order effects, e.g. optical rectification [20] and optical parametric frequency conversion [21, 22], that could exist in an ideal crystalline lattice. Interpretation of the photorefractivity involves intricate details of the material properties and has not yet been finalized in its present form. Therefore novel studies on the subject continue to frequently appear in the literature (see, e.g., [23–25]). Photorefractivity of lithium niobate, quite strong in the UV and visible parts of the spectrum, diminishes in the infrared and far infrared. Photorefractivity in infrared has been observed and studied using pulsed light to maintain a high enough infrared intensity [26, 27]. This effect has been also observed with a continuous wave 780 nm and 1 550 nm light in a WGR made of nominally pure as-grown congruent $LiNbO_3$ [16–18]. The experiments shed light on the long wavelength properties of the impurities of the material and support the point of view that the photorefractivity does not have a distinct red boundary in wavelength. The observation is also important for various above mentioned applications of lithium niobate resonators and it points out a possible source of "aging" of telecom devices that use lithium niobate elements. In this chapter we review recent advances in the applications of the lithium niobate WGM resonators in science and technology. We show that the resonators are powerful and very useful tools worthy of further research and development. We start from describing WGM-based modulators and filters, and review frequency doubling experiments with PPLN WGRs. We describe experiments on photorefractivity involving $LiNbO_3$ WGRs. Finally, we discuss theoretical limitations of the transparency as well as photorefractive properties of lithium niobate in the infrared, which is important for future applications of lithium niobate WGM resonators.

13.2 Modulators

The motivation for the optical resonator-based modulators stems from the relatively large operating powers required to drive the existing modulators. Both broadband integrated Mach–Zender modulators and free space microwave cavity-assisted narrow-band modulators typically require 0.5 to several Watts of microwave power to achieve a significant modulation. By utilizing high-Q resonances instead of zero-order interferometry or polarization rotation as the basis for electro-optic modulation, the driving power can be potentially reduce by many orders of magnitude.

This, of course, is at the expense of a limited bandwidth that, nevertheless, is still practical for many applications. WGR-based modulators have been recently realized [3–10]. The core of the modulators is a WGR made with LiNbO$_3$. Even a small voltage applied across the area of confinement of the optical WGMs is enough to induce a change in the frequency of the WGM resonance with a magnitude comparable to its bandwidth. This forms the basis for an efficient modulation in a single mode regime. The higher the Q-factor of the cavity modes in this regime, the stronger the nonlinear interactions. However, this relationship also creates a narrow spectral range where the nonlinear interactions take place. This restricts the application of the resonant nonlinearities for the enhancement of wide-band nonlinear phenomena, which include multi-wave mixing, photon up- and down-conversion, and broadband optical modulation. A multi-mode electro-optical modulator (EOM) is more attractive when compared with the single-mode one, because it can modulate light at high microwave frequencies even if the resonator used in the modulator has an extremely large Q-factor. However, modulation that involves several optical and one microwave modes must fulfill phase matching conditions. An approach to solve the problem has been proposed in [1, 2]. In that study, an efficient resonant interaction of three optical WGMs and a microwave mode was achieved by engineering the shape of a microwave resonator coupled to a micro-toroidal optical cavity. Based on this interaction a new kind of electro-optic modulator was suggested and realized [3–6]. A detailed experimental and theoretical study of the all-resonant optical-microwave interaction in a multi-mode EOM has been presented in [7]. It was shown that a very efficient three-wave mixing process can be realized based on the high quality factors of the optical WGMs and a microwave mode. This wave mixing approach has a high saturation threshold with respect to the optical fields, and a high sensitivity to the microwave field. A resonant EOMs have been realized in the X-band (at 9 GHz), and in the Ka-band (at 33 GHz). The feasibility of the modulators for photonic reception of microwave signals with direct upconversion into optical domain has been proven. A microphotonic receiver with 2.5 nW detectable microwave power with about 14 dB signal-to-noise ratio, corresponding to the noise floor at $\sim$0.1 nW, and $\sim$5 kHz analog bandwidth has been realized. It has been shown that the performance can be further enhanced by increasing the Q-factors and improving microwave and optical field overlap. A microwave-optical receiver based on nonlinear optical modulation in a LiNbO$_3$ WGM resonator has been presented in [9, 10]. This receiver does not use high-speed electronic components for transmitted carrier links; the second-order nonlinearity of a WGR is used instead. The nonlinearity results in the mixing process required to extract the baseband signal from the transmitted carrier microwave signal. Demodulation of 50 Mb/s digital data from a 8.7 GHz carrier frequency has been demonstrated in [9]. Demodulating of 100 Mb/s digital data from a 14.6 GHz carrier frequency has also been demonstrated more recently [10]. Results of employing this photonic microwave receiver in a short-distance Ku-band wireless link, was also reported in [10], demonstrating the potential of using high-Q optical WGRs in microwave photonics applications.

13.2.1 Principle of Operation

The scheme of the modulator described in [8, 7] is shown in Fig. 13.1. Light is sent into a LiNbO$_3$ WGR via a coupling diamond prism. The resonator is a disk with radius a ($a = 2.4$ mm in [7]) and thickness d ($d = 150$ μm in [7]). The side-wall of the disk is polished such that the resonator becomes a segment of an oblate spheroid or a sphere. The extraordinary axis of LiNbO$_3$ coincides with the symmetry axis of the cavity. The index of refraction of the prism (it is $n_p = 2.42$ for diamond) exceeds the index of refraction of the cavity whispering gallery modes $n = 2.14$ to create an effective coupling. The WGR is placed between two electrodes of a resonator that is pumped with an external microwave source. The modulator requires resonant input light. Because adjacent WGMs differ in their azimuthal field dependence by exactly one period added to the closed circular waves, the microwave field applied to the resonator must not be uniform along the rim. A a half-wave microstrip resonator consisting of a half-circular electrode along the rim of the WGR has been used. A typical quality factor of such a cavity is $Q_M = 100$, with a bandwidth $\sim$150 MHz, which is sufficiently close to the bandwidth of the optical resonances. By tuning the length of the stripline electrodes the microwave cavity can be tuned to a frequency equal the optical free spectral range.

13.2.2 Performance

A typical dependence of demodulated microwave power at the output of the modulator on the input microwave power is presented in Fig. 13.2. The saturation point at $\sim$30 μW corresponds to the limit imposed by multiplication of the harmonics. The optimal operational power within the linear regime is estimated at 10 μW.

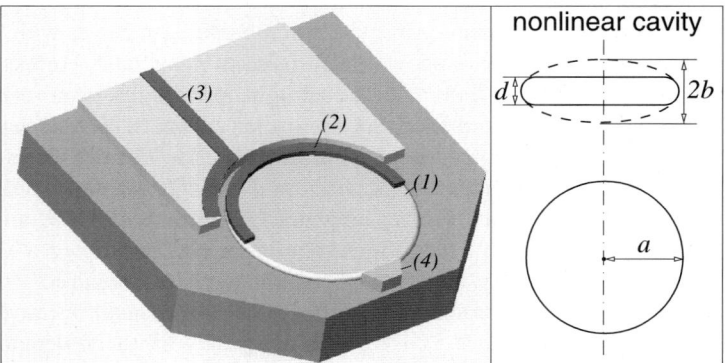

Fig. 13.1. Experimental setup of a WGM-based EOM: (1) is the LiNbO$_3$ WGM resonator, (2) is the microwave resonator, (3) is the microwave feeding stripline, and (4) the diamond coupling prism. *Inset*: geometrical characteristics of the resonator. The figure is reprinted from [7]

13.3 Tunable Filters

Crystalline LiNbO$_3$ WGM filters have high-Q and wide tunability, which makes them useful in optics and microwave photonics. A single WGM works as a first order Lorentzian filter [12]. Three properly coupled resonators form a third-order filter. A filter made with three LiNbO$_3$ WGM resonators [13] has 30 MHz linewidth and is tunable over a wider range of frequency (>20 GHz) (see Fig. 13.4). While increasing the number of coupled WGRs theoretically yields the required higher-order high-contrast filter function, the technical problems associated with the device fabrication pile up exponentially. These problems were analyzed in [28]. A demonstration of a fifth-order filter was also reported in the same work. In this section we review progress in fabrication of optical and microwave filters made of lithium niobate whispering gallery mode resonators.

13.3.1 First-Order Filter

A schematic diagram of the tunable Lorenzian filter configuration based on a disk cavity fabricated from lithium niobate wafer is shown in Fig. 13.3. A Z-cut MgO: LiNbO$_3$ disk resonator had 10 mm in diameter and 30 μm in thickness. The resonator perimeter edge was polished in the toroidal shape. We studied several nearly identical disks. The repeatable value of the quality factor of the main sequence of the resonator modes is $Q = 2 \times 10^7$ at 780 nm wavelength (the observed maximum was $Q = 5 \times 10^7$ at this wavelength), which corresponds to approximately 20 MHz bandwidth of the mode. We should note, that 780 nm laser was used to fabricate the filter/modulator suitable for spectroscopy of rubidium. The filter gives even a

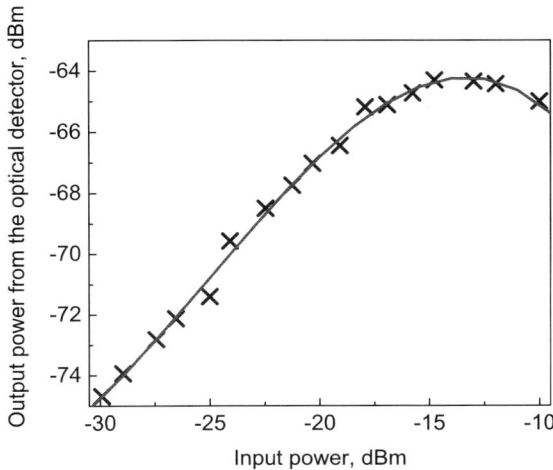

Fig. 13.2. Demodulated microwave power vs input microwave power in a WGM modulator (Fig. 13.1) operating at 5.2 GHz. Demodulated microwave power can be increased if the input optical power is increased

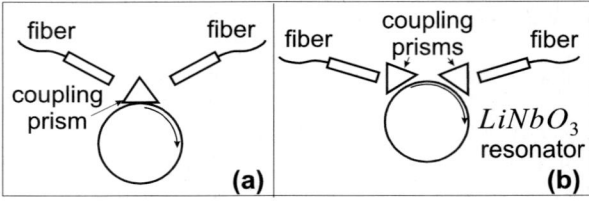

Fig. 13.3. (a) Single prism, and (b) double-prism Lorentzian filter setups

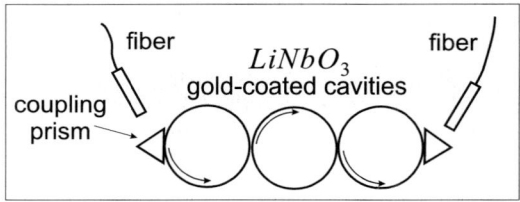

Fig. 13.4. A scheme of a third order tunable optical filter made of gold-coated lithium niobate discs. Reprinted from [13]

better performance at 1 550 nm. The typical fiber-to-fiber loss was approximately 5 dB when we used a single coupling prism (absorptive filter, Fig. 13.3(a). The repeatable value of fiber-to-fiber insertion loss was 12 dB when we used a two prism configuration (transmittive filter, Fig. 13.3(b). The minimum/maximum transmission was achieved when light was resonant with the modes in the filter shown in Fig. 13.3(a)/(b). Tuning of the filter is achieved by applying a DC voltage to the top and bottom disk surfaces, which are coated with metal. The coating is absent on the central part of the resonator perimeter edge where WGMs are localized. Experimentally measured electro-optic tuning of the filter's spectral response and tuning of center wavelength with applied voltage exhibit a linear voltage dependence in ±100 V tuning range, i.e. the total tuning span significantly exceeds the FSR of the resonator. Changing the tuning voltage from zero to 10 V shifted the spectrum of the filter by 4 GHz for the TM polarization, in agreement with the theoretically predicted value.

13.3.2 Third-Order Filter

Coupled optical fiber resonators are widely used as optical and photonic filters [29, 30]. WGRs also can be arranged in a coupled configuration. A miniature resonant electro-optically tunable Butterworth third-order filter was realized using three gold-coated WGRs fabricated from a Z-cut commercially available lithium niobate wafer. The filter, operating at the 1 550 nm wavelength, is an extension of the tunable filter design based on single lithium niobate resonator, described in the previous section. While tunable single-resonator filters are characterized by their finesse, which is equal to the ratio of the filter free spectral range and the filter bandwidth,

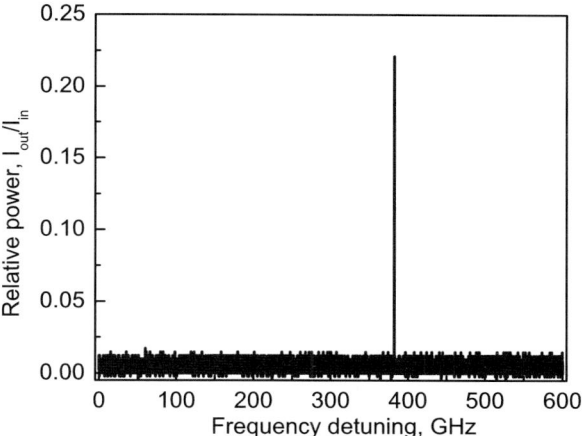

Fig. 13.5. Transmission curve of the filter consisting of three coupled WGM resonators. There is no other transmission line at the level of -20 dB in the frequency range of ± 300 GHz, at least

the three-resonator filter has a more sparse spectrum when compared with a single WGM resonator, as in the coupled fiber-ring resonators [29, 30]. The differences in the size of the WGRs is rather important for the device fabrication because it allows to rarify the overall transmission spectrum of the filter. On the other hand, the spectral widths of the interacting WGMs should be nearly identical. The obtained spectrum is so clean that the frequency separation between the filter transmission lines could not be observed with available techniques. It can only be asserted that this value exceeds 300 GHz (see Fig. 13.5). The tuning speed of the filter is approximately 10 ns, while the settling time is determined by the filter's bandwidth and does not exceed 30 ns. Figure 13.6 shows the spectrum obtained in the experiment with three gold-coated lithium niobate resonators. To highlight the filter performance the theoretical third-order Butterworth fit of the curve is also plotted. Obviously, the three-cavity filter has a much faster roll-off compared with the Lorentzian line of the same full width at half maximum. On the other hand, the filter function does not look exactly like a third order function because of small differences between each cavity's Q-factor and dimension. When three cavities are placed in series, the light amplitude transmission coefficient is

$$T = \frac{T_1 T_2 T_3}{(1 - R_1 R_2 \exp[i\psi_{12}])(1 - R_2 R_3 \exp[i\psi_{23}]) - R_1 R_3 |T_2|^2 \exp[i(\psi_{12} + \psi_{23})]}, \quad (13.1)$$

where T_j and R_j ($j = 1, 2$) are the transmission and reflection coefficients of the resonators, and ψ_{jk} is the phase shift introduced by the coupling of resonators j and k. Let us consider the case of resonators with slightly different resonance frequencies and tvalues of linewidth, and assume that $\exp(i\psi_{jk})$ are properly adjusted.

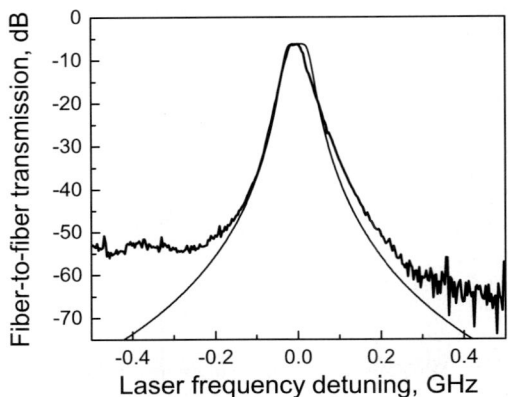

Fig. 13.6. Transmission curve of the filter and its fit with Butterworth profile function $\tilde{\gamma}^6/[(\nu - \tilde{\nu}_0)^6 + \tilde{\gamma}^6]$, where $\tilde{\gamma} = 29$ MHz, $\tilde{\nu}_0$ determines the center of the filter function and primarily depends on the resonators' geometrical dimensions. Voltages applied to the resonators vary near zero in 10 V range to properly adjust frequencies of each individual resonator and construct the collective filter function as shown. Reprinted from [13]

Under optimum conditions [31] power transmission through the system is

$$|T|^2 \simeq \frac{\tilde{\gamma}^6}{\tilde{\gamma}^6 + (\nu - \tilde{\nu}_0)^6}, \tag{13.2}$$

where $\tilde{\gamma}$ is the bandwidth of the filter, and $\tilde{\nu}_0$ is the central frequency of the filter. The transmission through the resonator system is small for any frequency when the resonant frequencies of the modes are far from each other ($|\nu_j - \tilde{\nu}_0|^2 \gg \tilde{\gamma}^2$). The transmission becomes close to unity when the mode frequencies are close to each other compared with the modes' width γ. Finally, the transmission for the off resonant tuning is inversely proportional to ν^6, not to ν^2, as for a single resonator Lorentzian filter. Those are the properties of the third order filters and they match the experimental observations. The filter tunability exhibits a linear voltage dependence in ±150 V tuning range, i.e. the total tuning span exceeds the FSR of the resonator. Changing the tuning voltage from zero to 10 V shifted the spectrum of the filter by 1.3–0.8 GHz for TM polarization, in agreement with theoretical value. Though theoretically this value does not depend on the resonator properties and is related to the fundamental limitations of optical resonator based high speed electro-optical modulators [32], the different results for different resonators measured in our experiment occur due to imperfections of the cavity metal coating as well as due to partial destruction of the coating during the polishing procedure.

13.3.3 Fifth-Order Filter

We have also realized a five-WGR filter, operating at the 1 550 nm wavelength (Fig. 13.7). The filter has approximately 10 MHz bandwidth and can be tuned in the

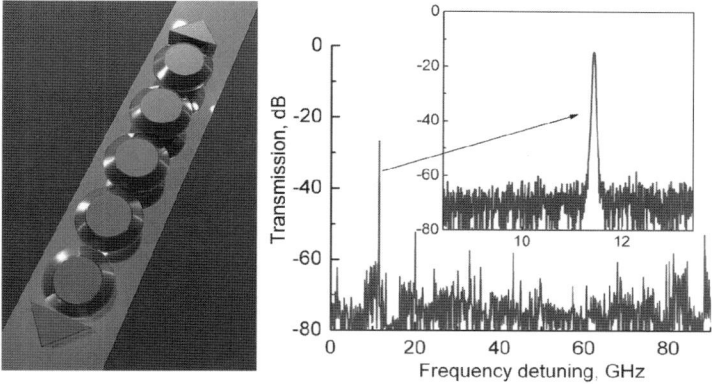

Fig. 13.7. Fabrication of the five-WGR filter. *Left*: Scheme. *Right*: Filter function. Reprinted from [28]

range of 10 GHz by applying a DC voltage of approximately 100 V to the graphite electrodes. The free spectral range of each resonator is approximately 10 GHz, and the filter may be tuned practically at any optical frequency in the transparency range of lithium niobate. We have demonstrated 12 dB of fiber-to-fiber insertion loss with this filter. The resonators were cut out from congruent Z-cut crystalline $LiNbO_3$ wafers and had approximately 4 mm in diameter and 100 μm thickness. Each resonator's rim was polished in the toroidal shape with a 50 μm curvature radius. The repeatable value of the loaded quality factor of the main sequence of the resonator modes was $Q = 3 \times 10^7$ (the unloaded maximum was $Q = 1.5 \times 10^8$). The resonators were arranged in the horizontal plane using flex manipulators. No vertical adjustment was required because the surface of the stage as well as the resonator was optically polished and all resonators were polished to have the same thickness (with 1 micron accuracy). The gaps between the resonators, and between the coupling prisms and the resonators, were less than 100 nm, which corresponds to the scale of the evanescent field. The gaps have been adjusted to maintain the appropriate system response.

13.3.4 Insertion Loss

Let us understand the nature of the insertion loss of the multiresonator filters. Using the formalism presented in [33] we describe the field accumulated in each resonator with complex transmission (T_j) and reflection (R_j) coefficients. The coupling between the resonators as well as the resonators and prism couplers is described by constants κ_j. The transmission and reflection coefficients are connected by a recurrent relation

$$\begin{bmatrix} T_{j+1} \\ R_{j+1} \end{bmatrix} = \frac{1}{i\sqrt{k_j} \exp[-(i\delta_j + \alpha_j)]} \Phi_j \begin{bmatrix} T_j \\ R_j \end{bmatrix},$$

$$\Phi_j = \begin{bmatrix} 1 & -\sqrt{1-k_j} \\ \sqrt{1-k_j}e^{-(i\delta_j+\alpha_j)} & -e^{-(i\delta_j+\alpha_j)} \end{bmatrix}, \qquad (13.3)$$

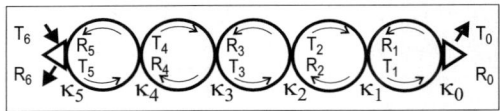

Fig. 13.8. Parameters used to describe the resonator chain. Reprinted from [28]

where δ_j and α_j represent the phase buildup and absorption per round trip in the resonator. For the sake of simplicity we assume that $\delta_j = \delta$ and $\alpha_j = 0$. Then the transfer functions for the five-resonator chain is (see Fig. 13.8)

$$H_{11} = \frac{T_0}{T_6} e^{(i\delta+\alpha)/2}, \qquad H_{21} = \frac{R_6}{T_6} e^{(i\delta+\alpha)}. \tag{13.4}$$

It is easy to find that $P_r/P_{in} = |H_{21}|^2$ and $P_{out}/P_{in} = |H_{11}|^2$. The relative power in resonator #1 is given by $P_1/P_{in} = |T_5/T_6|^2$, in resonator #2 is $P_2/P_{in} = |T_4/T_6|^2$, etc. To further simplify the problem we assume that $\kappa_0 = \kappa_5 = \kappa_a$, $\kappa_1 = \kappa_4 = \kappa_b$, and $\kappa_2 = \kappa_3 = \kappa_c$. The resonator chain becomes fifth order Butterworth filter with power transfer function

$$|H_{11}|^2 \simeq \frac{1}{1 + (2^{7/5}\delta/\kappa_a)^{10}}, \quad \text{if } \kappa_b \simeq \frac{\kappa_a^2}{8}, \ \kappa_c \simeq \frac{\kappa_a^2}{8}(1 - \kappa_a), \tag{13.5}$$

and κ_a is arbitrary. Finally, note that

$$\gamma_a = \frac{\alpha}{2\pi a}\frac{c}{n}, \qquad \gamma_c = \frac{\kappa_a}{2\pi a}\frac{c}{n}, \qquad \delta = \Delta\omega\frac{2\pi a n}{c},$$

where a is the radius of the resonators, n is the refractive index of the material. In reality $\alpha \neq 0$ and the filter is always absorptive ($|H_{11}|^2 < 1$ for $\delta = 0$). We calculate $|H_{11}(\delta = 0)|^2$ keeping the relative coupling between the resonators the same as was evaluated for the lossless case. The resultant absorption increases approximately exponentially with the increase of the relative intrinsic absorption Fig. 13.9. The resonators should be overcoupled to obtain small losses. In our experiment $\gamma_a/\gamma_c \sim 0.25$. The observed 12 dB fiber-to-fiber insertion loss stems from the absorption in the chain as well as other miscellaneous losses (reflection from the prism surface, not perfect phase matching, etc.).

13.4 WGRs Made of Periodically Poled Lithium Niobate

WGMs are attractive in nonlinear optics because of their small volumes and high quality factors [34, 35]. An efficient parametric nonlinear interaction among the modes is possible if the cavity supporting WGMs is fabricated from a low loss $\chi^{(2)}$ nonlinear material, like lithium niobate. On the other hand, the nonlinear interactions are usually strongly forbidden by the momentum conservation law (phase

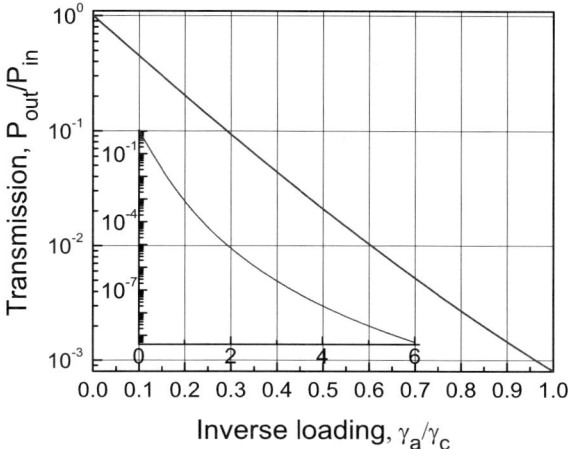

Fig. 13.9. Dependence of the transmission coefficient of the filter on the resonator loading. Reprinted from [28]

matching condition) because modes of a dielectric cavity possessing rotational symmetry are orthogonal to each other in the momentum space. An interaction could only be possible if the symmetry of the system is broken or modified. An approach for such a modification was suggested for a strongly-nondegenerate three-wave mixing process and was realized in the discussed above concerning electro-optical modulators. A resonant interaction of light confined in three optical WGMs, and a microwave field, was achieved by engineering the geometry of a microwave resonator coupled to a LiNbO$_3$ WGR. A parametric interaction among different optical WGMs was realized using a WGR fabricated from periodically poled lithium niobate (PPLN, [36]) [14]. In this section we review our achievements in fabrication of various devices involving WGRs made of PPLN.

13.4.1 Optical Frequency Doubling

Efficient frequency doubling at $\lambda = 1.55\,\mu m$ and $\lambda = 1.319\,\mu m$ was demonstrated using the same WGR made of PPLN. The resonator was doubly resonant, both at fundamental and second harmonic frequencies. Q-factor of WGMs did not depend on the poling of the resonator host material and was the same for WGMs of resonators made of congruent PPLN and congruent Z-cut lithium niobate. The basic trivial condition (energy conservation law) for the frequency doubling is

$$\lambda_p = 2\lambda_s, \tag{13.6}$$

where λ_p and λ_s are the pump and signal wavelengths in vacuum. This condition should be supplied with the momentum conservation law

$$2k_p - k_s = 0, \tag{13.7}$$

where k_p and k_s are the wave vectors of the pump and signal waves. Momentum conservation is not generally satisfied in crystals: $2k_p - k_s \neq 0$. This problem may be overcome using a quasi-phase matching technique [36], which is based on producing a modulation of the nonlinear susceptibility of the material with a period Λ such that

$$k_s - 2k_p = \frac{2\pi}{\Lambda} \rightarrow \Lambda = \frac{\lambda_s}{n_s - n_p}, \tag{13.8}$$

where n_s and n_p are indices of refraction for the signal and pump light. Spatial modulation of nonlinear susceptibility may be obtained by local flipping of the sign of the spontaneous polarization, i.e. by flipping the spins in ferroelectric domains. From the point of view of optical nonlinearity, oppositely polarized domains exhibit opposite signs of second-order susceptibility [36]. To achieve quasi-phase matching for parametric frequency doubling in a WGR, the frequency dependent dispersion of the host material of the dielectric cavity, as well as the dispersion introduced by the intrinsic modal geometry must be taken into account. Modes inside a WGR are localized close to the cavity rim. The shorter the wavelength, the closer is the mode to the cavity surface, and the longer is the mode path. This geometrical property of WGMs significantly changes the phase matching condition compared with bulk material. The frequency of the main sequence of high order TE WGMs can be estimated from

$$\frac{2\pi R}{\lambda} n(\lambda) + \sqrt{\frac{n^2(\lambda)}{n^2(\lambda) - 1}} \simeq \nu + \alpha_q \left(\frac{\nu}{2}\right)^{1/3} + \frac{3\alpha_q^2}{20}\left(\frac{2}{\nu}\right)^{1/3}, \tag{13.9}$$

where λ is the wavelength in vacuum, ν is the mode order, $n(\lambda)$ is the wavelength dependent index of refraction, R is the radius of the cavity, and α_q is the qth root of the Airy function, $\text{Ai}(-z) = 0$. The phase matching condition (13.8) should be rewritten here as

$$\nu_s - 2\nu_p = \nu_\Lambda, \tag{13.10}$$

where ν_s, ν_p, and ν_Λ are numbers of signal, pump, and the nonlinearity modulation (i.e. number of poling periods in the WGR) modes. In the experiments a WGR fabricated from a commercial flat Z-cut LiNbO$_3$ substrate was used, with TE modes corresponding to the extraordinary waves in the material. The resonator had 1.5 mm radius and 0.5 mm thickness. The curvature of the rim was approximately 1.2 mm. The lithium niobate substrate was periodically poled with $\Lambda = 14$ μm period. The poling was made in stripes (Fig. 13.10(b)). Because WGMs are localized close to the surface of the resonator, they encounter many periods of nonlinearity modulation, not just a single period as in the case of common "in stripes" poling. This provides the opportunity to achieve frequency doubling at a wide range of frequencies including those that do not correspond to the original planar poling period. We found that for resonators used in the experiment, the amplitude of the corresponding harmonics of the nonlinearity of the poled crystal was $\chi^{(2)}_{1.55} \simeq 1.6 \times 10^{-3} \chi^{(2)} \cos(\nu_{1.55}\phi)$ and $\chi^{(2)}_{1.319} \simeq 8 \times 10^{-5} \chi^{(2)} \cos(\nu_{1.319}\phi)$, where $\chi^{(2)}$ is the nonlinearity of the material. We observed doubling for both frequencies. The efficiency of the frequency

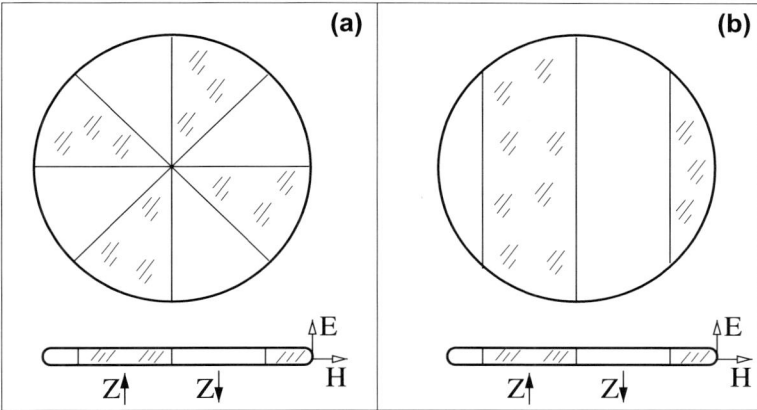

Fig. 13.10. Scheme of two possible periodical poling structures ((**a**) *"flower-like"*, and (**b**) *"striped"*) for a lithium niobate WGM resonator. Reprinted from [7]

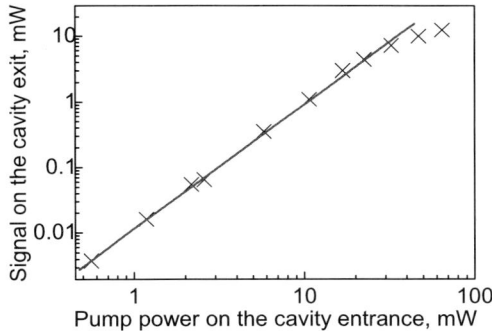

Fig. 13.11. Efficiency of the generation of the second harmonic in a WGM PPLN resonator pumped at 1.55 μm. Reprinted from [14]

doubling at $\lambda = 1.55\,\mu m$ is shown in Fig. 13.11. The absorption spectrum of the resonator measured by frequency scanning of the pump laser at 1.55 μm and the emission spectrum for the second harmonic at 775 nm are shown in Fig. 13.12. Light from a pump laser is sent into the resonator via a coupling diamond prism. The prism output is collected with a multimode fiber. Using spatial selection and a photodiode detecting the second harmonic only, we are able to filter out the residual pump light going out of the prism. The frequency of the pump was tuned in such a way that the second harmonic was also resonant. The phase matching conditions are not periodic with the FSR of the WGR. The FSR (13.6 GHz) is easily identifiable through the periodic pattern of the absorption peaks in Fig. 13.12. Only one mode for the second harmonic is efficiently coupled to the pump, within more than 80 GHz scan of the pump laser – roughly six times the free spectral range. The bandwidth of the second harmonic nonlinear resonance, in units of pump laser frequency tuning,

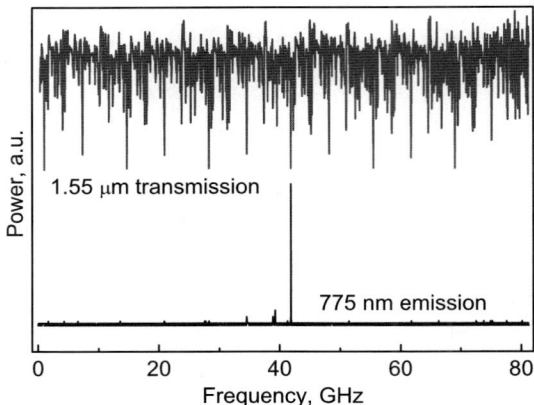

Fig. 13.12. Transmission and emission spectra of the WGR. Reprinted from [14]

is approximately 5 MHz. The maximum signal power on the prism output observed in the experiment was approximately 12.3 mW. The pump power at this point was 25 mW at the input port of the cavity. For the frequency doubling at $\lambda = 1.319\,\mu$m the maximum efficiency was 2×10^{-2} for the pump power available (30 mW). This relatively small efficiency at 1.319 μm may be explained by the small effective value of nonlinearity, or in other words, the poor efficacy of the chosen grating poling period. The conversion efficiency of the frequency doubling can be found theoretically and compared with the experiment (see, for example, [37, 38]).

13.4.2 Calligraphic Poling

Realization of an efficient parametric frequency conversion in a WGR made of a PPLN requires a specific "flower like" poling pattern (Fig. 13.10(a)). There exist a variety of techniques for poling lithium niobate, involving wet etching and photolithographic preformed masks [39–41] used to generate domain flipping. Unfortunately these techniques are not very convenient for WGRs. A technique for creating complex and arbitrary micron-scale domain patterns on ferroelectric materials, dubbed as calligraphic poling, has been developed and demonstrated [15]. A sharp 0.5 μm tungsten tip was used as a writing electrode. A bias voltage of ∼1.5 kV is applied to the electrode to pole a 100 μm thick congruent $LiNbO_3$ crystal of arbitrary shape at room temperature. The domain flipping generally occurs in one to ten seconds after the bias voltage is applied. This technique is convenient for WGR applications because it allows generating patterns on crystalline WGRs without degrading their optical quality. There are other advantages of calligraphic poling over traditional approaches. The method is flexible and allows creating an arbitrary complex domain pattern on a crystalline wafer without fabrication of an expensive mask. It is also possible to observe and manipulate the poling process in the real time. Furthermore, the method does not require extreme environment conditions.

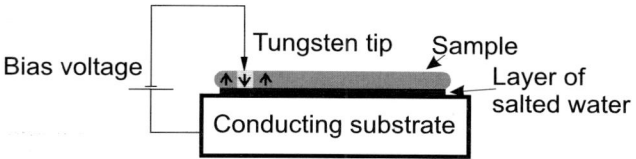

Fig. 13.13. Diagram of the calligraphic poling apparatus. The tungsten tip is free to move across the surface of a crystalline sample. When voltage is applied, domain reversal takes place locally under the position of the tip. The *arrows* represent the direction of the polarization for local regions on the crystal. To pole crystals close to their edge a hydrophobic, transparent, and insulating oil should be used. It prevents sparks from charge build up on the edge of the crystal. Reprinted from [15]

Domain reversal through the complete depth of the crystal occurs in a time span of less than one second with the bias voltages less than 800 V. The other poling techniques typically take longer time and require a higher voltage bias. With calligraphic poling (see Fig. 13.13) the entire process takes place at room temperature and at atmospheric pressure, unlike conventional techniques requiring high temperatures. Using the calligraphic poling technique the complex domain structures could be generated with two types of immediately available poling patterns: curves and hexagons. Narrow domains less than 2 μm across are achievable either as straight lines across the crystal or as smooth curves. Changes to the poling process due to alignment with the crystal axes have not been observed. Domains in the shape of hexagons (the crystal structure of $LiNbO_3$ is hexagonal) aligned along the x and y-axes of the crystal have been obtained. In other implementations of hexagon poling, the preformed mask had to be aligned to the crystal axes by the user. With calligraphic poling, the alignment is automatic. The technique is practical only for crystals that are less than 0.2 mm thick. Thicker crystals require a greater bias voltages than was available. Since the crystals are so thin, inhomogeneities in their composition play a decidedly large role in the poling process. This disadvantage, on the other hand, can be used as a non-destructive test for crystal purity.

13.4.3 Reconfigurable Filters

Calligraphic poling allows generating narrow-band, widely tunable optical filters with a reconfigurable spectrum [42]. A homogeneous electric field applied to a WGR made of an electro-optic material results in a homogeneous change of the refractive index for a single-domain crystalline resonator and, as a consequence, in a frequency shift of the whole resonator spectrum. However, various modes of a WGR have various spatial dependencies. Choosing the spatial dependence of the electro-optic coefficients in a correct manner by poling the crystal allows changing the tunability of the modes in different ways to realize a reconfigurable filter. Tunable filters generally do not change their FSRs significantly but only shift the comb of their optical modes making it overlap with itself for each frequency shift proportional to the FSR. Hence, such a filter can be tuned at any prescribed single frequency if the linear tunability exceeds the FSR. Filters possessing this property were discussed above. But some photonic applications require narrowband filters that simultaneously pass both the carrier and its sidebands. For example, this is important for generation of spectrally pure microwave signals in opto-electronic oscillators, where beating of the optical sidebands and the carrier on a fast photodiode generates microwave signals. Tunability of the microwave frequency of the oscillator requires that the frequency difference between the filter passbands changes controllably. The reconfigurable WGM filter complies with this requirement. Such reconfigurable filters have bee experimentally demonstrated [42]. Several disk-shaped resonators were fabricated from congruent lithium niobate. The resonators had 2.6 mm in diameter and 120 μm in thickness. The rims of the cylindrical disks were polished to resemble the surface of a sphere. A ferroelectric domain structure reminiscent of a set of rings concentric with the axis of the disk resonator was generated. This was done by dragging an electrode across the surface of the crystal while applying a 2.5 kV bias between the electrode and the bottom of the crystal, causing a permanent change in the structure of the material polarization. The top and bottom surfaces of the polished and poled disk resonator were placed into contact with metal electrodes. These electrodes were connected to a 0–150 V regulated DC power supply. A 1.55 μm probe laser light was coupled into the resonator through a diamond prism and scanned over 20 GHz. This allowed the observation of the absorption spectrum of the poled disk resonator and the motion of the modes as the voltage bias across the resonator axis was variede. The spacing between selected WGMs was manipulated in the linear regime with a tuning rate of approximately 21 MHz/V.

13.5 Photorefractive Damage

We have discussed some applications of WGRs made of lithium niobate. There is, however, an obvious problem associated with usage of lithium niobate as a resonator host material. The build up of the light intensity in modes with high-Q and small volume results in a significant enhancement of photorefractivity in $LiNbO_3$, which restricts the intended range of applications. The photorefractive effect leads to the so called photorefractive damage [19] that is detrimental to applications such as efficient parametric frequency conversion [43], switching [53] and modulation [80] of light. Photorefractive effects of as-grown and doped lithium niobate has been recently studied using WGM resonators [16–18]. We review those studies below. It was shown that the mode spectrum of WGRs fabricated with lithium niobate changes while the resonators interact with the infrared light. The observed modifications of the WGM spectra can be attributed to the photorefractive damage of the resonator host material because (i) the observed modification of the WGM spectrum does not disappear if the light is switched off; (ii) the changes can be removed by illuminating the resonators with UV light; (iii) the corresponding time scale of the observed effect is in the range of hours; and (iv) the observed effects are more pronounced for shorter wavelengths of light. The frequency shift of the WGM spectrum was monitored to find the photorefractive induced change of the index of refraction using the ordinarily polarized light that couples to TM WGMs. The TM mode family was selected because of its higher quality factor and lower photorefractive damage effect. The TM frequency spectrum of a spherical WGM resonator is given by

$$\omega_0 \simeq \frac{c\nu}{an_0}\left[1 + \alpha_q\left(\frac{1}{2\nu^2}\right)^{1/3} - \frac{1}{\nu n_0\sqrt{n_0^2 - 1}}\right], \qquad (13.11)$$

where n_0 is the ordinary refractive index, $\nu \gg 1$ is the angular momentum number of the WGM, a is the radius of the resonator, α_q is the qth root of the Airy function, Ai$(-z)$, which is equal to 2.338, 4.088, and 5.521 for $q = 1, 2, 3$, respectively (see [44] and Appendix). Photorefractivity results in a decrease of the index of refraction n_0 by Δn which leads to an increase of the WGM frequency by $\Delta\omega$ such that $\Delta n/n_0 \simeq \Delta\omega/\omega_0$. This relative frequency shift could be measured with a very high accuracy, better than Q^{-1}. The accuracy is given by the laser noise as well as the measurement time. The Q-factor of WGMs was measured to find the material absorption. Usually, the Q-factor is restricted because of the light absorption and scattering in the bulk and on the surface of the resonator. In WGRs, according to the experiment, the scattering and surface absorption are negligible, so the Q factor is inversely proportional to the bulk material absorption α ($Q_{max} = 2\pi n_0/(\lambda\alpha)$, where n_0 is the refractive index of the material and λ is the wavelength of the light in vacuum). Values of the change in the quality factor and the absolute and relative changes in the frequency of WGMs were both tracked. The mode motion in the frequency space results from the changing refractive index of the material due to the influence of light. To observe the mode motion two different techniques were used. In one technique the frequency of a laser was scanned across several free spectral ranges of a resonator and the light transmission through the resonator was recorded. This technique measures the change of the resonator's spectrum with time. However, since the frequency scan of the laser is much larger than the spectral width of the WGMs, the average amount of light penetrating into the resonator is quite small in this case and does not lead to a significant photorefractivity. To improve this technique and observe the effect of photorefractivity on the modal structures a pump/probe scheme was used with two lasers having nearly the same wavelength. One beam (the "pump") has a high power and the other one (the "probe") has a low power. The power of the probe was set so that the probe light itself did not introduce any observable modification of the material properties during the time of the observation. The probe was slowly scanned through a frequency span equal to about two FSRs of the resonator. The pump was manually tuned to match the frequency of the mode family under study, and was frequency-modulated at a much higher speed than the scan frequency of the probe. The modulation frequency span of the pump was as wide as several spectral widths of the WGMs. The pump and the probe were sent to the resonator using the same coupler and were both brought to the same spot on the photodiode. They could, however, be distinguished because of their different modulation frequencies; the pump light resulted in a constant background shift of the observed probe spectrum. Because of the high speed of modulation of the pump frequency, all fluctuations resulting from the modulation of the pump were averaged out.

13.5.1 Congruent LiNbO$_3$

The WGRs used in this study are identical to the resonators used in modulators and filters described above. The resonators were fabricated out of as-grown congruent Z-cut crystalline cylindrical preforms, mechanically cut from commercially available crystalline wafers. The WGRs had a disk shape with the radius of several millimeters and a thickness of a hundred microns. The side-walls of the preforms were polished in such a way to make the resonators a part of an oblate spheroid in the vicinity of their rim. Light from a pigtailed laser was sent into the resonator via a coupling diamond prism. The typical input optical power at the entrance of the coupling prism varied from 0.5 mW to 10 mW, and the coupling efficiency of light from the prism to the resonator was better than 40%. External cavity diode lasers with wavelengths tunable around 780 nm were used in the majority of experiments with the congruent material. The quality factors of the resonators were typically larger than 10^7, as was experimentally found from $Q = \omega/\Delta\omega$, where the observed full width at the half maximum (FWHM) $\Delta\omega$ of the resonance curves was less than $2\pi \times 40$ MHz. A time-dependent decrease of the Q-factor when the resonator was illuminated with 780 nm light was observed [18]. The largest detected change was approximately five times as large as the initial value. The Q-factor did not increase when the light was switched off. The quality factors were measured by detecting the FWHM of the selected mode subjected to cw laser radiation. The value of FWHM of the modes that had the strongest interaction with light was 17 MHz at the start of the experiment. This value changed to 82 MHz after a four-hour illumination of the resonator with the pump light. The modes that were not exposed to the pump light and were weakly coupled to the probe light nearly conserved their values of FWHM. For instance, modes with approximately 20 MHz FWHM at the start of the experiment finally ended up with 25 MHz FWHM. It was unclear, though, if the same weakly coupled modes were observed, because the WGM spectrum was modified significantly. The Q-factor obtained by measuring the FWHM of the modes could include a systematic error resulting from the inability to distinguish between inhomogeneous and homogeneous frequency broadening of the WGMs. For instance, the method is not valid if the spectrum of the resonator is very dense [45]. Hence, there are two possible reasons for the increase of FWHM of the modes. One is obviously related to an increase of the light absorption in the material. The other is related to the increase of interaction and/or overlap between resonator modes that occur because of photorefractivity. The second reason is more probable because the observed increase of the FWHM did not lead to a change in the coupling efficiency. A follow up detailed study of the effect involving ring-down measurements of the Q-factor [45] or measurement with resonators having a clean spectrum [46] is required. "Beam fanning" was not observed in the WGRs despite the fact that the intensity of circulating IR light in the modes was very high, and light induced modification of the refractive index was significant. The probable reasons could be the

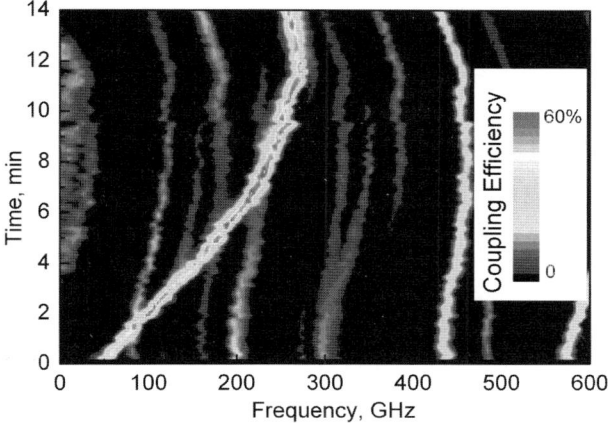

Fig. 13.14. A typical frequency map of the mode motion in the spectrum of the WGM resonator made of congruent as-grown lithium niobate. The coupling efficiency of the modes to the external 780 nm light source are shown by the color. *Red* corresponds to the maximum coupling (60%), *blue* – to the minimum. Reprinted from [18]

comparably small modal cross sections (beam fanning decreases with decrease of the cross section of the light beam [47]) and the specific symmetry of the system. Because the beam fanning is suppressed in these resonators, they are free from systematic errors present in the Z-scan technique [48], making them well suited for the study of photorefractivity. The most interesting effect observed in the congruent lithium niobate resonators is the relative motion of the modes in the frequency space (the motion of the resonator modes as a whole was also present). A fragment of a typical frequency map of the motion of the modes is shown in Fig. 13.14. The red color in Fig. 13.14 corresponds to a more efficient coupling; the maximum coupling efficiency was 60%. In this experiment a 780 nm laser was scanned across a two-gigahertz frequency span. The laser light had 8 mW of power. Using the value of the smallest observed FWHM of the WGM (20 MHz) that at the same time had the most efficient light coupling, the value of the time-averaged power of light coupled into the resonator mode was estimated to be 48 μW. To estimate the value of the light intensity in the mode we noted that mode localization area was of the order of 100 μm². Since the initial finesse of the resonator was equal to 220, the circulating intensity of the pumping light was approximately 10 kW/cm². The photorefractivity of congruent $LiNbO_3$ was also studied at 1550 nm using a single laser as well as a much smaller resonator [17]. The relative mode motion was observed as well, though it was slower than the relative mode motion at 780 nm (Fig. 13.15). No decrease of the Q-factors of the WGMs was observed. The laser was scanned in a 1.2 GHz frequency range, and its power was 18 mW; the maximum coupling efficiency was 65%. The best quality factor of the WGMs was $Q = 2 \times 10^8$ (approximately 1 MHz FWHM). The time averaged optical power coupled to the mode was ∼10 μW. The resonator had a 1.2 mm diameter and was approximately 5 μm

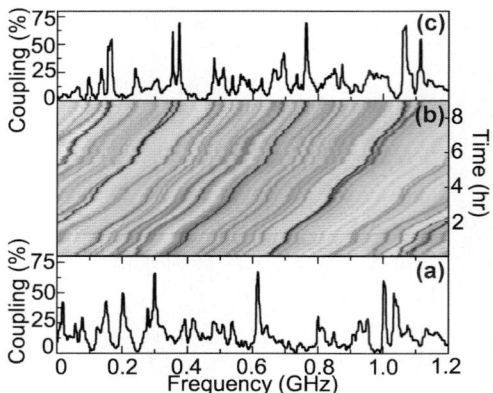

Fig. 13.15. (a) A segment of the initial spectrum of the WGM resonator at 1 550 nm. (b) A time/frequency map illustrating relative mode motion. (c) The segment of the spectrum after 8 hours of illumination with 1 550 nm light. Reprinted from [17]

in thickness in the geometrical area where the modes were localized. Hence, the intensity of light in the mode was approximately $9\,\mathrm{GW/m^2}$.

13.5.2 Magnesium Doped Congruent LiNbO$_3$

The strong photorefractive effect of a nominally pure lithium niobate crystal can be suppressed by doping the material with magnesium [49, 50]. It is important to determine the extent to which photorefractivity can be suppressed in order to understand the boundaries of the applicability of the material. The photorefractivity of highly Mg-doped LiNbO$_3$ in the infrared (780 nm) was studied with the WGM technique [16]. A single laser was used in the experiment. The laser power was 9 mW, and the total frequency span of the laser frequency scan was 9 GHz. Light from the laser was sent through a fiber pigtail into a disk-shaped optical WGR via a coupling diamond prism. The coupling efficiency was 75%. The resonator was fabricated from a commercially available flat Z-cut Mg:LiNbO$_3$ substrate and was temperature stabilized. As-grown congruent lithium niobate crystal doped with 5 mol % MgO was used. The TM mode family was studied because it has a lower absorption in the lithium niobate resonators. The free spectral range of our resonator was 3.42 GHz at 780 nm, and the smallest FWHM of a mode was approximately 20 MHz, corresponding to $Q_{\mathrm{max}} = 2 \times 10^7$. Therefore, the absorption coefficient of the material was less than $\alpha = 9 \times 10^{-3}\,\mathrm{cm}^{-1}$ (calculated for $n_0 = 2.25$). The loaded FWHM was approximately 50 MHz, and the value of the loaded finesse was 70. The length scale over which the optical intensity varies (the diameter of the crossection of the WGM channel) was approximately 30 μm. The average power coupled to the mode was 38 μW, and the maximum intensity in the mode was approximately $3\,\mathrm{GW/m^2}$. The laser light bandwidth was less than 300 kHz. The result of the measurement is shown in Fig. 13.16. The frequency change over the observation time exceeds

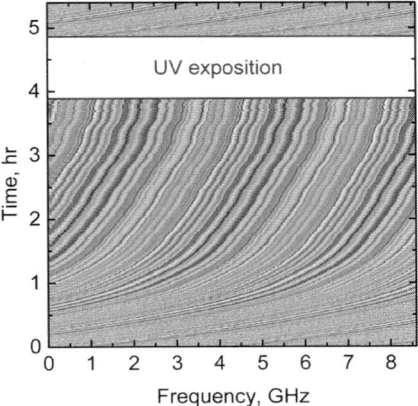

Fig. 13.16. A frequency map describing the motion of the WGM spectrum of a magnesium doped LiNbO$_3$ resonator. The plot density corresponds to the coupling efficiency of light to the mode. The *darker* the color, the stronger is the coupling. The resonator was exposed to ultraviolet light for an hour after the light induced mode motion was nearly saturated. The mode motion started again with the initial speed after the UV light illumination. The temperature of the resonator was actively stabilized. Reprinted from [16]

14 GHz. This corresponds to a change of the ordinary refractive index of the material by about 8×10^{-5}. The initial speed of the drift of the spectrum was 3.1 MHz/s, which corresponds to $\Delta(dn/dt)_{t=0}/(In_0) \simeq 1.3 \times 10^{-17}$ m^2/sW, where I is the intensity of the light. This is the feature of the photovoltaic effect. The charge diffusion effect is small because of the temperature stabilization of the resonator. Then light from a powerful UV lamp was sent to the resonator for one hour to examine the influences of UV light on the infrared light induced refractive index change. The laser was switched off during the UV exposure, and the measurement of the spectrum was continued after that. The change of the refractive index induced by infrared light was eliminated after UV exposure, and the mode motion regained the same speed as it had initially. Modification of the WGM spectrum with time was observed as well. The spectrum changes after the exposure to 780 nm light. A more careful observation of the mode spectrum shows that there is a relative motion of the WGMs, similarly to the mode motion in the undoped material. Subsequent illumination of the resonator with the UV lamp restores the spectrum completely. On the other hand, the modified spectrum was conserved in the dark. Clearly, the change in the spectrum results from photorefraction. It was also found that the absorption of the as-grown material is approximately three times less than that of the doped lithium niobate. The mode motion in the resonators made from nominally pure material is also different: WGMs of the undoped resonator move with significantly different speeds depending on the light power confined in the modes. This results in a significant relative mode motion after the power reaches a specific threshold value. It was predicted that the infrared photorefraction of pure LiNbO$_3$ originates form the interaction of light with deep and shallow electron traps [16]. The deep traps

generally include iron ions (Fe^{2+}) and bipolarons $Nb_{Li}^{4+}Nb_{Nb}^{4+}$, while small polarons (Nb_{Li}^{4+}) are the shallow traps [59, 77] (see Sect. 13.6 below for details). The presence of the two types of traps results in the superlinear behavior of the change of the refractive index with respect to the intensity of the infrared light [23]. The WGMs have different drift rates because of the superlinear photorefractivity, the incomplete mode overlap, and the different coupling efficiencies to the modes. Doping the crystals with Mg above 5 mol % threshold for congruent lithium niobate [50] eliminates the Nb_{Li} antisites forbidding the formation of the small polarons as well as bipolarons. In such a material weakly trapped polarons Nb_{Nb}^{4+} become possible. Those polarons have broadband optical absorption at 1.3 μm [63]. We expect, that those polarons are responsible for the observed effects. Magnesium doping also changes the structure of the deep iron traps shifting Fe^{3+} closer to the conduction band and making them ineffective as deep traps [84]. As a result, a significant superlinear behavior of the infrared photorefractivity was not observed in the doped samples. All WGMs experienced almost the same frequency shift. The residual relative mode motion can be explained by the presence of some number of uncompensated small polarons in the crystal. An interesting feature is that the change of the refractive index is very large (8×10^{-5}) in the magnesium doped sample illuminated with infrared light. As a comparison, the refractive index change of up to 2.8×10^{-4} was detected in iron-doped lithium niobate crystals interacting with 760 nm ordinarily polarized light [81]. An experimental study using green light was performed with nominally pure congruent as well as stoichiometric lithium niobate, which indicated an increase of the photorefractivity with the light intensity and the index change on the order of 10^{-3} [82]. Finally, it is easy to estimate the maximum change of the refractive index that could result from the electro-optic effect in the material:

$$\Delta n = \frac{n_0^3}{2} r_{13} E_Z, \qquad (13.12)$$

where $r_{13} = 10\,pm/V$ is the electro-optic constant, and E_Z is the amplitude of the electric field applied along the cavity axis. The domain reversal in a congruent $LiNbO_3$ crystal occurs at $E_Z \simeq 20\,kV/mm$ corresponding to refractive index change $\Delta n_{max} \simeq 1.1 \times 10^{-3}$, which is on the order of the observed shifts in the nominally pure as well as iron doped materials. The high doping of the material with magnesium reduces this index change by an order of magnitude only, which is rather unexpected. This means that the material will experience a strong photorefractive damage in the infrared, and points out the need for a theory that is able to give a quantitative description of the material properties.

13.5.3 Crossings and Anticrossings of the Modes

Ideally, the spectrum of a WGM resonator should be clean enough to allow a high accuracy measurement of the light induced change of the refractive index. Any interaction between the modes could change the measurement result and introduce an error. For example, the experiment described above was repeated for a resonator that has three close modes coupled to the 780 nm pump light almost with the same efficiency. During the measurement the pump laser was scanned across those three modes. The interaction between the modes resulted in 1% decrease (not increase) of the free spectral range for the modes. Generally, the photorefractive effect results in a decrease of the index of refraction and an increase of the free spectral range of the resonator. The opposite of this is observed because of the interaction between the WGMs. Some WGMs experience crossings, and other experience anticrossings with other modes due to the interaction effects. The coupling of light to some WGMs also changes with time. The mode crossing and anticrossing can be understood as follows. The frequency difference between the modes of different families is determined by their spatial overlap. If the modes have a small overlap they do not interact, even if they have the same frequency. In this case one observes mode crossing. On the other hand, if the overlapping modes begin to strongly interact at some point, then the frequency of the mode coupled to the pump laser may be shifted toward another mode. In this case one observes mode anticrossing.

13.5.4 Holographic Engineering of the WGM Spectra

To better illustrate the relative mode motion the following measurements were performed. Using the pump/probe technique described above, the drive laser was tuned to a mode that had a strong coupling with the light. Other conditions of the experiment were the same as indicated for the previous experiments. With this configuration, the selected mode began to move. It was able to move into the vicinity of another mode of approximately the same contrast. We then left it there by switching off the drive laser (Fig. 13.17). As a result, the spectrum of the resonator changed. The relative mode motion can be understood as a consequence of the superlinear decrease of the refractive index in the spatial area where the selected mode is localized. Because the modes do not entirely overlap in space, it is possible to have a mode family propagating in a channel where the refractive index is different from the refractive index of the nearby area. This also offers the possibility of engineering the resonator spectrum. As the experiments indicate, one can modify the refractive index of one mode family until the frequency of the family shifts close to the frequency of another mode family.

13.6 Infrared Transparency and Photorefractivity of Lithium Niobate Crystals: Theory

The question regarding processes that set the minimum level of light absorption in transparent optical crystals is fundamentally important. An ideal crystalline lattice has no inhomogeneities to cause Rayleigh scattering, which restricts transparency in amorphous materials. The blue- and red-wing fundamental absorption bands have a negligible effect on the absorption of light with wavelength that falls in the transparency window of the crystal. It was shown recently that Raman scattering introduces some tiny but measurable absorption in the majority of crystals [51]. The

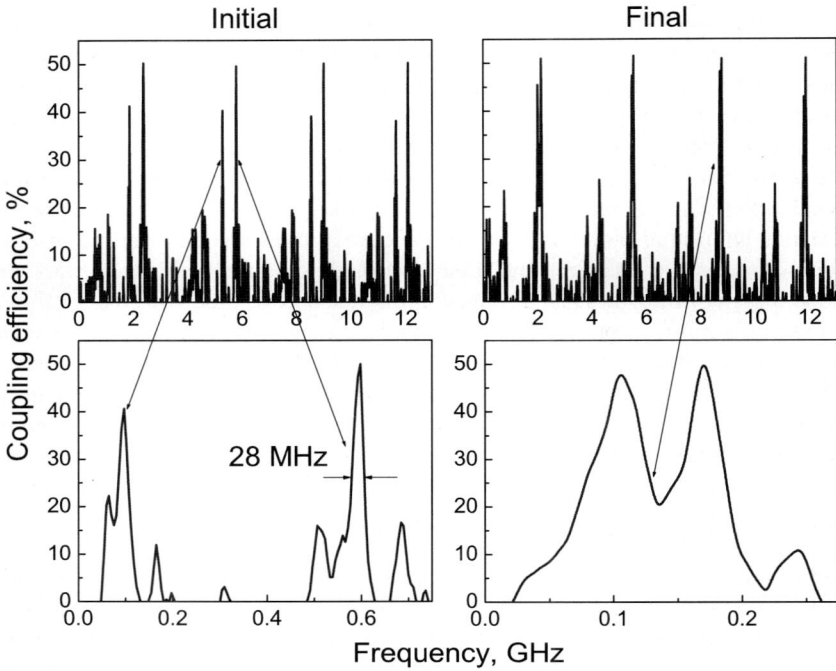

Fig. 13.17. An example of the engineering of the spectrum of a LiNbO$_3$ WGM resonator. The pump light is used to tune the eigenfrequency of one mode family toward the eigenfrequency of the other mode family. The pump light is switched off after the modes were close to each other. The initial and final spectra of the WGM resonator are shown in the plot. Reprinted from [18]

only significant reason for the absorption of light in real crystals is the presence of impurities. The goal of this contribution is to characterize the transparency window, and determine the maximum transparency of realistic as-grown lithium niobate (LiNbO$_3$) crystals. We are interested in optical properties of pure LiNbO$_3$ characterized with an ultra-low absorption in the infrared. The study of LiNbO$_3$ parameters in the infrared is particularly important because many applications of the crystal, especially those related to optical communications, are designed for this wavelength region. Even "pure" crystals always contain some defects, and the concentration of their intrinsic defects varies from several molar percent to tens of ppm (parts per million), depending on the level of stoichiometry; so they are usually called "nominally-pure". The concentration of extrinsic impurities of nominally-pure crystals rarely falls below 1 ppm [52]. Even such small concentrations significantly modify the optical properties, limiting the transparency as well as enhancing photorefractivity in these crystals. Photorefractivity is generally studied with the visible light, so a goal of the present contribution is to attract attention to the infrared photorefractivity occurring in the transparency window of lithium niobate. We show that this effect is not negligible, as is generally believed. The increase in the intensity of the infrared light, which usually results when the size of optical elements are significantly reduced, also results in the manifestation of infrared photorefractivity and the subsequent "aging" of optical elements [53, 54]. Photorefractivity can be treated as a property characterizing nonideality of electro-optic crystals. This is unlike other fundamental second-order effects such as optical rectification [20] and optical parametric frequency conversion [21, 22], which may be observed in crystals with ideal crystalline lattices. Photorefractivity requires the presence of acceptor impurities serving as electron traps in the forbidden energy gap of materials, which guarantee space charge distribution in the crystals. The space charge generates electric fields that modify the optical properties of the crystal via the linear electro-optic effect. Photorefractivity is also possible in the absence of donor impurities. The electron traps could be filled with direct ionization of electrons from the valence to the conduction band, with a subsequent decay into the traps. However, such a process is very weak for the visible and infrared light. Real crystals contain many donors that drastically change their optical properties. The concentration of these donor impurities can be varied by thermal annealing [55, 56]. Oxidization of lithium niobate also appreciably lowers its Fermi level, resulting in the removal of electrons from essentially all donors. This decreases the dark conductivity, and makes crystals more transparent. However, even such a treatment is unable to eliminate the donors completely. Our theoretical study is motivated by the lack of a complete experimental study of transparency and photorefractivity of $LiNbO_3$ in the infrared, a region of the spectrum where the crystals are generally treated as "transparent". This is partly because photorefractivity in the infrared is a slow process and its detection poses stringent requirements on the sensitivity of the experimental technique. Existing methods, including the holographic technique [23] and the free-beam approach (Z-scan technique) [57], are not very convenient for such a study. The first method generally does not deal with pure as-grown crystals because of a lack of sensitivity; however it allows spectroscopy of the impurities responsible for the photorefractivity in pretreated materials. The second method handles pure materials, but describes integrated effects, without a proper spectral selectivity. As a result, to the best of our knowledge, photorefractivity has not been previously detected for wavelengths exceeding 1.3 µm using conventional methods. The studied that involve WGRs as discussed above did result in the observation of some photorefractivity at 1.55 µm [17]. However, the those studies do not produce a coplete picture to give numerical explanation of the effect. Using the outcome of the existing experimental studies (see Appendix for a review) together with Kukhtarev's rate equations [58] we have approximated the influence of various impurities on the properties of the material. Namely, we find the concentration of iron, small polarons, and bipolarons in as-grown material, to approximate the spectral dependencies in the infrared, and predict photorefractive properties, as well as the minimum absorption of the material. We show that although light-induced change of the refractive index is slow in the infrared, the ultimate value of the change is approximately the same as that of a change induced by green light. We point out the possibility of an "infrared induced infrared light absorption", similar to the observed green light induced infrared absorption. It is believed that bipolarons contribute significantly to the photorefractive properties of lithium niobate [24]. However, to the best of our knowledge, there is no detailed description of the properties of bipolarons in this material. The concentration, as well as the absorption cross section of the defects, has not been previously studied. Also, the analytical expression for the spectral dependence of light absorption by the impurities has not yet been derived. Finally, the corresponding Kukhtarev's-type rate equation has not been worked out. We address all these issues below. This part of the chapter is organized as follows. A closed set of rate equations describing impurities (bipolarons, small polarons, and iron) of the lithium niobate crystals is presented in Sect. 13.6.1. The solution of the set of equations is discussed in Sect. 13.6.2. The dependence of the optical transparency of the crystal on the wavelength is presented in Sect. 13.6.3. Finally, a short review of the investigations of the optical properties of lithium niobate is given in Appendices.

13.6.1 Rate Equations

In this section we discuss a microscopic model for the interaction of the infrared light with lithium niobate, and obtain analytical expressions describing photorefractive and absorptive properties of this material. The formalism involved is based on Kukhtarev's rate equations [58]. The novel features of the equations we analyze include an explicit form of the rate equation for the bipolarons, as well as an analytical description of the population exchange between the small polarons and bipolarons. We show that a small concentration of iron results in much stronger photorefractive effects than was previously believed. In previous studies the role of iron in as-grown LiNbO$_3$ was underestimated because of several unjustified assumptions made while solving the rate equations. For instance, we show that it is not correct to neglect the recombination and thermal excitation rates for small polarons in the as-grown media with small concentration of iron. We utilize the model shown in Fig. 13.18(a) [59–61]. This includes shallow (X) and deep ($Y_{1,2}$) traps for electrons as well as conduction and valence bands. Even though we study nominally-pure materials, the model takes into account the direct exchange of charge carriers between the impurities. This exchange is important because the concentration of $Nb_{Li}^{4+/5+}$ defects is generally very large (about 1 mol % in a congruent material). Therefore, the localized electrons could directly migrate from a small polaron Nb_{Li}^{4+} (X) to an iron trap (Y_2) and vice versa. The same is true for the exchange between the small polarons and the bipolarons (Y_1). Moreover, following [62], we assume that such an exchange is inevitable because an ionized bipolaron becomes a small polaron. The excitation and recombination of electrons is described by the rate equations

$$\frac{\partial}{\partial t} N_{Y_1}^- = -(\beta_{Y_1} + I q_{Y_1} S_{Y_1}) N_{Y_1}^- + \Gamma_{Y_1 X} (N_X^-)^2, \tag{13.13}$$

$$\frac{\partial}{\partial t} N_{Y_2}^- = -(\beta_{Y_2} + I q_{Y_2} S_{Y_2} + I q_{Y_2 X} S_{Y_2 X} N_X^0) N_{Y_2}^-$$
$$+ (\Gamma_{Y_2} N_e + \Gamma_{Y_2 X} N_X^-) N_{Y_2}^0, \tag{13.14}$$

$$\frac{\partial}{\partial t} N_X^- = -\left(\beta_X + \Gamma_{XY_1} N_X^- + \Gamma_{XY_2} N_{Y_2}^0 + I q_X S_X\right) N_X^-$$
$$+ \left(\Gamma_X N_e + I q_{Y_2 X} S_{Y_2 X} N_{Y_2}^-\right) N_X^0, \qquad (13.15)$$

where N_e is the density of free electrons in the conduction band, N_X^- is the concentration of small polarons Nb_{Li}^{4+}, N_X^0 is the concentration of shallow traps Nb_{Li}^{5+}, $N_{Y_1}^-$ is the concentration of bipolarons, $N_{Y_2}^-$ is the concentration of Fe^{2+}, $N_{Y_2}^0$ is the concentration of Fe^{3+}, $\beta_{Y_{1,2}}$ and β_X are the rates of thermal excitation of the electrons from the traps into the conduction band, I is the intensity of the infrared light, $q_X S_X$, $q_{Y_{1,2}} S_{Y_{1,2}}$, and $q_{Y_2 X} S_{Y_2 X}$ are the absorption cross sections of the photons by the traps leading to the trap ionization, q_X, $q_{Y_{1,2}}$, and $q_{Y_2 X}$ are the quantum efficiencies of the absorption events, Γ_X and Γ_{Y_2} are the conduction band electron recombination coefficients to the corresponding traps, $\Gamma_{Y_{1,2} X}$ and $\Gamma_{X Y_{1,2}}$ are the coefficients of the electron recombination between the deep and shallow traps. Equation (13.13) was derived using results presented in [62], where it is stated that the recombination rate of the bipolaron concentration is proportional to the square of the concentration of small polarons. The total number of impurities of each kind is seriately conserved, i.e.

$$N_{Y_2}^- + N_{Y_2}^0 = N_{Y_2}, \qquad (13.16)$$
$$N_{Y_1}^- + N_X^- + N_X^0 = N_X, \qquad (13.17)$$

where N_{Y_2} and N_X are constants. The number of bipolarons is given by the number of small polarons. We assume that the ionization of a bipolaron generates a small polaron. Equations (13.13) and (13.15) are to be supplied with the continuity equation.

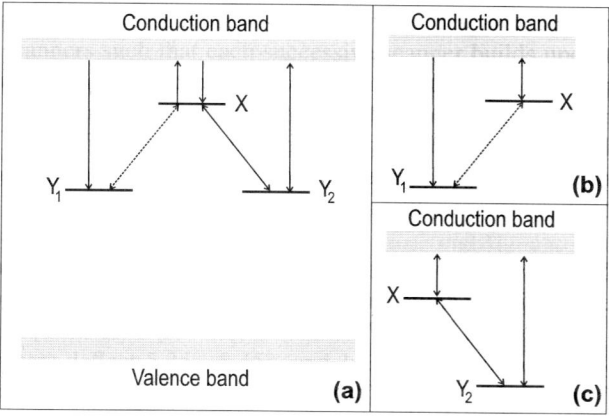

Fig. 13.18. Charge transport model for as-grown lithium niobate and lithium tantalate. (**a**) The model takes into account bipolaron (Y_1), iron (Y_2), and small polaron (X) traps. (**b**) The model involves bipolarons and small polarons only. (**c**) The model involves iron and small polarons

Assuming that the charge transport is associated primarily with the photovoltaic effect, we write a single dimension equation where the direction of the Z axis coincide with the direction of the crystalline axis

$$\frac{\partial}{\partial z} j = e \left(\frac{\partial}{\partial t} N_X^- + 2 \frac{\partial}{\partial t} N_{Y_1}^- + \frac{\partial}{\partial t} N_{Y_2}^- + \frac{\partial}{\partial t} N_e \right); \quad (13.18)$$

j is the current density defined as

$$j = e\mu N_e E + I \left(\kappa_{Y_1} N_{Y_1}^- + \kappa_{Y_2} N_{Y_2}^- + \kappa_X N_X^- \right), \quad (13.19)$$

where μ is the electron mobility in the conduction band, $\kappa_{Y_{1,2}}$ and κ_X are the bulk photovoltaic coefficients for excitation of electrons from the corresponding donor center to the conduction band, E is the space charge field that can be found from the Gauss' law

$$\frac{\partial}{\partial z} \epsilon \epsilon_0 E = -e \left(N_e + 2 N_{Y_1}^- + N_{Y_2}^- + N_X^- - N_C \right), \quad (13.20)$$

where ϵ is the dielectric coefficient of the material in the z direction, ϵ_0 is the electric permittivity of the vacuum, N_C is the immobile positive charge that compensates the charge carriers initially trapped by the photorefractive centers and maintains the overall charge neutrality of the sample. The self-induced absorption change of the light is determined by [59]

$$\alpha - \alpha_0 = \hbar \omega \left[S_X N_X^- + S_{Y_1} N_{Y_1}^- + \left(S_{Y_2} + S_{Y_2 X} N_X^0 \right) N_{Y_2}^- \right], \quad (13.21)$$

where ω is the frequency of light, and α_0 is the nonphotorefractive part of the absorption. Equation (13.21) shows that the absorption of light results from the ionization of the donor impurities or charge transfer from an iron ion Fe^{2+} to a small polaron. We assume that this kind of a charge transfer from a bipolaron to a small polaron is forbidden because the ionization of a bipolaron generates small polaron (see (13.17)). We will use (13.21) to find the minimal absorption of realistic lithium niobate crystals. The change of the extraordinary (n_e) and ordinary (n_o) indices of refraction are

$$\Delta n_e = -\frac{1}{2} n_e^3 r_{33} E, \quad (13.22)$$

$$\Delta n_o = -\frac{1}{2} n_o^3 r_{13} E, \quad (13.23)$$

where r_{33} and r_{13} are the electro-optic constants of the material. It is useful to find the rate of change of the refractive index at $t = 0$ under short-circuit conditions (no homogeneous fields in the sample). This value characterizes the efficiency of the photorefractive process. To do it, we use (13.18) and (13.20) and derive

$$\epsilon \epsilon_0 \frac{\partial E}{\partial t} = -j. \quad (13.24)$$

This is in fact the z-component of the Maxwell's equation describing Ampere's law for $\nabla \times \mathbf{H}|_z = 0$. Combining this equation and (13.23) we find

$$\epsilon\epsilon_0 \frac{\partial}{\partial t}\Delta n_o = -e\mu N_e \Delta n_o + \frac{1}{2}n_o^3 r_{13} I\left(\kappa_{Y_1} N_{Y_1}^- + \kappa_{Y_2} N_{Y_2}^- + \kappa_X N_X^-\right). \quad (13.25)$$

As follows from (13.25) the characteristic time constant of the photorefractive process is $\epsilon\epsilon_0/e\mu N_e$. The initial change of the index of refraction resulting from photorefractivity is

$$\left.\frac{\partial}{\partial t}\Delta n_o\right|_{t=0} \simeq \frac{n_o^3 r_{13}}{2\epsilon\epsilon_0} I\left(\kappa_{Y_1} N_{Y_1}^- + \kappa_{Y_2} N_{Y_2}^- + \kappa_X N_X^-\right), \quad (13.26)$$

where it is assumed that the population exchange in the deep traps is much faster than the photorefractive process itself.

13.6.2 Solution of the Rate Equations

To simplify the solution of (13.13–13.15) we consider two separate models. One model involves bipolarons and small polarons (Fig. 13.18(b)), and the other iron traps and small polarons (Fig. 13.18(c)). It is worth emphasizing that there has been no complete theoretical study of the first case, though it is generally assumed that the bipolarons play a significant role in the photorefractive properties of the material. In contrast, the photorefractivity due to the iron/small polaron interaction has been widely discussed (see e.g. [59–61]).

Bi- and Small Polaron Induced Photorefractivity

In this section we estimate the maximum change of the index of refraction of the material assuming that the influence of the iron traps (Y_2) is insignificant. It is generally accepted that the adiabatic approximation is satisfied with respect to the population of the conduction band [61], i.e. $\partial N_e/\partial t = 0$. Assuming that $\partial j/\partial z = 0$ and using (13.13), (13.15), and (13.18) we get

$$\frac{\partial}{\partial t} N_X^- = -2\Gamma_{Y_1 X}(N_X^-)^2 + (\beta_{Y_1} + I q_{Y_1} S_{Y_1})(N_C - N_X^-), \quad (13.27)$$

$$N_{Y_1}^- = \frac{1}{2}(N_C - N_X^-), \quad (13.28)$$

$$N_e \simeq \frac{(\beta_{Y_1} + I q_{Y_1} S_{Y_1}) N_{Y_1}^- + (\beta_X + I q_X S_X) N_X^-}{\Gamma_X N_X}, \quad (13.29)$$

where we used inequality $N_X \gg N_{Y_1}^- + N_X^-$. The solution of (13.27) is

$$N_X^- = \tilde{N}_X^- - \frac{(\tilde{N}_X^- - \bar{N}_X^-)\xi \exp(-\xi t)}{\xi - 2\Gamma_{Y_1 X}(\tilde{N}_X^- - \bar{N}_X^-)[1 - \exp(-\xi t)]}, \quad (13.30)$$

where

$$\xi = \beta_{Y_1} + I q_{Y_1} S_{Y_1} + 4\tilde{N}_X^- \Gamma_{Y_1 X}, \tag{13.31}$$

$$\tilde{N}_X^- = \frac{\beta_{Y_1} + I q_{Y_1} S_{Y_1}}{4\Gamma_{Y_1 X}} \left[\sqrt{1 + \frac{8\Gamma_{Y_1 X} N_C}{\beta_{Y_1} + I q_{Y_1} S_{Y_1}}} - 1 \right]. \tag{13.32}$$

For reasonably small light intensity $N_C \gg N_X^-$ and

$$\tilde{N}_X^- \simeq \sqrt{\frac{I q_{Y_1} S_{Y_1}}{2\Gamma_{Y_1 X}} N_C}. \tag{13.33}$$

Then, as follows from (13.18), (13.20), and (13.23), the steady state solution for the change of ordinary index of refraction, is

$$\Delta n_o = \frac{1}{2} n_o^3 r_{13} \frac{\Gamma_X N_X}{e\mu} \frac{I(\kappa_{Y_1} N_{Y_1}^- + \kappa_X N_X^-)}{(\beta_{Y_1} + I q_{Y_1} S_{Y_1}) N_{Y_1}^- + (\beta_X + I q_X S_X) N_X^-}. \tag{13.34}$$

Iron and Small Polaron Induced Photorefractivity

In this section we find the change of the refractive index resulting from small polarons and iron traps, assuming that the influence of bipolarons (Y_1) is insignificant for the as-grown nominally-pure materials. This problem was studied in detail in [59, 60] for a high concentration of iron impurities, where it is possible to assume that $\Gamma_X = 0$ and $\beta_X = 0$. This assumption does not hold for a low iron concentration. A simple scaling of the equations for the case of a high iron concentration applied to the case of low iron concentration leads to the conclusion that the presence of iron traps cannot explain the observed photorefractive effects. In what follows we prove that even small concentrations of iron traps is enough for large light-induced modification of the refractive index of congruent lithium niobate, if one properly takes the small polarons into account. In the adiabatic approximation from (13.14) we get, (13.15), and (13.18)

$$\frac{\partial}{\partial t} N_X^- = -(\beta_X + I q_X S_X) N_X^- + I q_{Y_2 X} S_{Y_2 X} N_X N_C, \tag{13.35}$$

$$N_{Y_2}^- = N_C - N_X^-, \tag{13.36}$$

$$N_e \simeq \frac{(\beta_{Y_2} + I q_{Y_2} S_{Y_2}) N_{Y_2}^- + (\beta_X + I q_X S_X) N_X^-}{\Gamma_X N_X}, \tag{13.37}$$

where we have used the numerical values of the parameters discussed above and neglected the influence of iron traps in the sum $\Gamma_X N_X + \Gamma_{Y_2} N_{Y_2}$ (the assumption $\Gamma_X N_X \gg \Gamma_{Y_2} N_{Y_2}$ is the opposite of the assumption made in [59, 60]). The solution of (13.35) is

$$N_X^- = \tilde{N}_X^-[1 - \exp(-\xi t)] + \bar{N}_X^- \exp(-\xi t), \tag{13.38}$$

where

$$\xi = \beta_X + I q_X S_X, \tag{13.39}$$

$$\tilde{N}_X^- = \frac{I q_{Y_2 X} S_{Y_2 X} N_X N_C}{\beta_X + I q_X S_X}. \tag{13.40}$$

The steady state solution $\tilde{N}_X^-$ saturates with the intensity of light. The saturated density is

$$\tilde{N}_X^- \simeq \frac{q_{Y_2 X} S_{Y_2 X}}{q_X S_X} N_X N_C. \tag{13.41}$$

As we have shown in Appendix, this value does not exceed $0.5 N_C$. We derive, similar to (13.34)

$$\Delta n_o = \frac{1}{2} n_o^3 r_{13} \frac{\Gamma_X N_X}{e\mu} \frac{I(\kappa_{Y_2} N_{Y_2}^- + \kappa_X N_X^-)}{(\beta_{Y_2} + I q_{Y_2} S_{Y_2}) N_{Y_2}^- + (\beta_X + I q_X S_X) N_X^-}. \tag{13.42}$$

The maximum saturated change of the refractive index does not depend on the concentration of the impurities. Because $|\kappa_X| \gg |\kappa_{Y_2}|$ in the infrared, this maximum is

$$\Delta n_o \to \frac{1}{2} n_o^3 r_{13} \frac{\Gamma_X N_X}{e\mu} \frac{\kappa_X}{q_X S_X}. \tag{13.43}$$

Substituting the corresponding values of the parameters presented in Appendix into (13.43) we get $\Delta n_o\,(780\,\text{nm})|_{\max} \simeq -3.8 \times 10^{-5}$. Equations (13.41) and (13.42) show that in steady state

$$\Delta n_o \simeq \frac{1}{2} \frac{n_o^3 r_{13}}{e\mu} \frac{\kappa_{Y_2} \Gamma_X}{q_{Y_2 X} S_{Y_2 X}} \left(1 + \frac{I \kappa_X q_{Y_2 X} S_{Y_2 X} N_X}{\kappa_{Y_2}(\beta_X + I q_X S_X)}\right). \tag{13.44}$$

The intensity dependence of the change in the photorefractive index becomes significant for $I > 10^6\,\text{W/m}^2$ for the above chosen parameters. However, the relative intensity dependence of light absorption that can be found from (13.21) is much smaller. Using (13.26) and (13.33) we arrive at

$$\left.\frac{\partial}{\partial t} \Delta n_o\right|_{t=0} \simeq \frac{n_o^3 r_{13}}{2 \epsilon \epsilon_0} I \left(\kappa_{Y_2} N_C + I N_X N_C \kappa_X \frac{q_{Y_2 X} S_{Y_2 X}}{\beta_X + I q_X S_X}\right). \tag{13.45}$$

It is easy to see that for $10^8\,\text{W/m}^2 > I > 10^6\,\text{W/m}^2$ the dependence becomes quadratic in the intensity of 780 nm light which corresponds to the experimental observation [23]. Using this result and results of measurements from [23] it is possible to find the wavelength dependence for the product $\kappa_X q_{Y_2 X} S_{Y_2 X}$, which is given by (13.63). For instance, the overall effect is 10^6 times less at 1 550 nm than at 780 nm. Therefore, to observe a similar refractive index variation at 1 550 nm as in the 780 nm case, one needs to use a thousand time larger intensity.

Discussion

According to (13.33) and (13.34), the bipolaron-small polaron model describes the first order photorefractive effect in lithium niobate satisfactorily. The iron-small polaron model is also the same way, (13.41) and (13.42). considering the first order effect only, it is impossible to say with full confidence if the bipolarons or the iron ions are responsible for the photorefractivity. Both models show that the concentration of shallow traps Nb_{Li}^{5+} (N_X) is important in the nominally-pure material. This value substitutes the concentration of iron ions Fe^{3+} (N_{Y_2}) responsible for the maximum light induced change of the refractive index in highly doped $LiNbO_3$. Our evaluation of the second order photorefractive effect suggests that it is likely that the iron impurities play a dominant role. The second order effect in the bipolaron-small polaron model, arising from the interaction of small polarons and bipolarons, is weak for the parameter values that we assigned to the bipolarons. The concentration and the influence of small polarons on the photorefractive properties of the material is small for reasonable values of the intensity of infrared light ($I \ll 10^{12}$ W/m^2). Moreover, according to (13.26) and (13.33) the superlinear increase of the time derivative of the refractive index $\partial \Delta n/\partial t$ scales as $I^{3/2}$, which was not observed experimentally (see (13.63)). Therefore, it is unlikely that the bipolarons play a significant role in the interaction. The model that includes iron and small polaron pair is more likely to properly describe the photorefractivity correctly because it gives its expected dependence on the light intensity.

13.6.3 Absorption of the Light and Initial Concentration of the Filled Traps

According to (13.21), the smallest measured absorption of light defines the number of filled deep traps. In this section we find the trap concentration and characterize the transparency window of lithium niobate. The spectral dependence of the absorption due to small polarons can be found using (13.57) and the absorption coefficient $S_X(633 \text{ nm}) = 1.3 \times 10^{-3}$ m^2/J [59]:

$$\alpha_{sp} \simeq 1.1 \times 10^{-26} \lambda \text{ (nm)} \exp\left[-10.7\left(1 - \frac{780}{\lambda \text{ (nm)}}\right)^2\right] N_X^-. \quad (13.46)$$

Calculating the absorption cross section for the bipolarons as for small Anderson polarons [64] and using the frequency dependence (13.60), we derive the expression for the material absorption per centimeter due to the bipolarons

$$\alpha_{bp}(\omega) \simeq 5.5 \times 10^{-27} \lambda \text{ (nm)} \exp\left[-9.6\left(1 - \frac{500}{\lambda \text{ (nm)}}\right)^2\right] N_{Y_1}^-. \quad (13.47)$$

This expression is in good agreement with the experimental results of [63] for the wavelengths ranging from 0.5 μm to 1 μm, but overestimates the absorption cross section two and six times at 1.3 μm and 1.5 μm, respectively. The cross section of the resonant absorption of the bipolarons is calculated using the similarity of the absorption spectra of small polarons (13.59) and small bipolarons (13.61) under the assumption of equality of their masses. We expect that (13.47) correctly gives the order of magnitude of the absorption value, only because of the equality of mass as well as identical-site polaron approximations made during the derivation. Similarly, using the results of our discussions of the iron traps (13.62) as well as the measurement data from [65], for the absorption per centimeter of ordinarily polarized light on the filled traps we get

$$\alpha_{fe}(\omega) \simeq 1.6 \times 10^{-23} \times \left\{\exp\left[-32\left(1 - \frac{480}{\lambda \text{ (nm)}}\right)^2\right] + \frac{1}{16}\exp\left[-15.4\left(1 - \frac{1100}{\lambda \text{ (nm)}}\right)^2\right]\right\} N_{Y_2}^-, \quad (13.48)$$

where the first term on the right hand side describes the absorption due to the trap ionization, and the second term stands for the absorption due to the ion excitation. To find the absorption of the material we have to add (13.47), (13.48), and (13.46) to the expressions for the blue and red wing absorption. Congruent undoped LiNbO$_3$ has an absorption edge that is shifted toward the lower energy by approximately 0.2 eV relative to the stoichiometric material [52, 66, 67]. The lithium niobate absorption coefficient reaches 10 cm^{-1} at 330 nm (congruent) and 309 (nearly stoichiometric, 49.7% Li). Using the experimental data [52, 66, 67] we approximate the blue wing absorption per centimeter wavelength dependence for the congruent LiNbO$_3$ crystals as

$$\alpha_{Bc} \simeq \exp\left[19\left(1 - \frac{\lambda \text{ (nm)}}{379}\right)\right]. \quad (13.49)$$

The fundamental red wing absorption arises primarily from multiphonon processes. We estimated the wavelength dependence of the absorption per centimeter coefficient for the congruent LiNbO$_3$ using data from [68]

$$\alpha_{Rc} \simeq \exp\left[16.7\left(1 - \frac{6800}{\lambda \text{ (nm)}}\right)\right]. \quad (13.50)$$

Presenting the total material absorption as

$$\alpha = \alpha_{Bc} + \alpha_{Rc} + \alpha_{bp} + \alpha_{fe} + \alpha_{sp}, \quad (13.51)$$

and fitting it to the experimental absorption data of the samples of as-grown congruent lithium niobate on which Fig. 13.19 is based [11, 14, 52, 69] we obtain the total number of the filled deep traps: $\bar{N}_{Y_1}^- = 1.5 \times 10^{21}$ m^{-3} and $\bar{N}_{Y_2}^- = 10^{21}$ m^{-3}. Here and below we indicate the initial values of the parameters by a bar on the top. The initial concentration of the small polarons is expected to be negligible and we can estimate $\bar{N}_X^- \ll 10^{19}$ m^{-3} from the absorption fitting. To find the light induced absorption we use (13.21) and substitute there the steady state concentrations of the small polarons and the iron traps. The residual absorption of the material (α_0)

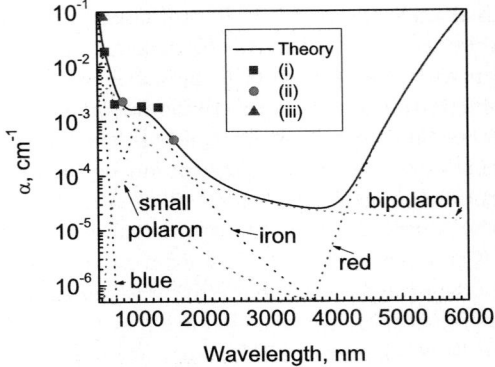

Fig. 13.19. Absorption of as-grown congruent lithium niobate for small intensity of light. The theoretical curve is obtained by selecting the values of the densities of the defects in (13.51). *Dotted lines* illustrate the contribution from the small polarons, bipolarons, iron, as well as fundamental *blue* and *red* absorption wings. Experimental points were taken from (i) [69]; (ii) [11, 14]; and (iii) [52]

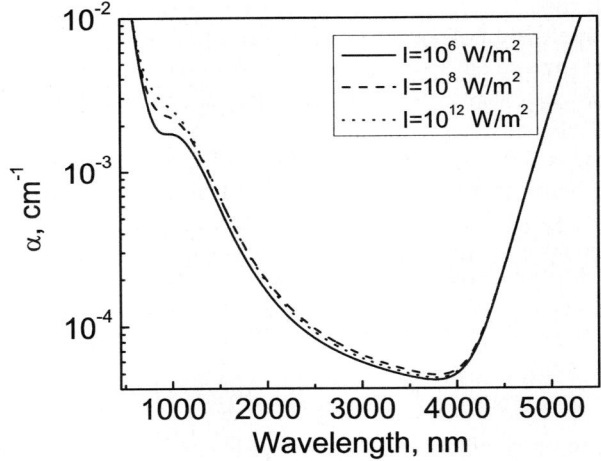

Fig. 13.20. Absorption of the as-grown congruent lithium niobate for various intensities of the light

includes the absorption at the blue and the red wings. The resulting spectral dependence of α at various light intensities is shown in Fig. 13.20. The induced absorption is insignificant in the far infrared. The strongest absorption change, up to several times, is localized in the vicinity of 780 nm wavelength. This absorption results from the filling of the shallow traps (small polarons) by the light. Finally, it is important to note that the optical transparency of as-grown crystals can be enhanced by oxidization of the samples [55]. This treatment results in a decrease of the donor/acceptor ratio, and consequently, an increase of the transparency of the material for the given concentration of the impurities. For example, the ratio $c_{Fe^{2+}}/c_{Fe^{3+}}$ can be made as small as 0.002 [56], an order of magnitude smaller than we have considered in our estimations. To conclude this section, we have theoretically studied the infrared transparency and photorefractivity of congruent as-grown lithium niobate. We have shown that the residual absorption of the material can be explained by the inevitable presence of a small concentration of extrinsic and intrinsic defects. This material has an absorption that is less than 10^{-4} in 2 μm–4.5 μm wavelength band. The same defects result in a significant photorefractivity. For instance, it is shown that the illumination of the crystal with high intensity monochromatic light at 780 nm will induce a change of approximately 4×10^{-4} in the refractive index, which is comparable to the index change realized with green light. Our study also points out the lack of experimental data available for this material. The theoretical, as well as an experimental, description of the complete spectral dependence for the photorefractive properties of the material, similar to the description of the material absorption presented in this part of the chapter, is an important task for future studies.

Appendix A: Basic Properties of WGMs

The electric field in a WGM resonator obeys the Maxwell equation

$$\nabla \times (\nabla \times \mathbf{E}) + \frac{\epsilon(r)}{c^2} \frac{\partial^2 \mathbf{E}}{\partial t^2} = 0, \quad (13.52)$$

where c is the speed of light in vacuum. Presenting the electric field as $\mathbf{E} = \int_0^\infty d\omega\, \mathbf{e}(\mathbf{r}) \exp(-i\omega t)$, we rewrite the above equation as

$$\nabla \times (\nabla \times \mathbf{e}) - k^2 \epsilon(\mathbf{r}) \mathbf{e} = 0, \quad (13.53)$$

where $k = \omega/c$ is the wave vector. Equation (13.53) may be solved in terms of TE and TM modes for a spherical or cylindrical WGM resonator. The latter represents the symmetry of the resonators used in our study. The radial field distribution for the modes is described by

$$\frac{\partial^2 \Psi}{\partial r^2} + \left[k^2 \epsilon(r) - \frac{\nu(\nu+1)}{r^2} \right] \Psi = 0, \quad (13.54)$$

where ν is an integer number $\nu = 0, 1, 2, 3, \ldots$ for a sphere ($\nu = 1/2, 3/2, 5/2, \ldots$ for an infinite cylinder). It is convenient to enumerate the discrete WGMs using this number. Hence, in what follows we refer to ν as the mode number for a spherical resonator (for a cylinder $\nu + 1/2 \approx \nu$ if $\nu \gg 1$). The electric field distribution has a dependence $\Psi(r)/r$ for a sphere, and $\Psi(r)/\sqrt{r}$ for a cylinder. Equation (13.54) has an exact solution for a homogeneous dielectric cavity $\epsilon(r) = \epsilon_0 = \text{const}$ if $r \leq a$. The same solution holds for a WGM resonator made of a material with (crystalline) axial symmetry, like lithium niobate, if the optical axis of the material coincides with the symmetry axis of a WGM. This solution reads $\Psi(r) = J_{\nu+1/2}(kr)$, where $J_{\nu+1/2}(kr)$ is the Bessel function of the first kind. The mode spectrum is determined by the boundary conditions $\Psi(r) \to 0$ for $r \to \infty$ and 0. We now consider lithium niobate WGM resonators and assume that the symmetry axis Z of the resonator coincides with the optical axis of the material. The spectrum of the resonator for a TE mode with large number $\nu \gg 1$ is given by (see, e.g., the first equation in [44])

$$\omega_e \simeq \frac{c\nu}{an_e}\left[1 + \alpha_q\left(\frac{1}{2\nu^2}\right)^{1/3} - \frac{n_e}{\nu\sqrt{n_e^2 - 1}}\right], \qquad (13.55)$$

where n_e is the extraordinary refractive index of the material, a is the radius of the resonator, α_q is the qth root of the Airy function, $\text{Ai}(-z)$, which is equal to 2.338, 4.088, and 5.521 for $q = 1, 2, 3$, respectively. The expression for the TM WGM modes looks similar

$$\omega_o \simeq \frac{c\nu}{an_o}\left[1 + \alpha_q\left(\frac{1}{2\nu^2}\right)^{1/3} - \frac{1}{\nu n_o\sqrt{n_o^2 - 1}}\right], \qquad (13.56)$$

where n_o is the ordinary refractive index of the material. Equations (13.55) and (13.56) show that relative change of the refractive index $\Delta n_e/n_e$ ($\Delta n_o/n_o$) results in the relative frequency shift of the mode $-\Delta\omega_e/\omega_e$ ($-\Delta\omega_o/\omega_o$). The frequency difference between the adjacent modes of the same family (the free spectral range, or FSR) is on the order of ten gigahertz for a resonator with a diameter of several millimeters, and is almost independent of the mode number ν. However the FSR depends on the refractive index of the material. This value can be easily measured, and the accuracy of the of the index of refraction derived from it depends on the accuracy of the measurement of the radius of the resonator, which is generally better than 1%. An accurate measurement of the radius is not required if one is interested in the relative value of ordinary/extraordinary index of refraction of the material. There are no pure TE or TM modes in generalized spheroids (the generalized geometry of our resonators), but the perturbation of the spectrum of the basic mode sequence is usually small if the spheroid's radius is large. In that case, (13.52) can be solved numerically and the mode spectrum of the spheroid can be found with high accuracy. An analytical approach to this problem has also been previously proposed [89].

Appendix B: Lithium Niobate Impurities: A Short Review of Existing Results

The photorefractivity of lithium niobate is determined by its intrinsic and extrinsic defects, namely donor and acceptor centers in the forbidden energy gap of the crystal. Intrinsic defects include perturbations of the crystalline lattice such as vacancies and polarons. Extrinsic defects are various dopants, such as Fe, Mg, Ti, Mn, etc. These defects generally influence each other and can not be considered independently. Our discussion involves three types of impurities important for infrared applications of LiNbO$_3$. Those impurities are small polarons, bipolarons, and iron ions. We are not aware of any strict theoretical description of the optical properties of the impurities in the infrared for as-grown nominally-pure LiNbO$_3$. Given formulae are devised from existing experimental data for reduced crystals.

B.1 Small Polarons

Lithium niobate is generally non-stoichiometric and has an excess of Nb ions. Those ions create intrinsic point defects Nb$_{Li}$ (niobium ion on the lithium site) to maintain the overall charge neutrality. Electrons trapped on the defects form small polarons. The small polaron bound to the Nb$_{Li}$ antisite defect is generally labeled as Nb$_{Li}^{4+}$ (or Nb$_{Li}^{4+}$Nb$_{Nb}^{5+}$ [70]). Small polarons result in broadband absorption of light centered at 780 nm [71]. The frequency dependence of the optical absorption due to small polarons is represented as [72]

$$\alpha_{sp}(\omega) = \frac{D_{sp}}{\hbar\omega} \exp\left[-\frac{(4U + E - \hbar\omega)^2}{8U\hbar\omega_0}\right] \quad (13.57)$$

where $\hbar\omega_0 \sim 0.1$ eV is the phonon energy, $U \sim 0.3$ eV is the polaron stabilization energy, $E \sim 0.4$ eV is the energy of the polaron transfer between non-equivalent sites, D_{sp} is the absorption amplitude scaling parameter. It is useful to rewrite (13.57) in terms of a model of a small polaron created on identical sites [64]

$$\alpha_{sp}(\omega) = \frac{D_{sp}}{\hbar\omega} \exp\left[-\frac{(\hbar\omega - 2E_b)^2}{4E_b\hbar\omega_0}\right]. \quad (13.58)$$

In this model a self-trapped carrier is excited from its localized state to a localized state at an identical site adjacent to the small polaron site. The difference of the electronic energy of the two equivalent sites is the lowering of the electronic energy associated with the small polaron formation ($-2E_b$). Equations (13.57) and (13.58) have the same functional form, and (13.57) seems to be more suitable for the description of small polarons in LiNbO$_3$ because it takes into account the non-equivalent sites. An advantage of the model obtained from (13.58) is that it allows finding an analytical form of the absorption amplitude scaling parameter

$$D_{sp} = c_X^- \frac{2\pi^{3/2} e^2}{m^* c} \frac{\hbar^3}{2m^* a^2} (4E_b \hbar\omega_0)^{-1/2}, \quad (13.59)$$

where c_X^- is the concentration of the particles per cubic centimeter, e is the charge of an electron, m^* is an effective electronic mass, and a is the lattice constant (0.3 nm for LiNbO$_3$). Unfortunately we cannot use (13.59) to estimate the absorption of light resulting from the polarons because m^* is unknown. A substitution of m^* by the mass of a free electron results in a value approximately thirty times less when compared to the experimentally measured value. On the other hand, this model is useful for approximating the relative absorption of small polarons and bipolarons. Small polarons have a finite average lifetime, so there are almost no small polarons in an as-grown crystal placed in the dark. The number of small polarons is determined by the thermal equilibrium between the small and bipolarons. The lifetime of small polarons ranges from milliseconds for congruent lithium niobate, to 400 ms for as-grown near stoichiometric lithium niobate [24, 52]. The actual lifetime varies by a wide range for the same sample (from nanoseconds to milliseconds), depending on the space location of the various deep traps with respect to the small polarons [73]. The lifetime increases with a temperature decrease, according to Arrhenius law ($\tau_{sp} \sim \exp(E_A/kT)$), where E_A is the thermal activation energy. For iron doped congruent lithium niobate $E_A \simeq 0.37$ eV [72] (see also [73–75]). The two-photon holographic data storage is possible due to the small polarons [24, 76], in which writing is realized by red/infrared light in the presence of a gating blue/green light, and reading is realized with the red/infrared light only.

B.2 Bipolarons

A bipolaron in lithium niobate is an intrinsic point defect $Nb_{Li}^{4+}Nb_{Nb}^{4+}$, containing two electrons with opposite spins [77]. Without the electrons the defect represents a two-electron trap. The optical resonance wavelength of the bipolaron is ~500 nm (~2.5 eV). The density of bipolarons, similar to the density of small polarons, decreases in the near-stoichiometric material. The bipolaron could be dissociated with light or temperature [62, 77]. The bipolaron dissociation energy is ~0.27 eV [62]. The bipolaron dissociates into a small polaron $Nb_{Li}^{4+}Nb_{Nb}^{5+}$ [70] and an electron. The electron produced with the bipolaron ionization enters the conduction band (leaves the cluster) and either becomes trapped on the Nb lattice sites, forming another Nb_{Li}^{4+} small polaron, or penetrates into another deep trap. Conversely, small polarons can nonradiatively transform to bipolarons by capturing electrons from the conduction band; so it is expected that the bipolarons play the role of deep traps in the nominally-pure lithium niobate [24]. Thermal induction of electrons from bipolarons results in the decay of the refractive index pattern formed due to the photorefractivity of the material. The decay ("dark decay") time could exceed a month in pure lithium niobate. We did not find any theory explaining the behavior of LiNbO$_3$ bipolarons in the infrared. To obtain clues about the concentration of these impurities in as-grown lithium niobate, as well as their infrared absorption, we assume that the bipolarons are free, small, Anderson bipolarons. The absorption of the light by the bipolarons is described by [64]

$$\alpha_{bp}(\omega) = \frac{D_{bp}}{\hbar\omega} \exp\left[-\frac{(\hbar\omega - 8E_b + U_b)^2}{16 E_b \hbar \omega_0}\right], \quad (13.60)$$

where

$$D_{bp} = c_{Y_1}^- \frac{2\pi^{3/2} e^2}{m^*c} \frac{\hbar^3}{2m^*a^2} (16 E_b \hbar \omega_0)^{-1/2}, \quad (13.61)$$

E_b has the same meaning as in the case of small polarons, U_b takes into account the mutual repulsion of the two electrons at one site in the lattice as well as the energy of the final state of the bipolaron. The energy of the final state is $-2E_b$ if we assume that the bipolaron decays to a free electron and a small polaron. Comparing (13.59) and (13.61), and assuming that the small polaron and bipolaron have equivalent effective masses we conclude that the absorption cross section of a small polaron is twice as much as that of the bipolaron. To find the wavelength dependence of the bipolaron induced light absorption we approximate the experimentally taken bipolaron absorption spectrum presented in [63] by varying E_b and U_b. We get $2E_b \sim 0.82\,\text{eV}$ and $U \sim 0.78\,\text{eV}$. It is interesting to note that a comparison of (13.57) and (13.58) gives $2E_b \sim 1.2\,\text{eV}$, which shows that the small polaron model developed for equivalent sites is not entirely suitable for the description of the small polarons and bipolarons in lithium niobate.

B.3 Iron

As-grown lithium niobate contains approximately a hundred times more acceptors Fe^{3+} than donors Fe^{2+}. The oxidization of the sample reduces this number even further [56]. Photoionization of acceptors Fe^{2+} is centered at 480 nm (2.6 eV) [78]. The absorption of light due to the impurities becomes important in the near infrared, and weakly depends on the polarization of the light. The photoionization of Fe^{2+} (an intervalence charge transfer from Fe^{2+} to Nb^{5+}) is characterized by the oscillator strength 1.3×10^{-2}. Absorption of light per centimeter due to Fe^{2+} at 532 nm is $\alpha_{Fe^{2+}}(532) = 1.1 \times 10^{-17} c_{Fe^{2+}}$, where the concentration $c_{Fe^{2+}}$ is taken as the number of particles per cubic centimeter [65]. The excitation of Fe^{2+} ions leads to a broadband absorption peak at 1 100 nm (1.1 eV). The excitation becomes possible due to the influence of the crystal field on the Fe^{2+} ions. This resonance can be seen with light polarized orthogonally to the crystal axis. However, this transition is weak. The oscillator strength measured in $LiNbO_3$ is 4×10^{-4} ($\alpha(2.6\,\text{eV})/\alpha(1.1\,\text{eV}) \simeq 16$) [78]. The equivalent absorption is $\alpha_{Fe^{2+}}(1\,100) \sim 10^{-18} c_{Fe^{2+}}$. Dipole-forbidden d–d transitions in Fe^{3+} ions result in absorption at 483 nm (2.55 eV) and 426 nm (2.95 eV) [65, 78]. These resonances can be seen with light polarized orthogonally to the crystal axis. The absorption of Fe^{3+} at 483 nm is $\alpha_{Fe^{3+}}(483) = 8.5 \times 10^{-20} c_{Fe^{3+}}$ [65]. The absorption is very low in existing as-grown materials so we neglect it. It could become important in strongly oxidized purified samples. To find the spectral dependence of the absorption we assume that there is a broadening of the transition resulting from thermal vibrations of the lattice. Following [79], we describe the absorption by Gaussian curves

$$\alpha(\omega) = \frac{D}{\delta E} \exp\left[-\left(\frac{\hbar\omega - E_0}{\delta E}\right)^2\right], \tag{13.62}$$

where E_0 is the energy corresponding to the absorption peak, δE is the energy width of the absorption band, D is a normalization constant. It was found that $\delta E(2.52\,\text{eV}) = 0.46\,\text{eV}$ and $\delta E(1.1\,\text{eV}) = 0.28\,\text{eV}$ for ordinarily polarized light propagating in $LiNbO_3$ at 30°C. The absorption peak shifts and broadens for extraordinarily polarized light $\delta E(2.63\,\text{eV}) = 0.48\,\text{eV}$. All the absorption peaks become broader with a temperature increase.

Appendix C: Photorefractivity in Red: A Short Review of the Existing Results

As is shown in the previous section, photorefractivity can be treated as a resonance effect. There are several selected frequencies where the maxima of photorefractive modification of the material properties can be achieved. The position as well as the strength of the maxima depends on both the extrinsic and intrinsic defects. As-grown nominally-pure LiNbO$_3$ crystals have almost no donor impurities that can be activated by the weak infrared light. The photorefractivity decreases with a wavelength increase, and it is generally assumed that far-infrared light causes no effect. Photorefractivity results in light-induced change of the refractive index of the material, as well as in the self-induced light absorption. We discuss both effects observed with the red (infrared) light in lithium niobate.

C.1 Light Induced Change of Refractive Index

It was found that photorefractivity enhances the crosstalk between LiNbO$_3$ waveguides at 633 nm, but no significant crosstalk was recorded in the same waveguide at 1 060 nm and light intensities up to 10 kW/cm^2 [53]. Higher infrared intensity revives the effect. Photorefractive damage was seen in titanium infused lithium niobate optical directional couplers at 1320 nm, when the light intensity exceeded 50 kW/cm^2. The saturation time constant was on the order of 100 hours [54]. Photorefractive damage revealed itself as the refractive index change not exceeding 3×10^{-4}, which modified the optical length in the coupler and caused the coupler imbalance. It was also shown that a proper annealing of titanium-infused LiNbO$_3$ modulators allows suppressing the photorefractive effect for several MW/cm^2 of light intensity at 1 320 nm [80]. To the best of our knowledge, no photorefractive damage was observed at 1 550 nm and longer wavelengths. A few observations of photorefractivity in the infrared were identified in the studies of LiNbO$_3$ with focus on the holographic data storage. Using data for the diffraction efficiency of the hologram created at 780 nm with 200 W/cm^2 light during 1 s exposure in nominally-pure LiNbO$_3$ [23] we infer the following important dependence describing the change of the ordinary index of refraction

$$\frac{\partial \Delta n_o}{\partial t} \simeq 10^{-21} \times \lambda \text{ (nm)} \times I^2 \text{ (W/m}^2\text{)} \times \exp\left[-49\left(1 - \frac{570}{\lambda \text{ (nm)}}\right)^2\right] \text{s}^{-1}.$$

Light-induced refractive-index changes up to 2.8×10^{-4} were detected in iron-doped lithium niobate crystals with 760 nm ordinarily polarized light. The intensity of the light was 4 mW/cm^2 and the concentration of the iron acceptor ions was $c_{Fe^{3+}} \simeq 2 \times 10^{19}$ cm^{-3}, which corresponds to the relative index change $\Delta n_o/c_{Fe^{3+}} \simeq 1.4 \times 10^{-23}$ cm^3. It is interesting to note that the changes are of the same order as changes for the green light illumination (7.2×10^{-4} at 514 nm, $\Delta n_o/c_{Fe^{3+}} \simeq 3 \times 10^{-23}$ cm^3) [81]. From the data we interpolate that iron-mediated maximum change of the refractive index in infrared is less than 2.8×10^{-5} in undoped lithium niobate because

concentration varies from $c_{Fe^{3+}} \sim 2 \times 10^{18}$ cm^{-3} [65] to $c_{Fe^{3+}} \sim 2 \times 10^{17}$ cm^{-3} [75] and $c_{Fe^{3+}} \sim 3 \times 10^{16}$ cm^{-3} [70] in such a crystal. The above conclusions are valid for low light intensities only. Higher intensities increase the photorefractivity, as shown in an experiment involving MW/cm^2 green light pump (532 nm) and red probe (633 nm) propagating in the iron doped lithium niobate [59]. However, the proposed theoretical model mentioned in the study is not necessarily valid for nominally pure crystals. An experimental study involving nominally-pure congruent as well as stoichiometric lithium niobate also has confirmed the increase of the photorefractivity with light intensity [82, 83]. Moreover, it was proven [83], that the change of the refractive index is proportional to the concentration of Nb$_{Li}$ defects. It was assumed that in the nominally-pure materials with small residual concentration of iron, the small polarons are created primarily by the dissociation of the bipolarons, while Fe^{3+} still acts as a deep trap for the formation of the long-lived refractive grating [84].

C.2 Light Induced Change of Absorption

Green light induced infrared absorption belongs to one of the basic photorefractive effects. Shining green light on a sample increases the absorption of the infrared light. This two-photon effect is unwanted in nonlinear optics experiments with the crystals that involve both infrared and visible light. It was shown that absorption of 1 064 nm light (21 kW/cm^2) in The undoped nearly-stoichiometric lithium niobate increases from 1.7×10^{-3} cm^{-1} to 7.5×10^{-3} cm^{-1} if 3.5 kW/cm^2 532 nm pump is switched on. Switching the pump off results in the restoration of the initial absorption value [85]. The effect is $\sim$1.5 times smaller in the congruent lithium niobate. The absorption increase is accompanied by an index of refraction change. This change was 2.7×10^{-5} for stoichiometric and 1.8×10^{-5} for congruent material illuminated with 2.5 kW/cm^2 532 nm pump. The difference was explained by the longer lifetime of the small polarons in the stoichiometric compared with the congruent crystal. Iron, having close values of concentration in the both materials, was expected to be a donor. No change in either the infrared light induced refractive index or absorption was detected in these experiments.

Appendix D: Numerical Values of the Basic Rates Characterizing the Impurities

To determine how best to handle the rate equations (13.13–13.15) it is useful to consider the experimentally measured numerical values for the parameters of lithium niobate, and adjust them to fit the conditions of the present study (infrared light and as-grown material). The results of such an adjustment are summarized in Tables 13.1–13.4. The values are obtained for the photorefractivity at 780 nm. We did not find enough data to evaluate the complete spectral dependence of all of the parameters. Moreover, many parameters are not known even for the selected wavelength.

Table 13.1. Parameters of the as-grown congruent LiNbO$_3$ at room temperature for 780 nm light as well as some related fundamental constants

Quantity	Value	Reference
$\bar{N}_e$	2.1×10^6 m^{-3}	[derived]
N_C	4×10^{21} m^{-3}	[derived]
μ	7.4×10^{-5} m^2/sV	[60]
ϵ	28	
n_o	2.25	
n_e	2.18	
α_o	2.7×10^{-3} cm^{-1}	[11, 14, 69]
α_e	8.5×10^{-3} cm^{-1}	[11, 14, 69]
r_{13}	11×10^{-12} m/V	
r_{33}	34×10^{-12} m/V	
ϵ_0	8.85×10^{-12} sA/Vm	
e	1.6×10^{-19} C	

Table 13.2. Parameters of the bipolarons in as-grown congruent LiNbO$_3$ at room temperature for 780 nm light

Quantity	Value	Reference
$\bar{N}_{Y_1}^-$	1.5×10^{21} m^{-3}	[derived]
β_{Y_1}	10^{-7} s^{-1}	[assumed]
$q_{Y_1} S_{Y_1}$	3×10^{-6} m^2/J	[assumed]
κ_{Y_1}	-10^{-33} m^3/V	[assumed]
$\Gamma_{Y_1 X}, \Gamma_{X Y_1}$	5.4×10^{-16} m^3/s	[derived]

Table 13.3. Parameters of the iron impurities in as-grown congruent LiNbO$_3$ at room temperature for 780 nm light. To estimate photovoltaic coefficient we assumed that the coefficient is proportional to the absorption of the light (see, e.g. [86])

Quantity	Value	Reference
N_{Y_2}	3×10^{22} m^{-3}	[71]
$\bar{N}_{Y_2}^-$	10^{21} m^{-3}	[derived]
β_{Y_2}	10^{-7} s^{-1}	[derived]
$q_{Y_2} S_{Y_2}$	1.4×10^{-7} m^2/J	Equation (13.62) and [59]
Γ_{Y_2}	10^{-15} m^3/s	[assumed]
κ_{Y_2}	-5×10^{-35} m^3/V	Equation (13.62) and [59]
$q_{Y_2 X} S_{Y_2 X}$	4.2×10^{-32} m^5/J	[derived]
$\Gamma_{Y_2 X}, \Gamma_{X Y_2}$	2×10^{-24} m^3/s	[assumed]

We find the initial density of the charges in the conduction band

$$\bar{N}_e = N_C - \bar{N}_X^- - 2\bar{N}_{Y_1}^- - \bar{N}_{Y_2}^- \quad (13.63)$$

Table 13.4. Parameters of the small polarons in as-grown congruent LiNbO$_3$ at room temperature for 780 nm light

Quantity	Value	Reference
N_X	10^{27} m^{-3}	[77, 60, 87]
$\bar{N}_{Y_1}^-/(\bar{N}_X^-)^2$	5.4×10^{-9} m^3	[derived]
$\bar{N}_X^-/N_X$	5.3×10^{-13}	[derived]
β_X	10^4 s^{-1}	[23, 24, 73]
$q_X S_X$	9×10^{-5} m^2/J	Equation (13.57) and [60]
Γ_X	2.5×10^{-15} m^3/s	[derived]
κ_X	-53.3×10^{-33} m^3/V	Equation (13.57) and [60]

from the expression for the dark conductivity $\sigma_d = \epsilon\epsilon_0/\tau_d = e\mu \bar{N}_e$. Taking the dark decay time to be equal to $\tau_d = 10^7$ s (several months, as per [24]), we get $\bar{N}_e = 2.1 \times 10^6$ m^{-3}. It is easy to see that $N_C \gg \bar{N}_e$ ($N_C = 4 \times 10^{21}$ m^{-3}). We evaluate the parameters of small polarons at the next step. The rate parameter Γ_X is estimated from the decay rate of the free carriers (N_e) excited by a sub-picosecond pulse of light. The decay rate, $\Gamma_X N_X$ exceeds 2.5×10^{12} s^{-1} [88] and, hence, $\Gamma_X = 2.5 \times 10^{-15}$ m^3/s. The initial concentration of small polarons is given by the expression (steady state solution of (13.14))

$$\frac{\bar{N}_X^-}{N_X} \simeq \frac{\Gamma_X \bar{N}_e}{\Gamma_X \bar{N}_e + \beta_X + \Gamma_{XY_1} \bar{N}_X^- + \Gamma_{XY_2} \bar{N}_{Y_2}}. \quad (13.64)$$

The number of empty traps Nb$_{Li}$ is $N_X \approx 10^{27}$ m^{-3}. Taking $\beta_X = 10^4$ s^{-1} (dark decay rate of the small polarons) and assuming $\beta_X \gg \Gamma_X \bar{N}_e + \Gamma_{XY_1} \bar{N}_X^- + \Gamma_{XY_2} \bar{N}_{Y_2}$ we find $\bar{N}_X^- \simeq 5.3 \times 10^{14}$ m^{-3}. We continue with the bipolaron parameters. We did not find any published information on the subject except an assumption that the bipolarons are responsible for the photorefractive effects in the as-grown materials [24]. We take the thermal ionization rate of the bipolarons to be equal to the inverse dark decay time $\beta_{Y_1} \approx \tau_d^{-1} = 10^{-7}$ s^{-1}. Recombination rate $\Gamma_{Y_1 X}$ of bipolarons and small polarons can be found from the results of [62] for relative concentration of the impurities, along with the steady state solution of the rate equations (13.13) ($I(t=0) = 0$)

$$\frac{\bar{N}_{Y_1}^-}{(\bar{N}_X^-)^2} = \frac{\Gamma_{Y_1 X}}{\beta_{Y_1}} \simeq 5.4 \times 10^{-9} \text{ m}^3. \quad (13.65)$$

Using (13.65) we derive $\Gamma_{Y_1 X} \simeq 5.4 \times 10^{-16}$ m^3/s. The absorption cross section of the bipolaron can be calculated from (13.21) and (13.47). It is equal to $S_{Y_1} = 5 \times 10^{-4}$ m^2/J at 780 nm. We did not find enough experimental data to infer quantum efficiency of the absorption process q_{Y_1} as well as the photovoltaic coefficient κ_{Y_1}, so we made the major assumption that the photorefractive properties of bipolarons are of the same order as those of the iron for the green light illumination. We also assume that $q_{Y_2} S_{Y_2} \approx q_{Y_1} S_{Y_1} \simeq 10^{-5}$ m^2/J at 500 nm. The expression (13.47) allows scaling of this value down to $3 \times 10^{-6}\,\text{m}^2/\text{J}$ at 780 nm, i.e. to get this result the quantum efficiency should be $q_{Y_1} \approx 6 \times 10^{-3}$. The photovoltaic coefficient of the bipolaron is chosen in a way to have the ratio $\kappa_{Y_1}/q_{Y_1} S_{Y_1}$ approximately equal to that of iron, i.e. $\kappa_{Y_1} \approx -10^{-33}\,\text{m}^3/\text{V}$. Though the parameters of the iron traps are tabulated much better than the parameters of all the other impurities, some assumptions are necessary here. To derive the thermal ionization rate β_{Y_2} for the iron ions we assume that $\tau_d \beta_{Y_2} \simeq 1$, which gives $\beta_{Y_2} \simeq 10^{-7}\,\text{s}^{-1}$. The same value can be obtained from $I_d = \beta_{Y_2}/q_{Y_2} S_{Y_2}$. It is known that I_d (532 nm) $\simeq 10^{-2}\,\text{W}/\text{m}^2$ and $q_{Y_2} S_{Y_2}$ (532 nm) $= 10^{-5}\,\text{m}^2/\text{J}$ [59], so $\beta_{Y_2} \simeq 10^{-7}\,\text{s}^{-1}$. The other parameters are estimated using tabulated values and spectral dependence of the light absorption by Fe^{2+}. We have assumed that the ratio $\kappa_{Y_2}/q_{Y_2} S_{Y_2}$ is independent of wavelength, which comes from the assumption that the saturated value of the maximum change of the refractive index of the material solely from the iron traps does not depend on the wavelength of the light. We found from (13.21) and (13.48) that $S_{Y_2} + S_{Y_2 X} N_X^0 = 5.8 \times 10^{-5}\,\text{m}^2/\text{J}$ at 780 nm. Rate $q_{Y_2} S_{Y_2}$ (780 nm) $= 1.4 \times 10^{-7}\,\text{m}^2/\text{J}$ is found by extrapolating $q_{Y_2} S_{Y_2}$ (532 nm) $= 10^{-5}\,\text{m}^2/\text{J}$ [59] using (13.48), q_{Y_2} is taken to be wavelength independent. Using the steady state solution of the rate equation (13.14), and assuming that $I(t=0) = 0$ and that $\bar{N}_{X,Y_2} \gg \bar{N}_{X,Y_2}^-$ we get

$$\frac{\bar{N}_{Y_2}^-}{N_{Y_2}} \simeq \frac{\Gamma_{Y_2} \bar{N}_e + \Gamma_{Y_2 X} \bar{N}_X^-}{\Gamma_{Y_2} \bar{N}_e + \Gamma_{Y_2 X} \bar{N}_X^- + \beta_{Y_2}}. \tag{13.66}$$

Utilizing (13.66) and assuming that the recombination rate $\Gamma_{Y_2} \simeq 1.65 \times 10^{-14}\,\text{m}^3/\text{s}$ and $\Gamma_{Y_2 X} \simeq 1.14 \times 10^{-21}\,\text{m}^3/\text{s}$ [59], we derive $\bar{N}_{Y_2}^-/N_{Y_2} \simeq 1$. This result suggests that the concentration of the filled iron traps is higher than we calculated ($\bar{N}_{Y_2}^- \simeq 10^{21}\,\text{m}^{-3}$). Hence, the recombination rates are likely to be overestimated because of the high iron concentration in the samples studied in [59] versus the small iron concentration in nominally-pure lithium niobate. We assume that $\Gamma_{Y_2} \simeq 10^{-15}\,\text{m}^3/\text{s}$ and $\Gamma_{Y_2 X} \simeq 2 \times 10^{-24}\,\text{m}^3/\text{s}$ to get expected $\bar{N}_{Y_2}^-/N_{Y_2} \simeq 0.03$.

References

1. V.S. Ilchenko, X.S. Yao, L. Maleki, Proc. SPIE **3930**, 154 (2000). Edited by A.V. Kudryashov and A.H. Paxton
2. V.S. Ilchenko, L. Maleki, Proc. SPIE **4270**, 120 (2001). Edited by A.V. Kudryashov and A.H. Paxton
3. D.A. Cohen, A.F.J. Levi, Electron. Lett. **37**, 37 (2001)
4. D.A. Cohen, M. Hossein-Zadeh, A.F.J. Levi, Electron. Lett. **37**, 300 (2001)
5. D.A. Cohen, A.F.J. Levi, Solid State Electron. **45**, 495 (2001)
6. D.A. Cohen, M. Hossein-Zadeh, A.F.J. Levi, Solid State Electron. **45**, 1577 (2001)
7. V.S. Ilchenko, A.B. Matsko, A.A. Savchenkov, L. Maleki, J. Opt. Soc. Am. B **20**, 1304 (2003)
8. V.S. Ilchenko, A.A. Savchenkov, A.B. Matsko, L. Maleki, IEEE Photonics Technol. Lett. **14**, 1602–1604 (2002)

9. M. Hossein-Zadeh, A.F.J. Levi, Solid State Electron. **49**, 1428 (2005)
10. M. Hossein-Zadeh, A.F.J. Levi, IEEE MTT **54**, 821 (2006)
11. A.A. Savchenkov, V.S. Ilchenko, A.B. Matsko, L. Maleki, Phys. Rev. A **70**, 051804 (2004)
12. A.A. Savchenkov, V.S. Ilchenko, A.B. Matsko, L. Maleki, Electron. Lett. **39**, 389 (2003)
13. A.A. Savchenkov, V.S. Ilchenko, A.B. Matsko, L. Maleki, IEEE Photonics Technol. Lett. **17**, 136 (2005)
14. V.S. Ilchenko, A.A. Savchenkov, A.B. Matsko, L. Maleki, Phys. Rev. Lett. **92**, 043903 (2004)
15. M. Mohageg, D. Strekalov, A. Savchenkov, A. Matsko, V. Ilchenko, L. Maleki, Opt. Express **13**, 3408 (2005)
16. A.A. Savchenkov, A.B. Matsko, D. Strekalov, V.S. Ilchenko, L. Maleki, Appl. Phys. Lett. **88**, 241909 (2006)
17. A.A. Savchenkov, A.B. Matsko, D. Strekalov, V.S. Ilchenko, L. Maleki, Phys. Rev. B **74**, 245119 (2006)
18. A.A. Savchenkov, A.B. Matsko, D. Strekalov, V.S. Ilchenko, L. Maleki, Opt. Commun. **272**, 257 (2007)
19. A. Ashkin, G.D. Boyd, J.M. Dziedzic, R.G. Smith, A.A. Ballman, J.J. Levinste, K. Nassau, Appl. Phys. Lett. **9**, 72 (1966)
20. M. Bass, P.A. Franken, J.F. Ward, G. Weinreich, Phys. Rev. Lett. **9**, 446 (1962)
21. P.A. Franken, A.E. Hill, C.W. Peters, G. Weinreich, Phys. Rev. Lett. **7**, 118 (1961)
22. J.A. Giordmaine, R.C. Miller, Phys. Rev. Lett. **14**, 973 (1965)
23. Y.S. Bai, R. Kachru, Phys. Rev. Lett. **78**, 2944 (1997)
24. L. Hesselink, S.S. Orlov, A. Liu, A. Akella, D. Lande, R.R. Neurgaonkar, Science **282**, 1089 (1998)
25. P. Herth, T. Granzow, D. Schaniel, Th. Woike, M. Imlau, E. Kratzig, Phys. Rev. Lett. **95**, 067404 (2005)
26. D. von der Linde, A.M. Glass, K.F. Rodgers, Appl. Phys. Lett. **25**, 155 (1974)
27. O. Beyer, I. Breunig, F. Kalkum, K. Buse, Appl. Phys. Lett. **88**, 051120 (2006)
28. A.A. Savchenkov, A.B. Matsko, V.S. Ilchenko, N. Yu, L. Maleki, in *6-th International Kharkov Symposium on Physics and Engineering of Microwaves, Millimeter, and Submillimeter Waves* (*MSMW'07*), Kharkov, Ukraine, June 25–30, 2007
29. P. Urquhart, J. Opt. Soc. Am. A **5**, 803 (1988)
30. K. Oda, N. Takato, H. Toba, J. Lightwave Technol. **9**, 728 (1991)
31. B.E. Little, S.T. Chu, H.A. Haus, J. Foresi, J.P. Laine, J. Lightwave Technol. **15**, 998 (1997)
32. J.-L. Gheorma, R.M. Osgood, IEEE Photonics Technol. Lett. **14**, 795 (2002)
33. C.K. Madsen, J.H. Zhao, J. Lightwave Technol. **14**, 437 (1996)
34. V.B. Braginsky, M.L. Gorodetsky, V.S. Ilchenko, Phys. Lett. A **137**, 393 (1989)
35. A.J. Campillo, J.D. Eversole, H.B. Lin, Phys. Rev. Lett. **67**, 437 (1991)
36. L.E. Myers, R.C. Eckardt, M.M. Fejer, R.L. Byer, W.R. Bosenberg, J.W. Pierce, J. Opt. Soc. Am. B **12**, 2102 (1995)
37. R.W. Boyd, *Nonlinear Optics* (Academic Press, New York, 1992)
38. V.S. Ilchenko, A.B. Matsko, A.A. Savchenkov, L. Maleki, J. Opt. Soc. Am. B **20**, 1304 (2003)
39. M. Yamada, N. Nada, M. Saitoh, K. Watanabe, Appl. Phys. Lett. **62**, 435 (1992)
40. R.G. Batchko, V.Y. Shur, M.M. Fejer, R.L. Byer, Appl. Phys. Lett. **75**, 1673 (1999)
41. V.Y. Shur, E.L. Rumyantsev, E.V. Nikolaeva, E.I. Shishkin, E.V. Fursov, R.G. Batchko, L.A. Eyres, M.M. Fejer, R.L. Byer, Appl. Phys. Lett. **76**, 143 (2000)

42. M. Mohageg, A. Savchenkov, D. Strekalov, A. Matsko, V. Ilchenko, L. Maleki, Electron. Lett. **41**, 91 (2005)
43. G.D. Boyd, A. Ashkin, Phys. Rev. **146**, 187 (1966)
44. C.C. Lam, P.T. Leung, K. Young, J. Opt. Soc. Am. B **9**, 1585 (1992)
45. A.A. Savchenkov, A.B. Matsko, L. Maleki, Opt. Lett. **31**, 92 (2006)
46. A.A. Savchenkov, I.S. Grudinin, A.B. Matsko, D. Strekalov, M. Mohageg, V.S. Ilchenko, L. Maleki, Opt. Lett. **31**, 1313 (2006)
47. J.-J. Liu, P.P. Banerjee, Q.W. Song, J. Opt. Soc. Am. B **11**, 1688 (1994)
48. L. Palfalavi, J. Hebling, G. Almasi, A. Peter, K. Polgar, K. Lengyel, R. Szipocs, J. Appl. Phys. **95**, 902 (2004)
49. G. Zhong, J. Jian, Z. Wu, J. Opt. Soc. Am. **70**, 631 (1980)
50. T. Volk, N. Rubinina, M. Woehlecke, J. Opt. Soc. Am. B **11**, 1681 (1994)
51. I.S. Grudinin, A.B. Matsko, L. Maleki, Opt. Express **15**, 3390 (2007)
52. H. Guenther, R. Macfarlane, Y. Furukawa, K. Kitamura, R. Neurgaonkar, Appl. Opt. **37**, 7611 (1998)
53. R.V. Schmidt, P.S. Gross, A.M. Glass, J. Appl. Phys. **51**, 90 (1980)
54. G.T. Harvey, G. Astfalk, A.Y. Feldblum, B. Kassahun, IEEE J. Quantum Electron. **QE-22**, 939 (1986)
55. G.E. Peterson, A.M. Glass, T.J. Negran, Appl. Phys. Lett. **19**, 130 (1971)
56. M. Falk, K. Buse, Appl. Phys. B **81**, 853 (2005)
57. F.Z. Henari, K. Cazzini, F.E. Akkari, W.J. Blau, J. Appl. Phys. **78**, 1373 (1995)
58. N.V. Kukhtarev, V.B. Markov, S.G. Odulov, M.S. Soskin, V.L. Vinetskii, Ferroelectrics **22**, 949 (1979)
59. F. Jermann, J. Otten, J. Opt. Soc. Am. B **10**, 2085 (1993)
60. A. Adibi, K. Buse, D. Psaltis, Phys. Rev. A **63**, 023813 (2001)
61. J. Carnicero, O. Caballero, M. Carrascosa, J.M. Cabrera, Appl. Phys. B **79**, 351 (2004)
62. J. Koppitz, O.F. Schirmer, A.I. Kuznetsov, Europhys. Lett. **4**, 1055 (1987)
63. G.K. Kitaeva, K.A. Kuznetsov, V.F. Morozova, I.I. Naumova, A.N. Penin, A.V. Shepelev, A.V. Viskovatich, D.M. Zhigunov, Appl. Phys. B **78**, 759 (2004)
64. D. Emin, Phys. Rev. B **48**, 13691 (1993)
65. S.A. Basun, D.R. Evans, T.J. Bunning, S. Guha, J.O. Barnes, G. Cook, J. Appl. Phys. **92**, 7051 (2002)
66. Y. Furukawa, M. Sato, K. Kitamura, F. Nitanda, J. Cryst. Growth **128**, 909 (1993)
67. Y. Furukawa, K. Kitamura, Y. Ji, G. Montemezzani, M. Zgonik, C. Medrano, P. Gunter, Opt. Lett. **22**, 501 (1997)
68. G.K. Kitaeva, K.A. Kuznetsov, A.N. Penin, A.V. Shepelev, Phys. Rev. B **65**, 054304 (2002)
69. D.J. Gettemy, W.C. Harker, G. Lindholm, N.P. Barnes, IEEE J. Quantum Electron. **24**, 2231 (1988)
70. I.S. Akhmadullin, V.A. Golenishchev-Kutuzov, S.A. Migachev, Phys. Solid State **40**, 1012 (1998)
71. L. Arizmendi, J.M. Cabrera, F. Agullo-Lopez, J. Phys. C **17**, 515 (1984)
72. P. Herth, D. Schaniel, Th. Woike, T. Granzow, M. Imlau, E. Kratzig, Phys. Rev. B **71**, 125128 (2005)
73. D. Berben, K. Buse, S. Wevering, P. Herth, M. Imlau, Th. Woike, J. Appl. Phys. **87**, 1034 (2000)
74. B. Faust, H. Muller, O.F. Schirmer, Ferroelectrics **153**, 297 (1994)
75. J. Rams, A. Alcazar-de-Velasco, M. Carrascosa, J.M. Gabrera, F. Agullo-Lopez, Opt. Commun. **178**, 211 (2000)

76. D. von der Linde, A.M. Glass, K.F. Rodgers, Appl. Phys. Lett. **25**, 155 (1974)
77. O.F. Schirmer, O. Thiemann, M. Woehlecke, J. Phys. Chem. Solids **52**, 185 (1991)
78. H. Kurz, E. Kratzig, W. Keune, H. Engelmann, U. Gonser, B. Dischler, A. Rauber, Appl. Phys. **12**, 355 (1977)
79. G. Panotopoulos, M. Luennemann, K. Buse, D. Psaltis, J. Appl. Phys. **92**, 793 (2002)
80. G.E. Betts, F.J. O'Donnell, K.G. Ray, IEEE Photonics Technol. Lett. **6**, 211 (1994)
81. K. Peithmann, A. Wiebrock, K. Buse, Appl. Phys. B **68**, 777 (1999)
82. F. Jermann, M. Simon, E. Kratzig, J. Opt. Soc. Am. B **12**, 2066 (1995)
83. S.M. Kostritskii, O.G. Sevostyanov, Appl. Phys. B **65**, 527 (1997)
84. A. Winnacker, R.M. Macfarlane, Y. Furukawa, K. Kitamura, Appl. Opt. **41**, 4891 (2002)
85. Y. Furukawa, K. Kitamura, A. Alexandrovski, R.K. Route, M.M. Fejer, G. Foulon, Appl. Phys. Lett. **78**, 1970 (2001)
86. K. Buse, Appl. Phys. **64**, 273 (1997)
87. M. Simon, S. Wevering, K. Buse, E. Kratzig, J. Phys. D **30**, 144 (1997)
88. O. Beyer, D. Maxein, T. Woike, K. Buse, Appl. Phys. B **83**, 527 (2006)
89. M.L. Gorodetsky, A.E. Fomin, IEEE J. Sel. Top. Quantum Electron. **12**, 33 (2006)

14 Applications of Domain Engineering in Ferroelectrics for Photonic Applications

D.A. Scrymgeour

14.1 Introduction

The advent of the laser in the early 1960's brought a surge of interest in techniques to modify, deflect, and change the frequency of laser light. These functions are extensively used today in such technological applications as displays, telecommunications, analog to digital conversion, printing, and data storage devices. Of the many competing technologies, optical devices fabricated in ferroelectric materials like lithium niobate and lithium tantalate offer a versatile solid-state platform to do all of these functions integrated seamlessly in the same device [1]. By patterning these crystals into periodic gratings, the wavelength of light can be converted to different wavelengths through nonlinear optical effects to create new laser sources not readily available [2]. If the domains are patterned into the shape of lenses or prisms, light passing through the crystal can be focused and deflected through the electro-optic effect [3, 4]. By precisely creating the domain structures in ferroelectric crystals, these functions and others can be combined in a single device offering large design flexibility, compactness, and utility. The chapter will be laid out as follows: a brief introduction to ferroelectric materials and domain creation will be in Sect. 14.2, applications of domain engineered structures for frequency generation and light manipulation will be presented in Sect. 14.3, a brief discussion of the challenges of domain engineered devices in Sect. 14.4, followed by conclusions in Sect. 14.5.

14.2 Ferroelectrics and Domain Engineering

Ferroelectrics are a class of materials that possess spontaneous polarization, P_s, in units of electrical dipole moments per unit volume, in the absence of an external electric field for specific temperature and pressure ranges [5]. This spontaneous polarization is a consequence of the atomic arrangements within the unit cell of the crystal. The spontaneous polarization directions are equivalent in energy, differing only in the direction of the polarization vector, and can be switched between at least two equivalent states by the application of an external electric field. Regions of the crystal with the same spontaneous polarization orientation are called domains, and a domain wall separates two domain states. The coercive field is the external electric field that must be exceeded to switch from one domain state to another. For a given temperature, a domain state is permanent in the crystal until an electric field greater than the coercive field is applied to the crystal.

This chapter will focus on the uniaxial single crystal ferroelectrics $LiNbO_3$ (LN) and $LiTaO_3$ (LT) that only possess two domain states which are orientated 180° to each other along the crystallographic c axis (also labeled as 3 or z axis). The techniques outlined in this chapter apply to all single crystal ferroelectric materials with appropriate electro-optic and nonlinear optical properties, although individual material systems may be less readily adapted to the discussed geometries. Lithium niobate and lithium tantalate have identical crystal structures and very similar optical properties, and as a consequence will be used interchangeably in this chapter. Each materials come in two varieties, a stoichiometric crystal with Li and Nb (or Ta) concentrations equal to each other [Li/(Li + Nb, Ta) = 0.5], and a congruent composition with a deficiency of lithium [Li/(Li + Nb, Ta) $\sim$ 0.485]. The congruent varieties, CLN and CLT, are much easier to grow and hence more common and commercially available than the stoichiometric varieties, SLN and SLT. A major result of the slight composition difference between the systems is the order of magnitude greater coercive field for the congruent materials compared to the stoichiometric varieties (for example $\sim$21 kV/mm versus 4 kV/mm for lithium niobate [6]).

There are several techniques for creating domain structures in LN and LT including poling by diffusion [7, 8], electron irradiation [9], and creation during growth using off-centered Czochraslski technique [10]. But the most common technique by far is electric field poling using structured electrodes. In this technique, an electric field exceeding the coercive field of the material is applied to the bulk of the crystal through patterned surface electrodes. Domains nucleate at regions of highest electric fields at the electrode edges or at defects in the crystal where the electric field is enhanced. Domains grow quickly along the z axis and then spread more slowly in the x–y plane. A schematic of this process is shown in Fig. 14.1, showing the continual balancing of nucleation and growth and the relative domain wall speeds both parallel and perpendicular to spontaneous polarization direction. Creating domains with complicated geometries that mirror lithographically patterning electrodes is termed *domain engineering* and offers a powerful way to create optical devices in a ferroelectric material.

The electrode used in electric field poling is either a patterned metal electrode or a patterned dielectric layer that is then covered by a liquid electrode. Either configuration is commonly used, with the patterned dielectric approach being the easiest in terms of processing, but metal electrodes offering more stablility especially when poling at higher temperatures. Critical to the operation of the device is that domain growth only occurs between the top and bottom electrodes. Unwanted growth in non-electroded regions, called domain overshoot, can result in diminished performance, or in the worst case, ruined devices. This domain overshoot is a particular problem for devices that utilize small domain sizes like periodically poled gratings.

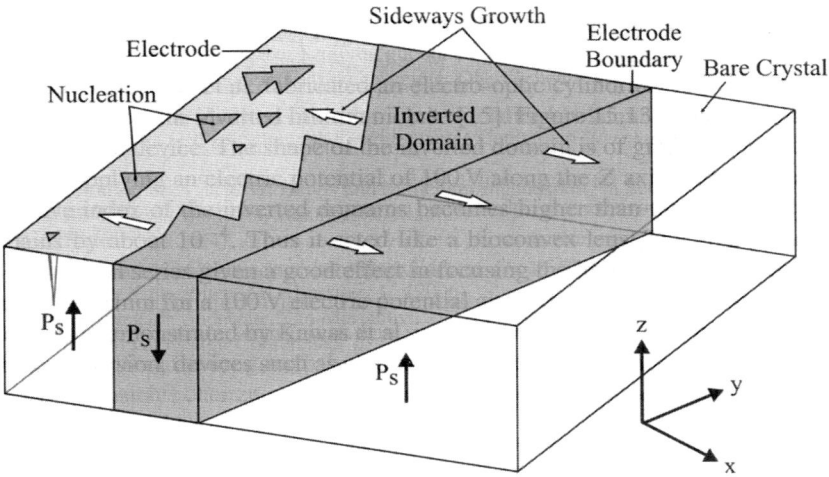

Fig. 14.1. Controlled growth of domains for domain micro engineering. The poling process is a competition between domain nucleation and sideways growth

Control over the applied electric field and resulting transient current is a key component in electric field poling. This transient current, which flows through the external circuit to compensate surface charges of the newly created domains, is related to the area of crystal undergoing domain reversal. Theoretically, knowledge of the electrode area, A, can be used to supply the exact amount of total charge, Q, through $Q = \int i \, dt = 2P_s A$, where P_s is the spontaneous polarization of the crystal, and i is the transient current curve. This can be used as a feedback loop in the poling process. Recently, a number of fundamental studies on the domain switching kinetics exploring the wall velocities, switching times, and stabilization mechanisms have been performed on lithium tantalate [11–13] and lithium niobate [14–16]. These basic studies can be used as guidelines to develop a particular poling waveform, however, most works on electrical poling have been empirically developed and include techniques like a single high voltage pulse [17], a series of short pulses [18], increasing electric fields required to maintain constant transient current flow [19], and short pulses following by low stabilization fields to exploit the backswitching phenomena [20].

14.3 Applications of Domain Engineered Structures

14.3.1 Frequency Conversion

Quasi phase matching (QPM) is a way to achieve a nonlinear optical conversion to create laser light at various frequencies and offers ways to generate coherent light in wavelength regions where compact and efficient lasers are unavailable. Varieties of nonlinear frequency generation include sum frequency generation [21], difference frequency generation [22], optical parametric oscillation and amplification [17], and the, most researched, second harmonic generation (SHG) [2]. There are many excellent reviews and textbooks which discuss in detail the theoretical aspects of quasi phase matching including Armstrong [23] and Yariv [24] and practical reviews about creating structures in nonlinear materials including ones by Rosenman [25] and Houe [26].

Because of the dispersive nature of solids, the speed at which light moves through a medium is dependent upon the wavelength of the light and the index of refraction, n, which is generally not the same at two different wavelengths. Therefore, when a material exhibits nonlinear frequency generation, the fundamental frequency and the generated frequency travel at different speeds though the crystal. This will cause the fundamental and generated secondary waves to go into and out of phase as they propagate through the crystal, and will cause constructive and destructive interference of the beams. This significantly reduces and limits the efficiency of the frequency conversion, as energy flows back and forth between the fundamental and secondary waves.

Quasi phase matching works around this problem by periodically reversing the sign of the nonlinear coefficient to offset the wavevector mismatch between the interacting waves by the reciprocal vectors of the lattice. It was first proposed by Armstrong et al. [23], and has advantages of generating wavelengths of light impossible though traditional birefringence phase matching, accessing the highest nonlinear coefficient, and creating specific frequencies of light which are dependent upon the period of the grating. The most effective QPM technique is the periodic reversal of the domain state in a ferroelectric material as shown in Fig. 14.2. For second harmonic generation, where light at frequency ω is converted to light at 2ω, the necessary domain period, Λ, is given by

$$\Lambda = \frac{m\lambda_\omega}{2(n_{2\omega} - n_\omega)} \quad (14.1)$$

where λ_ω is the wavelength of input light, n_ω and $n_{2\omega}$ are the index of refraction at ω and 2ω, respectively, and m is an integer greater than or equal to 1 [5]. It is

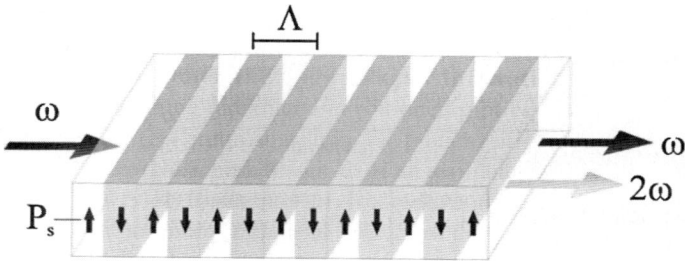

Fig. 14.2. SHG device formed by a grating of alternating polarization where light at ω is converted to light at 2ω

important to point out that for a given material and input light, the domain period, Λ, can be engineered to generate light at specific wavelengths.

Lithium niobate and lithium tantalate are particularly well-suited for quasi phase matching due to the relatively large nonlinear coefficient (25.7 in LN and 13.8 in LT at 1 064 nm) [27], large transmission regions (360 nm–4.5 μm) [28], and the ability to modulate the nonlinear coefficient through domain engineering. QPM has allowed frequency generation from the UV through the visible and up into the IR range. State of the art QPM structures are moving away from congruent varieties, which have high coercive fields (>21 kV/mm), significant photorefractive damage (PRD) and green-induced infrared absorption (GRIIRA) [29, 30], toward stoichiometric and MgO doped crystals. Stoichiometric crystals offer an order of magnitude reduction in the coercive fields [6], shorter wavelength transparency [31], higher thermal conductivity [32], enhanced PRD threshold [33], and reduced GRIIRA [30, 34]. The addition of 1–5% MgO to either congruent or stoichiometric varieties provide similar benefits [35–37].

As the QPM technology matures, the continual frontier is the creation of higher power devices and smaller domain gratings for UV generation. High power levels can introduce thermal lensing, thermal dephasing, and even surface damage that can be avoided by making larger aperture devices. An accurate metric of the required domain period sizes that reflects the challenges is to consider the aspect ratio of the domains, which is the ratio of crystal thickness to grating period. Very fine periods are generally limited to thinner crystals and coarser domain structures are only possible in thicker crystals due spreading of the domains in the x–y plane accompanying growth along the z axis as discussed in Sect. 14.2. A summary of advances and state of the art devices is shown in Table 14.1. As can be seen from this table, the reduction of the coercive field in stoichiometric and MgO doped crystals allow thicker crystals to be poled.

Table 14.1. Domain Periods and Aspect ratios in QPM structures in LiNbO$_3$ and LiTaO$_3$

Domain period	Thickness*	Aspect ratio	Material	Year	Source
15.5 μm	500 μm	32	CLN	1995	[17]
4 μm	500 μm	125	CLN	1999	[38]
2.6 μm	200 μm	76	CLT	1997	[39]
1.5 μm	200 μm	133	CLT	2001	[40]
29 μm	2 mm	70	SLT	2000	[41]
8 μm	830 μm	103	SLT	2005	[42]
8 μm	1 mm	125	Mg:SLT	2004	[32]
6.5 μm	1 mm	154	Mg:SLT	2006	[43]
32 μm	5 mm	156	Mg:CLN	2005	[44]
1.78 μm	300 μm	169	Mg:CLN	2004	[45]

* Thickness of crystal or effective depth of useful domain structures

Quasi phase matched devices require precise control of the domain growth in both the thickness (z direction) and grating directions (x–y plane), which becomes problematic as the grating period decreases or the crystal thickness increases. Current challenges are to reliably create domain structures with aspect ratios greater than ∼150 uniformly over large areas of the crystal. New domain control techniques are needed, as growth along the z axis is accompanied by broadening of the domains in the in x–y plane, which allows domains to move past the boundaries of the electrode.

14.3.2 Electro-Optic Devices

The electro-optic effect is the change in the index of refraction of a material with an applied electric field. Like all second order nonlinearities, this effect is present due to the break in symmetry of the crystal system (specifically along ⟨001⟩ for LiNbO$_3$). The electro-optic effect, for materials with $3m$ symmetry like LN and LT is defined with respect to the positive z axis (or in Voigt notation the 3 direction) and is given as

$$\Delta n_e = -1/2 n_e^3 r_{33} E_3 \qquad (14.2)$$

where n_e is the extraordinary refractive index, r_{33} is the electro-optic coefficient, and E_3 is the field applied parallel to the 3 axis. An electric field parallel to the $+z$ axis sees an extraordinary index change of $-\Delta n_e$ and a change of $+\Delta n_e$ for a field antiparrallel to the $+z$ axis. Under a uniform electric field in the presence of a domain wall, the index increases and decreases on either side of the domain wall to give a total change of $\pm 2\Delta n_e$ at the domain boundary that is directly proportional to the electric field.

This change in indices can be utilized to create optical devices by pattering the domains into specific geometries as shown in Fig. 14.3–14.5. To fabricate such a device, domains are created in the material via patterned electrodes and electric field poling as discussed in Sect. 14.2. After poling, the ends are polished to optical grade, and the entire top and bottom surface are electroded. As long as the field applied to the electrodes does not exceed the coercive field of the material, the domains are permanent. Light traveling through the crystal perpendicular to the domain walls can then be manipulated by the electric field tunable index profile created through domain engineering. Because the change in index is small, ∼10^{-4} even at high fields of 15 kV/mm, clever design of the domains is necessary to create useful devices.

While LN and LT do not have the strongest electro-optic responses, especially when compared to liquid crystals, it is the ability to create domain structures that give these materials utility. The electro-optic coefficients in LiTaO$_3$ and LiNbO$_3$ are summarized in Table 14.2. A figure of merit (FOM) for domain-engineered devices is the coercive field, E_c, times the r_{33} coefficient, which is the effective change of index achievable before domain reversal and destruction of domain patterns. Notice from Table 14.2, that CLN and CLT have much higher FOM than competing materials. The higher coercive fields in the congruent crystals allow a much higher operating field to be applied to the crystal and therefore, a greater index change and

Table 14.2. Figure of merit for domain engineered devices

Material	r_{33} (pm/V)*	E_c (kV/mm)	FOM ($E_c r_{33} \times 10^{-4}$)
CLN	31.5 [46]	21 [6]	6.4
SLN†	38.3 [46]	4 [6]	1.5
CLT	30.5 [28]	21 [47]	6.4
SLT†	31.0 [48]	1.7 [49]	0.52
$(Sr_{0.75}Ba_{0.25})Nb_2O_6$ (SBN)	1340 [50]	0.25 [51]§	3.4
$KTiOPO_4$ (KTP)	35 [52]	~4.1 [53, 54]	1.4

* Measured 633 nm
† Near stoichiometric compositions
§ For composition $(Sr_{0.61}Ba_{0.39})Nb_2O_6$

more effective devices. It must be pointed out that most electro-optic crystals, even those with more than two domain states or where creation of domains is difficult or impossible, can be used in electro-optic devices discussed in this chapter. However, they will rely on patterned electrodes instead of patterned domain shapes, and efficiencies are reduced by a factor of 2 compared to domain engineered crystals (only a change in Δn). Additionally, the non-uniform electric fields near the patterned boundaries of the electrodes can create stray index gradients. Restrictions on the coercive field are removed if the devices are operated in a unipolar direction where the electric field applied in the same direction as spontaneous polarization only limited only by the breakdown strength of the material.

Single Domain Wall Devices

Light deflection using a field tunable refractive index contrast at an interface via the electro-optic effect was first proposed by Lotspeich [55]. In the early 1970's, this technique was accomplished using electro-optically and diffusion-induced optical gratings or polished and stacked prisms of opposite orientation [56, 57]. Only recently have domain engineered devices been demonstrated [58]. Solid-state electro-optic devices and deflectors based on domain inverted ferroelectrics have several advantages over other systems (i.e. galvanometric mirrors and acouto-optic modulators) including non-inertial scanning, small device sizes, durability, and high operating speeds. Additionally, domain engineering allows several device functions to be integrated into a single device.

The simplest domain engineered architecture is a single domain wall shown in Fig. 14.3. At a single domain interface, light travels through the domain wall at

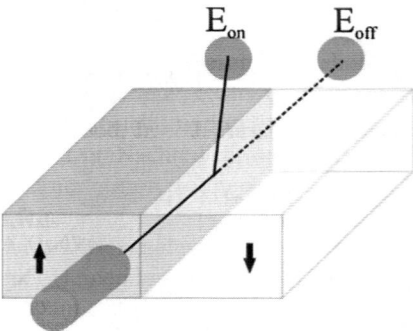

Fig. 14.3. Single interface device with light incident from bottom either passes straight through when no field is applied, or is deflected through total internal refraction when an electric field is applied

an angle (typically at grazing incidence $>80°$ from normal for maximum performance). An applied electric field changes the refractive index on both sides of the wall and results in deflection and/or total internal reflection, which has been demonstrated by a number of groups [59, 60]. This scheme can been used as an analog deflector or a digital switch, and has the advantage of single simple interface and small electrode size which reduces electrode capacitance and increases device speed [61]. A digital switch as pictured in Fig. 14.3 takes advantage of the bias induced total internal reflection (TIR) present at a single interface, allowing light to pass through the boundary or to be reflected totally at a given bias, giving two directions of output light.

Dynamic Focusing Lens Stacks

Patterning domains in the shapes of lenses can create lens stacks to focus and diverge beams passing through the crystal. This can be used to collimate highly divergent beams, e.g. at the output from a fiber, or to change the focal point of a beam in the far field as shown for the domain engineered lens is shown in Fig. 14.4. Because the change in index in each lens is small ($\sim 10^{-4}$), a stack of many lenses is needed to achieve focusing. If the domains are patterned into a stack of simple thin lenses like those pictured in Fig. 14.4(a), the power, ϕ, of the combined lens is given by

$$\phi = \frac{1}{f} = N\left(\frac{n_2 - n_1}{n_1}\right)\left(\frac{2}{R_c}\right) = N\left(\frac{2\Delta n_e}{(n_e - \Delta n_e)}\right)\left(\frac{2}{R_c}\right) \qquad (14.3)$$

where N is the number of lenses in the stack, f is the focal length, R_c is the radius of curvature of the front and back face, and the indices inside and outside the lenses are given by $n_1 = (n_e - \Delta n_e)$ and $n_2 = (n_e + \Delta n_e)$, respectively. The lens power is then tunable by an electric field. Several different groups have demonstrated simple plano-convex lens stacks in $LiTaO_3$ [3, 62].

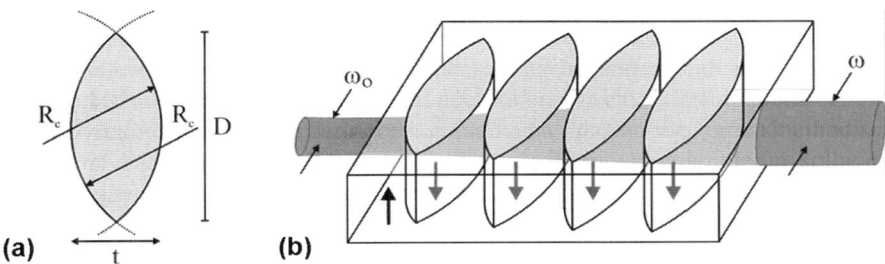

Fig. 14.4. (a) A biconvex lenses formed by two hemispherical surfaces with radius of curvature, R_c. (b) A collimating lens stack composed of cylindrical domain inverted lenses collimating input ω_o to output ω

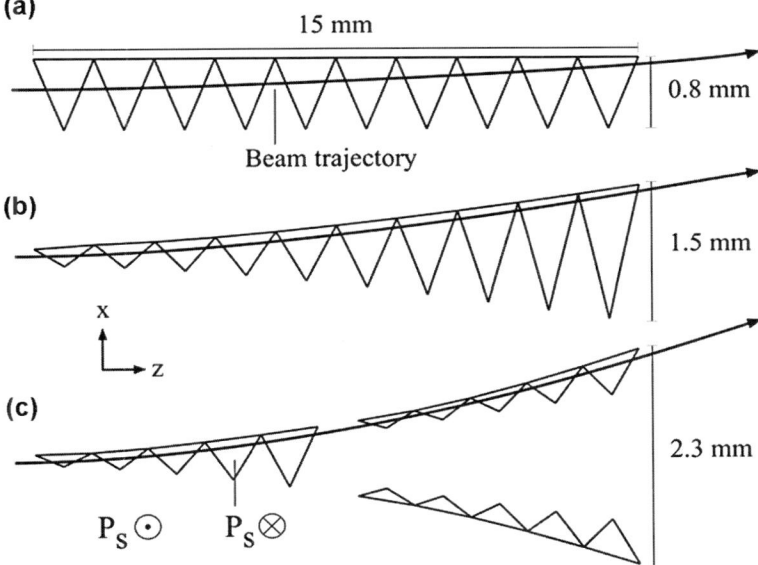

Fig. 14.5. Schematics of (a) rectangular, (b) horn-shaped, and (c) cascaded horn-shaped prism type deflectors with same length and number of interfaces (10). Deflection angles (one way) are 5.25°, 8.45°, and 14.77° respectively. Domain orientation inside the triangles is 180° opposite to that of the surrounding areas

Prism Type Deflectors

A more complicated geometry is the so-called analog or prism type deflector shown in Fig. 14.5. This scanner geometry was theoretical proposed by Lotspeich [55] and first demonstrated in domain inverted waveguides in LiTaO$_3$ by Chen et al. [58]. Since then, there have been continual refinements and improvements in the device architecture and domain engineering to allow greater scanning angles and deflection of larger diameter beams.

In paraxial approximation, where the deflection angles are small, one can use a series of prisms in sequence, with each prism successively deflecting the light beam further from the optical axis. The simplest design, a *rectangular scanner*, consists of a sequence of N identical prisms of base l and height W. The total length of the rectangular scanner is $L = Nl$ and width is W [55]. Such a scanner is shown in Fig. 14.5(a).

The total deflection angle observed external to the crystal, θ_{ext}, of the rectangular scanner in a domain engineered ferroelectric is given by

$$\theta_{ext} = 2\Delta n_e \frac{L}{W} \tag{14.4}$$

where $2\Delta n_e$ is the index difference given in (14.2). Note that the scan angle is electric field tunable. Equation (14.4) shows that the longer the device, L, and narrower the width, W, the higher the deflection angle, θ_{ext}. However, one cannot simply reduce the width and increase the length of the scanner to boost the scan angle, as the width needs to be wide enough to completely enclose the trajectory of the beam. Since the pivot point for deflection in a rectangular scanner lies at the center of the rectangle, the maximum deflection angle inside the crystal is limited to $\theta_{int,\,max} = \tan^{-1}[(W - 2\omega)/L]$ where ω is the radius of the light beam at the output of the scanner, which imposes the geometry for maximum deflection [63].

An improvement to this design is the *horn-shaped scanner* shown in Fig. 14.5(b). The deflection angle for a given length L is increased by keeping the width W of the scanner just wide enough to accommodate the trajectory of the beam [63]. The lateral position of the beam $x(z)$ from the optic axis is determined as a function of propagation distance z as

$$\frac{d^2 x}{dz^2} = \frac{\Delta n}{n_e} \frac{1}{W(z)} \tag{14.5}$$

given $L \gg W$ and $\Delta n \ll n_e$. The scanner width $W(z)$ for a Gaussian beam is given as

$$\frac{W(z)}{2} = x(z) + w_o \sqrt{1 + \left(\frac{\lambda_o (z - z_o)}{\pi n_e \omega_o^2}\right)^2} \tag{14.6}$$

where λ_o is the free space wavelength, ω_o is the beam waist, and z_o is the position of the beam waist [64].

Although the horn shaped scanner significantly improves deflection angles, it too has limitations on beam deflection. The amount of deflection becomes diminished near the end of the scanner, as shown in the displacement and its derivative plots of Fig. 14.6(a) and (b). As noticed in Fig. 14.6(b) for the horn shaped scanner, the derivative curve, which shows the change in deflection angle per length, flattens out near the end of the scanner where the beam is deflected less.

To further increase the deflection angle for a given length, an extension of the horn shaped scanner, called the *cascaded scanner*, is utilized. This design takes advantage of the efficient deflection near the beginning of a horn shaped scanner where the width of the scanner is narrow compared to the beam waist. The cascaded

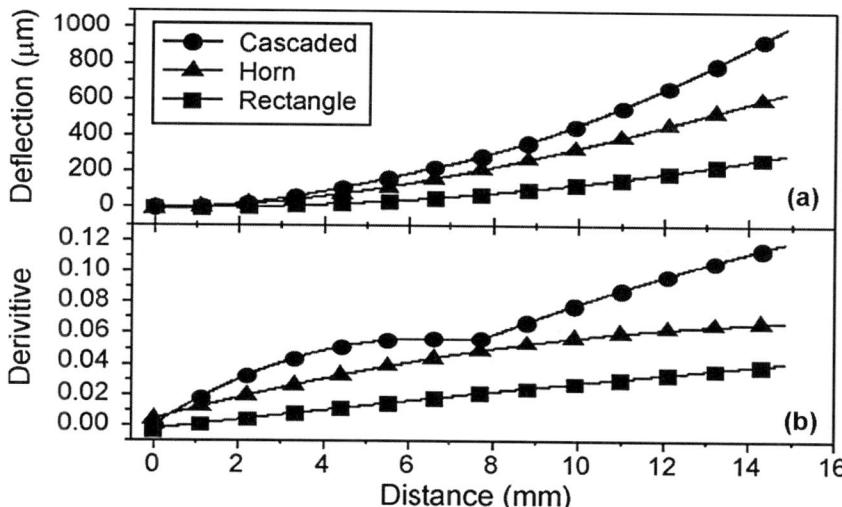

Fig. 14.6. (a) Comparison of the beam trajectories and (b) derivative of trajectories of the three scanner designs pictured in Fig. 14.5 determined from the analytical models

scanner consists of a series of individual scanners aligned so that subsequent scanners in the stack are aligned to the previous scanner's peak deflection. Each subsequent scanner is fabricated in the same piece of electro-optic crystal but has different power supply. A schematic of a 2-stage cascaded scanner is shown in Fig. 14.5(c). At peak operating field, the first stage scanner deflects the beam along a given direction. A second set of scanners is oriented along this maximum deflection direction, so that when the first stage is at its maximum deflection, a bias applied to the second stage scanners will further deflect the beam. By cascading a series of rectangular or horn-shaped scanners such that each successive scanner builds upon the deflection of the previous scanner, one can effectively increase the scan angle arbitrarily. The advantage of this design are that, for a given deflection angle, additional stages can be appended until a final target deflection is achieved, or that the peak field can be reduced for a given scan angle by increasing the number of stages. The trade off in this design is more complex drive controllers, as each scanner must work in tandem with the others.

A comparison of the three scanner designs shown in Fig. 14.5 is given in Fig. 14.6. For the same scanner length (15 mm), number of interfaces (10), input beam size (100 µm), and operating field (15 kV/mm), the rectangular, horn, and cascaded scanners deflect 5.25°, 8.45°, and 14.77°, respectively as determined by analytical models. Scan angle is increased by a factor of 2.8 by going from a rectangular to a cascaded scanner.

A comparison among the different types of electro-optic deflectors is given in Table 14.3. Scanners are characterized in terms of the maximum total angular displacement also called the scan angle. However, this value can be manipulated by

Table 14.3. Analog deflectors in domain-engineered LiNbO$_3$ and LiTaO$_3$

Type	Material	λ (nm)	Resolvable spots	Deflection Max	(°/kV)	(°/kV/mm)	Source
Single interface	LN	632.8	–	1°	1.54	1.18	[60]
Single interface	LN	632.8	50	4.8°	1.92	0.57	[61]
Prism–rectangular	LT	632.8	11	0.46°	0.38	0.19	[58]
Prism–rectangular	LT	632.8	–	1.4°	0.72	0.16	[65]
Prism–horn	LT	632.8	13	10.4°	5.21	0.78	[64]
Prism–split-horn	LT	632.8	–	18.9°	2.09	0.61	[66]
Prism–cascaded	LT	632.8	29	26.08°	11.5	1.2	[67]
Prism–beamlet cascaded	LT	1064	–	10.3°	–	1.9	[68]

focusing optics. A true measure of a deflector is given by the number of resolvable spots, N, in the far field which is given by

$$N = \frac{\theta_{\text{tot}}}{2\theta_{\text{div}}} \qquad (14.7)$$

where θ_{tot} is the total deflection angle (peak-to-peak) and θ_{div} is the far-field divergence angle given by

$$\theta_{\text{div}} = \frac{\lambda}{\pi \omega_0} \qquad (14.8)$$

where λ is the wavelength of light and ω_0 is the waist at the focal point of the beam [55]. In this sense, single interface deflectors have resolution advantages compared to prism-type deflectors. However, due to the finite angular divergence of a focused beam, performance of a single interface device will always show changes in transmitted intensity near TIR. A single interface device does not necessarily have deflection that is linearly dependent on voltage, but can be designed as such. Prism deflectors, however, are inherently linear and capable of high resolution. The increased interfaces increase reflection losses of the devices, but, because Δn is small, maximum Fresnel reflection losses are only $\sim 10^{-6}$ per interface. Prism deflectors offer design flexibility and can be optimized for different geometries and beam sizes [68]. The type of deflector design chosen for a particular device depends upon the device's application.

14.4 Challenges of Domain Engineered Ferroelectric Devices

The design flexibility of domain-engineered devices makes them attractive for many optical applications. However, barriers must be overcome before widespread application. One challenge is the poling of smaller and smaller domain features over larger areas or in thicker crystals as discussed in Sect. 14.2. Another is the high voltage requirements of such devices, which is a particular problem for high frequency operation. While the intrinsic response of the electro-optic effect is known to be in the GHz range [69, 70], the ultimate speed performance of the device is limited by the large voltage requirements (neglecting piezoelectric resonances). When driven at high frequencies, these devices are essentially parallel plate capacitors with current requirements (when driven by a sine wave) given by

$$i = \omega C_\omega V_{\text{peak}} \cos(\omega t) \tag{14.9}$$

where ω is the angular frequency of the applied signal, C_ω is the capacitance of the device at the drive frequency, V_{peak} is the peak voltage applied to the device. The required current output of the driver scales linearly with both frequency and drive voltage, which makes driving these devices at high frequency primarily a driver issue. The voltage requirements, V_{peak}, can be reduced through better optimization of device structures as demonstrated for the prism-type deflector [67] or by moving completely to traveling wave geometries [71, 72]. Another method is to reduce the crystal thickness or to make the devices in thin film or waveguide geometries [73, 74]. Reducing the crystal thickness, however, limits the power handling capability of the devices. Alternative materials with higher electro-optic coefficients can be used, however, as seen Table 14.2, the coercive field of the ferroelectric must be taken into account. The high coercive field of the congruent compositions makes their use in electro-optic devices ideal.

14.5 Conclusions

Despite these limitations, domain engineered devices offer many exciting design possibilities. The true utility and power of domain engineered ferroelectric devices come from the ability to integrate several different functions into a single piece of material. Any of the various applications discussed here – gratings, scanners, lenses – can all be created in a single optical device. Additionally, device elements not based on domain-engineered structures, such as electro-optic phase shifting regions [75], and waveguides to confine the light into near surface channels [76, 77] can be seamlessly integrated into a single piece of crystal. The further development of these devices can be used in next-generation optical devices and systems.

Acknowledgments

The author would like to thank the Gopalan group at The Penn State University where the domain engineered device work was performed. The writing of this document was supported in part by an appointment to the Sandia National Laboratories Truman Fellowship in National Security Science and Engineering, sponsored by Sandia Corporation (a wholly owned subsidiary of Lockheed Martin Corporation) as

Operator of Sandia National Laboratories under its US Department of Energy Contract Number DE-AC04-94AL85000. The submitted manuscript has been authored by a contractor of the US Government under contract No. DE-AC04-94AL85000. Accordingly, the US Government retains a nonexclusive, royalty-free license to publish or reproduce the published form of this contribution, or allow others to do so, for US Government purposes.

References

1. V. Gopalan, T.E. Mitchell et al., Interg. Ferroelectr. **22**, 465 (1998)
2. M.M. Fejer, G.A. Magel et al., IEEE J. Quantum Electron. **28**, 3631 (1992)
3. M. Yamada, M. Saitoh et al., Appl. Phys. Lett. **69**, 3659 (1996)
4. K.T. Gahagan, V. Gopalan et al., Proc. SPIE **3620**, 374 (1999)
5. M.E. Lines, A.M. Glass, in *Principles and Applications of Ferroelectrics and Related Materials*, pp. 59–126 (Clarendon Press, Oxford, 1977)
6. V. Gopalan, T.E. Mitchell et al., Appl. Phys. Lett. **72**, 1981 (1998)
7. S. Miyazawa, J. Appl. Phys. **50**, 4599 (1979)
8. S. Makio, F. Nitanda et al., Appl. Phys. Lett. **61**, 3077 (1992)
9. M. Fujimura, T. Suhara et al., Electron. Lett. **28**, 721 (1992)
10. F. Duan, M. Nai-Ben et al., Appl. Phys. Lett. **37**, 607 (1980)
11. V. Gopalan, T.E. Mitchell, J. Appl. Phys. **83**, 941 (1998)
12. V. Gopalan, T.E. Mitchell et al., Appl. Phys. Lett. **72**, 1981 (1998)
13. S. Kim, V. Gopalan et al., J. Appl. Phys. **90**, 2949 (2001)
14. L.-H. Peng, Y.-C. Fang et al., Appl. Phys. Lett. **74**, 2070 (1999)
15. Y.L. Chen, J.J. Xu et al., Opt. Commun. **188**, 359 (2001)
16. J.H. Ro, T.-H. Kim et al., J. Korean Phys. Soc. **40**, 488 (2002)
17. L.E. Myers, R.C. Eckardt et al., J. Opt. Soc. Am. B **12**, 2102 (1995)
18. W.K. Burns, W. McElhanon et al., IEEE Photonics Technol. Lett. **6**, 252 (1994)
19. D.A. Scrymgeour, Y. Barad et al., Appl. Opt. **40**, 6236 (2001)
20. R.G. Batchko, G.D. Miller et al., Proc. SPIE **3610**, 36 (1999)
21. P. Baldi, C.G. Trevifio-Palacios et al., Electron. Lett. **31**, 1350 (1995)
22. C.Q. Xu, H. Okayama et al., Appl. Phys. Lett. **63**, 1170 (1993)
23. J.A. Armstrong, N. Bloembergen et al., Phys. Rev. **127**, 1918 (1962)
24. A. Yariv, *Introduction to Optical Electronics*, 2nd edn. (Holt Rinehart and Winston, New York, 1976)
25. G. Rosenman, A. Skliar et al., Ferroelectr. Rev. **1**, 263 (1999)
26. M. Houe, P.D. Townsend, J. Phys. D **28**, 1747 (1995)
27. I. Shoji, T. Kondo et al., J. Opt. Soc. Am. B **14**, 2268 (1997)
28. T. Yamada, Group III condensed matter LiNbO$_3$ family oxides. In *Landolt–Bornstein*, ed. by T. Mitsui, S. Nomura (Springer, Berlin, 1981), pp. 149–163
29. R.G. Batchko, G.D. Miller et al., OSA Tech. Digest Ser. **12**, 75 (1998)
30. Y. Furukawa, K. Kitamura et al., Appl. Phys. Lett. **78**, 1970 (2001)
31. Y. Furukawa, K. Kitamura et al., J. Cryst. Grow. **197**, 889 (1999)
32. N.E. Yu, S. Kurimura et al., Jpn. J. Appl. Phys. **43**, L1265 (2004)
33. Y. Furukawa, K. Kitamura et al., Appl. Phys. Lett. **77**, 2494 (2000)
34. M. Katz, R.K. Route et al., Opt. Lett. **29**, 1775 (2004)
35. R. Sommerfeldt, L. Holtmann et al., Phys. Status Solidi A **106**, 89 (1988)

where the directions x, y, and z are the principal dielectric axes that the directions in the crystal along which D and E are parallel. $1/n_x^2$, $1/n_y^2$, and $1/n_z^2$ are the principal values of the impermeability tensor.

According to the quantum theory of solids, the optical dielectric impermeability tensor depends on the distribution of charges in the crystal. The application of an electric field will result in a redistribution of the bond charges and possibly a slight deformation of the ion lattice. The net result is a change in the optical impermeability tensor. This is known as the electro-optic effect. The electro-optic coefficients are defined as

$$\eta_{ij}(E) - \eta_{ij}(0) \equiv \Delta\eta_{ij} = r_{ijk}E_k + s_{ijkl}E_k E_l, \quad (15.2)$$

where E is the applied electric field. The constants r_{ijk} are the linear (or Pockels) electro-optic coefficients, and s_{ijkl} are the quadratic (or Kerr) electro-optic coefficients. Since the higher-order effects are too small, they are neglected in the above expansion. In most practical applications of the electro-optic effect, the applied electric field is small compared with the electric field in atom, which is typically of order 10^8 V/cm. As a result, the quadratic effect is small compared to the linear effect and is often neglected when the linear effect is present. Since η_{ij} is a second-rank tensor, $\Delta\eta_{ij}$ is also a second-rank tensor describing the change in the impermeability tensor. Thus r_{ijk}, the linear electro-optic coefficient tensor, is a third-rank tensor with 27 elements. According to the definition of η_{ij}, it must be a symmetric tensor, provided that the medium is lossless and optically inactive. Consequently, the indices i and j in (15.2) can be permuted. Because of this permutation symmetry, the independent elements of r_{ijk} can be reduced from 27 to 18, and it is convenient to introduce contracted indices to abbreviate the notation. They are defined as

$$\begin{aligned}
1 &= (11), \\
2 &= (22), \\
3 &= (33), \\
4 &= (23) = (32), \\
5 &= (13) = (31), \\
6 &= (12) = (21).
\end{aligned} \quad (15.3)$$

Using these contracted indices, we can write

$$\begin{aligned}
r_{1k} &= r_{11k}, \\
r_{2k} &= r_{22k}, \\
r_{3k} &= r_{33k}, \\
r_{4k} &= r_{23k} = r_{32k}, \\
r_{5k} &= r_{13k} = r_{31k}, \\
r_{6k} &= r_{12k} = r_{21k}, \\
k &= 1, 2, 3.
\end{aligned} \quad (15.4)$$

The linear electro-optic effect exists only in crystals with noncentrosymmetric structure. Since the effect of an electric field on the propagation is expressed most conveniently by giving the changes in the constants $1/n_x^2$, $1/n_y^2$, $1/n_z^2$ of the index ellipsoid. Using the above contracted indices, we can take the equation of the index ellipsoid in the presence of an electric field as

$$\left(\frac{1}{n^2}\right)_1 x^2 + \left(\frac{1}{n^2}\right)_2 y^2 + \left(\frac{1}{n^2}\right)_3 z^2 + 2\left(\frac{1}{n^2}\right)_4 yz + 2\left(\frac{1}{n^2}\right)_5 xz + 2\left(\frac{1}{n^2}\right)_6 xy = 1. \tag{15.5}$$

If we choose x, y, z to be parallel to the principal dielectric axes of the crystal, then with zero applied field, (15.5) must reduce to (15.1). In the presence of an arbitrary electric field $E(E_x, E_y, E_z)$, the linear change in the coefficients due to this field is defined by

$$\Delta\left(\frac{1}{n^2}\right)_i = \left(\frac{1}{n^2}\right)_i\bigg|_E - \left(\frac{1}{n^2}\right)_i\bigg|_{E=0} = \sum_{k=1}^{3} r_{ik} E_k, \tag{15.6}$$

where E_k ($k = 1, 2, 3$) is a component of the applied electric field. Here 1, 2, 3 correspond to the principal dielectric axes x, y, z. Equation (15.3) can be expressed in a matrix form as

$$\left(\Delta\frac{1}{n^2}\right)_i = \begin{pmatrix} r_{11} & r_{12} & r_{13} \\ r_{21} & r_{22} & r_{23} \\ \vdots & \vdots & \vdots \\ r_{61} & r_{62} & r_{63} \end{pmatrix} \begin{pmatrix} E_x \\ E_y \\ E_z \end{pmatrix}. \tag{15.7}$$

The 6×3 matrix with elements r_{ij} is called the electro-optic tensor. The form of the tensor r_{ij} can be derived from symmetry considerations indicating which of the 18 r_{ij} coefficients are zero, as well as the relationships that exist between the remaining coefficients. Thus the equation of the index ellipsoid in the presence of an electric field can be written as

$$\left(\frac{1}{n_x^2} + r_{1k} E_k\right) x^2 + \left(\frac{1}{n_y^2} + r_{2k} E_k\right) y^2 + \left(\frac{1}{n_z^2} + r_{3k} E_k\right) z^2$$
$$+ 2yz r_{4k} E_k + 2xz r_{5k} E_k + 2xy r_{6k} E_k = 1. \tag{15.8}$$

In general, the principal axes of the ellipsoid (15.8) do not coincide with the unperturbed axes (x, y, z) given by the unperturbed ellipsoid (15.1). A new set of principal axes can be found by a coordinate rotation.

15.2.2 Electro-Optical Effect for Crystals of 3m Symmetry Group

For a crystal possess a crystal symmetry of $3m$ such as lithium niobate and lithium tantalite, the electro-optic tensors are in the form

$$r_{ij} = \begin{pmatrix} 0 & -r_{22} & r_{13} \\ 0 & r_{22} & r_{13} \\ 0 & 0 & r_{33} \\ 0 & r_{51} & 0 \\ r_{51} & 0 & 0 \\ -r_{22} & 0 & 0 \end{pmatrix}.$$

Thus, when an arbitrary external electric field $E(E_x, E_y, E_z)$ is applied, the linear change in the coefficients is defined by

$$\left(\Delta\frac{1}{n^2}\right)_i = \begin{pmatrix} 0 & -r_{22} & r_{13} \\ 0 & r_{22} & r_{13} \\ 0 & 0 & r_{33} \\ 0 & r_{51} & 0 \\ r_{51} & 0 & 0 \\ -r_{22} & 0 & 0 \end{pmatrix} \begin{pmatrix} E_x \\ E_y \\ E_z \end{pmatrix}. \quad (15.9)$$

We can obtain the equation of the index ellipsoid in the presence of a field $E(E_x, E_y, E_z)$ as

$$\left(\frac{1}{n_o^2} - r_{22}E_y + r_{13}E_z\right)x^2 + \left(\frac{1}{n_o^2} + r_{22}E_y + r_{13}E_z\right)y^2$$
$$+ \left(\frac{1}{n_e^2} + r_{33}E_z\right)z^2 + 2r_{51}E_y yz - 2r_{51}E_x xz - 2r_{22}E_x xy = 1, \quad (15.10)$$

where $n_x = n_y = n_o$, $n_z = n_e$, since the crystals of 3m symmetry group are uniaxial.

Electric Field Along the z Axis

We now consider the case where the electric field is applied along the z axis of the crystal only. In this case, the equation of the index ellipsoid can be written as

$$\left(\frac{1}{n_o^2} + r_{13}E_z\right)x^2 + \left(\frac{1}{n_o^2} + r_{13}E_z\right)y^2 + \left(\frac{1}{n_e^2} + r_{33}E_z\right)z^2 = 1. \quad (15.11)$$

Since no mixed terms appear in (15.11), the principal axes of the new index ellipsoid remain unchanged. The lengths of the new semiaxes are

$$\begin{cases} n'_x = n_o - \frac{1}{2}r_{13}n_o^3 E_z, \\ n'_y = n_o - \frac{1}{2}r_{13}n_o^3 E_z, \\ n'_z = n_e - \frac{1}{2}r_{33}n_e^3 E_z. \end{cases} \quad (15.12)$$

Notice that under the influence of the electric field in the direction of z axis, the crystal remains uniaxially anisotropic. If a light beam is propagating along the x axis, the birefringence seen by it is

$$n_z - n_y = (n_e - n_o) - \frac{1}{2}(r_{33}n_e^3 - r_{13}n_o^3)E_z. \quad (15.13)$$

Electric Field along the y Axis of the Crystal

While there is only an electric field along the y axis of the crystal, that is to say $E_x = E_z = 0$. Thus the index ellipsoid equation in the presence of a field E_y can be written, according to (15.10), as

$$\left(\frac{1}{n_o^2} - r_{22}E_y\right)x^2 + \left(\frac{1}{n_o^2} + r_{22}E_y\right)y^2 + \frac{1}{n_e^2}z^2 + 2r_{51}E_y yz = 1. \quad (15.14)$$

In this case, it is clear that the new principal axis x' will coincide with the x axis, because the "mixed" term in (15.14) involves only y and z. A rotation in the yz plane is therefore required to put in a diagonal form. Let θ be the angle between the new coordinate $y'z'$ and the old coordinate yz. The transformation from x, y, z to x', y', z' is given by

$$\begin{cases} x = x', \\ y = y'\cos\theta - z'\sin\theta, \\ z = y'\cos\theta + z'\sin\theta. \end{cases} \quad (15.15)$$

We now substitute (15.15) for x, y, z in (15.14) and require that the coefficient of the $y'z'$ vanishes. This yields

$$\left(\frac{1}{n_o^2} - r_{22}E_y\right)x'^2 + \left(\frac{1}{n_o^2} + r_{22}E_y + r_{51}E_y \tan\theta\right)y'^2 + \left(\frac{1}{n_e^2} - r_{51}E_y \tan\theta\right)z'^2 = 1 \quad (15.16)$$

with θ given by

$$\tan 2\theta = \frac{2r_{51}E_y}{\frac{1}{n_o^2} - \frac{1}{n_e^2}}. \quad (15.17)$$

Thus, the new principal refractive indices, according to (15.13), are given by

$$\begin{cases} n'_x = n_o + \frac{1}{2}r_{22}E_y n_o^3, \\ n'_y = n_o - \frac{1}{2}r_{22}E_y n_o^3 - \frac{1}{2}r_{51}E_y n_o^3 \tan\theta, \\ n'_z = n_e + \frac{1}{2}r_{51}E_y n_e^3 \tan\theta. \end{cases} \quad (15.18)$$

As indicated by (15.18), $n_{x'} \neq n_{y'}$. Thus, the crystal is no longer uniaxial, and the new index ellipsoid equation (15.16) has its principal axes rotated at an angle θ about the x axis with respect to the unperturbed principal axes when a filed E_y is applied along the y direction. This angle is very small, even with moderately high electric field. For LiNbO$_3$ with an applied field $E_y = 10^6$ V/m, this angle is only $0.1°$. So with a moderate field E_y, θ is very small and is almost linear proportional to $r_{51}E_y$, and we have the following:

$$\theta \approx \frac{r_{51}E_y}{\frac{1}{n_o^2} - \frac{1}{n_e^2}}. \quad (15.19)$$

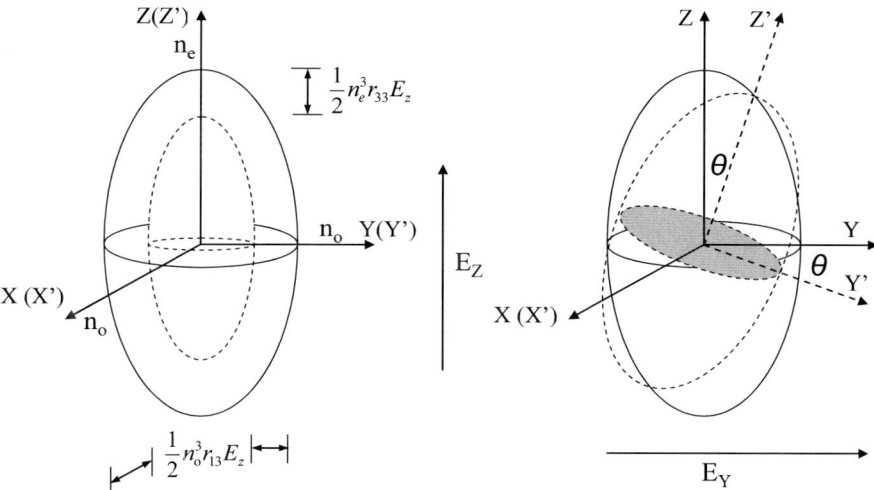

Fig. 15.1. Deformation of the index ellipsoid when an external electric field is applied along the crystal with 3*m* symmetry

Figure 15.1 shows an illustrated diagram of the index ellipsoid deformation when an external electric field is applied. When an electric field is applied along the Z axis, the crystal remains uniaxial with the same principle axes, only the length of the principle axes changes according to the applied electric field. Whereas in the presence of an external electric field along the Y axis, the index ellipsoid deforms to make the Y and Z axes rotate a small angle about the X axis.

15.2.3 Electro-Optical Effect in Periodically Domain-Inverted Crystals with 3*m* Symmetry

As mentioned in the introduction part, in the periodically domain-inverted crystals, the spontaneous polarization is periodically reversed by domain-inversion. Besides the nonlinear optical coefficient, other third-rank tensors, such as the electro-optic (EO) coefficient, are also modulated periodically because of the periodically reversed ferroelectric domains.

Figure 15.2 shows a simple schematic diagram of a *z* cut periodically domain-inverted crystals with 3*m* symmetry. X, Y, and Z represent the principal axes of the original index ellipsoid. The arrows inside the crystal indicate the spontaneous polarization of the original domain and inverted domain. This type of domain-inverted crystal can be fabricated by room temperature electric field poling technique easily with a periodic electrode on the surface of the crystal. For a ferroelectric crystal with the group symmetry of 3*m*, in the inverted domain, the crystal structure rotates 180° about the X axis, thus the electro-optic coefficients change subsequently under this operation. It is easy to demonstrate that all elements of the electro-optic tensor have

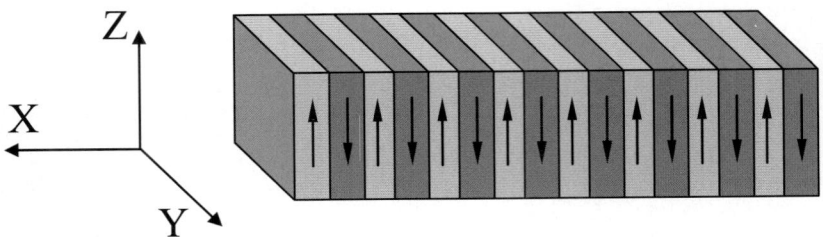

Fig. 15.2. Schematic diagram of a periodically domain-inverted crystals

different signs in different domains. Since the sign of the electro-optic coefficients for the original domain and inverted domain is opposite to each other, a periodic layered media with a periodically refractive index modulated structure is induced by applying an electric field on the periodically domain-inverted crystal.

Now we discuss two basic structures based on the electro-optic effect of the periodically domain-inverted crystals with $3m$ symmetry. One is Bragg diffraction grating structure, and the other one is a so-called Solc layered structure.

Bragg-Diffraction Grating Structure

By applying an electric field E_z along the Z axis of the crystal of $3m$ symmetry with periodically inverted domain, a periodically refractive index modulated structure is induced. The refractive index of original domain and inverted domain are given as follows:

$$\begin{cases} n_{x\pm} = n_o \mp \frac{1}{2} r_{13} n_o^3 E_z, \\ n_{y\pm} = n_o \mp \frac{1}{2} r_{13} n_o^3 E_z, \\ n_{z\pm} = n_e \mp \frac{1}{2} r_{33} n_e^3 E_z, \end{cases} \quad (15.20)$$

where $n_{x,y,z\pm}$ correspond to the refractive index of the original and inverted domain, respectively. E_z is the applied electric field. Thus, the induced refractive index change is periodic in sign. Its amplitude is given by

$$\begin{aligned} \Delta n_o &= -r_{13} n_o^3 E_z/2, \\ \Delta n_e &= -r_{33} n_e^3 E_z/2, \end{aligned} \quad (15.21)$$

where Δn_o is the refractive index change for the o-polarized light and Δn_e for the e-polarized light.

Figure 15.3 shows the schematic diagram of the Bragg-diffraction grating structure based on periodically domain-inverted crystals. Assume that the period of the periodically inverted domain is Λ, the length and the thickness of the crystal are L and d, respectively.

While a uniformed electric field is applied along the Z axis of the crystal, an incident e-polarized or o-polarized light, which propagates through the crystal at the angle θ_B called the Bragg angle to the periodically index modulated structure,

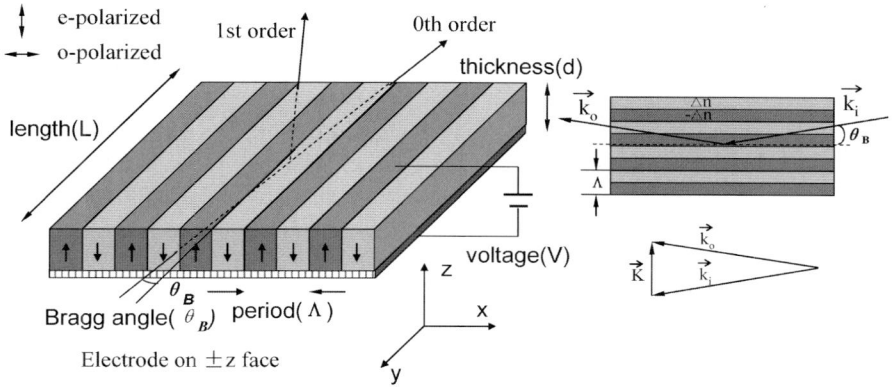

Fig. 15.3. Schematic diagram and principle of a Bragg grating structure based on periodically domain-inverted crystals

is strongly diffracted. The 0th-order is the nondiffracted light, and the 1st-order is the first-order diffracted light. As shown in Fig. 15.3, the relation between the wave number vectors of an incident light k_i, a diffracted light k_o, and a periodic grating K is shown in the following equation:

$$\begin{cases} k_i + K = k_o, \\ |k_i| = |k_o| = 2\pi n/\lambda, \\ |K| = 2\pi/\Lambda. \end{cases} \qquad (15.22)$$

Then the light Bragg angle θ_B of a incident light with wavelength of λ, which diffracted strongly by the Bragg-diffraction grating with a period of Λ, is given by

$$\theta_B = \sin^{-1}\left(\frac{\lambda}{2n\Lambda}\right). \qquad (15.23)$$

In this equation, we can see that, for a given grating period, a large range of wavelengths can satisfy this equation by alternation of the launch angle such that the only constraints on the operating condition are set by the device geometry. Using a simple model based on a thick sinusoidal grating, the first-order diffraction efficiency of such a thick Bragg grating is given by the equation [14]

$$\eta = \sin^2(\pi \Delta n L / \lambda \cos \theta_B), \qquad (15.24)$$

where Δn is the change of the refractive index between original domains and inverted domains when an electric field is applied along Z axis. Δn equals to Δn_o for the o-polarized light and Δn_e for the e-polarized light which is given by (15.21). Since Δn is proportional to the applied electric filed amplitude as described in (15.21), the intensity of the diffracted light which determined by the diffraction efficiency is proportional to the field intensity. Thus, this grating structure can produce intensity modulation directly and can be used as a modulator. According to (15.24),

100% diffraction efficiency is obtained when

$$\pi \Delta n L / \lambda \cos \theta_B = (2m + 1)\pi/2, \quad m = 0, 1, 2 \ldots. \quad (15.25)$$

It should be noted that in the real case the induced refractive index modulated profile is far from the sinusoidal and follows a pattern of square wave. While this refractive index structure can be Fourier decomposed into many fine period sinusoidal gratings, which would in turn cause scattering of the incident beam into several higher orders, these higher-order diffracted spots are not Bragg-matched and therefore contain only a small fraction of power from the incident beam.

Solc Layered Structure

As mentioned in Sect. 15.2.2, when an electric field is applied along the Y axis of a crystal with $3m$ symmetry, the new index ellipsoid deforms to make its principal axes rotated at an angle θ about the x axis with respect to the unperturbed principal axes. Since θ is proportional to the product of the electric field and the electro-optic coefficient, when a uniformed electric field is applied along the Y axis of the crystal, the rotation angle of the original domain is opposite to that of the inverted domain because of the reversal of the spontaneous polarization. Thus, in a periodically domain-inverted crystal with a uniformed electric field applied along the Y axis, a structure with alternating left and right rotation angle θ will be formed due to the periodic EO coefficient. This structure is similar to the well-known folded Solc filter structure with alternating azimuth angles of the crystal axes. So we call it the Solc layered structure.

Figure 15.4 shows the schematic diagram of a Solc layered structure based on electro-optic periodically domain-inverted crystals. X, Y and Z are the principal axes of the unperturbed index ellipsoid, Λ is the period of this structure, and N is the period number. The arrows inside the structure indicate the spontaneous polarization directions. When an electric field is applied along the Y axis, the index ellipsoid deforms. $Y_{o,i}$ and $Z_{o,i}$ are the new perturbed principal axes of the original domain and the inverted domain, respectively.

In this structure, by applying a uniformed electric field along the Y axis, alternating rotation of the crystal axes is realized due to the periodic electro-optic coefficient change. Thus this structure can be viewed as a periodic medium. The alternating rotation of the Y, Z axes constitute a periodic perturbation to the propagation of Z-polarized and Y-polarized wave. This perturbation couples these two waves.

A coupled-mode theory can be used to study the properties of such a structure. The original dielectric tensors in the principal axes X, Y, and Z is

$$\varepsilon(0) = \varepsilon_0 \begin{bmatrix} n_o^2 & 0 & 0 \\ 0 & n_o^2 & 0 \\ 0 & 0 & n_e^2 \end{bmatrix}. \quad (15.26)$$

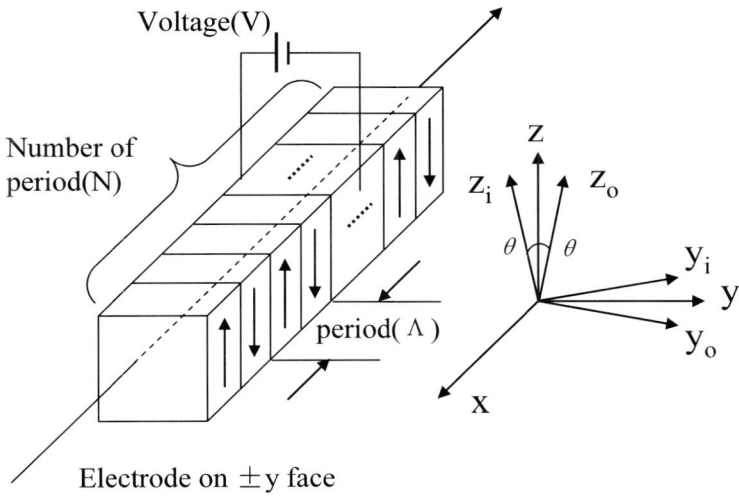

Fig. 15.4. Schematic diagram of a Solc layered structure based on electro-optic periodically domain-inverted crystals

With an electric field applied along the Y axis of the crystal, the new principal axes rotated alternately left and right about the x axis with respect to the unperturbed principal axes due to the periodic electro-optic coefficients. The rotation angle is given by

$$\theta = \frac{r_{51} E_y}{(1/n_o^2 - 1/n_e^2)} f(x), \tag{15.27}$$

where

$$f(x) = \begin{cases} +1, & \text{if } x \text{ is in the original domains,} \\ -1, & \text{if } x \text{ is in the inverted domains,} \end{cases} \tag{15.28}$$

is a factor due to the periodic electro-optic coefficients in the periodically domain-inverted crystal. Thus the dielectric tensors of periodically domain-inverted crystal in the original XYZ coordinate system with an electric field applied along the Y axis is given by

$$\varepsilon = \varepsilon(0) + \Delta\varepsilon, \tag{15.29}$$

where

$$\Delta\varepsilon = \varepsilon_0 \begin{bmatrix} 0 & 0 & 0 \\ 0 & 0 & -\frac{1}{2}(n_o^2 - n_e^2)\sin 2\theta \\ 0 & -\frac{1}{2}(n_o^2 - n_e^2)\sin 2\theta & 0 \end{bmatrix}$$

$$= -\varepsilon_0 r_{51} E_Y n_o^2 n_e^2 \begin{bmatrix} 0 & 0 & 0 \\ 0 & 0 & 1 \\ 0 & 1 & 0 \end{bmatrix} f(x) \tag{15.30}$$

is the dielectric tensor change. Since θ is very small and is given by (15.19), the diagonal matrix elements of the dielectric tensor change are neglected. Assuming in one period that the thickness of the original domain is $k\Lambda$ ($0 < k < 1$), the thickness of the inverted domain is $(1-k)\Lambda$. Thus this period function $f(x)$ can be written as the Fourier series

$$f(x) = \sum_m \frac{i(1 - e^{-i2mk\pi})}{m\pi} \exp\left[-im\frac{2\pi}{\Lambda}x\right], \quad m = 1, 2, 3, \ldots. \tag{15.31}$$

The change of the dielectric tensor could be viewed as a disturbance, thus the coupled wave equations of the Z-polarized and Y-polarized plane wave may be obtained as [13]

$$\begin{cases} dA_2/dx = -i\kappa A_3 e^{i\Delta\beta x}, \\ dA_3/dx = -i\kappa^* A_2 e^{-i\Delta\beta x}, \end{cases} \tag{15.32}$$

with

$$\Delta\beta = k_1 - k_2 - G_m, \quad G_m = m\left(\frac{2\pi}{\Lambda}\right),$$

$$\kappa = -\frac{\omega}{2c} \frac{n_o^2 n_e^2 r_{51} E_Y}{\sqrt{n_o n_e}} \frac{i(1 - e^{-i2mk\pi})}{m\pi}, \quad m = 1, 2, 3, \ldots, \tag{15.33}$$

where A_2 and A_3 are the amplitudes of the Y-polarized and Z-polarized waves, respectively; k_1 and k_2 are the corresponding wave vectors, and G_m is the mth reciprocal vector. The general solution of the coupled mode equation is

$$\begin{aligned} A_2(x) &= e^{-i(\Delta\beta/2)x}\left\{\left[\cos sx - i\frac{\Delta\beta}{2s}\sin sx\right]A_2(0) - i\kappa^*\frac{\sin sx}{s}A_3(0)\right\}, \\ A_3(x) &= e^{i(\Delta\beta/2)x}\left\{-i\kappa\frac{\sin sx}{s}A_2(0) + \left[\cos sx - i\frac{\Delta\beta}{2s}\sin sx\right]A_3(0)\right\}, \end{aligned} \tag{15.34}$$

where s is given by $s^2 = \kappa\kappa^* + (\Delta\beta/2)^2$. From (15.34) one can see that the fraction of power, which is coupled from Z-polarized to Y-polarized or vice versa in the length of the crystal which equals to $N\Lambda$, is

$$\frac{|\kappa|^2}{|\kappa|^2 + (\Delta\beta/2)^2} \sin^2 N\Lambda\sqrt{|\kappa|^2 + (\Delta\beta/2)^2}. \tag{15.35}$$

The maximum conversion is achieved when $\Delta\beta = k_1 - k_2 - G_m = 0$, which means that the reciprocal vector may also compensate for the wave vector mismatch similar to QPM frequency conversion. This condition could be also called the QPM condition. This condition can be expressed as

$$\lambda = \Delta n \Lambda/m, \quad m = 1, 2, 3, \ldots, \tag{15.36}$$

where Δn is the difference of the refractive index between Z-polarized and Y-polarized light. This means that when the period Λ of the structure is given, the QPM condition can be satisfied for a given light with wavelength λ according to this equation. Beside the QPM condition, the dynamical condition

$$|\kappa|N\Lambda = (2n+1)\pi/2, \quad n = 0, 1, 2, \ldots,$$

should also be satisfied for 100% conversion to occur.

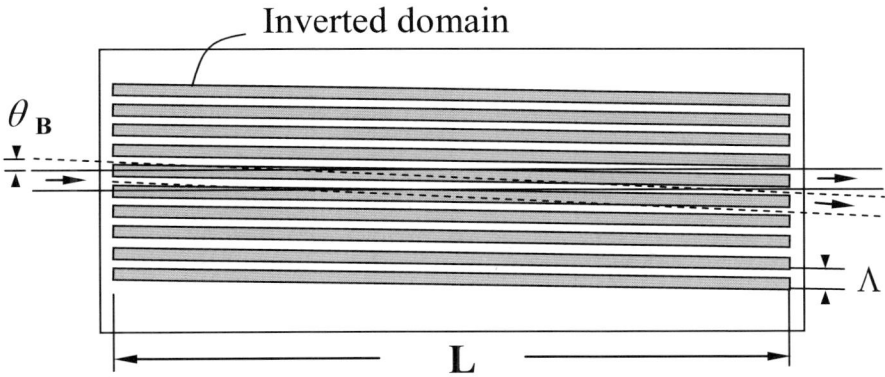

Fig. 15.5. Topview of the switching devices [15]

15.3 Applications

Based on these two basic structures in periodically domain-inverted crystal discussed hereinbefore, many application devices such as electro-optic Bragg modulator, electro-optic Solc filter, and so on can be realized. Some other novel application devices based on the Electro-optic effect in periodically domain-inverted crystal with $3m$ symmetry are also introduced in the below paragraphs.

15.3.1 Devices Based on Bragg Diffraction Grating Structure

As we discussed in Sect. 15.2.3, when an electric field is applied along the Z axis of the periodically domain crystal with $3m$ symmetry, a Bragg-diffraction grating structure is induced. The Bragg diffraction grating will deflect the incident light to the diffracted light. The intensity of the diffracted light determined by the diffraction efficiency is proportional to the field intensity. Thus, this structure can act as an electro-optic deflector, electro-optic switching device, or electro-optic modulator.

Figure 15.5 shows the top view of a switching device which is the first demonstration of a electro-optically induced Bragg grating switch in periodically domain-inverted lithium niobate by Yamada et al. in 1996 [15]. This device was fabricated in a 200 μm thick lithium niobate of length 4 mm and grating period 8 μm and was tested using light of wavelength 633 nm. The power ratio of the light in the diffracted order with applied voltage (V) is shown in Fig. 15.6. This data shows a good fit to a sinusoidal curve, and the on/off voltage is approximately 75 V. But the power ratio of the diffracted light is not 0 as shown in Fig. 15.6, which is treated as a voltage offset. In both works, a voltage offset is seen, which is attributed to a residual grating after poling caused by the existence of an internal field, and this phenomena is also observed in other applications of the periodically domain-inverted lithium niobate.

The next Bragg grating switch in periodically domain-inverted lithium niobate (PPLN) was developed by Gnewuch et al. in 1998 [16]. This is a bulk-optical Bragg

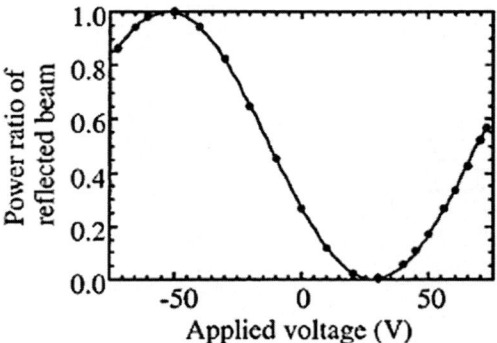

Fig. 15.6. The power ratio of the diffracted light plotted against applied voltage [15]

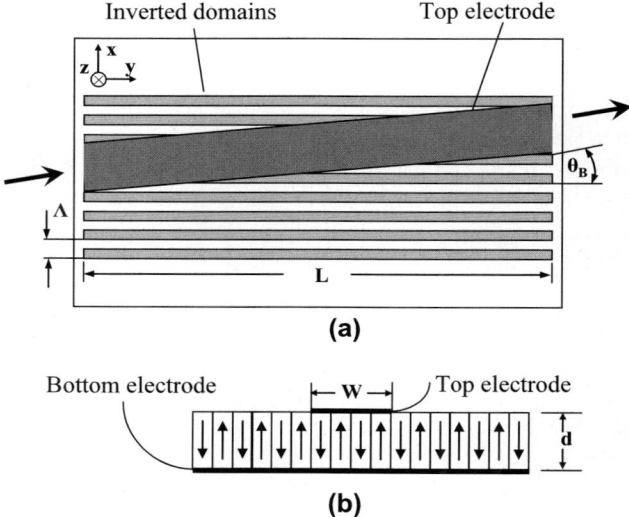

Fig. 15.7. Schematic of the grating deflector. (a) Top view where x, y, z are the main axes of the one-domain LiNbO$_3$ crystal, L and Λ are the grating length and period, respectively, and θ_B is the Bragg angle. (b) View of the x–z plane in the domain-inverted region, where d is the wafer thickness, and w is the top electrode width [16]

deflectors in PPLN with, in comparison to AO Bragg-deflectors, similar diffraction efficiency but fivefold reduction in switching time. They demonstrated both static operation and its nanosecond response. Figure 15.7 shows the schematic of the grating deflector.

The device was 300 μm thick, with an area of periodically domain-inverted regions forming a grating of length 20 mm and period 40 μm, and was tested at a wavelength of 633 nm. Response of the device to 100 ns, 20 ns, and 5 ns pulses are obtained. The static efficiency of the diffracted (first-order) and nondiffracted (zero-

surfaces of the crystals. It is easy to fabricate periodic structures that are special in the domain shape.

In 1996, Yamada et al. fabricated an electro-optic cylindrical lens based on the periodically domain-inverted lithium niobate [15]. Figure 15.13 is the schematic of the cylindrical device. The shape of the inverted domain is of grains of rice 0.2 mm wide R. Applying an electric potential of 100 V along the Z axis of the crystal, the refractive index of the inverted domains becomes higher than that of the adjacent domains by about 10^{-4}. Thus it acted like a bioconvex lens. The arrangement of 50 domains in series given a good effect in focusing the light beam with a focusing length of 8.5 mm for a 100 V electric potential applied. A similar electro-optic lens stacks also demonstrated by Kawas et al. in lithium tantalite. Based on other shapes domain-inversion, devices such as electro-optic scanner are realized [25, 26].

References

1. W. Martienssen, H. Warlimont, Ferroelectric and antiferroelectric. In *Springer Handbook of Condensed Matter And Materials Data* (Springer, Berlin, 2005), p. 903.
2. J. Valasek, Phys. Rev. **15**, 537 (1920)
3. J. Valasek, Phys. Rev. **17**, 475 (1921)
4. T.H. Maiman, Stimulated optical radiation in ruby masers. Nature **187**, 493–494 (1960)
5. P.A. Franken, A.E. Hill, C.W. Peters, G. Weinreich, Generation of optical harmonics. Phys. Rev. Lett. **7**, 118–119 (1961)
6. J.A. Armstrong, N. Bloembergen, J. Ducuing, P.S. Pershan, Interaction between light waves in a nonlinear dielectric. Phys. Rev. **127**, 1918 (1962)
7. R.C. Miller, Optical harmonic generation in single crystal $BaTiO_3$. Phys. Rev. **134**, A1313–A1319 (1964)
8. S. Thaniyavarn, T. Findakly, D. Booher, J. Moen, Domain-inversion effects in $Ti–LiNbO_3$ integrated optical devices. Appl. Phys. Lett. **46**, 933 (1985)
9. W. Wenshen, Z. Qun, G. Zhaohua, F. Duan, Study of $LiTaO_3$ crystals grown with modulated structure. J. Cryst. Growth **79**, 706 (1986)
10. A. Feisst, P. Koidl, Current induced periodic ferroelectric domain structures in $LiNbO_3$ applied for efficient nonlinear optical frequency mixing. Appl. Phys. Lett. **47**, 1125–1127 (1985)
11. M. Yamada, K. Kishima, Fabrication of periodically reversed domain structure for SHG in $LiNbO_3$ by direct electron beam lithography at room temperature. Electron. Lett. **26**, 188 (1991)
12. M. Yamada, N. Nada, M. Saitoh, K. Watanabe, First-order quasi-phase matched $LiNbO_3$ waveguide periodically poled by applying an external field for efficient blue second-harmonic generation. Appl. Phys. Lett. **62**(5), 435 (1993)
13. A. Yariv, P. Yeh, *Optical Waves in Crystal: Propagation and Control of Laser Radiation* (Wiley, New York, 1984)
14. H. Kogelnik, Bell Syst. Tech. J. **48**, 2909 (1969)
15. M. Yamada, M. Saitoh, H. Ooki, Electric-field induced cylindrical lens, switching and deflection devices composed of the inverted domains in $LiNbO_3$ crystals. Appl. Phys. Lett. **69**, 3659–3661 (1996)

16. H. Gnewuch, C.N. Pannell, G.W. Ross, P.G.R. Smith, H. Geiger, Nanosecond response of Bragg deflectors in periodically poled LiNbO$_3$. IEEE Photonics Technol. Lett. **10**, 1730–1732 (1998)
17. M. Yamada, Electrically induced Bragg-diffraction grating composed of periodically inverted domains in lithium niobate crystals its application devices. Rev. Sci. Instrum. **71**(11), 4010–4016 (2000)
18. Y.Y. Lin, S.T. Lin, G.W. Wang, A.C. Chiang, Y.C. Huang, Y.H. Chen, Electro-optic periodically poled lithium niobate Bragg modulator as a laser Q-switch. Opt. Lett. **32**, 545–547 (2007)
19. Y.Q. Lu, Z.L. Wan, Q. Wang, Y.X. Xi, N.B. Ming, Electro-optic effect of periodically poled optical superlattice LiNbO$_3$ and its applications. Appl. Phys. Lett. **77**, 3719–3721 (2000)
20. Y.H. Chen, Y.C. Huang, Actively Q-switched Nd:YVO$_4$ laser using an electro-optic periodically poled lithium niobate crystal as a laser Q-switch. Opt. Lett. **28**, 1460–1462 (2003)
21. X.F. Chen, J.H. Shi, Y.P. Chen, Y.M. Zhu, Y.X. Xia, Y.L. Chen, Electro-optic Solc-type wavelength filter in periodically poled lithium niobate. Opt. Lett. **28**, 2115–2117 (2003)
22. J.H. Shi, X.F. Chen, Y.P. Chen, Y.M. Zhu, Y.X. Xia, Y.L. Chen, Observation of Solc-like filter in periodically poled lithium niobate. Electron. Lett. **39**, 224–225 (2003)
23. J. Shi, X. Chen, Y. Xia, Y. Chen, Polarization control by use of the electro-optical effect in periodically poled lithium niobate. Appl. Opt. **42**, 5722–5725 (2003)
24. M.J. Kawa, D.D. Stancil, T.E. Schlesinger, V. Gopalan, Electro-optic lens stacks on LiTaO$_3$ by domain-inversion. J. Lightwave Technol. **15**, 1716–1719 (1997)
25. V. Gopalan, M.J. Kawas, M.C. Gupta, T.E. Schlesinger, D.D. Stancil, Integrated quasi-phase-matched second harmonic generator and elctrooptic scanner on LiTaO$_3$ single crystal. IEEE Photonics Technol. Lett. **8**, 1704–1706 (1996)
26. K.T. Gahagan, V. Gopalan, J.M. Robinson, Q.X. Jia, T.E. Mitchell, M.J. Kawas, T.E. Schlesinger, D.D. Stancil, Integrated electro-optic lens/scanner in a LiTaO$_3$ single crystal. Appl. Opt. **38**, 1186–1189 (1996)

Index

$3m$ symmetry group, 404, 405

absorption coefficient, 238
absorption length, 238

beat-note frequency, 290
birefringence, 174, 178, 187, 201, 202
Borrmann effects, 233
Bragg conditions, 232, 233
Bragg-diffraction grating, 408
Bravais lattice, 266
bright soliton, 110–113, 115
buried waveguide, 85, 87, 90

cantilever, 14–17
charged point defects, 240
classical electronic polarization mechanism, 235
coercive field, 177, 191, 193
confocal microscopy, 141
constructive interference, 232
cross-talk, 213
crystal lattice parameters, 230

dark soliton, 106, 107
deflector, 393, 395, 396
difference-frequency generation, 288
differential etching, 5, 8, 11, 17
diffractometers, 242
diffuse scattering peak, 240
digital holography, 173, 174
dislocations, 240
displacement field, 237
domain engineering, 22, 23, 27–29, 40, 44, 385, 386, 390, 391, 393
domain inversion, 153–158
domain wall width, 222
dual grid method, 273

effective absorption coefficient, 238
effective penetration length, 238
elastic scattering phenomena, 232
electric field distribution, 103–105
electric field poling (EFP), 59, 172

electro-optic deflector, 413
electro-optic effect, 178, 192, 311, 322, 390, 402, 403
electro-optic modulator, 413
electro-optic switching device, 413
emission energies, 144
emission spectrum, 139, 154
entangled photon, 298, 300
etch-rate, 6–8, 10, 17

ferroelectric crystal, 124, 401, 407
ferroelectric domain, 210
fiber alignment groove, 13
finite difference time domain (FDTD), 309
finite element method (FEM), 323
focused ion beam (FIB), 313
frequency comb, 292
frequency conversion, 387, 388, 402
fundamental grating vector, 240

gas lasers, 287

high resolution X-ray diffraction, 232
holographic lithography, 61

incoherent scattering processes, 247
interdigital transducer (IDT), 327
internal field, 182, 192, 194
ion beam milling, 4

kinematical scattering theory, 233
KTP, 190, 191

laser ablation, 4
lateral electrostatic force microscopy, 223
lens stacks, 392
lithium niobate ($LiNbO_3$), 101, 118, 130, 171, 174, 198, 210, 338, 341–343, 346–348, 350, 352, 359, 360, 362–364, 369, 370, 372–377, 401, 404
lithium tantalate, 81, 404

mid-IR metrology, 288
modulator, 338–340, 413, 415, 416
Moiré effect, 67

Index

nonlinear photonic crystal, 22, 32, 263–265, 272, 278, 280
nonlinear waveguide, 85
not conventional Czochralski technique, 230

optical damage, 81, 85, 88, 94, 95
optical parametric oscillators, 288
overpoling, 59, 60, 63, 65, 66

parametric amplification, 124
parametric down-conversion, 298
perfectly matched layer (PML), 310
periodic poling, 85
periodically domain-inverted, 402, 407
periodically poled $LiNbO_3$, 27, 29, 32, 34, 39, 187, 288
periodically poled $LiTaO_3$, 30
PFM background, 216
PFM calibration, 213
PFM frequency dependency, 216
PFM resolution, 222
phase matching, 286
phononic band structure, 324
phononic crystal, 307
photonic band structure, 309
photonic waveguide, 319
photorefractive effect, 103
photorefractivity, 338, 352, 353, 355, 356, 359–361, 365, 366, 372, 376
photovoltaic effect, 238
phoXonic crystal, 333
piezoelectric effect, 326
piezoelectric interaction, 240
piezoresponse force microscopy (PFM), 209
pockels effect, 311, 322
polarizability, 237

Q-switch, 416
quantitative PFM, 221
quantum cascade lasers, 287
quasi-periodic lattice, 273–276, 279
quasi-phase-matching, 22, 263, 279, 287, 401
quasicrystal, 273, 279

rare earth ions, 137–139, 144
reactive ion etching (RIE), 4, 314, 326
reciprocal lattice, 262, 265, 266, 273
reciprocal map, 236

refractive index, 166–171
resonator, 338–347, 349, 352–358, 360, 371
reverse-proton-exchange (RPE), 85

scanning force microscopy (SFM), 211
scanning near-field optical microscope (SNOM), 320
scattered intensity, 236
second harmonic generation, 84, 85, 93, 94, 96
second order susceptibility, 261
self-assembly, 40, 42
self-deflection, 119, 120
single crystal tip, 17
single domain wall, 391
single-photon detectors, 301
slow light, 309, 312
Solc filter, 410, 413, 417
Solc layered structure, 410, 411, 416
soliton, 105, 118
soliton waveguide, 116, 117, 119
sound line, 330
spectroscopic source, 296
spontaneous polarization, 230, 232, 239
squeezed light, 297
static displacements, 247
stoichiometric lithium tantalate, 83
strain gradient, 251
structural modifications, 237
structure factor, 235
sub-micrometric PPLN crystal, 246
sub-micron structuring, 63
surface acoustic waves (SAW), 324
surface roughness, 8, 9

Takagi approximation, 240
thin films, 332
total reflection, 121
tunable filters, 341

ultra-sharp tip, 11

waveguide, 101, 102, 117
wet etching, 4
whispering gallery modes, 340

X-rays apparatus, 242

y-face etching, 11

Springer Series in
MATERIALS SCIENCE

Editors: R. Hull   R. M. Osgood, Jr.   J. Parisi   H. Warlimont

50 **High-Resolution Imaging and Spectrometry of Materials**
Editors: F. Ernst and M. Rühle

51 **Point Defects in Semiconductors and Insulators**
Determination of Atomic and Electronic Structure from Paramagnetic Hyperfine Interactions
By J.-M. Spaeth and H. Overhof

52 **Polymer Films with Embedded Metal Nanoparticles**
By A. Heilmann

53 **Nanocrystalline Ceramics**
Synthesis and Structure
By M. Winterer

54 **Electronic Structure and Magnetism of Complex Materials**
Editors: D.J. Singh and D. A. Papaconstantopoulos

55 **Quasicrystals**
An Introduction to Structure, Physical Properties and Applications
Editors: J.-B. Suck, M. Schreiber, and P. Häussler

56 **SiO_2 in Si Microdevices**
By M. Itsumi

57 **Radiation Effects in Advanced Semiconductor Materials and Devices**
By C. Claeys and E. Simoen

58 **Functional Thin Films and Functional Materials**
New Concepts and Technologies
Editor: D. Shi

59 **Dielectric Properties of Porous Media**
By S.O. Gladkov

60 **Organic Photovoltaics**
Concepts and Realization
Editors: C. Brabec, V. Dyakonov, J. Parisi, and N. Sariciftci

61 **Fatigue in Ferroelectric Ceramics and Related Issues**
By D.C. Lupascu

62 **Epitaxy**
Physical Principles and Technical Implementation
By M.A. Herman, W. Richter, and H. Sitter

63 **Fundamentals of Ion-Irradiated Polymers**
By D. Fink

64 **Morphology Control of Materials and Nanoparticles**
Advanced Materials Processing and Characterization
Editors: Y. Waseda and A. Muramatsu

65 **Transport Processes in Ion-Irradiated Polymers**
By D. Fink

66 **Multiphased Ceramic Materials**
Processing and Potential
Editors: W.-H. Tuan and J.-K. Guo

67 **Nondestructive Materials Characterization**
With Applications to Aerospace Materials
Editors: N.G.H. Meyendorf, P.B. Nagy, and S.I. Rokhlin

68 **Diffraction Analysis of the Microstructure of Materials**
Editors: E.J. Mittemeijer and P. Scardi

69 **Chemical–Mechanical Planarization of Semiconductor Materials**
Editor: M.R. Oliver

70 **Applications of the Isotopic Effect in Solids**
By V.G. Plekhanov

71 **Dissipative Phenomena in Condensed Matter**
Some Applications
By S. Dattagupta and S. Puri

72 **Predictive Simulation of Semiconductor Processing**
Status and Challenges
Editors: J. Dabrowski and E.R. Weber

73 **SiC Power Materials**
Devices and Applications
Editor: Z.C. Feng

Springer Series in
MATERIALS SCIENCE

Editors: R. Hull   R. M. Osgood, Jr.   J. Parisi   H. Warlimont

74 **Plastic Deformation
in Nanocrystalline Materials**
By M.Yu. Gutkin and I.A. Ovid'ko

75 **Wafer Bonding**
Applications and Technology
Editors: M. Alexe and U. Gösele

76 **Spirally Anisotropic Composites**
By G.E. Freger, V.N. Kestelman,
and D.G. Freger

77 **Impurities Confined
in Quantum Structures**
By P.O. Holtz and Q.X. Zhao

78 **Macromolecular Nanostructured
Materials**
Editors: N. Ueyama and A. Harada

79 **Magnetism and Structure
in Functional Materials**
Editors: A. Planes, L. Mañosa,
and A. Saxena

80 **Micro- and Macro-Properties of Solids**
Thermal, Mechanical
and Dielectric Properties
By D.B. Sirdeshmukh, L. Sirdeshmukh,
and K.G. Subhadra

81 **Metallopolymer Nanocomposites**
By A.D. Pomogailo and V.N. Kestelman

82 **Plastics for Corrosion Inhibition**
By V.A. Goldade, L.S. Pinchuk,
A.V. Makarevich, and V.N. Kestelman

83 **Spectroscopic Properties of Rare Earths
in Optical Materials**
Editors: G. Liu and B. Jacquier

84 **Hartree–Fock–Slater Method
for Materials Science**
The DV–X Alpha Method for Design
and Characterization of Materials
Editors: H. Adachi, T. Mukoyama,
and J. Kawai

85 **Lifetime Spectroscopy**
A Method of Defect Characterization
in Silicon for Photovoltaic Applications
By S. Rein

86 **Wide-Gap Chalcopyrites**
Editors: S. Siebentritt and U. Rau

87 **Micro- and Nanostructured Glasses**
By D. Hülsenberg and A. Harnisch

88 **Introduction
to Wave Scattering, Localization
and Mesoscopic Phenomena**
By P. Sheng

89 **Magneto-Science**
Magnetic Field Effects on Materials:
Fundamentals and Applications
Editors: M. Yamaguchi and Y. Tanimoto

90 **Internal Friction in Metallic Materials**
A Handbook
By M.S. Blanter, I.S. Golovin,
H. Neuhäuser, and H.-R. Sinning

91 **Ferroelectric Crystals
for Photonic Applications**
Including Nanoscale Fabrication
and Characterization Techniques
Editors: P. Ferraro, S. Grilli, and P. De Natale

92 **Solder Joint Technology**
Materials, Properties, and Reliability
By K.-N. Tu

93 **Materials for Tomorrow**
Theory, Experiments and Modelling
Editors: S. Gemming, M. Schreiber,
and J.-B. Suck

94 **Magnetic Nanostructures**
Editors: B. Aktas, L. Tagirov,
and F. Mikailov

95 **Nanocrystals
and Their Mesoscopic Organization**
By C.N.R. Rao, P.J. Thomas,
and G.U. Kulkarni

96 **Gallium Nitride Electronics**
By R. Quay

97 **Multifunctional Barriers
for Flexible Structure**
Textile, Leather and Paper
Editors: S. Duquesne, C. Magniez,
and G. Camino